Principles of Biochemistry

Geoffrey L. Zubay
Columbia University

William W. Parson
University of Washington

Dennis E. Vance
University of Alberta

Wm. C. Brown Publishers
Dubuque, Iowa • Melbourne, Australia • Oxford, England

Book Team

Editor *Elizabeth M. Sievers*
Developmental Editor *Robin P. Steffek*
Publishing Services Coordinator *Julie Avery Kennedy*
Photo Editor *Lori Hancock*
Permissions Coordinator *Karen L. Storlie*

Wm. C. Brown Publishers
A Division of Wm. C. Brown Communications, Inc.

Vice President and General Manager *Beverly Kolz*
Vice President, Publisher *Kevin Kane*
Vice President, Director of Sales and Marketing *Virginia S. Moffat*
Vice President, Director of Production *Colleen A. Yonda*
National Sales Manager *Douglas J. DiNardo*
Marketing Manager *Patrick E. Reidy*
Advertising Manager *Janelle Keeffer*
Production Editorial Manager *Renée Menne*
Publishing Services Manager *Karen J. Slaght*
Royalty/Permissions Manager *Connie Allendorf*

Wm. C. Brown Communications, Inc.

President and Chief Executive Officer *G. Franklin Lewis*
Senior Vice President, Operations *James H. Higby*
Corporate Senior Vice President, President of WCB Manufacturing *Roger Meyer*
Corporate Senior Vice President and Chief Financial Officer *Robert Chesterman*

Cover Images:
Main Cover and Volume 1: Image copyright 1994 by the Scripps Research Institute/ Molecular Graphics Images by Michael Pique using software by Yng Chen, Michael Connolly, Michael Carson, Alex Shah, and AVS, Inc. Endonuclease III by Kuo, McRee, Fisher, O'Handley, Cunningham, and Tainer

Volume II: Courtesy of Dr. Klaus Piontek

Volume III: Courtesy of Dr. John Kuriyan

Freelance Permissions Editor Karen Dorman
Copyediting and Production by York Production Services
Design by York Production Services
Composition by York Graphic Services

The credits section for this book begins on page C-1 and is considered an extension of the copyright page.

Contributors to the End-of-Chapter Problems

Chapters 2–24
Hugh Akers
Lamar University

Chapters 25–31
Caroline Breitenberger
Ohio State University

A Times Mirror Company

Library of Congress Catalog Card Number: 94–70034 Complete
94–70098 Volume One, Two, and Three

ISBN 0–697–24169–6—Volume One
0–697–24170–X—Volume Two
0–697–24171–8—Volume Three
0–697–14275–2—Complete

Printed in the United States of America by Wm. C. Brown Communications, Inc., 2460 Kerper Boulevard, Dubuque, IA 52001

10 9 8 7 6 5 4 3 2 1

This text is dedicated to:

Bongsoon
Polly, and
Jean

Publisher's Note

Principles of Biochemistry, first edition, is available as a single, full-length casebound text or in three paperback volumes, which may be used separately or together in any combination that best suits your course needs.

Binding Option	**Description**	**ISBN**
Principles of Biochemistry, 1/E (Casebound)	The full-length text, Chapters 1–31, plus four Supplements.	0-697-14275-2
Principles of Biochemistry, 1/E Volume 1: Energy, Proteins, and Catalysis (Paperback)	Chapters 1–10	0-697-24169-6
Principles of Biochemistry, 1/E Volume 2: Metabolism (Paperback)	Chapters 11–24, plus Supplements 1 and 2.	0-697-24170-X
Principles of Biochemistry, 1/E Volume 3: Molecular Genetics (Paperback)	Chapters 25–31, plus Supplements 3 and 4.	0-697-24171-8

Brief Contents

Volume One

Energy, Proteins, and Catalysis

Part 1

Part 2

Part 3

Volume Two

Metabolism

Part 4

Part 5

Metabolism of Lipids 379

Part 6

Metabolism of Nitrogen-Containing Compounds 485

Volume Three

Molecular Genetics

Part 7

Storage and Utilization of Genetic Information 625

Extended Contents

Part 2

Protein Structure and Function 47

Chapter 3

Chapter 4

Chapter 5

Chapter 6

Part 3

Catalysis 133

Chapter 7

Chapter 13

Chapter 14

Chapter 15

Chapter 16

Structures and Metabolism of Oligosaccharides and Polysaccharides 356

Part 5

Metabolism of Lipids 379

Chapter 17

Structure and Functions of Biological Membranes 381

Chapter 18

Metabolism of Fatty Acids 411

Part 6

Metabolism of Nitrogen-Containing Compounds 485

Chapter 21

Amino Acid Biosynthesis and Nitrogen Fixation in Plants and Microorganisms 436

Chapter 22

Amino Acid Metabolism in Vertebrates 511

Chapter 23

Nucleotides 533

Part 7

Storage and Utilization of Genetic Information 625

Chapter 25

Chapter 26

Chapter 27

List of Supplementary Text Elements

Boxes

Supplemental Chapters Principles of Physiology and Biochemistry

Methods of Biochemical Analysis*

* *Please Note:* Most of the material on **Methods of Biochemical Analysis** and **Biochemical Experiments** presented in this textbook is **integrated** into the main part of the text. This material is **"flagged"** throughout the text by the **lab door icon. Supplemental biochemical methods material which can be identified within the text by the blue-tabbed pages is listed below.**

Preface

As the subject of biochemistry has grown, the need for texts that are appropriate for different types of courses has also grown. There is a need for texts at three distinct levels: the ''comprehensive'' level that includes a great deal of detail, the ''middle'' level, and the ''short course'' level. Most up-to-date texts are written at the ''comprehensive'' level despite the fact that most students would benefit more from a less detailed, ''middle'' level text that emphasizes *principles.* Currently, there are very few up-to-date ''middle'' level texts available. The main goal of our text, as indicated by the title, is to serve the needs of the *principles-oriented,* ''middle'' level audience.

In addition to presenting material that we think is important for the ''middle'' level audience, we have tried to write this text with the following goals in mind:

- to give a balanced presentation that explains the biochemistry of both prokaryotes and eukaryotes;
- to indicate the biomedical significance of defects in metabolism;
- to describe metabolism with special emphasis on how pathways are organized and how they are regulated;
- to give the student a feel for the process of scientific research by describing how some major biochemical discoveries were made; and finally,
- to present the techniques used by biochemists during the course of their research.

Unique Features of *Principles of Biochemistry*

Some of the unique features of our text can be classified as *general features* in that they apply to the entire text, while other *content-specific features* apply to an individual chapter or group of chapters.

General Features

1. **The text is available as a single, full-length (31 Chapters; plus 4 Supplements) casebound text or in three paperback volumes (Volume One: Chapters 1–10, Volume Two: Chapters 11–24, plus Supplements 1 & 2; and Volume Three: Chapters 25–31, plus Supplements 3 & 4).** It is generally useful for students to bring their texts to class, but students of biochemistry are not inclined to do this because of the large size of the texts. The three volume binding format makes it convenient for students, and professors alike, to carry only one third of the textbook around at any one time. An additional advantage of the three volume binding format is that each of the volumes may be used separately or together in any combination that best suits the needs of the biochemistry course. The three volumes therefore allow greater flexibility for professors and greater manageability and affordability for students.
2. ***Eight icons****, graphic images that are intended to help the textbook reader remember and mentally cross-reference key biochemical concepts and principles through image association,* are used throughout this textbook to ''flag'' sections of chapters that discuss the following: Methods of Biochemical Analysis, Biomedical Applications, Regulatory Aspects of Biochemistry, Plant Biochemistry, Neurochemistry, the Biochemistry of Vision, Immunochemistry, and the Biochemistry of Cancer Cells. Additional information on the use of the eight icons in this textbook is given in the ''Learning Aids'' section of this preface.

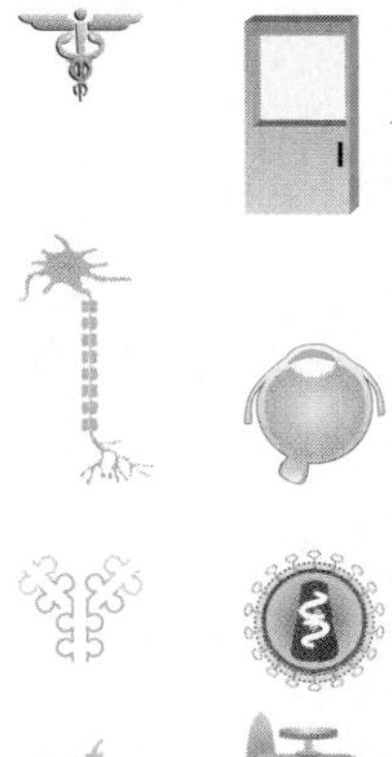

Content-Specific Features

1. Chapter 6: Detailed step-by-step descriptions of the purification schemes for two proteins are presented to give the student an understanding of how different purification procedures may be effectively combined.

2. Chapters 7–9: Emphasis on basic enzymology—enzyme kinetics, mechanisms and regulation.
3. Chapter 10: An exclusive chapter devoted to vitamins and coenzymes.
4. Chapter 11: A comprehensive description of basic metabolic strategies that govern substrate and energy flow in metabolic pathways.
5. Chapter 12: A single cohesive chapter on all major aspects of glucose metabolism—glycolysis, gluconeogenesis, and regulation.
6. Chapters 14 and 18: Correct values for ATP production resulting from catabolism. Other texts do not take into account the energy required for ATP, ADP, and P_i transport across the inner mitochondrial membrane. This makes a substantial difference in the calculated energy yields.
7. Chapter 20: A chapter exclusively devoted to cholesterol metabolism.
8. Chapter 22: A chapter exclusively devoted to amino acid metabolism in vertebrates.
9. Chapters 12–16, 17–23: Organization of carbohydrate metabolism, lipid metabolism, and the metabolism of nitrogen-containing compounds, respectively, into self-contained units.
10. Chapters 30 and 31: Separate chapters devoted to the regulation of gene expression in prokaryotes and eukaryotes, respectively.

Learning Aids for the Student

In order for students to succeed in their study of biochemistry, they must be able to understand the material presented, utilize the text as a tool for learning, and enjoy reading the text. Therefore, we have included many aids to make the study of biochemistry efficient and enjoyable.

The text chapters contain the following:

Chapter Opening Outline and Overview The chapter opening outline and overview were written to help students preview how the chapter is organized and what major concepts are to be covered in the chapter.

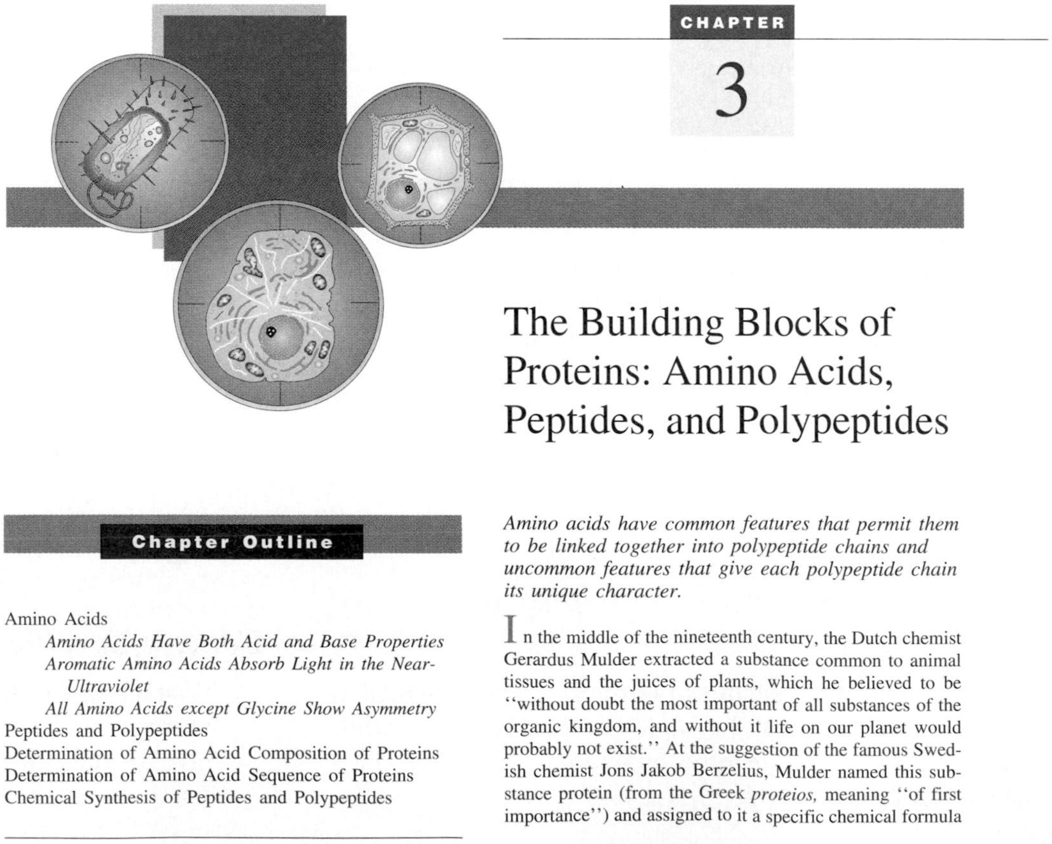
CHAPTER 3

The Building Blocks of Proteins: Amino Acids, Peptides, and Polypeptides

Chapter Outline

Amino Acids
Amino Acids Have Both Acid and Base Properties
Aromatic Amino Acids Absorb Light in the Near-Ultraviolet
All Amino Acids except Glycine Show Asymmetry
Peptides and Polypeptides
Determination of Amino Acid Composition of Proteins
Determination of Amino Acid Sequence of Proteins
Chemical Synthesis of Peptides and Polypeptides

Amino acids have common features that permit them to be linked together into polypeptide chains and uncommon features that give each polypeptide chain its unique character.

In the middle of the nineteenth century, the Dutch chemist Gerardus Mulder extracted a substance common to animal tissues and the juices of plants, which he believed to be "without doubt the most important of all substances of the organic kingdom, and without it life on our planet would probably not exist." At the suggestion of the famous Swedish chemist Jons Jakob Berzelius, Mulder named this substance protein (from the Greek *proteios,* meaning "of first importance") and assigned to it a specific chemical formula

Declarative Statement Headings Clear, informative headings help introduce students to each important topic.

Underlined Terms and Key Concepts Underlined important terms and key concepts are easy for students to locate.

> The second law of thermodynamics is that the universe inevitably proceeds from states that are more ordered to states that are more disordered. This phenomenon is measured by a thermodynamic function called entropy, which is denoted

Numbered Equations Important equations within a chapter are numbered so that they can be easily located and referenced by students.

Summary Tables Summary tables are designed to help students more easily understand and utilize important facts or characteristics presented for a specific topic.

Eight Concept and Applications Icons Throughout the text, the student will find text sections "flagged" by eight different graphic icons or images. These icons are intended to help the student remember and mentally cross-reference the following concepts and applications:

Methods of Biochemical Analysis

Biomedical Applications

Regulatory Aspects of Biochemistry

Plant Biochemistry

Neurochemistry

Biochemistry of Vision

Immunochemistry

Biochemistry of Cancer Cells

Boxed Readings Thought-provoking boxed readings on relevant topics are featured in various chapters—for example, Box 20A, Metabolism and Heart Disease.

BOX
20A Cholesterol Metabolism and Heart Disease

Molecular Graphics and Illustration Program Molecular graphics images of key molecules enable students to "see" and interpret three-dimensional structures.

Figure 8.6a

Trypsin, Chymotrypsin, Elastase

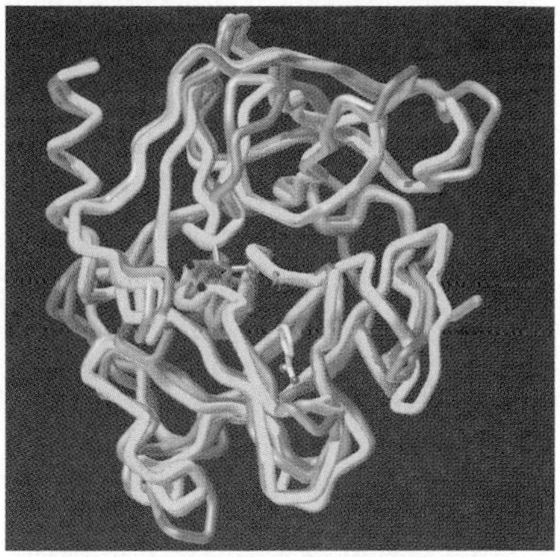

End-of-Chapter Enumerated Summary This concise summary of chapter concepts is designed to serve as a guide for chapter study.

Summary

In this chapter we discussed the ways in which biochemistry parallels ordinary chemistry and those in which it is quite different. The chief points to remember are the following.

1. The basic unit of life is the cell, which is a membrane-enclosed, microscopically visible object.
2. Cells are composed of small molecules, macromolecules, and organelles. The most prominent small molecule is water, which constitutes 70% of the cell by weight. Other small molecules are present only in quite small amounts; they are precursors or breakdown products of macromolecules or coenzymes. There are four types of macromolecules: lipids, carbohydrates, proteins, and nucleic acids.
3. Noncovalent intermolecular forces largely determine the folded structure adopted by a macromolecule, particularly the relative affinity of different groupings on the macromolecule for water. In general, hydrophobic groupings are buried within the folded macromolecular structure, whereas hydrophilic groupings are located on the surface, where they can interact with water.
4. Biochemical reactions utilize a limited number of elements, most prominently carbon, hydrogen, oxygen, nitrogen, sulfur, and phosphorus. Many biochemical reactions are simple organic reactions.

End-of-Chapter Selected Readings Each chapter concludes with carefully selected references that contain further information on the topics covered in that chapter.

End-of-Chapter Problems These problems (over 400) are designed to help students access their understanding of the chapter's basic concepts. Brief solutions for the odd-numbered problems are found in the back of the text.

End-of-Chapter Methods of Biochemical Analysis Supplemental information on experimental techniques and procedures are presented at the end of some chapters. The "Methods of Biochemical Analysis" can be easily located throughout the text as their text pages are tabbed with a blue band. The lab door icon is used to cross-reference the

''Methods of Biochemical Analysis'' to pertinent text sections within a chapter.

Methods of Biochemical Analysis **3 A**

Measurement of Ultraviolet Absorption in Solution

The general quantitative relationship that governs all absorption processes is called the Beer-Lambert law:

$$I = I_0 10^{-\epsilon c d}$$

where I_0 is the intensity of the incident radiation, I is the intensity of the radiation transmitted through a cell of thickness d (in centimeters) that contains a solution of concentration c (expressed either in moles per liter or in grams per 100 ml), and ϵ is the extinction coefficient, a characteristic of the substance being investigated (see fig. 3.7).

Light absorption is measured by a spectrophotometer as shown in the illustration. The spectrophotometer usually is capable of directly recording the absorbance A, which is related to I and I_0 by the equation

$$A = \log_{10}\left(\frac{I_0}{I}\right)$$

Hence $A = \epsilon c d$, and A is a direct measure of concentration. We can see from figure 3.7 that the ϵ values are largest for tryptophan and smallest for phenylalanine.

Because protein absorption maxima in the near-ultraviolet (240–300 nm) are determined by the content of the aromatic amino acids and their respective values, most proteins have absorption maxima in the 280-nm region. By contrast, absorption in the far-ultraviolet (around 190 nm) is shown by all polypeptides regardless of their aromatic amino acid content. The reason is that absorption in this region is due primarily to the peptide linkage.

Figure 1

Schematic diagram of a spectrophotometer for measuring light absorption. Laboratory instruments for making measurements are much more complex than this, but they all contain the same basic components: a light source, a monochromator, a sample, and a detector. λ is the wavelength of the light, I_0 and I are the incident light intensity and the transmitted light intensity, respectively, and d is the thickness of the absorbing solution.

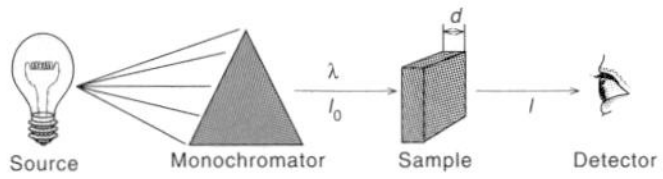

In addition to these chapter aids, this text also features the following:

Four ''Principles of Physiology and Biochemistry'' Supplements These supplements cover the topics of **neurotransmission, vision, immunobiology, and carcinogenesis.** The Neurotransmission and Vision supplements, Supplements 1 and 2, are located at the end of Part 6 of the text. The Immunobiology and Carcinogenesis supplements, Supplements 3 and 4, are located at the end of Part 7 of the text. Physiological discussions within the text that relate to these topics are also ''flagged'' by the corresponding concept icon.

Page-Referenced Listing of Supplementary Learning Aids This helpful ready reference tool is located in the frontmatter of this text.

Illustrated Guide to the ''Comparative Sizes of Biomolecules, Viruses, and Cells'' This highly visual guide is located at the end of the text.

Common Abbreviations in Biochemistry These abbreviations can be found at the end of the text.

Organic Chemistry for Biochemistry This appendix, which describes the similarities and the differences between organic chemistry and biochemistry, is also located at the end of this text.

Appendix A: Some Major Discoveries in Biochemistry Summarizes major research discoveries from the past and present.

Appendix B: Career Paths in the Biological Sciences A useful starting point for career information for biochemistry students.

Appendix C: Answers to Selected Problems Brief solutions are provided to help the student determine if they are on the right track.

Glossary Over 700 important biochemical terms are defined in this end-of-book glossary.

Index An easy-to-use, comprehensive index is provided.

Endsheets with Reference Material The endsheets of the text contain useful easily accessible reference information.

Organization of the Text

This text is divided into seven parts: part 1, An Overview of Biochemical Structures and Reactions That Occur in Living Systems; part 2, Protein Structure and Function; part 3, Catalysis; part 4, Metabolism of Carbohydrates; part 5, Metabolism of Lipids; part 6, Metabolism of Nitrogen-Containing Compounds; and part 7, Storage and Utilization of Genetic Information.

The subjects presented in each of the seven parts, the level of the discussion, the organization of the parts, and the organization within each of the chapters are all designed to satisfy the teaching needs of introductory biochemistry courses.

We may think of biochemistry as being divided into three main areas: biomolecular structure, intermediary metabolism, and nucleic acid and protein metabolism. It would be possible to segregate the coverage of these topics in a biochemistry text and present them as three consecutive blocks. However, we did not take this approach in the development of this text because we felt it would not be good pedagogy for two reasons: (1) students get bored if they sit through a third of a year where they only hear about structure; and (2) students often forget what was said about the structure of a biomolecule by the time they read the section on its metabolism and therefore have to review it before going on to the metabolism material.

Our general approach in writing *Principles of Biochemistry* was to integrate the presentation of biomolecular structure and metabolism throughout the text. We were successful in applying this integrative approach to presenting biomolecular structure and metabolism in five of the seven parts of the text. Specifically, amino acid structure and protein structure are discussed in part 2, well before considering their metabolism in parts 6 and 7, respectively. This deviation from our general approach is essential because one must understand amino acid structure before one can understand the function of proteins and enzymes, which is discussed in early chapters (chapters 5, 7, 8, and 9 of parts 2 and 3). An additional advantage to discussing protein structure early in the text is that most students find it exciting. Notwithstanding the early discussion of basic protein structure in chapter 4, additional discussions pertaining to protein structure are included within the text where they are most appropriate to furthering the student's understanding protein function. Such discussions can be found in chapter 17 where membrane proteins are discussed and in chapters 30 and 31 where gene regulatory proteins are discussed.

The structures of sugars and polysaccharides are covered in the appropriate chapters within part 4 just prior to discussing their metabolism. Similarly, the structures of lipids are presented in the lipid metabolism chapters found in part 5. Nucleotide structures are addressed in chapter 23 before considering their metabolism. Finally, nucleic acid and nucleoprotein structures are examined in the first chapter (chapter 25) of part 7 prior to the discussion of the roles these molecules play in nucleic acid and protein metabolism in the six subsequent chapters.

Metabolism is regulated at the level of the gene and at the level of the enzyme. Regulation is emphasized in this text both because of its importance and because it reflects some of the most unique characteristics of biochemical reactions. Of necessity, our treatment of the subject of regulation at the gene level is confined primarily to part 7, as this is where the metabolism of nucleic acids is discussed. We cover metabolic regulation at the level of the enzyme throughout this text. Specifically, the regulation of the oxygen binding capacity of hemoglobin and the process of muscle contraction is discussed in chapter 5. An entire chapter, chapter 9, is devoted to regulatory enzymes. With respect to the metabolism of carbohydrates (chapter 12) and fatty acids (chapter 18), the regulation of the catabolic and anabolic pathways of these biomolecules is presented together within the respective chapters. We feel that this approach simplifies the discussion of the regulation of these opposing pathways.

Detailed Chapter-by-Chapter Description of the Text

Having indicated some aspects of the overall organization of this text, we now turn to a more detailed chapter-by-chapter description of the text. Part 1, An Overview of Biochemical Structures and Reactions That Occur in Living Systems, contains only two chapters. Chapter 1, Cells, Biomolecules and Water, gives an introductory presentation of the structure of cells and the molecules from which cells are composed. The central role of water in determining the structures that are formed is emphasized. A general description of the principles that govern cellular organization and determine the course of these reactions is presented. An introductory survey of the types of molecules that make up the cell is given. Finally, the overall strategy of living systems and the evolution of living systems is briefly discussed.

The main function of chapter 1 is to present an overview of the principles and facts that are presented in greater depth in various parts of the text. Most of the material in this chapter is presented in a good introductory biology course. However, there are a number of students coming from a chemistry background that may not have had the benefits of such a course. Reference should also be made at this point to the appendix ''Organic Chemistry for Biochemistry'' (located in the endmatter of the text), which gives a brief review of the similarities and differences between organic chemistry and biochemistry. It is assumed that all students reading this text will have had a year of general chemistry and one term of organic chemistry. Or, at the very least, that they will be taking organic chemistry concurrently with biochemistry.

In chapter 2, thermodynamics is presented in a way that is most relevant to the consideration of biochemical phenomena. We chose to present the rather difficult topic of thermodynamics in the beginning of the text because thermodynamics is an important consideration in all aspects of biochemistry. The central role of ATP as the main carrier of free energy in biochemical systems is emphasized.

The discussion of thermodynamics presented in chapter 2 gives a perspective of bioenergetic considerations, which are discussed at many points throughout the text.

Thermodynamic applications that relate to biomolecular structure and biochemical reactions are also elaborated on in specific sections of subsequent chapters.

Part 2, Protein Structure and Function, contains four chapters that relate to the structures and functions of proteins. In chapter 3, The Building Blocks of Proteins: Amino Acids, Peptides, and Polypeptides, we discuss basic structural and chemical properties of amino acids, peptides and polypeptides. In chapter 4, The Three-Dimensional Structures of Proteins, we describe how and why polypeptide chains fold into long fibrous molecules in some cases, or into compact globular molecules in other cases. In chapter 5, Functional Diversity of Proteins, we turn to the question of how protein structure relates to protein function. To explore this question, two protein systems, hemoglobin and the actin-myosin complex are examined in detail. In chapter 6, Methods for Purification and Characterization of Proteins, the primary goal is to acquaint the reader with the techniques used for protein purification. The first part of chapter 6 presents methods for protein fractionation. In the second part of this chapter, purification procedures for two proteins, UMP synthase and lactose carrier protein, are presented so that the student can see how different purification steps are combined for maximum effectiveness.

Part 3, Catalysis, is divided into four chapters. The first three chapters in part 3 deal with the properties and functions of enzymes. The last chapter explains the essential properties of coenzymes and cofactors.

Enzyme kinetics is studied for two reasons: (1) it is a practical concern to determine the activity of the enzyme under different conditions; (2) frequently the analysis of enzyme kinetics gives information about the mechanism of enzyme action. Chapter 7, Enzyme Kinetics, begins with an introductory section on the discovery of enzymes, basic enzyme terminology and a description of the six main classes of enzymes and the reactions they catalyze. The remainder of the chapter deals with basic aspects of chemical kinetics, enzyme-catalyzed reactions and various factors that affect the kinetics.

Chapter 8, How Enzymes Work, starts with a description of the basic chemical mechanisms that are exploited by enzymes. The latter half of this chapter presents a detailed description of how three enzymes—chymotrypsin, RNase, and triose phosphate isomerase—exploit these basic mechanisms of enzyme catalysis.

Most enzymes spontaneously process substrates when present, as long as inhibitory factors do not prevent this from happening. A few enzymes, known as regulatory enzymes, do not react spontaneously with their substrates unless signaled to do so by overiding metabolic conditions. Chapter 9, Regulation of Enzyme Activities, describes a wide range of mechanisms that are used to control the activity of regulatory enzymes. This chapter concludes with a detailed description of how the activity of three enzymes—phosphofructokinase, aspartate transcarbamylase and glycogen phosphorylase—are regulated.

Frequently enzymes act in concert with small molecules, coenzymes or cofactors, which are essential to the function of the amino acid side chains of the enzyme. Coenzymes or cofactors are distinguished from substrates by the fact that they function as catalysts. They are also distinguishable from inhibitors or activators in that they participate directly in the catalyzed reaction. Chapter 10, Vitamins and Coenzymes, starts with a description of the relationship of water-soluble vitamins to their coenzymes. Next, the functions and mechanisms of action of coenzymes are explained. In the concluding sections of this chapter, the roles of metal cofactors and lipid-soluble vitamins in enzymatic catalysis are briefly discussed.

Part 4, Metabolism of Carbohydrates, begins with a general overview in chapter 11, Metabolic Strategies. This chapter also relates strongly to parts 5 and 6. Chapter 11 starts with an explanation of how biochemical reactions are organized into energy-generating and energy requiring pathways and a handful of principles that explain the design of metabolic pathways. The next five chapters describe the metabolism of carbohydrates insofar as they relate to energy-generating catabolism and energy-consuming biosynthesis. Chapter 12, Glycolysis, Gluconeogenesis, and the Pentose Phosphate Pathway, deals with all aspects of glucose metabolism. After a brief section describing structures, glycolysis, the pathway for the breakdown of glucose, is examined. This is followed by a shorter section on gluconeogenesis, the synthesis of glucose from three carbon compounds. While glycolysis produces energy, gluconeogenesis consumes energy. These two processes are discussed side-by-side so that we can consider the closely related question of the regulation of these pathways. This chapter concludes with a short section on an oxidative route for glucose catabolism, the pentose phosphate pathway which serves multiple purposes.

Under anaerobic conditions the breakdown of glucose stops at the three carbon compound stage. Further catabolism requires oxygen as described in chapter 13, The Tricarboxylic Acid Cycle. The tricarboxylic acid cycle is an energy-producing process although this is not immediately obvious because the major products of the cycle outside of CO_2—H_2O and a single ATP molecule—are the reduced forms of the coenzymes NAD^+ and FAD. These coenzymes contain the potential chemical energy for ATP production, but an elaborate process of membrane-associated electron transport and proton transport must precede the synthesis of ATP. This process is described in chapter 14, Electron Transport and Oxidative Phosphorylation.

Most of the energy that is used to drive biochemical processes originates from the sun. The way in which solar energy is harnessed to produce chemical energy and to fix atmospheric carbon dioxide into reduced organic compounds is described in chapter 15, Photosynthesis.

Up to this point in part 4, the focus has been on the roles that sugars and carbohydrates play in energy metabolism. Polymeric carbohydrates are major structural components in plant and bacterial cell walls and in the extracellular matrix of vertebrates. Branched-chain oligosaccharides covalently linked to proteins are used to give proteins unique signatures that guide them to their final destination within a cell and facilitate specific interactions between free proteins (ligands) and proteins attached to cells (receptors). In chapter 16, Structures and Metabolism of Oligosaccharides and Polysaccharides, the structures and functions of some of these polymeric carbohydrates are discussed.

Part 5, Metabolism of Lipids, comprises four chapters that deal with the structure and metabolism of lipids. In chapter 17, Structure and Function of Biological Membranes, we start by examining the constituents of membranes with the aim of developing a general model for membrane structure. We then turn to the question of how cells transport materials across membranes.

In chapter 18, Metabolism of Fatty Acids, we discuss the synthesis and breakdown of fatty acids. The chapter starts with a discussion of fatty acid breakdown. A second section covers the pathway for fatty acid biosynthesis. Finally, we consider the regulatory mechanisms that determine the conditions under which each of these processes occurs. As in the case of glucose metabolism, it is convenient to discuss the synthesis and breakdown in the same chapter so that the closely related topic of regulation can be considered alongside.

Thus far we have been concerned with the metabolism of fatty acids in relationship to the storage and release of energy. In chapter 19, Biosynthesis of Membrane Lipids, we focus on the metabolism of lipids that serve other roles. Many types of lipids are essential membrane components. A number of lipids also function as metabolic signals in response to hormonal signals. These lipid molecules are known as second messengers.

The final chapter in part 5, chapter 20, Metabolism of Cholesterol, deals with the synthesis of cholesterol and some of its derivatives, the steroid hormones and the bile acids. This chapter considers the structure, function and metabolism of these molecules. Also, the health-related concerns associated with cholesterol excess are addressed.

Part 6, Metabolism of Nitrogen-Containing Compounds, is concerned mostly with the metabolism of amino acids and nucleotides. Chapter 24, the last chapter in this part, deals with the integration of metabolism.

Amino acid metabolism is divided into two chapters. Amino acids are best known as the building blocks of proteins. In addition to this role, amino acids serve as precursors to many important low molecular weight compounds including nucleotides, porphyrins, parts of lipid molecules and precursors for several coenzymes. Amino acids also serve as the ''vehicles'' for converting inorganic forms of nitrogen and sulfur to organic forms. As an alternative energy source amino acid catabolism can be coupled to the regeneration of ATP from ADP of AMP. In chapter 21, Amino Acid Biosynthesis and Nitrogen Fixation in Plants and Microorganisms, we focus on the biosynthesis of amino acids and their role in bringing inorganic nitrogen and sulfur into the biological world. In addition, nonprotein amino acids are discussed briefly.

Amino acid metabolism in vertebrates contrasts sharply with amino acid metabolism in plants and microorganisms. Most striking is the fact that plants and microorganisms can synthesize all twenty amino acids required for protein synthesis whereas vertebrates can only synthesize about half this number. This leads to complex nutritional needs for vertebrates, which are discussed in chapter 22, Amino Acid Metabolism in Vertebrates. Vertebrate amino acid degradation pathways are also discussed in chapter 22 along with the existence of many pathological states that result from enzyme deficiencies in the degradative pathways.

Chapter 23, Nucleotides, deals with the biosynthesis of ribonucleotides, deoxyribonucleotides, the roles of these biomolecules in metabolic processes, and the pathways for their degradation. Medically related topics such as nucleotide metabolism deficiencies or the use of nucleotide analogs in chemotherapy are also considered.

Chapter 24, Integration of Metabolism and Hormone Action, explains the organization strategies used to integrate metabolic processes in a multicellular organism. Like the first chapter in part 4, the content of chapter 24 relates to all of the chapters on metabolism (chapters 11–24). This chapter emphasizes the fact that hormones and closely related growth factors play a dominant role in regulating metabolic activities in different tissues.

Part 6 concludes with two brief, informative supplements that integrate physiological and biochemical principles as they apply to the ''nonmetabolic'' functions of amino acids and lipids: Supplement 1—Principles of Physiology and Biochemistry: Neurotransmission; and Supplement 2—Principles of Physiology and Biochemistry: Vision.

Part 7, Storage and Utilization of Genetic Information, is composed of seven chapters, plus two supplements. In this part, we examine the means by which genetic information is replicated and expressed in the cell. This discussion includes a description of the relevant structures, their function, and their metabolism. In addition, this part includes a chapter that explains the procedures by which DNA is experimentally manipulated to build new genes or new combinations of genes. In chapter 25, Structures of Nucleic Acids and Nucleoproteins, we begin by considering the key experiments that revealed the genetic significance of DNA. At every stage in this chapter and the subsequent chapters of this section, explanations of genetic observations essential to the understanding of biochemical phenomena are given.

From the complementary duplex structure of DNA described in chapter 25, it is a short intuitive hop to a model for replication that satisfies the requirement for one round of DNA duplication for every cell division. In chapter 26, DNA Replication, Repair, and Recombination, key experiments demonstrating the semiconservative mode of replication in vivo are presented. This is followed by a detailed examination of the enzymology of replication, first for how it occurs in bacteria and then for how it occurs in animal cells. Also included in this chapter are select aspects of the metabolism of DNA repair and recombination. The novel process of DNA synthesis using RNA-directed DNA polymerases is also considered. First discovered as part of the mechanisms for the replication of nucleic acids in certain RNA viruses, this mode of DNA synthesis is now recognized as occurring in the cell for certain movable genetic segments and as the means whereby the ends of linear chromosomes in eukaryotes are synthesized.

Before going into the processes for the utilization of genetic information in the cell, DNA manipulation and some of its applications are considered. In chapter 27, DNA Manipulation and Its Applications, DNA sequencing is the first subject to be explained. Different approaches for amplifying and isolating specific genes are also explored. Following this, methods for restructuring existing DNA sequences are described. Finally, some applications of new technologies are considered. These applications focus on the mapping of the human globin gene family and the gene responsible for the genetically inherited disease, cystic fibrosis.

In chapter 28, RNA Synthesis and Processing, the DNA-directed synthesis of RNA is considered. As with the other chapters in part 7, a balanced presentation of how these processes occur in bacteria and eukaryotes is given. First, the structures of different classes of RNA are described. The RNA classes include messenger RNA, transfer RNA, and ribosomal RNA. A single enzyme is responsible for the transcription of the RNAs in bacteria. The initial transcripts for transfer RNA and ribosomal RNA undergo extensive modification after synthesis, while the messenger RNA is used without modification. In eukaryotes, even the messenger RNA undergoes extensive modifications before it can be utilized. Also, in eukaryotes the process of transcription is much more complicated, involving several RNA polymerases and many more protein subunits which are associated with the polymerases. Every attempt is made to treat these complications without presenting an overwhelming amount of detail. RNA synthesis in certain viruses is also discussed and the fascinating subject of RNA enzymes is briefly considered. Lastly, selective inhibitors of RNA polymerases that have diagnostic value or medical value in chemotherapy are discussed.

In chapter 29, Protein Synthesis, Targeting, and Turnover, the processes of protein synthesis and transport are described. First the process whereby amino acids are ordered and polymerized into polypeptide chains is described. Next, posttranslational alterations of newly synthesized polypeptides is considered. This is followed by a discussion of the targeting processes whereby proteins migrate from their site of synthesis to their target sites of function. Finally, proteolytic reactions that result in the return of proteins to their starting materials, the amino acids, are considered.

Regulation of gene expression is on the cutting edge of research in molecular biology and biochemistry. This topic is so expansive that we chose to divide our coverage of it into two chapters: one that considers bacteria and another that considers eukaryotes. In chapter 30, Regulation of Gene Expression in Prokaryotes, the classical systems of the *lac* and *trp* operons and regulation of bacteriophage lambda are considered, with particular emphasis on the nature of gene regulatory proteins and how they interact with DNA. While most forms of regulation of gene expression that are understood involve regulation at the transcriptional level, one excellent example of translational control, the regulation of ribosomal protein synthesis, is also presented in chapter 30. Chapter 30 ends with a comprehensive summary section on the different types of regulatory proteins that are used to regulate transcription in prokaryotes.

The subject of regulation of gene expression in eukaryotes is complex and diffuse because so many different types of systems have been studied and the level of research effort in this field has reached unprecedented heights. Studies on this subject are truly at the cutting edge of modern biological investigations. In chapter 31, Regulation of Gene Expression in Eukaryotes, regulation of gene expression is first examined in yeast, a unicellular organism. We then examine regulatory mechanisms prevalent in multicellular eukaryotes. A section on the types of regulatory proteins most frequently found in eukaryotic systems is presented next. Finally, modes of regulation specific to developmental processes are considered.

Part 7 concludes with two brief, informative supplements that integrate key physiological and biochemical principles as they apply to two specialized processes that utilize genetic information: Supplement 3—Principles of Physiology and Biochemistry: Immunobiology and Supplement 4—Principles of Physiology and Biochemistry: Carcinogenesis and Oncogenes.

Ancillary Materials

For the Instructor

An *Instructor's Manual with Test Item File* contains suggestions on how to utilize the text in different course situations and detailed, worked-out solutions for the even-numbered problems found in the text chapters. These answers are **not** included in the *Student's Solutions Manual* that accompanies the text. In addition, this manual offers an average of 30 objective test questions for each chapter which can be used to generate exams. (ISBN 14276)

Classroom Testing Software is offered free upon request to adopters of this text. The software provides a database of questions for preparing exams. No programming experience is required. The software is available in IBM and Macintosh formats: IBM DOS 3.5 (ISBN 14278), IBM DOS 5.25 (ISBN 14279), WINDOWS 3.5 (ISBN 14280), and MAC 3.5 (ISBN 14282).

A set of *150 full-color acetate transparencies* is available free to adopters. These acetates feature key illustrations that can be used to enhance your classroom lectures. (ISBN 14277)

A set of *150 full-color projection slides* derived from the transparency illustrations is also available free to adopters. (ISBN 22871)

Electronic acetates, computerized image files for a majority of the text illustrations, will also be available free to adopters upon request. The electronic acetates will be available in a Mac/Windows (ISBN 26204). These electronic acetates can be clearly projected on large lecture hall screens using a LCD projection system.

A set of *175 transparency masters* is available free to adopters upon request. These black-and-white versions of in-text tables and illustrations can be used to prepare course handouts or additional transparency acetates. (ISBN 22872)

For the Student

A *Student's Solutions Manual* by Hugh Akers, Lamar University, and Caroline Breitenberger, Ohio State University, provides detailed, worked-out solutions for the odd-numbered problems found in the text. This solutions manual can help students to better understand how to solve the problems in the text and prepare for exams. (ISBN 22870)

A *Student Study Art Notebook,* a lecture companion containing the illustrations from the text that correspond to the acetate transparency images, is designed to help students spend more of their lecture time listening to the professor and less time copying down art from the overhead transparencies. A copy of this student study art notebook is packaged FREE with each new text. (ARTPAK ISBN 27016)

Computer Software and CD-ROM for Instructors and Students

NOTE: PLEASE ALSO REFER TO THE CD-SAMPLER THAT WAS PACKAGED WITH YOUR COPY OF *PRINCIPLES OF BIOCHEMISTRY*

Gene Game Software, by Bill Sofer of The State University of New Jersey–Rutgers, is an interactive software game that tests students' critical-thinking skills and knowledge of the scientific method as they attempt to work through "dry" lab protocols to clone a fictitious "Fountain of Youth" gene. The software provides direct feedback and hints to the student as protocols are completed. Protocols used and the results obtained are automatically recorded in the program's "lab notebook." Contact your bookstore, or call Wm. C. Brown Publishers at 1-800-338-5578 to place an order or request more information on this challenging, interactive software game. (ISBN 24893)

Biochemical Pathways Software, also created by Bill Sofer, is an easy-to-use tutorial review software program for Macintosh that provides quizzes/memory exercises that test the students' knowledge of Glycolysis and the TCA Cycle. Contact your bookstore, or call Wm. C. Brown Publishers at 1-800-338-5578 to place an order or request more information on this software. (ISBN 25100)

Molecules of Life CD-ROM, developed by a talented team of professionals from Purdue University, is a four CD-ROM set that allows the user to visualize key biomolecular structures through *interactive* animations, simulations, drills and tutorials. The set includes material on (1) Amino acids and Proteins, (2) Carbohydrates and Lipids, (3) Nucleic acids, and (4) Metabolism and Photosynthesis. The four CD-ROMs are available individually, or as a set. Contact your bookstore, or call Wm. C. Brown Publishers at 1-800-338-5578 to place an order or request more information on this software (Four CD set/MAC ISBN 27235) (Four CD set/WINDOWS ISBN 27264).

Acknowledgments

We wish to thank our biochemistry colleagues and contributing authors to the third edition of *Biochemistry;* Dr. Raymond L. Blakley, Dr. James W. Bodley, Dr. Ann Baker Burgess, Dr. Richard Burgess, Dr. Perry Frey, Irving Geis, Dr. Lloyd L. Ingram, Dr. Gary R. Jacobson, Dr. Julius Marmur, Dr. Richard Palmiter, Dr. Milton H. Saier, Jr., Dr. Pamela Stanley and Dr. H. Edwin Umbarger, for providing us with a wealth of comprehensive information from which to draw from in the development of *Principles of Biochemistry.* Special recognition is also due to Dr. Raymond L. Blakley for his work in writing chapter 23, and to Dr. Perry A. Frey, Dr. Gary R. Jacobson, Dr. Pamela Stanley, and Dr. H. Edwin Umbarger for their invaluable feedback and proofing efforts during the many phases of writing and publishing this text. In addition, we thank Hugh Akers, Lamar University, and Caroline Breitenberger, Ohio State University, for their insightful contributions to the end-of-chapter problems in this text. We are also indebted to Michael Pique, The Scripps Research Institute, and Holly Miller, Wake Forest University Medical Center, for their special contributions to the development and generation of many of the molecular graphics images in this text.

My fellow authors and I would also like to extend a special thank you to our many colleagues across the country for reviewing the text manuscript and making many helpful suggestions. The reviewers included:

Hugh Akers
Lamar University

Richard M. Amasino
UW–Madison

Dean R. Appling
The University of Texas at Austin

John N. Aronson
University of Arizona

Paul Austin
Hanover College

Derek Baisted
Oregon State University

Terry A. Barnett, Ph.D.
Southwestern College

E. J. Behrman
The Ohio State University

Paul Arthur Berkman
Ohio State University

Frank O. Brady, Ph.D.
University of South Dakota School of Medicine

Caroline A. Breitenberger
Ohio State University

Ronald W. Brosemer
Washington State University

Oscar P. Chilson
Washington University

David P. Chitharanjan
University of Wisconsin, Stevens Point

Alan D. Cooper
Worcester State College

Rick H. Cote
University of New Hampshire

Mukul C. Datta
Tuskegee University

Lawrence C. Davis
Kansas State University

Dr. Paul H. Demchick
Barton College

Michael W. Dennis
Montana State University-Billings

Kathleen A. Donnelly, Ph.D.
Russell Sage College

Lawrence K. Duffy
University of Alaska Fairbanks

John R. Edwards
Villanova University

Alfred T. Ericson
Emporia State University

Robert J. Evans
Illinois College

David Fahrney
Colorado State University

H. Richard Fevold
University of Montana

Christopher Francklyn
University of Vermont College of Medicine

Edward A. Funkhouser
Texas A & M University

Edwin J. Geels
Dordt College

Darrel Goll
University of Arizona

Dr. Eugene Gooch
Elon College

Milton Gordon
University of Washington, Seattle

Joan M. Griffiths
Cornell University

Lonnie J. Guralnick
Western Oregon State College

James H. Hageman
New Mexico State University

B. A. Hamkalo
University of California, Irvine

Kenneth D. Hapner
Montana State University

Gerald W. Hart
University of Alabama at Birmingham

Terry L. Helser
S.U.N.Y College at Oneonta

Pui Shing Ho
Oregon State University

Joel Hockensmith
University of Virginia

Daniel Holderbaum
Case Western Reserve University

Charles F. Hosler, Jr.
University of Wisconsin-La Crosse

Larry Jackson
Montana State University

Ralph A. Jacobson
CAL POLY

John R. Jefferson
Luther College

Colleen B. Jonssen
New Mexico State University

Floyd W. Kelly
Casper College

Mary B. Kennedy
California Institute of Technology

R. L. Khandelwal
University of Saskatchewan

Nazir A. Khatri
Franklin College of Indiana

Ramaswamy Krishnamoorhi
Kansas State University

James I. Lankford
St. Andrews Presbyterian College

Daniel J. Lavoie
Saint Anselm College

Franklin R. Leach
Oklahoma State University

Carol Leslie
Union University

Michael Leung
State University of New York/Old Westbury

Randolph V. Lewis
University of Wyoming

Albert Light
Purdue University, West Lafayette

Donald R. Lueking
Michigan Technological University

Dr. Celia L. Marshak
University of San Diego 1994
(Emeritus, San Diego State University)

Lynn M. Mason
Lubbock Christian University

Harry R. Matthews
University of California at Davis

Martha McBride
Norwich University

William L. Meyer
University of Vermont

Holly Miller
Wake Forest University Medical Center

Michael J. Minch
University of the Pacific

Debra M. Moriarity
University of Alabama in Huntsville

Mary E. Morton
College of the Holy Cross

Melvyn W. Mosher
Missouri Southern State College

Stephen H. Munroe
Marquette University

Richard M. Niles
Marshall University School of Medicine

Jennifer K. Nyborg
Colorado State University

William R. Oliver
Northern Kentucky University

Dr. Richard Steven Pappas
Georgia State University

Raymond Earl Poore, Ph.D.
Jacksonville State University

William T. Potter
The University of Tulsa

Gary L. Powell
Clemson University

Michael Eugene Pugh
Bloomsburg University of Pennsylvania

Paul D. Ray
University of North Dakota

Philip Reyes
University of New Mexico School of Medicine

John M. Risley
The University of North Carolina at Charlotte

H. Alan Rowe
Norfolk State University

John E. Robbins
Montana State University

Norman G. Sansing
University of Georgia

Roy A. Scott, III
The Ohio State University

Steven E. Seifried
John A. Burns School of Medicine
University of Hawaii

Ralph Shaw
Southeastern Louisiana University

J. M. Shively
Clemson University

Roger D. Sloboda
Dartmouth College

Deborah Kay Smith
Meredith College

Thomas Sneider
Colorado State University

Wesley E. Stites
University of Arkansas

Eric R. Taylor
University of Southwestern Louisiana

Martin Teintze
Montana State University

Arrel D. Toews
University of North Carolina at Chapel Hill

H. Edwin Umbarger
Purdue University

Harry van Keulen
Cleveland State University

Robert J. Van Lanen
Saint Xavier University

Charles Vigue
University of New Haven

William H. Voige
James Madison University

Raymond E. Waldner
Palm Beach Atlantic College

Arthur C. Washington
Tennessee State University

Daniel Weeks
University of Iowa

Steven M. Wietstock, Ph.D.
Alma College

Steven Woeste
Scholl College of Podiatric Medicine

Robert Zand
University of Michigan

We are grateful for the assistance of the editorial staff at Wm. C. Brown Publishers, especially Kevin Kane, publisher, Liz Sievers, our editor, Robin Steffek, developmental editor, Julie Kennedy, in-house production services coordinator, and Lori Hancock, photo editor. In taking this text from the raw manuscript stage to a production-ready stage, the authors have received tremendous assistance from Robin Steffek. Her input covers a wide range of activities from the meticulous checking of the manuscript for accuracy to numerous suggestions for modifications and special learning aids. She has pursued this project with great enthusiasm, skill, and dedication.

Once the project was ready for the production stage, we knew that a highly technical multicolor text of this magnitude would require an outstanding production team and a leader to see that every element in the final product would attain the highest quality. Laura Skinger directed the York Production Services team with confidence and determination to turn out the best possible final product. Whenever we the authors complained that something was not quite the way we wanted it, Laura would most willingly see that the concern was properly addressed without hesitation. In addition, she went way beyond our concerns to produce a text in which we could take great pride.

We hope very much that this text will be interesting and educational for students and a help to their instructors. We would appreciate any comments and suggestions from our readers. If you should find errors, please notify us so that we can make corrections in the second printing.

Geoffrey L. Zubay William W. Parson Dennis E. Vance

PART 7

Storage and Utilization of Genetic Information

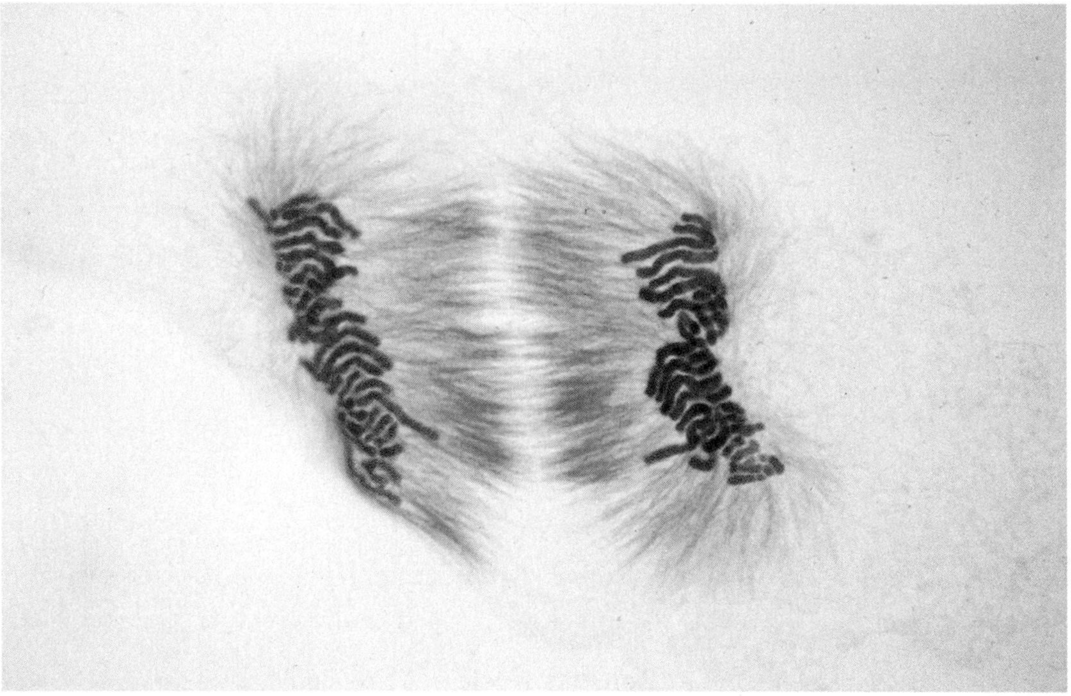

Light micrograph of late mitosis in a plant. Microtubules are stained red and chromosomes are counterstained blue. (Courtesy of Andrew Bajer.)

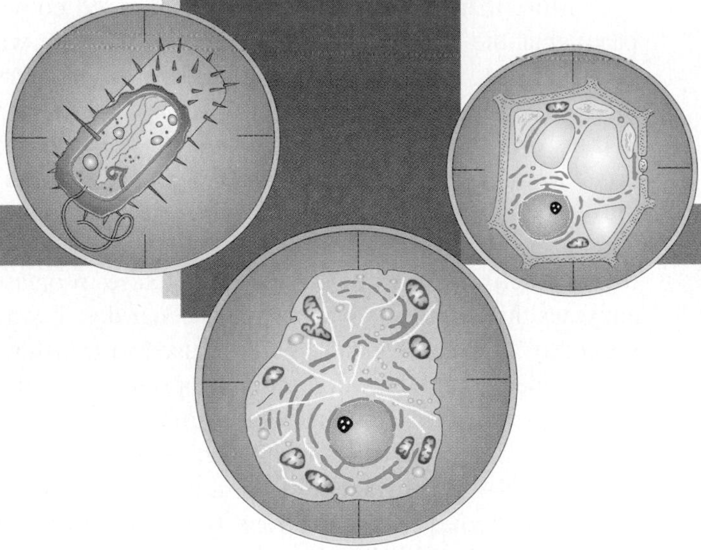

CHAPTER

25

Structures of Nucleic Acids and Nucleoproteins

Chapter Outline

Most DNAs form a double stranded helical structure in which nitrogen bases from opposing chains form hydrogen bonds with one another.

Nucleic acids occupy a unique position in the biochemical world. Not only are they involved in many important reactions, but they carry genetic information, which must be faithfully duplicated so that it can be passed from one cell generation to the next and from one organism to the next. Nonetheless this information must be mutable to produce the variability on which the evolutionary selection process feeds. Finally the DNA must be selectively transcribed so that each cell can synthesize the proteins it needs.

In the following five chapters we focus on the biochemistry of DNA, RNA, and protein, with occasional reminders of the genetic significance of these closely related processes. Appropriately, this section on nucleic acids and protein metabolism begins with a description of key experiments that demonstrated the genetic significance of nucleic acids. Following this we turn to a consideration of the structural properties of DNA and chromosomes. Next, reactions involving DNA are discussed, first purely biochemical reactions (chapter 26) and then *in vitro* and *in vivo* reactions in which DNA is manipulated (chapter 27). Chapter 28 deals with information transfer from DNA to RNA, and, finally, chapter 29 covers the mechanism of information transfer from RNA to protein.

The Genetic Significance of Nucleic Acids

After the discovery around the turn of the century that genes are carried by chromosomes, a great deal of effort went into characterizing the sizes and shapes of chromosomes. But it was not until much later that significant progress was made in elucidating the chemical nature of the gene. Biologists were aware, as early as 1900, that chromosomes are composed of both nucleic acids and proteins. However, the seemingly simple chemical composition of nucleic acids misled early investigators into believing that nucleic acids were a purely structural component of the chromosome. The favored theory was that the arrangement of specific proteins along the chromosome accounted for gene specificity. This notion was dispelled in 1944 when Oswald T. Avery and his colleagues at the Rockefeller Institute (now University) demonstrated that purified deoxyribonucleic acid (DNA) contains the genetic determinants of the bacterium *Diplococcus pneumoniae* (now called *Streptococcus pneumoniae*). In this section we discuss Avery's results, as well as some other historically important observations that led up to his experiments.

Transformation Is DNA-Mediated

In 1928, Fred Griffith was experimenting with two different strains of pneumococcus. Type S bacteria (S for smooth, from the appearance of bacterial colonies on agar plates) are encapsulated by polysaccharide. The capsules protect them from the host immune system, making the S bacteria pathogenic; even when small numbers of S bacteria are injected into mice, death results. By contrast, R bacteria (R for rough colonies) are nonencapsulated; they are readily attacked by the mouse's immune system and consequently are nonpathogenic. Although heat-killed S bacteria by themselves are nonvirulent, Griffith found that when heat-killed S bacteria were mixed with live R bacteria and introduced into a susceptible laboratory mouse, death of the animal frequently occurred (fig. 25.1*a*). S bacteria could then be detected in the blood. Apparently the genetic factor required for encapsulation was transferred from killed S cells to live R cells.

Griffith's result was duplicated outside of the animal by Dawson and Sia in 1930. Both bacterial strains were grown in liquid growth medium and distinguished by the distinctive appearance of their colonies on plates (see fig. 25.1*b*). In parallel with the results obtained in the mouse, heat-treated extracts of S cells transformed R cells into S cells.

More than 10 years later, Avery provided convincing proof that the active agent in the S cell extracts was the cellular DNA. He and his co-workers did this by purifying the DNA from S cells and showing that it had the capacity to transform R cells into S cells *in vitro* (see fig. 25.1*b*). The transforming activity in the purified extract was destroyed if the extract was first incubated with the enzyme DNase which specifically degrades DNA. By contrast the transforming activity was not affected by RNase, proteases, or enzymes that degrade capsular polysaccharides. It was subsequently found that DNA could be used to transfer many other genetic traits between the appropriate pairs of donor and recipient bacterial strains. For example, resistance to the antibiotics streptomycin or penicillin could be transferred, with the DNA of resistant cells, to sensitive cells. The transformation studies with different traits conferred by the donor DNA showed that the active genetic material being transferred faithfully reflected the genetic patterns of the donor strains. The procedure of altering the genetic composition of one cell strain by exposing it to DNA of another strain is termed transformation.

Transformation occurs naturally in only a small number of bacterial species other than *Streptococcus pneumoniae.* In organisms that are not transformed naturally, laboratory procedures are available to make the cell envelope partially permeable (e.g., of *E. coli*) and permit uptake of DNA. This is termed artificial transformation. It can be brought about by Ca^{2+} treatment and temperature shock; but the most efficient method found to be useful with most organisms is electroporation, or electric shock. For instance, by exposing yeast to electric field pulses it is possible to obtain transformation efficiencies that are 10^4 times greater than those obtained by conventional methods. Transformation is an essential step in the cloning of genes (see chapter 27). Transformation with self-replicating plasmids that may be carrying different genes has made it possible to select cells carrying the gene(s) of interest from a large population of cells.

Avery's experiments on cells were followed by additional experiments on viruses, demonstrating that the genetic information of a virus is carried by nucleic acids. In the case of a cell the genetic information is always carried by DNA. In the case of viruses the genetic information can be carried by either DNA or RNA, depending on the virus.

Structural Properties of DNA

The early work equating genetic material with DNA plunged genetics into an entirely new vocabulary of chemical terms. The genetic consequences of these early studies

Figure 25.1

In vivo and *in vitro* evidence that DNA causes transformation of *Streptococcus pneumoniae.* (*a*) Transformation experiment by Griffith. R bacteria are nonvirulent. S bacteria are virulent. A mixture of R bacteria and heat-killed S bacteria is also virulent if transformation has occurred. (*b*) Transformation experiment by Avery and co-workers. When bacteria from a liquid culture are spread on a semisolid medium, each cell adheres to the medium at random. As time passes, the cells and their offspring grow and divide, leading to visible clones or colonies, each of which arose from a single cell. R and S cells each have a distinct clonal morphology. Whereas R cells produce small rough colonies, S colonies are smooth. R cells exposed to DNA from S cells produce a mixture of both types of colonies: untransformed R colonies and transformed S colonies.

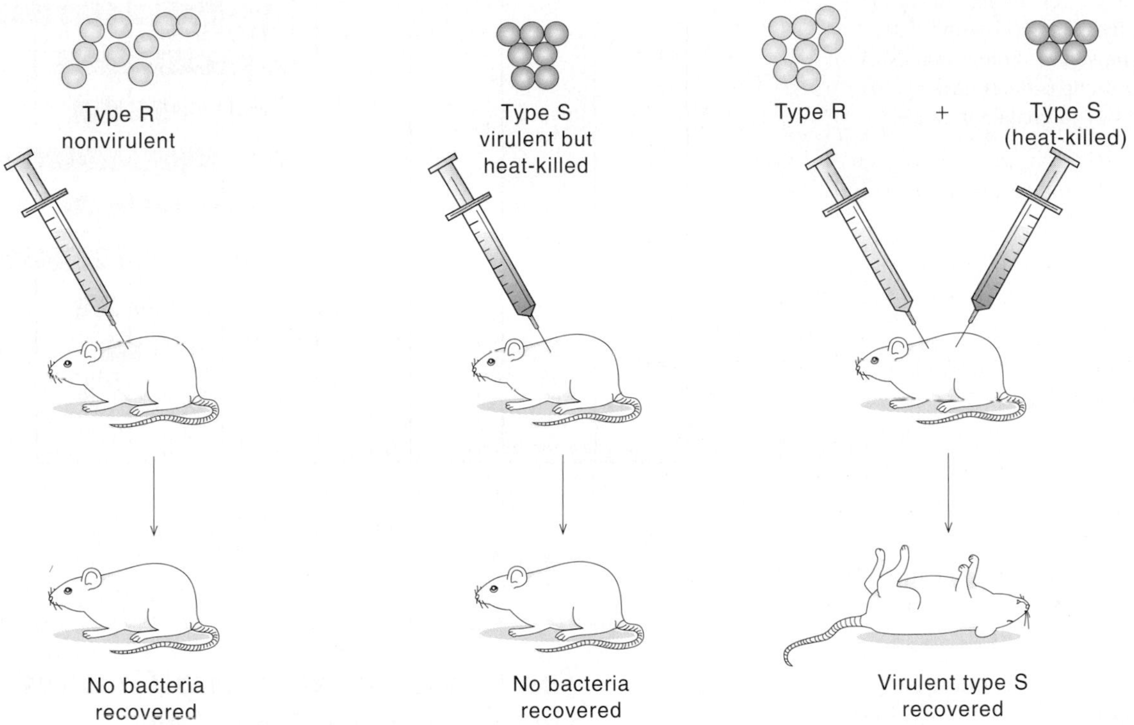

(a) *In vivo* results

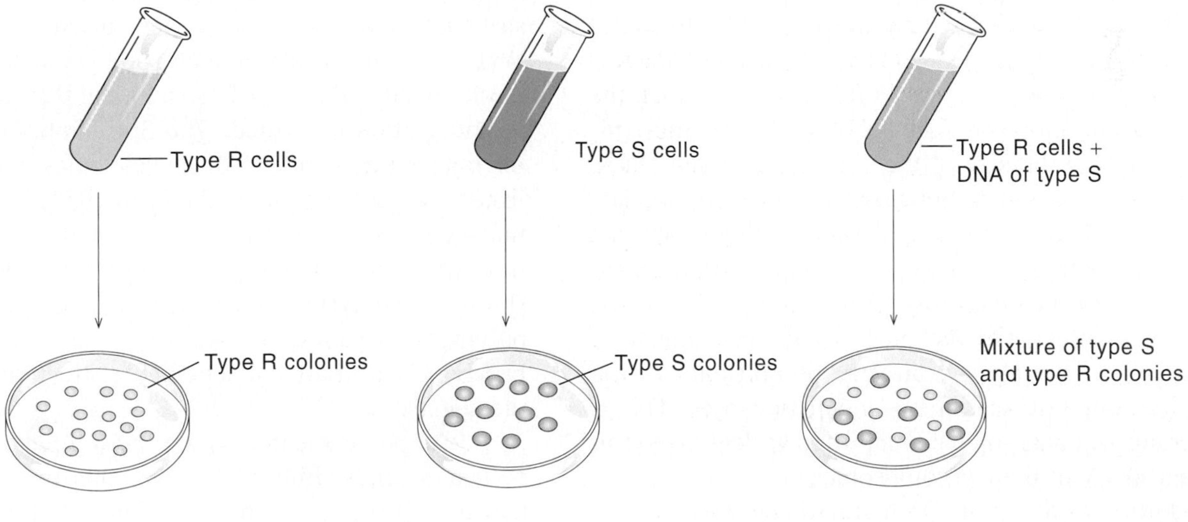

(b) *In vitro* results

Figure 25.2

Genome size in different cells, viruses, and plasmids. Plasmids are small circular DNA molecules that replicate autonomously in cells harboring them. Unlike viruses, they do not form any complex nucleoprotein structures. In the case of plasmids, most viruses, and bacteria, the genome size is equivalent to the size of the chromosomal DNA because there is only one chromosome. For all of the remaining eukaryotic organisms listed here, the genome is subdivided into two or more chromosomes. Some organisms contain more than 100 chromosomes. All chromosomes are believed to contain a single DNA molecule.

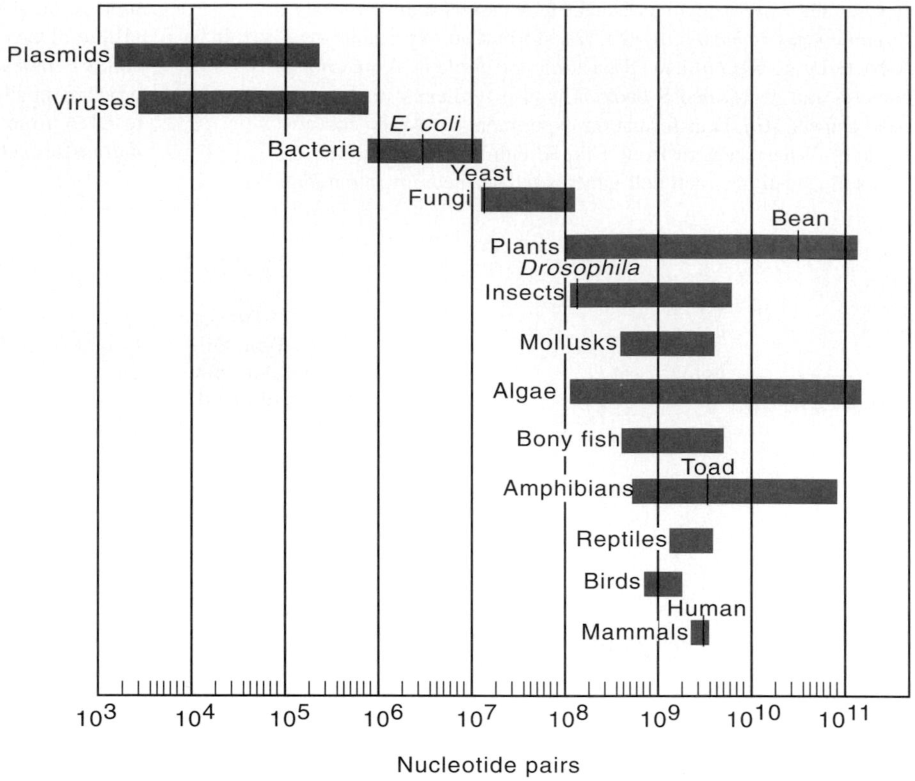

could be understood, but the chemistry was new. The remainder of this chapter focuses on the chemical and structural properties of DNA.

The amount of DNA per cell differs widely among different organisms (fig. 25.2). Mammalian cells contain about 1,000-fold more DNA than bacterial cells. Bacterial viruses such as the T type (T1–T7) bacteriophages (phages) that infect *E. coli* contain 10- to 20-fold less DNA than the bacterial host chromosome. The DNA of the smallest viruses is about 1/10 the size of the smallest T phage DNA, containing barely enough genetic material to accommodate about 10 genes. This finding is consistent with the fact that viruses do not contain sufficient genetic information for independent growth, but can only grow parasitically in the host cells they infect. On the other hand, the amount of DNA per cell is not always directly proportional to the amount of genetic information an organism carries. This is because complex eukaryotes contain a great deal of noninformational DNA in their chromosomes.

In addition to the main DNA associated with the cell or the virus, informational DNA is found in special organelles, such as chloroplasts and mitochondria. This DNA carries genes whose products are exclusively associated with organelle function.

The Polynucleotide Chain Contains Mononucleotides Linked by Phosphodiester Bonds

Nucleotides are the building blocks of nucleic acids; their structures and biochemistry were discussed in chapter 23. When a 5′-phosphomononucleotide is joined by a phosphodiester bond to the 3′-OH group of another mononucleotide, a dinucleotide is formed. The 3′-5′-linked phosphodiester internucleotide structure of nucleic acids was firmly established by Lord Alexander Todd in 1951. Repetition of this linkage leads to the formation of polydeoxyribonucleotides in DNA or polyribonucleotides in RNA. The structure of a short polydeoxyribonucleotide is shown in figure 25.3. The polymeric structure consists of a sugar phosphate diester backbone with bases attached as distinctive side chains to the sugars.

The polynucleotide chain has a directional sense with 5′ and 3′ ends. Either of these ends may contain a free hydroxyl group or a phosphorylated hydroxyl group. The structure shown in figure 25.3 contains a phosphate group on the 5′ end but none on the 3′ end. By convention, one writes a nucleic acid sequence from the 5′ to the 3′ end so that a comparable structure is written pTpApCpG. With no

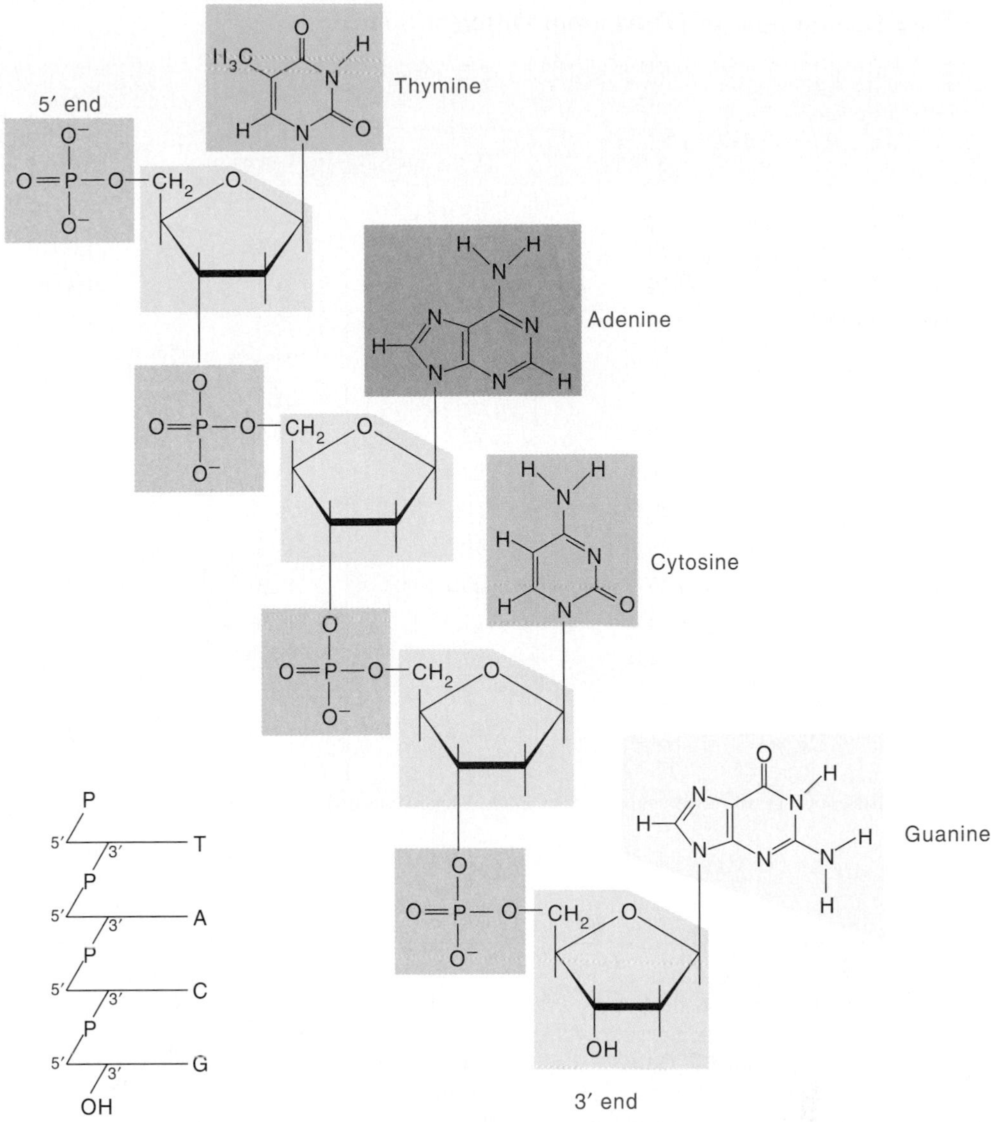

Figure 25.3

The structure of a deoxyribonucleotide. Drawn in abbreviated form at lower left. The illustrated structure is written pTpApCpG.

phosphate on the 5′ end, the structure is designated TpApCpG; alternatively, if the terminal phosphate is on the 3′ end rather than the 5′ end, the structure is written TpApCpGp. When the phosphates are not indicated, the structure is indicated by dashes on either end: -TACG-. The letters ''d'' or ''r'' sometimes precede the capital letter of the nucleotide to indicate a deoxyribo- or a riboderivative.

Most DNAs Exist as Double-Helix (Duplex) Structures

Like most other types of biological macromolecules, nucleic acids adopt highly organized three-dimensional structures. The dominant factors that determine nucleic acid conformations are the limitations imposed by the stereochemistry of the polynucleotide chains, the high negative charge resulting from the regularly repeating phosphate groups, and the noncovalent affinities between purine and pyrimidine bases.

A body of chemical information that proved vital to understanding DNA structure came from Erwin Chargaff's analyses of the nucleotide composition of duplex DNAs from various sources (table 25.1). Although the base compositions varied over a wide range, Chargaff found that within the DNA of each source that he examined, the amount of A was very nearly equal to the amount of T, and the amount of G was very nearly equal to the amount of C. The C is present as both unmodified C and, to a lesser extent, 5-methyl-cytosine, which results from postreplicative

Table 25.1

Base Composition of DNAs from Different Sources

	(A) Adenine	(G) Guanine	(C) Cytosine	(5-MC) 5-Methyl-cytosine	(T) Thymine	$\frac{A + T}{G + C + 5\text{-MC}}$
Human	30.4	19.6	19.9	0.7	30.1	1.53
Sheep	29.3	21.1	20.9	1.0	28.7	1.38
Ox	29.0	21.2	21.2	1.3	28.7	1.36
Rat	28.6	21.4	20.4	1.1	28.4	1.33
Hen	28.0	22.0	21.6		28.4	1.29
Turtle	28.7	22.0	21.3		27.9	1.31
Trout	29.7	22.2	20.5		27.5	1.34
Salmon	28.9	22.4	21.6		27.1	1.27
Locust	29.3	20.5	20.7	0.2	29.3	1.41
Sea urchin	28.4	19.5	19.3		32.8	1.58
Carrot	26.7	23.1	17.3	5.9	26.9	1.16
Clover	29.9	21.0	15.6	4.8	28.6	1.41
Neurospora crassa	23.0	27.1	26.6		23.3	0.86
Escherichia coli	24.7	26.0	25.7		23.6	0.93
T4 Bacteriophage	32.3	17.6		16.7[a]	33.4	1.91

[a] In the T4 bacteriophage all of the cytosine exists in the 5-hydroxymethyl form 5-HMC.

Figure 25.4

Dimensions and hydrogen bonding of (*a*) thymine to adenine and (*b*) cytosine to guanine. Note that two hydrogen bonds are formed in the A-T base pair and three in the G-C base pair. The overall dimensions of the base pairs are the same. Consequently they fit at any position in an otherwise regular polymeric structure. (Source: Adapted from S. Arnott, M. H. F. Wilkins, L. D. Hamilton, and R. Langridge, Fourier synthesis studies of lithium DNA, part III: Hoogsteen models, *J. Mol. Biol.* 11:391, 1965.)

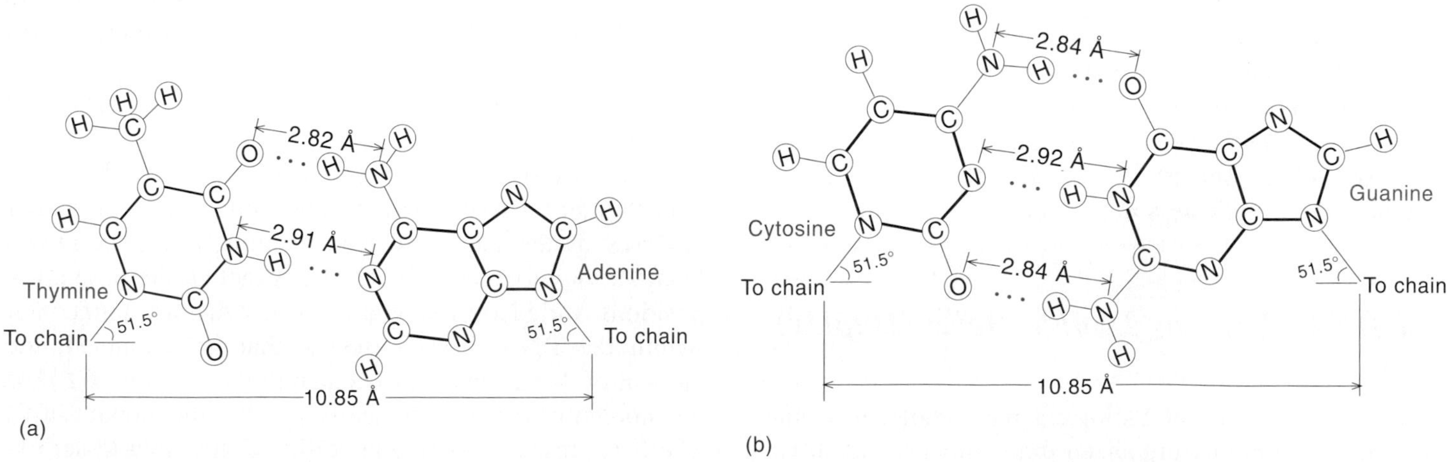

Figure 25.5

Segment of DNA, drawn to emphasize the hydrogen bonds formed between opposing chains. Each type of base is represented by a different color, with the sugar-phosphate backbones in blue and yellow. Note the three hydrogen bonds in the G-C pairs and the two in the A-T pairs (A, red; T, green; G, yellow; C, blue). The two strands are antiparallel: One strand (left side) runs 5′ to 3′ from top to bottom, and the other strand (right side) runs 5′ to 3′ from bottom to top. The planes of the base pairs are turned 90° to show the hydrogen bonds between the base pairs. (Source: Adapted from A. Kornberg, The Synthesis of DNA, *Scientific American*, October 1968.)

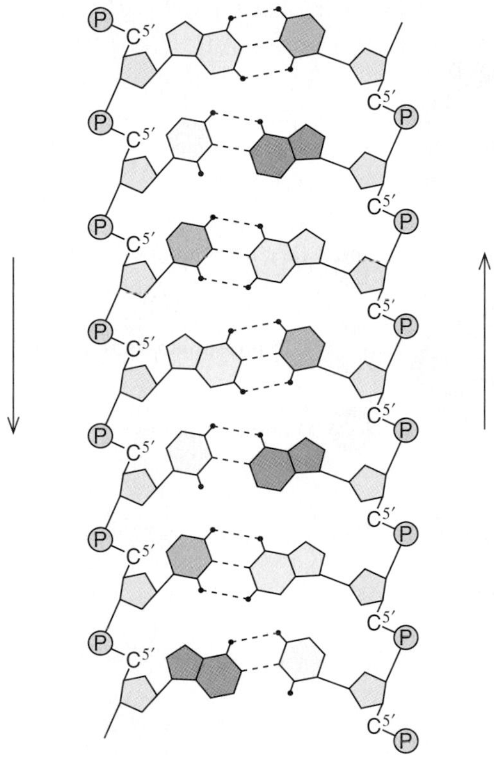

modification. The two equalities were the first indication that regular complexes occur between A and T and between G and C in DNA.

While searching for the meaning of these equalities, James Watson noted that hydrogen-bonded base pairs with the same overall dimensions could be formed between A and T and between G and C (fig. 25.4). The A-T base-paired structure has two hydrogen bonds, whereas the G-C base pair has three. The hydrogen-bonded pairs are formed between bases of opposing strands and can only arise if the directional senses of the two interacting chains are opposite or antiparallel (fig. 25.5). With this notion in mind Francis Crick took a closer look at the x-ray diffraction pattern produced by DNA and was able to interpret the diffraction pattern in terms of a helix (see Methods of Biochemical Analysis 25A) composed of two polynucleotide strands. In this structure the planes of the base pairs are perpendicular to the helix axis, and the distance between adjacent pairs along the helix axis is 3.4 Å, bringing them into close contact (fig. 25.6). The structure repeats itself after 10 residues, or once every 34 Å along the helix axis; the repeating distance is referred to as the pitch length or just the pitch. An average-sized bacterial gene, which encodes the information to make a single protein, is about 1,000 bp in length, equivalent to 100 helical turns. As we see (chapters 26 and 28) the complementary structure of duplex DNA hints at how the genetic material is faithfully replicated as well as how it is expressed.

An important feature of the helical structure is the grooved nature of the surface resulting from the helical twist. Alternating wide (major) and narrow (minor) grooves are displayed in a side view of the helix structure (see fig. 25.6*a* and *b*). Different sections of the purine and pyrimidine bases are exposed in these two grooves as indicated in figure 25.6*c*. Many different proteins interact with DNA; most of the interactions occur with the phosphoryl groups on the outer surface of the structure and with the purine and pyrimidine bases in the major groove because of its greater accessibility. Specific instances of DNA-protein interactions are considered in later chapters (chapters 30 and 31).

Hydrogen Bonds and Stacking Forces Stabilize the Double Helix

Several factors account for the stability of the double-helix structure. The negatively charged phosphoryl groups are all located on the outer surface, where they have a minimum effect on one another. The repulsive electrostatic interactions generated by these charged groups are often partly neutralized by interaction with cations such as Mg^{2+}, basic polyamines (such as putrescine and spermidine), and the positively charged side chains of chromosomal proteins. The core of the helix is composed of the base pairs held together by the specific hydrogen bonds and also by favorable stacking interactions between the planes of adjacent base pairs. These stacking interactions are complex, involving dipole–dipole interactions and van der Waals forces; this results in a stacking energy comparable in magnitude to the stabilizing energy generated by the hydrogen bonds between the base pairs. The result is that stacking is maximized in most nucleic acid structures. In this connection it is noteworthy that two fully extended polynucleotide strands can form a hydrogen-bonded base-paired complex, leading to a stepladderlike structure (fig. 25.7). In this structure the chains do not form a helix but lie straight, with a distance of

Figure 25.6

The most common form of the double-helix DNA. The base-paired structure shown in figure 25.5 forms the helix structure shown in (*a*) and (*b*) by a right-handed twist. The two strands are antiparallel as indicated by the curved arrows in (*a*). (Reprinted with permission from *Nature* (171:737, 1953) Copyright 1953 Macmillan Magazines Limited.) In (*b*) a space-filling model depicts the sugar-phosphate backbones as strings of mostly gray, red, white, and yellow spheres, and the base pairs are rendered as horizontal flat plates composed of dark blue spheres. (Reprinted with permission from *Nature* (175:834, 1955) Copyright 1955 Macmillan Magazines Limited.) In (*c*) the orientation of the groups in the base pairs with respect to the major and minor grooves is indicated.

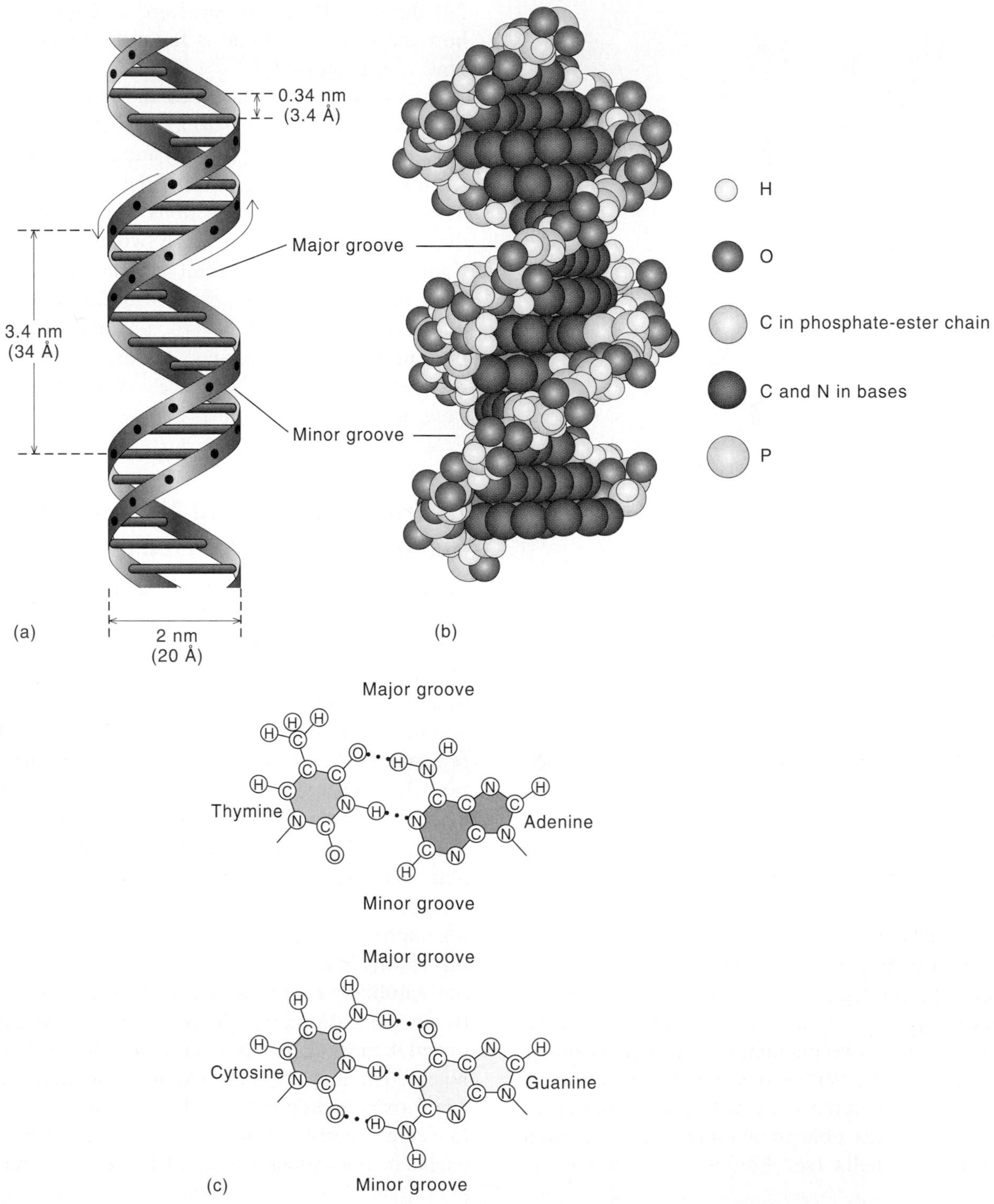

Figure 25.7

Different conformations of base-paired DNA: (*a*) the untwisted straight ladder, (*b*) the normal spiral ladder. The stepladder structure is unstable; it can be converted into a spiral ladder by a right-handed twist, a change that permits the planes of the base pairs to come into close contact.

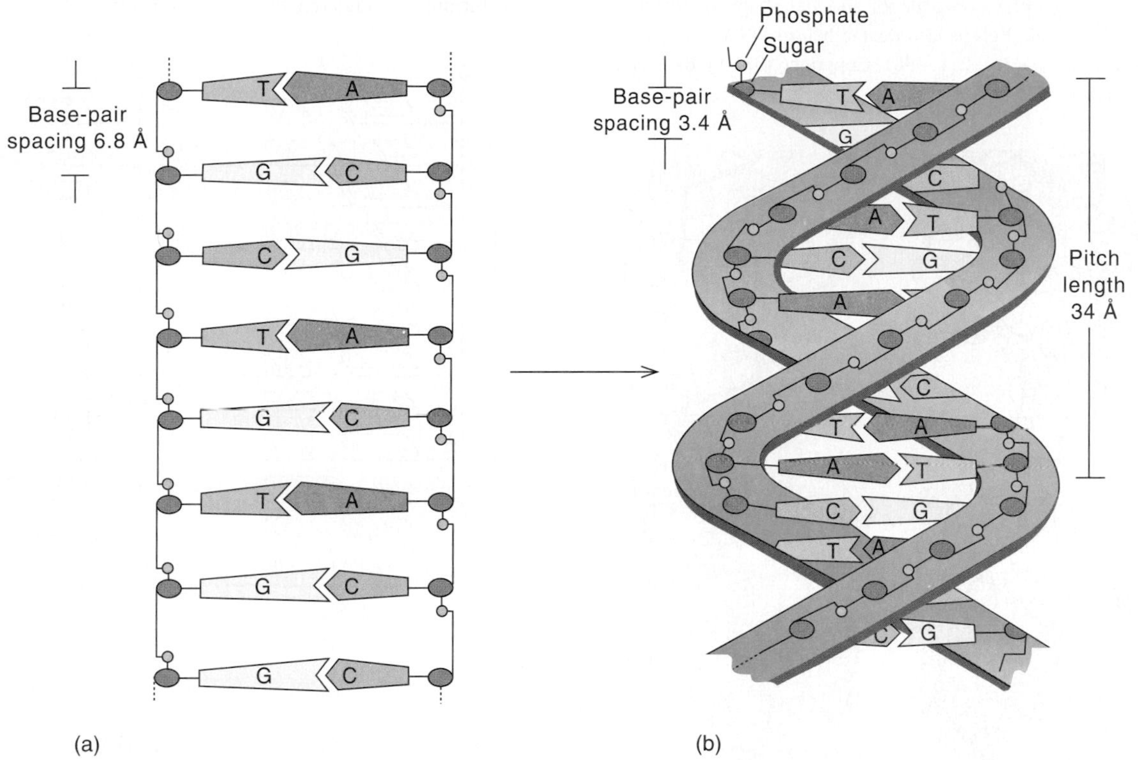

6.8 Å between adjacent nucleotides in the direction of the long axis. This 6.8-Å distance between adjacent base pairs produces a gap that would presumably be filled by water. Such a conformation is unstable because the planes of the bases prefer close contacts with one another as does water. The stepladder structure is related to the helix structure by a simple right-handed (clockwise) twist (see fig. 25.7*a*). Following this operation the distance between base pairs decreases until they are in close contact, with a spacing of 3.4 Å.

Conformational Variants of the Double-Helix Structure

The same base-pairing arrangement is found in all naturally occurring double-helix structures. However, the inherent flexibility in the ribose sugar and the degrees of freedom generated by several rotatable single bonds per residue, six in the sugar-phosphate backbone and one in the C-1′–N-glycosidic linkage, lead to considerable variation in the conformations adopted by double-helix structures.

The most striking conformational variant observed for a DNA double helix with Watson-Crick base pairing is referred to as the Z form. In Z DNA the backbone is twisted in the left-handed (counterclockwise) direction. This structure was first detected by Alex Rich and his co-workers (fig. 25.8). The Z form is a considerably slimmer helix than the B form and contains 12 bp/turn rather than 10. In the Z form, the planes of the base pairs are rotated approximately 180° with respect to the helix axis from their orientation in the B form (fig. 25.9).

Because of the different orientations of the bases in Z DNA, this DNA conformation requires that the sequence of purine and pyrimidine bases in each chain strictly alternate. An alternating sequence of G and C, or T and G or A and C residues can adopt a Z conformation. Of course, in all cases the opposing strand must contain a sequence of bases that is complementary as in all DNA duplex structures. An alternating A-T DNA sequence cannot adopt the Z confor-

Figure 25.8

Space-filling models of (*a*) B DNA and (*b*) Z DNA. The irregularity of the Z DNA backbone is illustrated by the heavy lines that go from phosphate to phosphate residue along the chain. In contrast, B DNA has a smooth line that connects the phosphate groups and the two grooves, neither one of which extends into the helix axis of the molecule. The space-filling model is excellent for displaying the volume occupied by molecular constituents and the shape of the outer surface. (From A. Wang et al., Left-handed double helical DNA: Variations in the backbone conformation, *Science,* 211, 1981. Copyright 1981 by the AAAS. Reprinted by permission.)

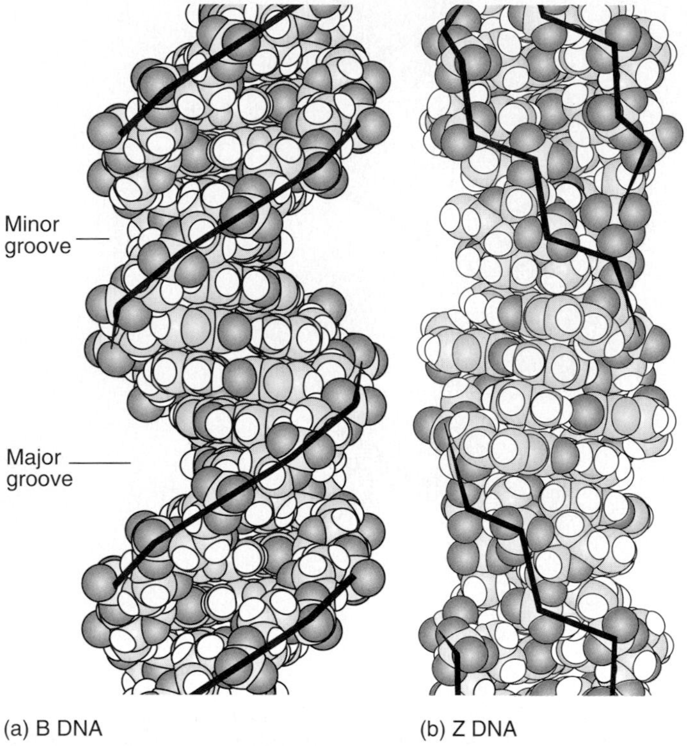

Figure 25.9

The change in topological relationship if a four-base-pair segment of B DNA is converted into Z DNA. Such a conversion could be accomplished by rotation of the bases relative to those in B DNA. This rotation is shown diagrammatically by coloring one surface of the bases. All of the colored areas are at the bottom in B DNA. In the segment of Z DNA, however, four of them are turned upward. The turning is indicated by the curved arrows.

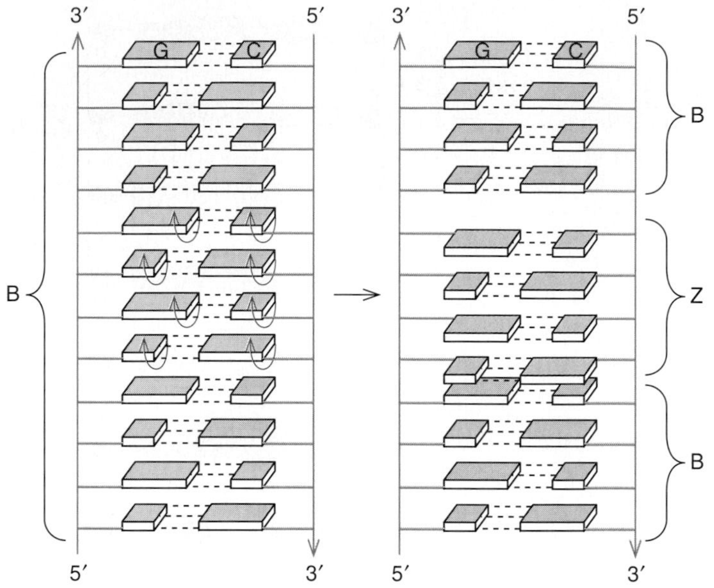

mation. This is believed to have something to do with the way that water molecules orient around an A-T base pair in the Z helix.

The biologic significance of Z DNA is currently unclear. However, several cellular proteins that bind specifically to Z DNA have been isolated from the nuclei of *Drosophila* fruit flies. The mere existence of such proteins suggests that they may function in some specific role when they encounter stretches of DNA that can adopt a Z conformation.

Duplex Structures Can Form Supercoils

The detection of different conformations of DNA underscores the inherent flexibility built into the DNA duplex. All the conformations discussed thus far involve regular linear duplexes. Energetically favorable interactions with other molecules, particularly proteins, can induce additional conformations that do not result in major changes in either pairing or stacking. Several conformations are believed to play important roles in different situations. Bends are known to be important in structures formed by chromosomes (discussed later in this chapter). Cruciforms, in which a single chain folds back on itself into a hairpinlike duplex, are formed at the ends of eukaryotic chromosomes (see the discussion of telomeres in chapter 26). Supercoiled DNA is a very common type of tertiary structure in which the double-helix segments twist around each other. Supercoiled DNA is topologically constrained by being covalently closed and circular, or by being complexed to proteins so that the ends of the DNA cannot rotate freely.

DNA can form right-handed (negatively supercoiled) or left-handed (positively supercoiled) supercoils. A right-handed supercoiled structure is shown in the middle of figure 25.10. Negative supercoiling imparts a torsional stress to the DNA that favors unwinding, whereas positive supercoiling favors tighter winding of the double helix.

Supercoiling of circular duplex DNA is quantitatively considered in terms of the linking number (L), an integer that specifies the number of complete turns made by one strand around the other. The linking number can change only if a covalent linkage in the DNA backbone is broken and reformed. Enzymes called topoisomerases (see chapter

Figure 25.10

A circular duplex molecule in different topological states. (Adapted from a diagram supplied by M. Gellert.) The linking number can be changed only by breakage and re-formation of the phosphodiester linkages, as shown in the conversion of the relaxed circular form (1) to the strained negatively supercoiled form (2). The strain in the negatively supercoiled form can be partitioned in different ways between twist (T) and supercoiling (S) as shown in the interconversion between (2) and (3). No phosphodiester linkages are broken in making this interconversion, and consequently no change occurs in the linking number (L) as indicated.

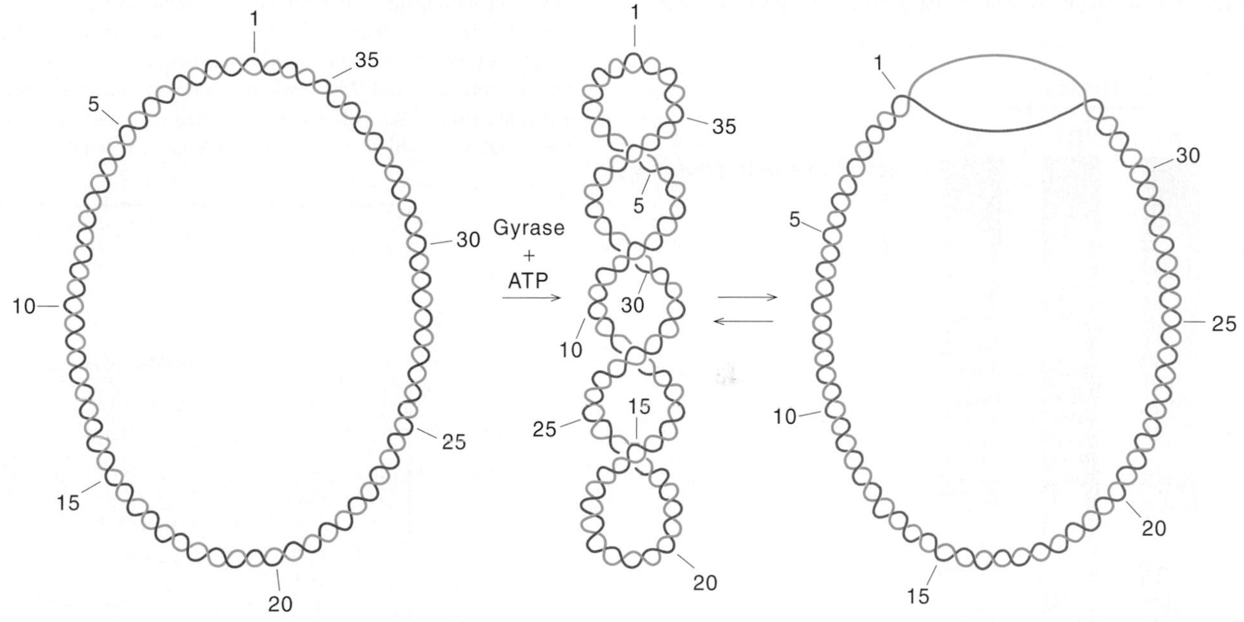

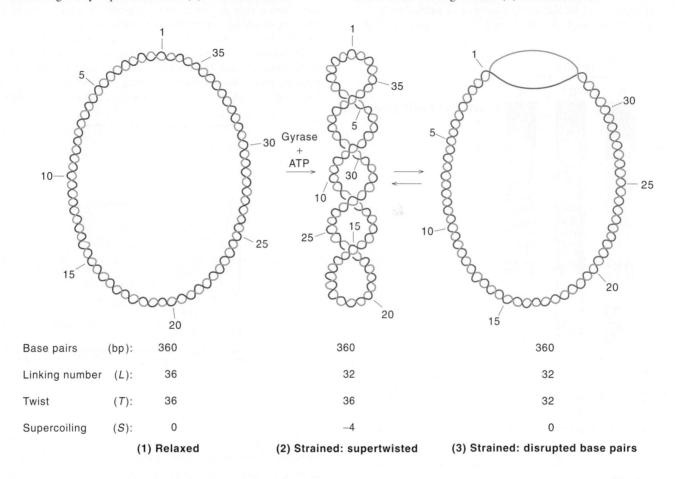

		(1) Relaxed	(2) Strained: supertwisted	(3) Strained: disrupted base pairs
Base pairs	(bp):	360	360	360
Linking number	(L):	36	32	32
Twist	(T):	36	36	32
Supercoiling	(S):	0	–4	0

26) catalyze changes in the linking number. If a molecule of DNA is projected onto a two-dimensional surface, the linking number is defined as the excess of right-handed over left-handed crossings of one strand over the other. Linear duplex DNA with free ends adopts a conformation in solution close to the B form, with about 10 bp/turn. Therefore, a closed circular duplex with this extent of twist is presumed to be under no torsional strain, and is said to be relaxed. Because B DNA is a right-handed helix, the linking number is normally positive by the sign convention we have adopted. The values of the linking number of relaxed DNA, $L°$, are distributed over a narrow range of integral values centered around 1 per 10 base pairs. DNA with a mean linking number smaller than this is termed negatively supercoiled, or underwound; DNA with a larger linking number is termed positively supercoiled, or overwound. The deviation of the linking number from its relaxed value, $\Delta L = L - L°$, can be partitioned between twist (altered double-helix coiling) and supercoiling.

$$L = \text{twist } (T) + \text{supercoiling } (S)$$

At present we do not know precisely how L partitions between twist and supercoiling for helices under torsional stress. For example, consider the hypothetical situation illustrated in figure 25.10. A 360-bp structure in the circular relaxed form ($L = +36$, $T = +36$, $S = 0$) is indicated on the left. Exposure to the bacterial topoisomerase known as DNA gyrase introduces negative supercoils into such a structure in a reaction that requires ATP. If four negative supertwists are introduced, L is reduced to $+32$. Barring other changes, T remains fixed and S becomes -4. In fact, the torsional strain introduced by the four negative supertwists tends to reduce T, causing a partial unwinding of the duplex. At one extreme, this effect can lead to the unwind-

Figure 25.11

Electrophoretic patterns of highly supercoiled or partially supercoiled DNA. Strip A represents a sample of circular duplex DNA obtained by deproteinization of the animal virus SV40. In strips B and C the DNA has been exposed for increasing times to an enzyme (topoisomerase) that catalyzes relaxation. Adjacent bands differ by 1 in linking number. (From W. Keller, Characterization of purified DNA-relaxing enzyme from human tissue culture cells, *Proc. Natl. Acad. Sci. USA* 72:2553, 1975.)

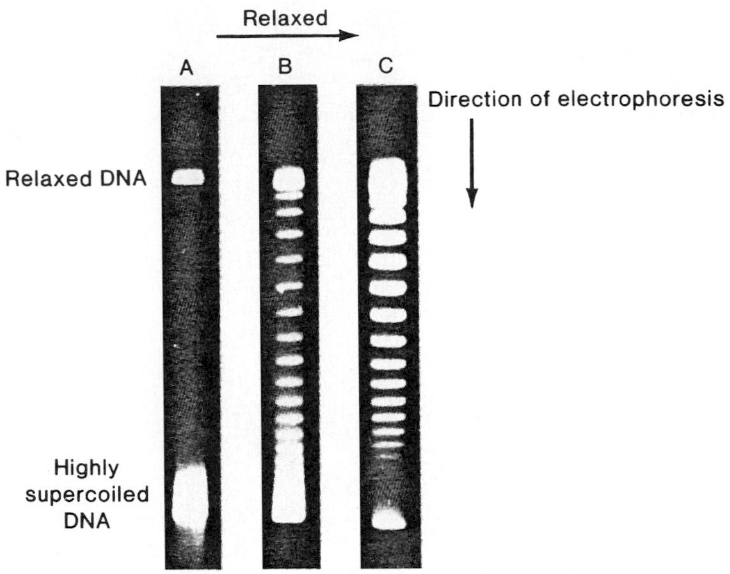

Figure 25.12

Effect of temperature on the relative absorbance of native, renatured, and denatured DNA. When native DNA is heated in aqueous solution, its absorbance does not change until a temperature of about 80°C is reached, after which the absorbance rises sharply, by about 40% (curve *a*). On cooling the absorbance falls, but along a different curve, and it does not return to its original value (curve *b*). Renatured DNA, in which the two strands have been brought back into perfect register, shows a sharp melting curve similar to native DNA (curve *c*). Renatured DNA is prepared from denatured DNA by holding the temperature at about 25°C below the denaturation temperature for an extended time. This subject is discussed in detail later in the text. The temperature at which the native DNA is half denatured is labeled T_m.

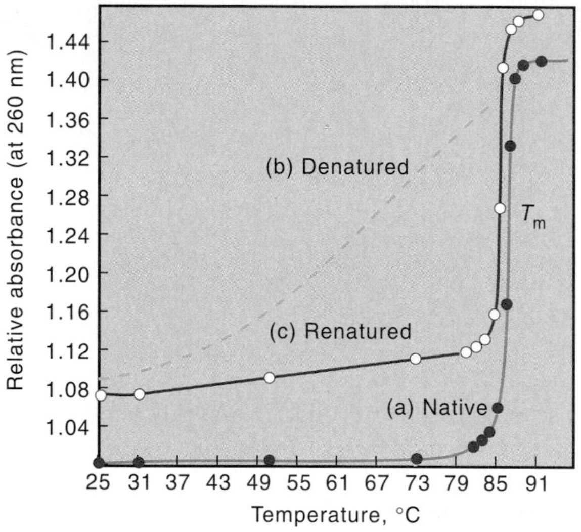

ing of four helical turns, in which case the supercoiling disappears entirely ($L = +32$, $T = +32$, $S = 0$). The actual situation probably leads to a reduction in the negative value of S and a concomitant reduction in T that is spread over the entire duplex as pictured. Note that the linkage number L changes only when covalent bonds are broken, as in the case of gyrase treatment.

For DNA with a molecular weight of less than 10^7, agarose gel electrophoresis is a most effective method for assessing the extent of supercoiling. DNA isomers differing by 1 in linking number form separate bands in the gel (fig. 25.11). The more highly supercoiled molecule migrates more rapidly through the gel as a result of its more compact structure.

DNA Denaturation Involves Separation of Complementary Strands

In the laboratory, double-stranded DNA can be separated into single strands. The process of separating the polynucleotide strands of duplex nucleic acid structures is called denaturation. Denaturation disrupts the secondary binding forces that hold the strands together. Recall that the secondary binding forces include the edge-to-edge hydrogen bonds between the base pairs of opposing strands and the face-to-face stacking forces between the planes of adjacent base pairs. Individually, these secondary forces are weak, but when they act cooperatively they give rise to a DNA duplex that is highly stable in aqueous solution. The conditions required to denature DNA provide us with a measure of the strength of these interactions.

One of the simplest ways to denature DNA is by heating. The extent of denaturation at any temperature can be measured by the change in ultraviolet absorbance of a solution of DNA. A rise in absorbance coincides with the disruption of the regular base-paired structure, and the separation of the two polynucleotide strands from one another. The sharpness of the disruption of the regularly hydrogen-bonded base-paired native structure may be likened to the melting of a pure organic compound. It is customary to refer to this ultraviolet absorption temperature profile as a melting curve (curve *a* in fig. 25.12).

The melting temperature, T_m, of DNA is defined as the temperature at the midpoint of the absorption increase. This is about 85°C for the example shown in figure 25.12. Rapid cooling of the denatured DNA solution leads to re-formation of intrastrand hydrogen bonds but in a nonspecific, irregular manner. The absorbance decreases, but only by about three-fourths of the total original increase, and the decrease occurs over a much broader range of temperatures, as shown by curve *b* in figure 25.12. On subsequent reheating and cooling, the absorbance follows this cooling curve in the appropriate direction, indicating that denaturation results in an irreversible change.

The melting curve is an excellent tool for detecting DNAs that occur naturally in the single-stranded conformation. For example, in certain bacteriophages that infect *E. coli,* such as ϕX174, fd, or M13, the DNA exists as a single, circular strand. The ultraviolet absorption temperature curve for the DNA of these phages is similar in shape to that observed for denatured DNA (curve *b* in fig. 25.12). Broad melting curves also are characteristic of most RNAs, which rarely have regions of regular base pairing that extend for more than 10 or 20 residues.

Melting curves also have provided evidence that the stability of the double-helix structure is a function of its base composition. The midpoint of denaturation (T_m) of naturally occurring DNAs is precisely correlated with the average base composition of the DNA: The higher the mole percent of G-C base pairs, the higher the T_m (fig. 25.13). This seems reasonable, because the G-C base pair contains three hydrogen bonds, whereas the A-T base pair contains only two (see fig. 25.4); thus DNA with a greater G-C content is expected to be more stable. As indicated above, base stacking is also believed to contribute to the stability of the duplex structure. In general the interaction energy gained by stacking between adjacent G-C base pairs is greater than that gained by interaction between A-T base pairs.

Other factors present in aqueous solution can affect the stability of the double-helix structure in a positive or a negative way. For example, salt has a stabilizing effect, which is mainly due to the repulsive electrostatic interactions between the negatively charged phosphate groups. Salt shields this charge interaction and therefore stabilizes the duplex structure. Thus, DNA in 0.15 M NaCl denatures at a T_m about 20°C higher than DNA in 0.01 M phosphate. In pure water (no salt present) DNA denatures at room temperature. Extremes of pH also have a destabilizing effect on the double-helix structure. When the pH is above 11.5 or below 2.3, extensive deprotonization or protonization, respectively, occurs in the hydrogen-bonding groups of the bases, which in turn disrupts the hydrogen-bonded structure. Alkali is an excellent DNA denaturant; it permits rapid separation of the strands without degradation. Alkali both denatures and degrades RNA to 2′(3′)-mononucleotides.

Many solutes that can form hydrogen bonds also lower the melting temperature (decrease the stability) of double-helix structures. The organic compounds formamide and urea are frequently used to lower the denaturation temperature as well as to prevent reaggregation in DNA manipulations, when it is important to avoid nonspecific aggregation. Reagents that increase the solubility of the DNA bases (e.g., methanol) or disrupt the water shell around them (e.g., trifluoracetate) reduce the hydrophobic interactions between the bases and lower the T_m. Most proteins that bind to DNA inhibit denaturation. However, some DNA-binding proteins destabilize the native state. Proteins of this class usually bind preferentially to single-stranded DNA, thereby favoring separation of the double strands. We discuss so-called single-strand binding proteins in the next chapter because they play an important role in DNA synthesis.

Figure 25.13

Dependence of the temperature midpoint (T_m) of DNA on the content of guanine and cytosine. As the percentage of G + C increases, the T_m increases. Two curves are shown to illustrate the point that the denaturation temperature is shifted to lower values when the ionic strength is lowered.

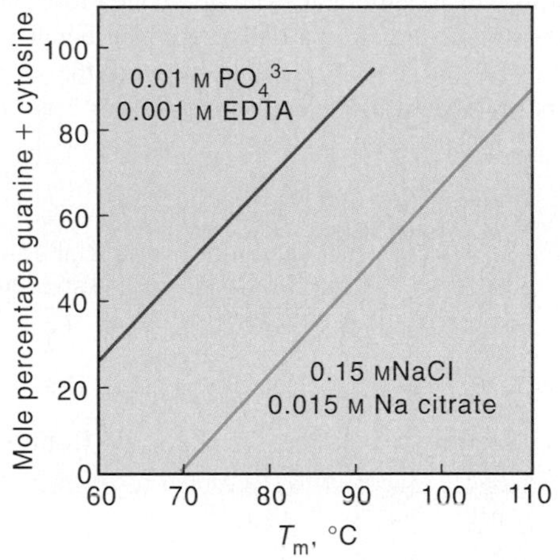

DNA Renaturation Involves Duplex Formation from Single Strands

We have seen that when a solution of heat-denatured DNA is allowed to cool rapidly, the regularly hydrogen-bonded structure does not reform. However, reassembly of the two separated polynucleotide strands into the native structure, called renaturation, is possible under certain specialized conditions.

The first indication that renaturation was possible came from Paul Doty's laboratory in experiments performed by Julius Marmur. They observed that when transforming DNA was heated and rapidly cooled it

Figure 25.14

Steps in denaturation and renaturation of a DNA duplex. In step 1 the temperature is raised to the point where the two strands of the duplex separate. If denatured DNA is slowly cooled, the events depicted as steps 2 and 3 follow. In step 2 a second-order reaction occurs in which two complementary strands of DNA must collide and form interstrand hydrogen bonds over a limited region. Step 3 is a first-order reaction in which additional hydrogen bonds form between the complementary strands that are partially hydrogen-bonded (zippering). Once complementary strands are partially bonded, the zippering reaction occurs rapidly. In the overall process, step 2 is rate-limiting.

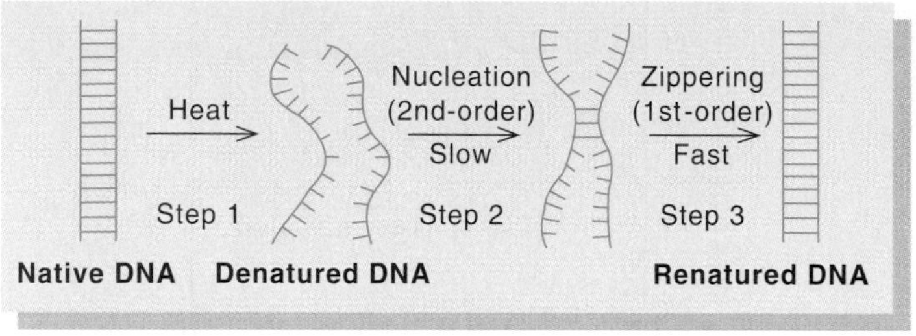

was biologically inactive; however, when denatured DNA was slowly cooled, a small percentage of the initial transforming activity was recovered. Further experiments showed that the optimal temperature for this recovery of activity, or renaturation, was about 25°C below the T_m. Similar to denaturation, renaturation can be followed spectrophotometrically. If the temperature of a denatured DNA solution is maintained at $T_m - 25°C$ for a long time, the absorbance of the solution gradually decreases until it approaches a value close to that of native DNA (see curve *c* in fig. 25.12). The optimum temperature for renaturation is called the annealing temperature. At the annealing temperature, irregularly hydrogen-bonded structures are unstable, but regularly hydrogen-bonded structures are stable. Consequently, prolonged exposure of denatured DNA at this temperature favors the formation of regularly base-paired structures.

Kinetic analysis indicates that renaturation is a two-step process. In the slow step effective contact is made between two complementary regions of DNA originated from separate strands. This rate-limiting step called nucleation is a function of the concentration of complementary strands. Nucleation is followed by a relatively rapid zippering up of adjoining base residues into a duplex structure. The steps involved in denaturation and renaturation are depicted in figure 25.14.

Since nucleation involves interaction between two molecules, it should occur at a rate proportional to the square of the concentration of single strands. If c is the concentration of single-stranded DNA at time t, then the rate equation for the loss of single-stranded DNA is

$$-\frac{dc}{dt} = k_2c^2 \quad \textbf{(1)}$$

where k_2 is the rate constant for a second-order reaction. Starting with a concentration c_0 of completely denatured DNA, the amount of single-stranded DNA left after renaturation for time t is given by

$$\frac{c}{c_0} = \frac{1}{1 + k_2c_0t} \quad \textbf{(2)}$$

At time $t_{1/2}$, when half of the DNA is renatured, $c/c_0 = 0.5$ and $t = t_{1/2}$, from which it follows that

$$c_0t_{1/2} = \frac{1}{k_2} \quad \textbf{(3)}$$

The rate of renaturation is also a function of chain length, but this effect is usually eliminated as a variable by shearing the starting DNA down to a uniform size. For a typical renaturation experiment the values of c/c_0 are plotted as a function of c_0t, and the resulting curve is referred to as a "cot" curve.

In figure 25.15 cot curves for four DNA samples and one synthetic polyribonucleotide sample are presented. The mouse satellite DNA is a fraction of the DNA from the mouse that contains highly repetitious sequences. The T4 DNA is the total DNA isolated from the bacteriophage T4. Similarly the *E. coli* DNA is the total DNA isolated from *E. coli* bacteria. The calf nonrepetitive fraction represents a fraction of calf nuclear DNA that contains mostly sequences that are represented only one time per haploid genome. At the top of this semilogarithmic plot is an additional scale indicating the nucleotide complexity, *N*, which is defined as the number of nucleotides in a nonrepeating sequence. If no sequences in the cellular DNA repeat, then *N* is equal to the number of nucleotides in the genome. It can be seen that $c_0t_{1/2}$ is proportional to *N* for these samples.

This family of curves has been used to calibrate more complex situations in which the test DNA is a mixture of unique sequence and repetitive DNA. In such cases the repetitive DNA fraction tends to anneal more rapidly. Bacterial DNA (*E. coli*) contains very little repetitive DNA (0.3%). This is mainly accounted for by the eight genes for *E. coli* ribosomal RNA that have nearly identical sequences. In complex eukaryotes single-copy DNA (i.e., nonrepetitive DNA) accounts for 40%–70% of the DNA, most of the remainder being roughly divided between middle repetitive ($<10^4$ copies/genome) and highly repetitive ($>5 \times 10^4$ copies/genome). Further analyses of eukaryotic gene struc-

Figure 25.15

Reassociation of double-stranded nucleic acids from various sources. The genome size is indicated by arrows near the upper nomographic scale. Over a factor of 10^{10}, this value is proportional to the c_0t (the "cot") required for half-reaction. All DNAs were sheared so that they have approximately the same fragment size (about 400 nucleotides, single-stranded). Correction has been made to give the rate that would be observed at 0.18 M sodium ion concentration. No correction for temperature has been applied because it was approximately optimum in all cases. The labels for the different DNAs should not concern the average reader. Mouse satellite and calf (nonrepetitive fraction) are fractions of the genome obtained from the indicated animals. (Source: Adapted from R. J. Britten and D. E. Kohne, Repeated sequences in DNA, *Science,* 161:529, 1968.)

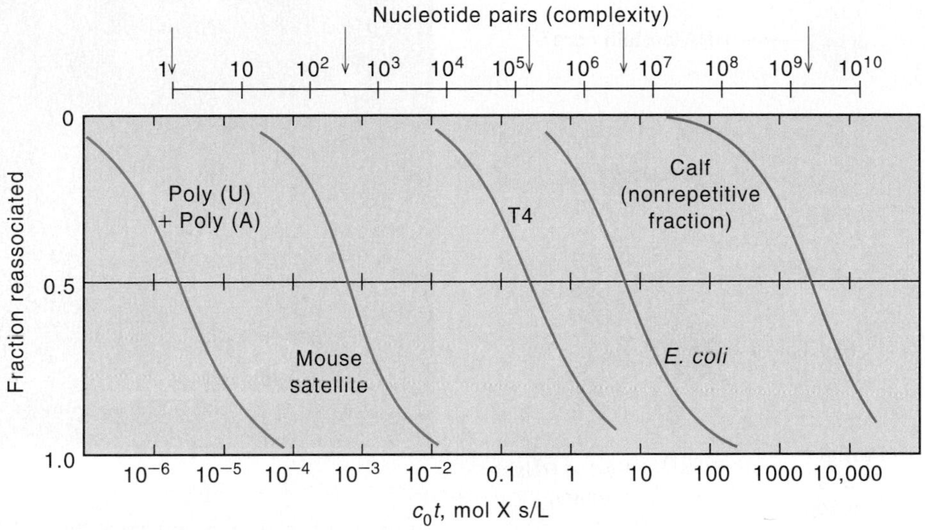

ture by a variety of other techniques have shown that most genes encode proteins that belong to the unique (single-copy) class. The middle repetitive class includes transfer RNA genes and ribosomal RNA genes that form part of the biochemical machinery for protein synthesis (chapter 29), as well as the genes encoding histones, the main chromosomal proteins. Some highly repetitive sequences occur in tandem (see mouse satellite DNA in fig. 25.15), and still other repetitive DNA elements are distributed at random throughout the genome. It has been argued that some families of repetitive DNA, referred to as "selfish DNA," represent "parasitic" sequences that replicate together with the genome without conferring any positive or negative characteristics to the host cells that have these sequences.

Chromosome Structure

All types of nucleic acids interact with proteins. Chromosomal DNA forms stable nonspecific complexes with structural proteins that stabilize their tertiary structure; it also forms transient complexes with enzymes and regulatory proteins that modulate DNA and RNA metabolism. The gross tertiary structure of DNA in *E. coli* and a typical eukaryotic chromosome is described in the next section.

Physical Structure of the Bacterial Chromosome

A single chromosome of *E. coli* contains about 3×10^9 daltons or about 4.5×10^6 bp of DNA. If all of this DNA were in a duplex structure stretched end to end, it would be 1.5 mm long, which is about 75 cell diameters. In fact, the chromosome is circular, centrally located in the cell, and highly folded, so that it is only about 2 μm across. No dramatic change in chromosome morphology is seen prior to cell division as in eukaryotes. Clearly the degree of compaction observed throughout the cell cycle does not interfere with transcription to any great extent because all the genes in *E. coli* are readily expressible.

Electron micrographs of the *E. coli* chromosome suggest a folded circular structure containing about 40–100 supercoiled loops (diagrammatically indicated in fig. 25.16). It is believed that the folded structure is held together by an RNA-protein core, although the manner in which this is done is not well understood. The structure is further stabilized because the core forms a complex with positively charged polyamines and certain basic proteins. D. E. Pettijohn and his co-workers have provided evidence of such a core. They first showed that the individual super-

Figure 25.16

The *E. coli* chromosome exists as a circular, folded, supercoiled duplex (*a*). This can be converted to a partially unfolded structure by brief treatment with RNase (*c*). There are 50–100 loops in the structure; supercoiling may be selectively eliminated from individual loops by single-strand nicking of the DNA within the loop (*b*).

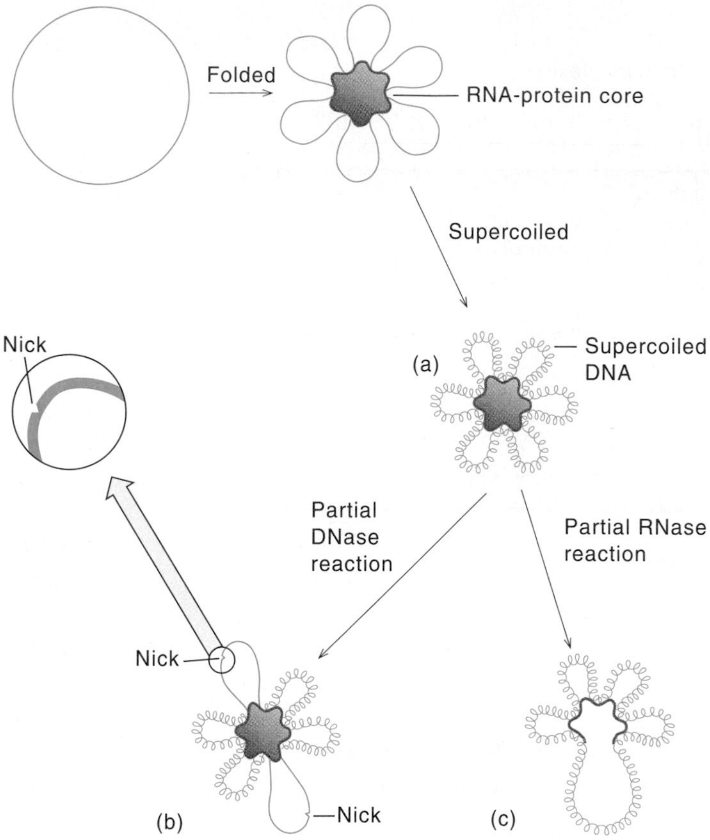

coiled loops maintain their supercoiling independently of one another. Thus, if a single nick is introduced into one of the loops by limited DNase action, that loop adopts an expanded relaxed conformation, but supercoiling in the other loops is maintained (see fig. 25.16). Limited RNase or protease treatment causes the partial breakdown of the looped structures without interfering with the supercoiling (see fig. 25.16). These results have led to the conclusion that each of the loops is a domain, the lateral motion of which is restricted by an RNA-protein core complex.

The Genetic Map of Escherichia coli

The circular *E. coli* chromosome contains enough base pairs to make about 3,000 average-sized genes. The relative positions of over half of these genes are known. In regions where our understanding is reasonably complete the impression is obtained of a tightly organized genome. Coding regions are interspersed with regulatory regions; there is no evidence for significant stretches of DNA with no function. Frequently, genes with a related function are clustered. These clustered genes are usually transcribed into single expression units (messenger RNAs) containing the information for the synthesis of several functionally related proteins (see chapters 28 and 29).

Figure 25.17

Swollen fibers of chromatin from the nucleus of the chicken red blood cell. The electron micrograph is enlarged about 325,000× and negatively stained with uranyl acetate. (Micrograph courtesy of A. L. Olins and D. E. Olins.)

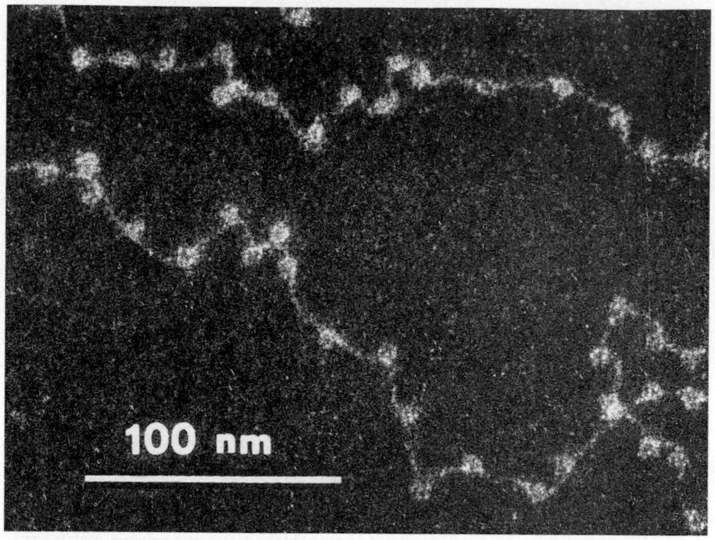

Eukaryotic DNA Is Complexed with Histones

DNA in eukaryotic chromosomes exists in a highly compacted form known as chromatin, a complex of DNA with a great variety of proteins. Five proteins called histones are present in large amounts (see table 25.2) and are believed to form a regularly repeating structural motif. The remaining proteins are present in smaller amounts and are irregularly distributed.

Electron-microscopic and x-ray diffraction studies on chromatin suggest that the DNA forms a coiled-coil structure with the histones of chromatin. A breakthrough in our understanding of the nucleohistone complex came when D. E. Olins and A. L. Olins observed that chromatin viewed after sudden swelling in water had a beaded structure (fig. 25.17). The beads, called nucleosomes, contain four of the five histones; they are about 10 nm in diameter, and the spacing between the beads is about 14 nm. Brief enzymatic digestion of chromatin with micrococcal nuclease fragments this structure. The DNA–histone frag-

Table 25.2

Characteristics of Histones

Name	Ratio of Lysine to Arginine	M_r	Copies per Nucleosome
Histone H1[a]	20	21,000	1 (not in bead)
Histone H2a	1.2	14,500	2 (in bead)
Histone H2b	2.5	13,700	2 (in bead)
Histone H3	0.7	15,300	2 (in bead)
Histone H4	0.8	11,300	2 (in bead)

[a] Not found in lower eukaryotes such as yeast.

ments from this partial digestion give rise to a banded pattern on agarose gel electrophoresis that suggests nucleoprotein structures containing 200 bp of DNA or multiples thereof (400, 600, 800 bp, etc.). Electron-microscopic examination of the individual fractions isolated after gel electrophoresis confirms the suggested correlation between size of the DNA estimated on gels and the number of nucleosomes. Thus, the most rapidly moving DNA band seen on gels was derived from a structure containing one nucleosome, and the second fastest migrating species contains nucleosome dimers, and so forth. Evidently, the brief treatment with endonuclease preferentially cleaves DNA in the internucleosomal region, where the DNA is least likely to be protected from enzyme attack. More extensive nuclease treatment results in a single band on gels that contain a single nucleosome with 140 bp of DNA. It appears that exhaustive nuclease digestion has removed all of the DNA that is not in direct contact with the nucleosome. From these results it was deduced that nucleosomes contain a core of histone with 140 base pairs of DNA wrapping; an additional 60 bp of more exposed DNA connects adjacent nucleosomes.

The histones present in chromatin are of five major types: H1, H2a, H2b, H3, and H4 (table 25.2). The lysine-rich histone H1 is not present in the nucleosome core particles, as evidenced by its release on extensive nuclease treatment and the finding that H1 is the only histone that readily exchanges between free and DNA-bound histone. H1 may play a key role in the conversion of chromatin to the highly compacted chromosome that occurs immediately before cell division. The other eight histones, two each of the other four histones, form the protein core of the nucleosome. These protein octamers do not come apart even when chromosomes duplicate.

An illustration of the coiled-coil structure of the nucleosome is presented in figure 25.18. The 140 bp of DNA make about one and three-quarters superhelical turns about the histone octamer. An additional 60 bp of spacer DNA (not shown) connect adjacent nucleosomes. A survey of chromatins from different species and in different tissues of the same species has shown that the spacer DNA actually varies in length from about 20–95 bp.

Salt bridges between positively charged basic amino acid side chains of histones and the negatively charged DNA phosphates play a major role in stabilizing the DNA–histone complex. Indeed, treatment of chromatin with concentrated NaCl (1–2 M), which is known to disrupt electrostatic bonds, causes a complete dissociation of DNA and histone in the nucleohistone complex.

Higher order structures beyond that of the nucleosome are less well understood. Electron-microscopic investigations indicate two types of fibers with diameters of 10 nm and 30 nm. To account for the 10 nm fiber the nucleosomes can be arranged edge to edge in a zigzag fashion to produce a fibril that is 10 nm wide. When the ionic strength is raised on an isolated preparation of chromatin, the fibrils reversibly condense into an irregularly supercoiled fiber about 30 nm in diameter. The nucleosome particles are thought to have their cylindrical axes approximately perpendicular to the long axis of the 30 nm fiber with six to seven nucleosomes per turn (fig. 25.19). Further coiling of these structures is necessary to explain the much larger structures seen in mitotic chromosomes.

Organization of Genes within Eukaryotic Chromosomes

The organization of genes within a typical eukaryotic chromosome is far more complex and less well understood than in prokaryotes. It is highly likely that a much lower percentage of the DNA is informational in complex eukaryotes than in prokaryotes. *E. coli* contains about 3,000 genes; the

Figure 25.18

(*a*) Path of DNA that can account for the bipartite structure of the nucleosome core is a superhelix with an external diameter of 110 Å and a pitch of 27 Å; the turns of the 20-Å-wide DNA helix are nearly in contact. About 80 bp of DNA occur per turn; the nucleosome core, an enzymatically reduced form of the nucleosome consisting of some 140 bp, has about one and three-quarter turns wrapped on it. The histone octamer complex, containing two each of histones H2a, H2b, H3, and H4, is packed on the inside of the DNA coiled-coil structure. (© Scientific American, Inc., George V. Kelvin. Reprinted with permission.) In (*b*) we see this histone octamer inserted into the nucleosome core. The H3-H4 tetramer is shown in yellow, and an H2a-H2b dimer is shown at each end in purple.

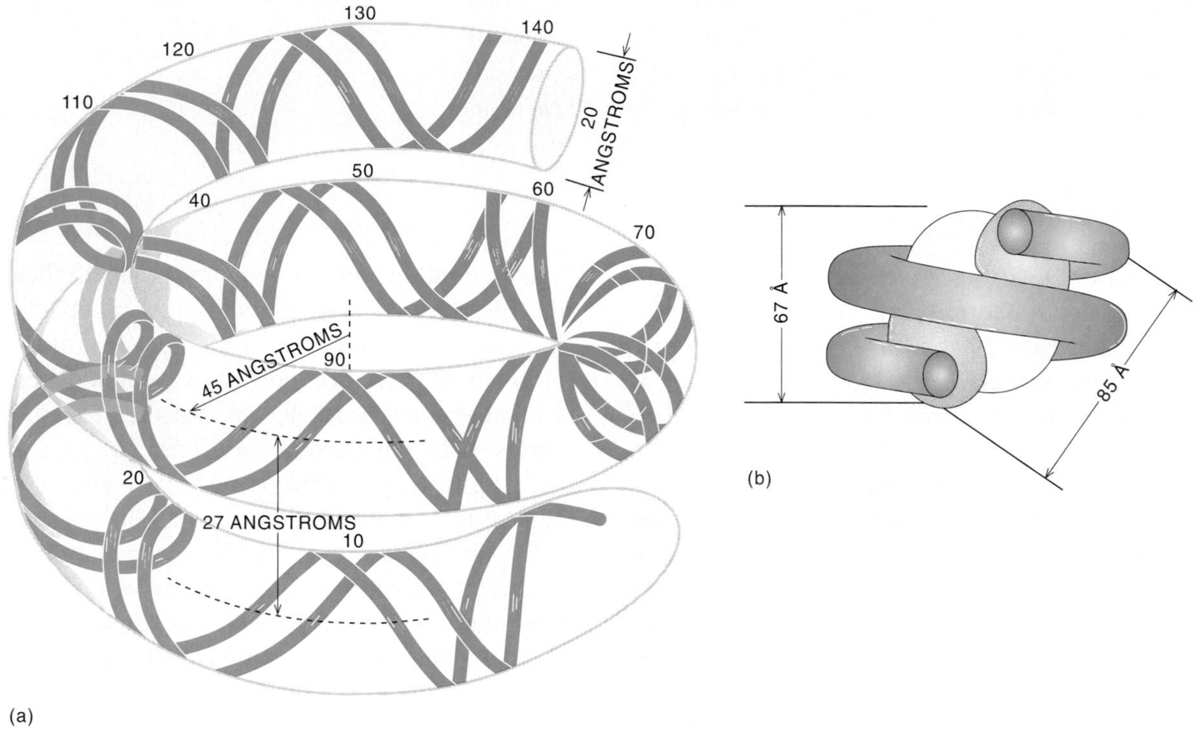

human genome is 1,000 times larger but it is very unlikely to contain 1,000 times the number of functional genes. Most estimates hover around 50,000. Even if there were 200,000 human genes, this would still leave approximately 90% of the DNA with no coding function. Measurements on the fruit fly *Drosophila melanogaster* indicate that the coding information for proteins in most genes probably accounts for no more than one-tenth of the base pairs within the gene. The discovery that noncoding regions (introns) occur between coding regions (exons) helps to explain the excessive amounts of DNA that appear to be present in eukaryotes (see chapter 28), although it is probably not the whole story. Noncoding regions serving as control loci may be larger on the average in eukaryotes. Nonfunctioning genes that have lost their initiation sites for being expressed (promoters) and repetitive genetic elements with no apparent coding function also help to account for the large amount of noncoding DNA.

As we have already discussed, not all repetitive DNA in eukaryotes is noncoding. Thus, histone genes, for example, are typically reiterated many times. The five different histone genes are usually clustered, and this cluster is then tandemly repeated many (up to 100 or more) times. Ribosomal RNA genes are also tandemly clustered. Other nonidentical but functionally related genes that show clustering include the globin genes and the immunoglobulin genes.

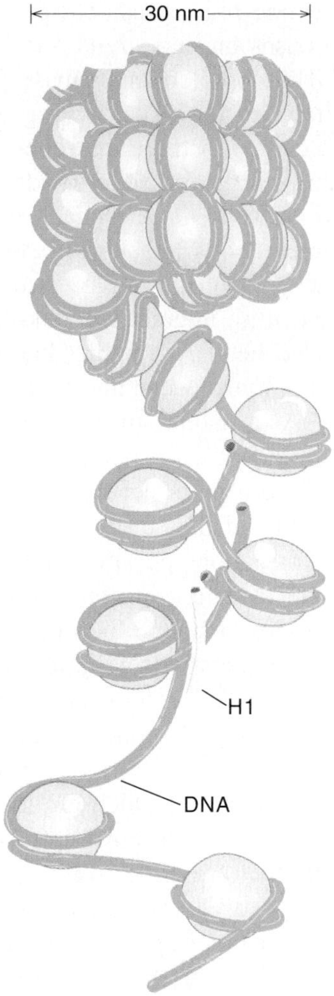

Figure 25.19

Helical superstructures might be formed with increasing salt concentration (*bottom to top*) as is suggested here. The zigzag pattern of nucleosome (1, 2, 3, 4) closes up, eventually to form a solenoid, a helix with about six nucleosomes per turn. (The helix is probably more irregular than it is in this drawing.) Cross-linking data indicate that H1 molecules on adjacent nucleosomes make contact. Extrapolation from the zigzag form to the solenoid suggests (but does not prove) that the aggregation of H1 at higher ionic strengths gives rise to a helical H1 polymer (not shown) running down the center of the solenoid. In the absence of H1 (*bottom*) no ordered structures are formed. The details of H1 associations are not known at this time; the drawing is meant to indicate only that H1 molecules contact one another and link DNA. (© Scientific American, Inc., George V. Kelvin. Reprinted with permission.)

Summary

1. The genetic material of cells and viruses consists of DNA or RNA. That DNA bears genetic information was first shown when the heritable transfer of various traits from one bacterial strain to another was found to be mediated by purified DNA.
2. All nucleic acids consist of covalently linked nucleotides. Each nucleotide has three characteristic components: (1) a purine or pyrimidine base; (2) a pentose; and (3) a phosphate group. The purine or pyrimidine bases are linked to the C-1′ carbon of a deoxyribose sugar in DNA or a ribose sugar in RNA. The phosphate groups are linked to the sugar at the C-5′ and C-3′ positions. The purine bases in both DNA and RNA are always adenine (A) and guanine (G). The pyrimidine bases in DNA are thymine (T) and cytosine (C); in RNA they are uracil (U) and cytosine. The bases may be post-replicatively or posttranscriptionally modified by methylation or other reactions in certain circumstances.
3. DNA exists most typically as a double-stranded molecule, but in rare instances it exists (in some phages and viruses) in a single-stranded form. The continuity of the strands is maintained by repeating 3′,5′-phosphodiester linkages formed between the sugar and the phosphate groups; they constitute the covalent backbone of the macromolecule. The side chains of the covalent backbone consist of the purine or pyrimidine bases. In double-stranded, or duplex, DNA the two chains are held together in an antiparallel arrangement. The base composition of DNA varies character-

istically from one species to another in the range of 25%–75% guanine plus cytosine. Specific pairing occurs between bases on one strand and bases on the other strand. The complementary base pairs are either A and T, which can form two hydrogen bonds, or G and C, which can form three hydrogen bonds. The duplex is stabilized by the edge-to-edge hydrogen bonds formed between these planar base pairs and face-to-face interactions (stacking) between adjacent base pairs. Twisting of the duplex structure into a helix makes stacking interactions possible.

4. The right-handed helical structure, known as B DNA, is the most commonly occurring conformation of linear duplex DNA in nature. In this structure, the distance between stacked base pairs is 3.4 Å, with approximately 10 base pairs per helical turn. The inherent flexibility of the structure, however, makes a variety of conformations possible under different conditions. In some instances, nucleotide sequence and degree of hydration dictate which conformations are favored. DNA interacts with a variety of proteins inside the cell, and these proteins can also have a significant influence on its secondary and tertiary structure.
5. Circular DNA molecules, which are topologically confined so that their ends are not free to rotate, can form supercoils that are either right-handed (negative) or left-handed (positive). Negative supercoiling exerts a torsional tension favoring the untwisting of the primary right-handed double helix, whereas positive supercoiling has the opposite effect. Negatively supercoiled DNAs are most commonly observed in prokaryotes, which contain an enzyme that generates the supercoiled structure.
6. When duplex DNA (or RNA) is heated, it dissociates (denatures) into single strands. The temperature at which denaturation occurs (the melting temperature) is a measure of the stability of the duplex and is a function of the G-C content of the DNA. A preparation of denatured DNA may be renatured in the native duplex structure by maintaining the temperature about 25°C below the melting temperature. The rate of renaturation is a measure of the sequence complexity of the DNA. In prokaryotes, which consist predominantly of unique sequences, the complexity (the number of base pairs) is approximately equal to the genome size. However, eukaryotic cells contain DNAs of varying sequence complexity that renature at quite different rates. The fastest renaturing fractions are present in many copies per nucleus, whereas the slowest renaturing fractions are present in single copies. Analyses by other techniques have shown that some of the repetitive DNA sequences exist as tandemly repeated structures, while other types of repetitive sequences are dispersed throughout the genome.
7. In the chromatin of eukaryotic cells DNA forms a coiled-coil structure with an approximately equal weight of a mixture of five basic proteins known as histones. Four of these histones in pairs form an octamer around which the DNA duplex occurs in a left-handed helix. The DNA octamer complex is called a nucleosome. Each nucleosome contains about 140 base pairs of DNA in a nuclease-resistant ''nucleosome core'' and approximately 60 base pairs of spacer between core particles. Histone H1 binds to the chromatin independently of the octamer and is the first histone to dissociate from the chromatin when the ionic strength is raised. Beyond the nucleosome the higher order structure of the chromosome involves coiled-coil structures with varying degrees of regularity.

Selected Readings

Avery, O. T., C. M. MacLeod, and C. McCarthy, Studies on the chemical nature of the substance inducing transformation of pneumococcal types. *J. Exp. Med.* 79:137–158, 1944.

Camerini-Otero, R. D., and P. Hsieh, Parallel DNA triplexes, homologous recombination and other homology-dependent DNA interaction. *Cell* 73:217–224, 1993.

Cantor, C. R., C. L. Smith, and M. K. Mathew, Pulsed-field gel electrophoresis of very large molecules. *Ann. Rev. Biophys. Chem.* 17:287–304, 1988.

Daniels, D. L., G. Plunkett III, V. Burland, and F. R. Blattner, Analysis of the *Escherichia coli* genome: DNA sequence of the region from 84.5 to 86.5 minutes. *Science* 257:771–778, 1992.

Dervan, P. B., Reagents for the site-specific cleavage of megabase DNA. *Nature* 359:87–88, 1992.

Dickerson, R. E., The DNA helix and how it is read. *Sci. Am.* 249(6)d:94–111, 1983.

Hershey, A. D., and M. Chase, Independent functions of viral proteins and nucleic acid in growth of bacteriophage. *J. Gen. Physiol.* 36:39–56, 1952.

Hillary, C. M., J. T. Finch, B. F. Luisi, and A. Klug, The structure of an oligo(dA), oligo(dT) tract and its biological implications. *Nature* 330:221–236, 1987.

Kang, C., X. Zhang, R. Ratliff, R. Moyzis, and A. Rich, Crystal structure of four-stranded Oxytricha telomeric DNA. *Nature* 356:126–131, 1992.

Kim, S. H., Three-dimensional structure of transfer RNA. *Prog. Nuc. Acid Res. Mol. Biol.* 17:181–216, 1973.

Kornberg, R. D., and A. Klug, The nucleosome. *Sci. Am.* 244(2):52–64, 1981.

Lerman, L. S., S. G. Fischer, I. Hurley, K. Silverstein, and N. Lumelsky, Sequence-determined DNA separations. *Ann. Rev. Biophys. Bioeng.* 13:399–423, 1983.

Morse, R. H., and R. T. Simpson, DNA in the nucleosome. *Cell* 54:285–287, 1988.

Nadeau, J. G., and D. M. Crothers, Structural basis for DNA bending. *Proc. Natl. Acad. Sci.* 86:2622–2626, 1989.

Ramakrishnan, V., J. T. Finch, V. Graziano, P. L. Lee, and R. M. Sweet, Crystal structure of globular domain of histone H5 and its implications for nucleosome binding. *Nature* 362:219–223, 1993.

Rich, A., A. Nordheim, and A. H.-J. Wang, The chemistry and biology of left-handed Z DNA. *Ann. Rev. Biochem.* 53:791–846, 1984.

Richmond, T. J., J. T. Finch, B. Rushton, D. Rhoades, and A. Klug, Structure of the nucleosome core particle at 7 Å resolution. *Nature* 311:532–537, 1984.

Roberts, R. W., and D. M. Crothers, Stability and properties of double and triple helices: Dramatic effects of RNA or DNA backbone composition. *Science* 258:1463–1466, 1992.

Saenger, W., *Principles of Nucleic Acid Structure.* New York: Springer-Verlag, 1984.

Schmid, M. B., Structure and function of the bacterial chromosome. *Trends Biochem. Sci.* 13:131–135, 1988.

Schwartz, D. C., and C. R. Cantor, Separation of yeast chromosome-sized DNAs by pulsed field gradient gel electrophoresis. *Cell* 37:67–75, 1984.

Strobel, S. A., L. A. Doucette-Stamm, L. Riba, D. E. Housman, P. B. Dervan, Site-specific cleavage of human chromosome 4 mediated by triple-helix formation. *Science* 254:1639–1642, 1991.

Structures of DNA, *Cold Spring Harbor Symp. Quant. Biol.* 47, 1983.

van Holde, K. E., *Chromatin.* New York: Springer-Verlag, 1988.

Watson, J. D., and F. H. C. Crick, Molecular structure of nucleic acids. *Nature* 171:737–738, 1953.

Wells, R. D., D. A. Collier, J. C. Hanvey, M. Shimizu, and F. Wohlrab, The chemistry and biology of unusual DNA structures adopted by oligopurine · oligopyrimidine sequences. *Faseb J.* 2:2939–2949, 1988.

Problems

1. Briefly describe how Avery was able to show that DNA is the genetic material in cells.
2. Why is the bond holding nucleotides together in nucleic acids called a phosphodiester bond?
3. Give the structure of the DNA strand complementary to pTpApCpG (see structure shown in fig. 25.3) in the abbreviated form.
4. How many different base-paired structures with two hydrogen bonds can be made using guanine (G) and thymine (T)? Which one of these is most similar to the standard Watson-Crick (G-C) base pair?
5. (a) List the hydrogen bond donors and acceptors available in the major and minor grooves of the DNA double helix.
   (b) The intermolecular forces that stabilize DNA-protein complexes often involve hydrogen bonds between specific amino acids and the exposed surfaces of the bases. Explain why interaction with the major groove is more common among DNA-binding proteins than interaction with the minor groove.
6. How many base pairs are found in the DNA pictured at the right in figure 25.10 if two negative supercoils are introduced without breaking the backbone? Describe the effect of a short stretch of Z DNA on the overall conformation of this DNA.
7. Describe a physical method that can be used to estimate the base composition of DNA. Describe the data obtained with two DNA samples: One with high G-C content and another with high A-T content? (Assume that the concentration of the samples is equal.)
8. Why is DNA denatured at either low pH (pH 2) or high pH (pH 11), and why is DNA stable at pH 7? (*Hint:* See p*K* values in table 23.2.)
9. You are given a sample of nucleic acid extracted from a virus. How would you determine whether the virus has an RNA or DNA genome and whether it is single- or double-stranded?
10. A column packed with the material hydroxyapatite preferentially binds double-stranded DNA over single-stranded DNA. The double-stranded DNA can

be eluted by changing the salt concentration. Using a hydroxyapatite column and the information in figure 25.15, propose a method for separating a mixture of equal amounts of T4 bacteriophage and *E. coli* DNAs into the respective components.

11. Why can't RNA duplexes or RNA–DNA hybrids adopt the B conformation?
12. Some DNA-binding proteins specifically bind to Z DNA. How could these proteins help stabilize DNA in the Z configuration? (*Hint:* How do single-stranded DNA-binding proteins destabilize duplex structures?)
13. Linear duplex DNA can bind more ethidium bromide than covalently closed circular DNA of the same molecular weight. Why? (*Hint:* Ethidium bromide molecules bind between adjacent base pairs of DNA, causing the duplex to unwind in the region of binding.)
14. Give the relative times for 50% renaturation of the following pairs of denatured DNAs, starting with the same initial DNA concentrations.
    (a) T4 DNA and *E. coli* DNA, each sheared to an average single-strand length of 400 nucleotides.
    (b) Unsheared T4 DNA and sheared T4 DNA.
15. When histone proteins are isolated from chromatin their mass is equal to the DNA, and the molar ratio of four of the histones is 1:1:1:1 (H2a:H2b:H3:H4), while H1 is found in half the amount (0.5). Discuss whether or not these data fit the bead-and-string model for nucleosomes.
16. Your supervisor shows you two test tubes and announces: "The labels fell off of these tubes in the freezer! One was supposed to contain DNA from *E. coli* and one was DNA from *Mycobacterium tuberculosis.* I can't figure out which is which, so you will just have to grow cells and prepare more DNA." Because *M. tuberculosis* is a dangerous pathogen, you wish to avoid culturing the cells. Can you devise a simple strategy to determine which sample is from which organism, based on the fact that *E. coli* DNA contains 52% G + C and *M. tuberculosis* DNA is 70% G + C? Show an example of the expected results.

X-Ray Diffraction of DNA

An x-ray pattern of DNA (fig. 1) is obtained by holding a stretched fiber containing many DNA molecules in a vertical direction and exposing it to a collimated monochromatic beam of x-rays. Only a small percentage of the x-ray beam is diffracted. Most of the beam travels through the specimen with no change in direction. A photographic film is held in back of the specimen; a hole in the center of the film allows the incident undiffracted beam to pass through (fig. 2). Coherent diffraction occurs only in certain directions, specified by Bragg's law: $2d \sin \theta = n\lambda$. Here d is the distance between identical repeating structural elements: θ is the angle between the incident beam and the regularly spaced diffracting planes. λ is the wavelength of x-rays used, and n is the order of diffraction, which may equal any integer but is usually strongest for $n = 1$. The most important point is that $\sin \theta \approx \theta$ and $d \approx 1/\theta$, so that a spot far out on the photographic film is indicative of a repeating element of small dimension and vice versa.

Watson and Crick were the first to appreciate the significance of strong 3.4-Å and 34-Å spacings and the central crosslike pattern, which reflects a helix structure in the x-ray diffraction pattern of DNA. They interpreted this as arising from the hydrogen-bonded antiparallel double-helix structure.

Figure 1

Diffraction pattern of a fibrous sample of DNA. (© M. H. F. Wilkins.)

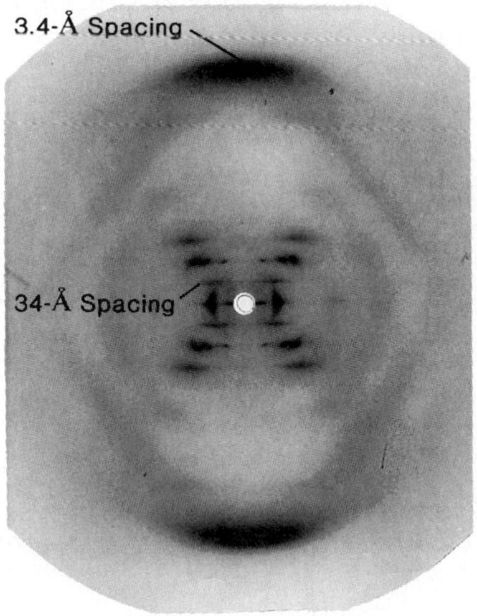

Figure 2

Camera setup for obtaining DNA diffraction pattern.

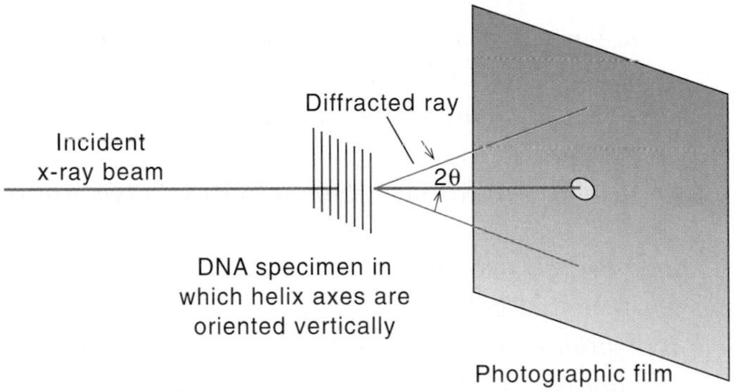

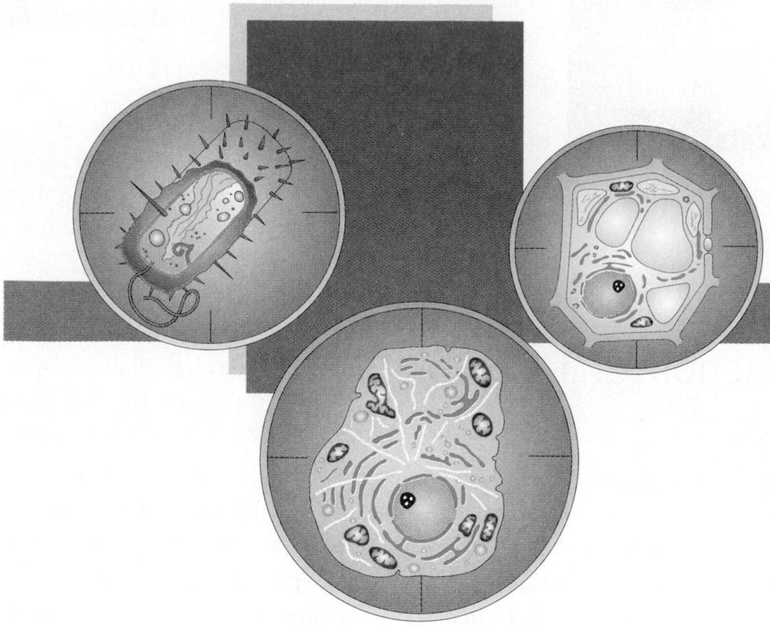

CHAPTER

26

DNA Replication, Repair, and Recombination

Chapter Outline

The genetic message carried by the sequence of bases in the DNA is replicated by a template mechanism.

From the complementary duplex structure of DNA it is a short intuitive hop to a model for replication that satisfies the requirement for one round of DNA duplication for every cell division. Such a proposal was made by Watson and

Figure 26.1

Watson-Crick model for DNA replication. The double helix unwinds at one end. New strand synthesis begins by absorption of mononucleotides to complementary bases on the old strands. These ordered nucleotides are then covalently linked into a polynucleotide chain, a process resulting ultimately in two daughter DNA duplexes.

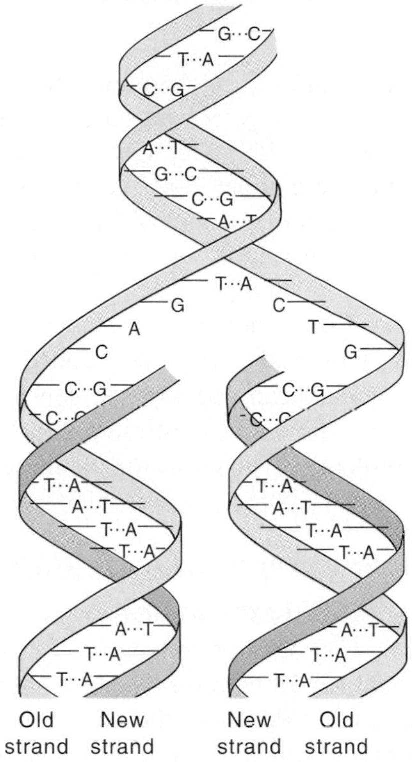

Crick when they proposed the duplex structure for DNA (fig. 26.1). First, the double helix unwinds; next, mononucleotides are absorbed into complementary sites on each polynucleotide strand; and finally these mononucleotides become linked to yield two identical daughter DNA duplexes. What could be simpler! Subsequent biochemical investigations showed that in many respects this model for DNA replication was correct, but they also indicated a much greater complexity than was initially suspected. Part of the reason for the complications is that replication must be very fast to keep up with the cell division rate, and it must be very accurate to ensure faithful transfer of information from one cell generation to the next.

In addition to the replication process there are two other major areas of DNA metabolism. One is concerned with repair of damaged DNA, and the other is concerned with recombination between DNA molecules. In this chapter we deal mainly with the replication process, but we also consider chromosome repair and recombination.

The Universality of Semiconservative Replication

The Watson-Crick model for DNA replication is called semiconservative because the daughter duplexes arising from replication each contain one old (conserved) strand and one new strand.

Matthew Meselson and Frank Stahl conceived a way of demonstrating the semiconservative mode of replication that involved the use of isotopes that result in DNAs of different densities after replication. For this purpose *E. coli* bacteria were grown for several generations on a medium in which all the nitrogen was of the heavy ^{15}N isotope type (normal nitrogen is ^{14}N). The resulting DNA had a greater than normal density because the ^{15}N was incorporated into the bases of the DNA. Following this step, the bacteria were transferred to a growth medium containing normal (^{14}N) nitrogen, and the cells were allowed to go through one or more rounds of duplication. The DNA from these cells was then isolated and analyzed by density-gradient centrifugation. Pure ^{15}N DNA produces a single band of DNA (fig. 26.2, frame 1). The same is true for ^{14}N DNA (see fig. 26.2, frame 2). The only difference is that denser DNA produces a band further down the tube. Thus, the location of the DNA in the centrifuge tube can be used as a measure of the density of the DNA. Cells containing pure ^{15}N DNA were allowed to grow in ^{14}N medium for one or more generations, and the DNA from them was similarly analyzed. After precisely one generation in ^{14}N medium, the only band visible in the isolated DNA was that corresponding to ^{15}N–^{14}N-hybrid DNA (see fig. 26.2, frame 4). These data argue strongly against the conservative mode of replication, but they do not discriminate between the semiconservative and other modes of replication in which both strands might be labeled approximately equally. For this purpose the results in further generations were examined. Only in the semiconservative mode would equal amounts of DNA with the hybrid density and the light density be expected after two generations. This result is shown in figure 26.2, frame 5. In the third generation we still see a band with the hybrid density, but the amount of pure light DNA has increased (see fig. 26.2, frame 6). These results strongly support the semiconservative mode for DNA replication. Indeed it is hard to think of another mode of replication that could give rise to these results.

Subsequent studies initiated by others have shown that the DNA in all chromosomes regardless of their size replicate in this semiconservative fashion. Thus, the semiconservative mode of DNA replication is well established and essentially universal. The genetic information implanted in

Figure 26.2

The Meselson-Stahl experiment demonstrating semiconservative replication for *E. coli* chromosomal DNA. Cesium chloride (CsCl) density-gradient centrifugation is used to discriminate between DNAs of different densities. When a concentrated solution of CsCl is centrifuged at high speed (50,000–100,000 times gravity), a stable concentration gradient of CsCl develops, with the concentration increasing along the direction of the centrifugal force. Since CsCl is much denser than water, this concentration gradient produces a density gradient. Macromolecules of DNA present in the solution are driven by the centrifugal field into the region where the solution density is equal to their own density. If the DNA has a uniform density, then after many hours of centrifugation (about 36 h) equilibrium is established, with all of the DNA concentrated in a single band. *E. coli* DNA has different densities when cells are grown in ^{14}N or ^{15}N medium (frames 1 and 2). When cells containing pure heavy DNA (^{15}N–^{15}N DNA) are grown in ^{14}N medium for one generation, all of the DNA is of intermediate density (^{14}N–^{15}N). After two generations of growth in ^{14}N medium, the cells contain equal amounts of light (^{14}N–^{14}N) DNA and intermediate density DNA (frame 5). In subsequent generations the hybrid DNA reappears in constant amounts, but the amount of light DNA increases. These results support the model of a semiconservative mode of DNA replication.

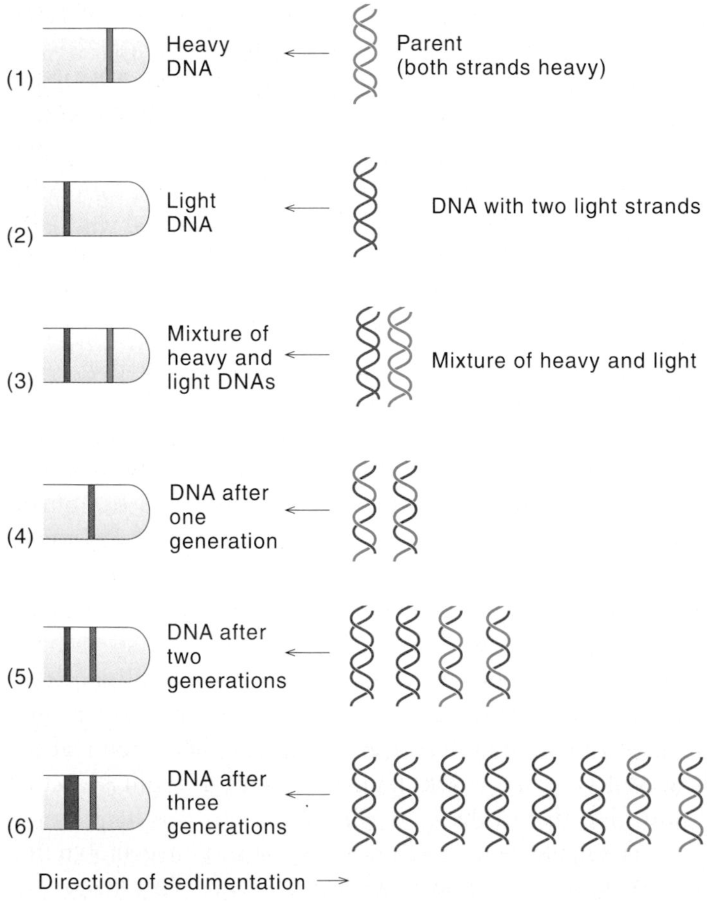

the base sequence is directly transferred from one generation of DNA to the next by the ability of DNA single strands to serve as templates for the absorption of complementary mononucleotides.

Overview of DNA Replication in Bacteria

More is known about the replication of DNA in *E. coli* than in any other system. The *E. coli* bacterium contains a single circular chromosome with about 4.5×10^6 bp. Within this chromosome there are about 2,000 genes, each gene being represented by a unique sequence of bases and encoding the genetic information essential for the synthesis of a specific protein. Exactly how many of these genes are required for DNA synthesis is not known; so far about 30 have been found, and in many cases the gene-encoded proteins have been extensively characterized. First we consider the overall strategy for bacterial DNA replication; then we discuss some of the proteins involved in the synthesis.

Growth during Replication Is Bidirectional

Replication of the *E. coli* chromosome can be made visible by very delicate techniques developed by John Cairns and Ric Davern for isolating intact ^{3}H-labeled chromosomes and subjecting them to autoradiography. After one round of replication in labeled medium, chromosomes appear circular and uniformly labeled.

Initiation of a second round of replication leads to a replication "eye" at the initiation site of replication (fig. 26.3). As synthesis proceeds the size of the replication eye becomes larger; at this stage the replicating chromosome is referred to as a theta structure because it has the appearance of the Greek letter θ. Semiconservative replication is consistent with the density of the autoradiographic tracks made by different parts of the chromosome after one and two rounds of replication in [^{3}H]thymidine (see fig. 26.3).

It seems clear that the replication eye must contain two partially separated parental DNA strands that are base-paired with short strands of newly synthesized DNA. Not resolved by these autoradiographs is the question of whether replication occurs in one or both directions about the origin of replication. If growth is unidirectional we expect one growth point (fig. 26.4*a*), called a growth fork; if growth is bidirectional we expect two growth points or two growth forks (see fig. 26.4*b*). Although examples of both types of replication occur, it is believed that most bacterial chromosomes including *E. coli* replicate bidirectionally.

Evidence for bidirectional replication is supported by additional experiments. Growing bacteria were briefly ex-

Figure 26.3

Simulated autoradiographs of the *E. coli* chromosome after one or more replications in the presence of [^{3}H]thymidine. After one round of replication, the autoradiograph shows a circular structure that is uniformly labeled. The second round of replication begins with the formation of a replication "eye." One branch in the replication eye is twice as strongly labeled as the remainder of the chromosome, indicating that this branch contains two labeled strands. This structure is consistent with semiconservative replication for the *E. coli* chromosome.

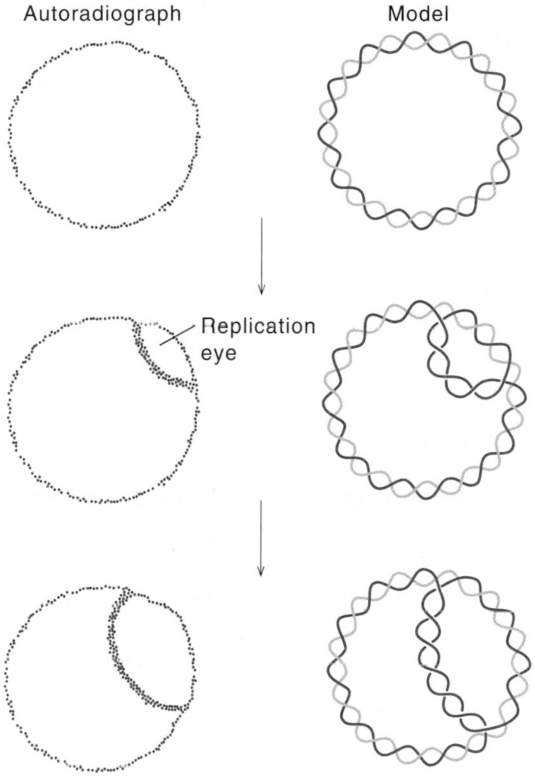

Figure 26.4

Schematic diagrams of two different modes of DNA synthesis at the growth fork(s). In unidirectional replication (*a*) one growth fork occurs; in bidirectional replication (*b*) two occur. Red indicates regions containing newly synthesized DNA.

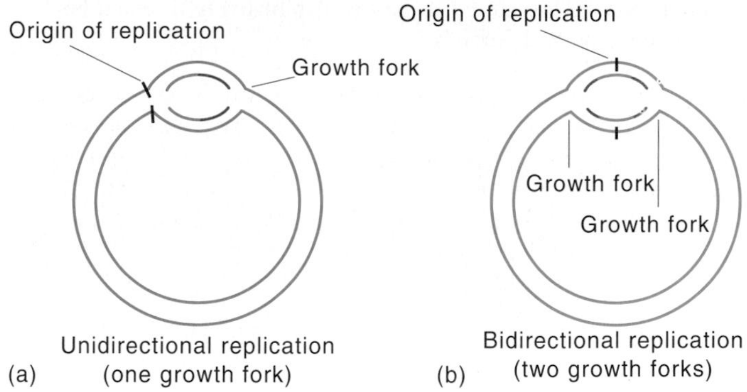

posed to radioactive thymidine after which the replicating chromosomes were examined by autoradiography; both of the forks in the replicating structures were intensely labeled (see fig. 26.4*b*). These results show that both forks must be active in replication, a finding consistent with bidirectional growth.

Growth at the Replication Forks Is Discontinuous

Continuous synthesis on both strands of the replication fork would require synthesis in the 5′ → 3′ direction on one strand and in the 3′ → 5′ direction on the other because of the antiparallel nature of the DNA duplex (fig. 26.5*a*). Continuous synthesis on both strands seems unlikely since the only known enzymes that catalyze DNA synthesis add bases to the growing chain in the 5′ → 3′ direction (see below). For this reason it was hypothesized that replication is discontinuous on one of the branches at the replication fork (see fig. 26.5*b*). Meticulous electron-microscopic examination of replication forks in bacterial viruses did in fact show that transient gaps sometimes are apparent on one of the DNA strands near the replicating fork. Observations such as this led to the notion of leading-strand and lagging-strand synthesis (see fig. 26.5*b*). The 5′ → 3′ synthesis of the leading strand can occur continuously in the direction of unwinding at the replication fork. But synthesis of the lagging strand in the 5′ → 3′ direction only occurs in discontinuous spurts in the opposite direction.

This concept of discontinuous synthesis was supported by the finding that about half of the newly synthesized DNA is first made in small pieces that subsequently become incorporated into larger units of DNA. Small replication fragments were first detected in the laboratory of Okazaki. This was done by exposing growing cells to ^{3}H-labeled thymidine for a very short time (2–10 s), followed by rapid isolation of the radioactively labeled DNA. After longer labeling times, 1–2 min, most of the labeled DNA was found in much larger segments of DNA.

A closer examination of the Okazaki fragments led to detection of short stretches of ribonucleotides at the 5′ ends. From this and many other observations on different systems it was determined that a new DNA chain can be initiated only by attaching the first deoxynucleotide through its 5′-phosphate to the 3′-OH of a short RNA polynucleotide. An RNA strand that functions in this capacity is called a primer. Primers are formed at points along the chromosome; they base-pair with the single-stranded template DNA in the regions where they are formed.

Figure 26.5

Models for synthesis at the replication fork. (*a*) Continuous synthesis on both strands. Note that both growth arrows are pointing in the same direction, which would require growth in the $5' \rightarrow 3'$ direction on one strand and in the $3' \rightarrow 5'$ direction on the other. If growth occurs only in the $5' \rightarrow 3'$ direction, synthesis would have to be discontinuous on one strand, as in (*b*). Alternatively, it could be discontinuous on both strands (*c*).

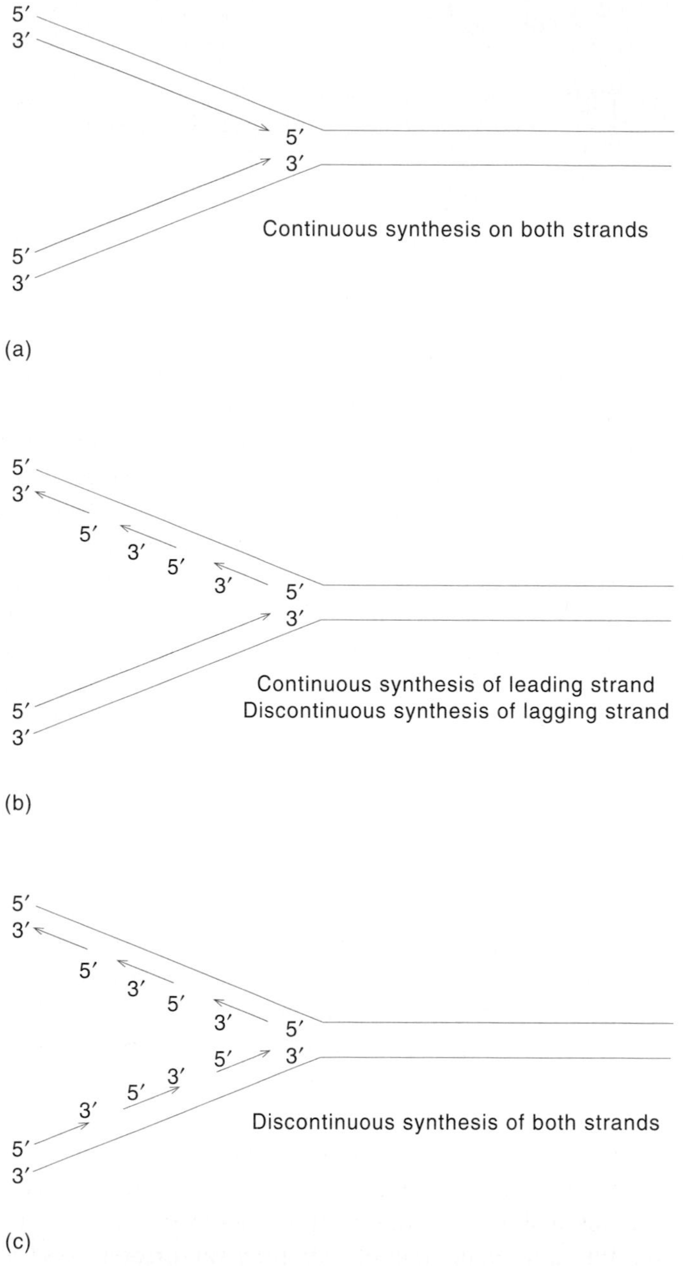

Figure 26.6

A model for discontinuous DNA synthesis. Synthesis occurs in a region that has been partially single-stranded. First, RNA primers are formed at various points on the single-stranded region (1). DNA synthesis starts at the $3'$ ends of the primers (2). Primers are removed (3). Gaps between DNA fragments are filled in by further DNA synthesis (4). Fragments are ligated to make one long, continuous piece of DNA (5). Newly synthesized RNA and DNA are indicated in red.

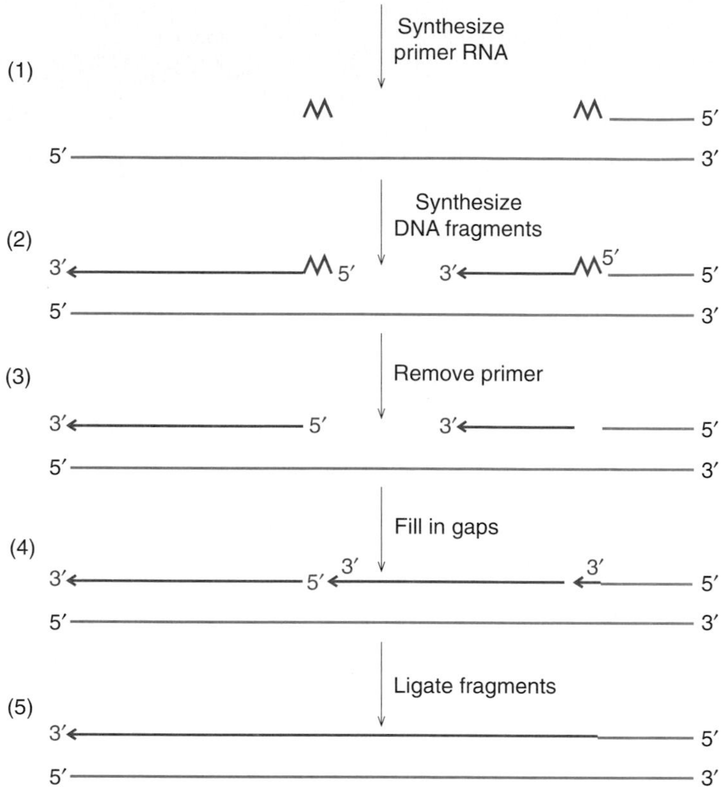

A detailed model for discontinuous synthesis could now be proposed (fig. 26.6). First, RNA primers are made on the single-strand region of the template; then DNA is synthesized. Finally, the RNA is removed from the fragments, and the gaps are filled in and ligated. An understanding of the detailed steps of this process has come largely from analysis of simpler viral replicating systems and more recently from examination of a system intended to resemble the *E. coli* chromosome (see below).

Proteins Involved in DNA Replication

Table 26.1 contains a list of some proteins involved in DNA replication and their functions. How does one go about analyzing such a complex situation? Historically, three general methods have been used for the identification and characterization of the proteins involved in DNA replication: Purification, reconstitution, and mutation. Insofar as possible, all three methods are used together.

The first method involves isolation of proteins with enzymatic activities that are logically related to the replication process, such as DNA polymerases and ligases. This is

Table 26-1

Proteins Involved in DNA Replication

Protein	Gene(s)	Function
DnaA	*dnaA*	Initiator protein: Binds *oriC;* promotes double-helix opening; DnaB loading
DnaC	*dnaC*	Complexes with DnaB; delivers DnaB to DNA
DNA polymerase III holoenzyme		
α	*dnaE*	Polymerase
ϵ	*dnaQ*	3′ to 5′ exonuclease
θ	*holE*	Unknown
τ	*dnaX*	
γ	*dnaX*	
δ	*holA*	
δ'	*holB*	
χ	*holC*	
ψ	*holD*	
β	*dnaN*	Processivity factor: ''Sliding clamp''
PolI	*polA*	Prime removal and gap filling; Initial leading strand synthesis on ColE1
Ligase	*lig*	Joins nascent DNA fragments
Gyrase	*gyrA* *gyrB*	Type II topoisomerase; replication swivel, DNA supercoiling, decatenation
DnaB	*dnaB*	5′ to 3′ helicase and activator of primase
Primase	*dnaG*	Primer synthesis
SSB	*ssb*	Binds single-stranded DNA

Source: Adapted from T. A. Baker and S. H. Wickner, Genetics and enzymology of DNA replication in *Escherichia coli. Ann. Rev. Genetics,* 26:447, 1992.

the classical biochemical approach and can be applied to any biological system. After isolation and characterization, several approaches may be used to demonstrate that the purified enzyme is active in the replication process *in vivo.* Sometimes this can be done by using inhibitors that act on both the purified protein and the cellular process. The concentration of inhibitor required to inhibit DNA replication *in vivo* should be approximately the same as that required to inhibit the purified enzyme *in vitro.* In prokaryotes, mutations have been very useful for confirming the functions of isolated proteins in the replication process. The induction of a new enzyme activity associated with a biological process, such as virus infection or cell proliferation, also provides useful evidence.

A second method used to identify proteins needed for replication involves reconstitution. Whole-cell lysates containing all of the components necessary for replication are fractionated, and the DNA replication process is then reconstituted with various combinations of the purified or partially purified proteins. Components of the replication system are recognized on the basis of their ability to restore overall activity *in vitro.* This procedure can be applied to any organism, even when relevant genetic mutants are not available.

The third method uses genetic mutation as the primary tool. This method requires the isolation of conditional mutants, that is, mutants that behave normally under one set of conditions but abnormally under another. Most commonly, temperature-sensitive DNA replication mutants are used. Such mutants have been isolated in *E. coli,* and they grow normally at a low (permissive) temperature (33°C) but poorly or not at all at a high (nonpermissive) temperature (41°C). Preliminary analysis of the temperature-sensitive step provides clues to the stage of replication affected. For example, the length of time required for DNA synthesis to stop after cells are shifted from permissive to nonpermissive temperatures can indicate whether the mutation occurs in a protein involved in initiation or elongation. If the mutation is in the gene for a protein required for elongation, most DNA synthesis stops immediately at the nonpermissive temperature because the majority of cells are usually at some stage in the elongation process. If the mutation is in a gene required only for initiation of replication, most DNA synthesis continues for some time and stops when the rounds of replication in progress are completed. *In vitro* assays are then used to aid in purifying the corresponding proteins from wild-type cells. Extracts from cells with the temperature-dependent defect are not active in DNA synthesis at the elevated temperature, but activity can be restored by adding the corresponding protein from normal wild-type cells. This complementation assay can be used to aid in the purification of particular replication proteins. To prove that the correct protein has been purified from wild-type cells, the proteins from the temperature-sensitive mutant also must be purified and shown to be abnormal, frequently exhibiting unusual instability at elevated temperatures.

Characterization of DNA Polymerase I in Vitro

We look at *E. coli* DNA polymerase I as an example of how biochemical and genetic studies are used to characterize a DNA replication enzyme. Then we consider the other major proteins involved in DNA replication in *E. coli* before examining the proteins that participate in the synthesis of DNA in different types of organisms.

The Watson-Crick proposal of a complementary duplex structure for DNA stimulated a search for a DNA polymerase enzyme with certain implied properties. The enzyme should require an intact DNA chain to serve as a template for the absorption of complementary bases, and the newly synthesized DNA should be a complement of one of the template DNA chains. Arthur Kornberg and his co-workers isolated a DNA-synthesizing enzyme from cells of *E. coli* that satisfied these requirements and named it DNA polymerase; it is now known as DNA polymerase I (PolI), or the Kornberg enzyme. This enzyme requires a DNA template, the four commonly occurring deoxynucleotide triphosphates, and Mg^{2+} ions for making DNA. The enzyme catalyzes the addition of mononucleotides to the 3′-OH end of a growing chain (fig. 26.7). Simultaneously with the formation of this linkage, the linkage between the two phosphates is broken, releasing a pyrophosphate group. The energy produced by the cleavage provides the energy necessary for linking the mononucleotide to the growing DNA chain. Subsequent cleavage of the pyrophosphate to orthophosphates ensures the irreversibility of the reaction. Bases added to the growing chain are determined by the sequence of bases in the DNA template. As nucleotides complementary to those on the template are added, the single-stranded DNA template gradually becomes converted to a double helix. Structure studies have shown that the enzyme has a complex surface with specific attachment sites for the template chain, the growing chain, and monomer nucleoside triphosphate. The enzyme is highly selective, because the only nucleotides it links to the growing chain are those that form Watson-Crick base pairs with the template strand. As the new chain lengthens by synthesis, the enzyme moves along the template one base at a time.

In addition to the characteristic template-directed polymerization activity, PolI contains an activity that results in the removal of mononucleotides from the 3′ end of a polynucleotide strand. This enzyme activity leads to cleavage of the bond between the 5′-phosphate of the terminal residue and the 3′-OH group of the penultimate residue. Since this is the same linkage that is made during synthesis, the degradation reaction may be thought of as a reversal of the polymerization process. For net chain elongation beyond the 3′-OH end of the DNA strand, the polymerization rate must exceed the depolymerization rate. Polymerization is much faster than depolymerization if the correct base-paired nucleotide is inserted into the growing chain. If a mismatched base is accidentally inserted, the opposite is true, and the mismatched base is usually removed. The combined polymerization–depolymerization reaction has been viewed as a "proofreading," error-reducing mechanism in DNA synthesis.

E. coli DNA polymerase I has a second associated activity catalyzing $5' \rightarrow 3'$ degradation of DNA. Whereas $3' \rightarrow 5'$ degradation activity of the enzyme is much more effective on unpaired or mispaired bases, the $5' \rightarrow 3'$ activity cleaves preferentially at base-paired regions. The ability of PolI to degrade DNA (or RNA) in the $5' \rightarrow 3'$ direction as well as to carry out a polymerization reaction suggests a role in removing RNA primer during gap filling.

Crystallography Combined with Genetics to Produce a Detailed Picture of DNA PolI Function

The question arises as to whether the three activities of PolI just described all originate from a single active site or from more than one site. Cleavage of PolI by the protease subtilisin leads to a small fragment (M_r = 30,000) with $5' \rightarrow 3'$-nuclease activity and a large fragment (M_r = 70,000), called the Klenow fragment, exhibiting the polymerization and $3' \rightarrow 5'$ depolymerization activities. A bril-

Figure 26.7

Template and growing strands of DNA. (*a*) Nucleotides are added one at a time to the 3′-OH end of the growing chain. Only residues that form Watson-Crick H-bonded base pairs with the template strand are added. (*b*) Covalent bond formation between the 3′-OH end of the growing chain and the 5′-phosphate of the mononucleotide is accompanied by pyrophosphate removal from the substrate nucleoside triphosphate.

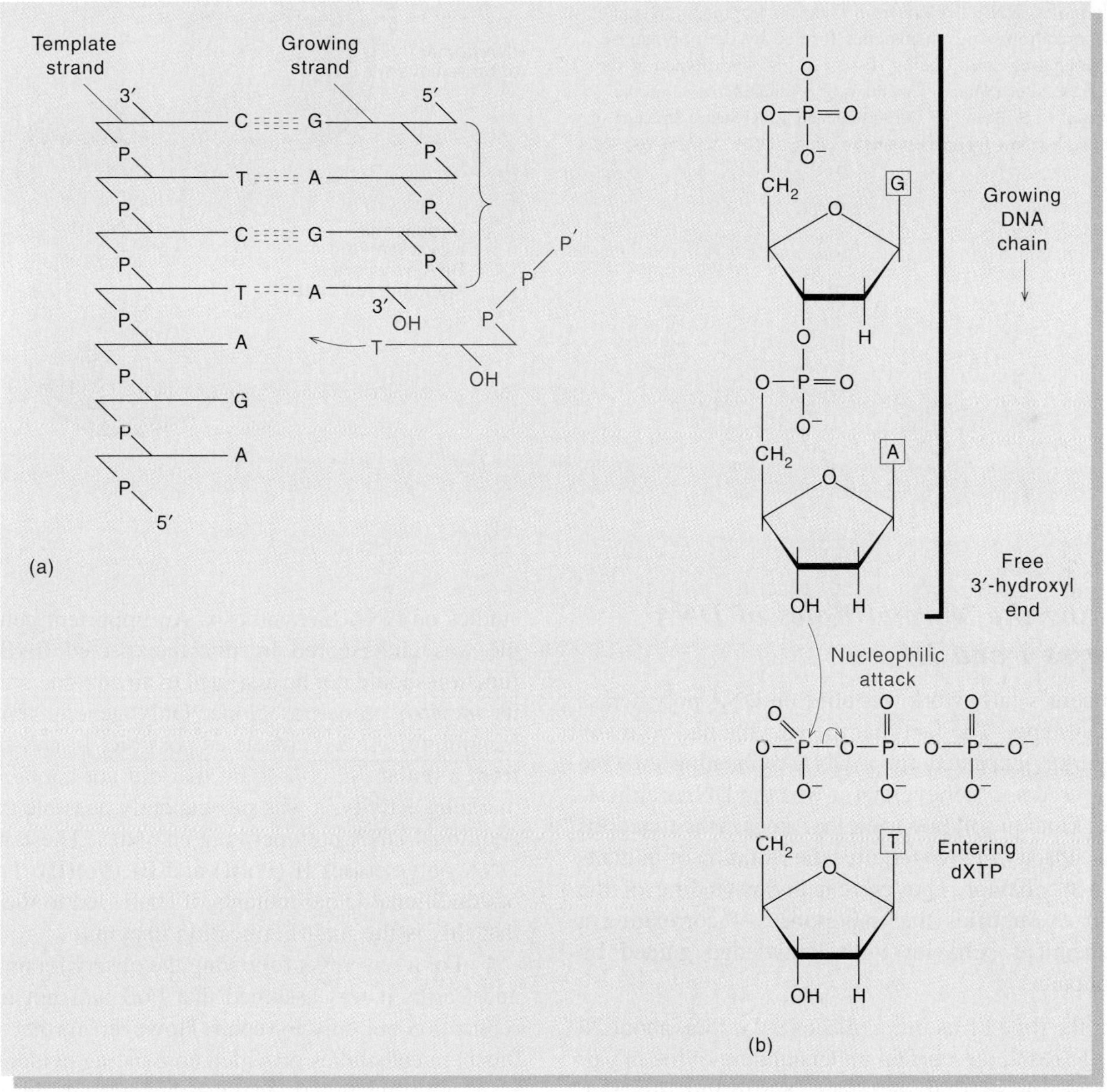

liant series of investigations by Tom Steitz and his coworkers led to a detailed understanding of the Klenow fragment. These studies were done on wild-type and mutant enzyme lacking the nuclease activity. From their studies Steitz and his colleagues concluded that within the Klenow fragment the polymerase and 3′ → 5′ exonuclease activities were located in different regions. Thus, the polymerase has three enzyme activities all resulting from different parts of the same protein molecule. The approximate locations of these activities are indicated relative to a chain undergoing synthesis in figure 26.8. Recent investigations on cocrystals of DNA and PolI indicate that the DNA makes a sharp bend between the 3′ → 5′ exo and the polymerase active sites. This somewhat clumsy-looking structure may facilitate the detection of imperfections in newly synthesized DNA and the backing up process that must accompany proofreading excision by the 3′ → 5′ exo.

Figure 26.8

A schematic drawing of the possible structure of the DNA-PolI complex in an elongating complex of DNA. The Klenow fragment contains the 5′–3′ polymerase site and the error correcting 3′–5′ exonuclease site. The fragment cleaved from the PolI enzyme by limited proteolysis contains the 5′–3′ exonuclease site. The 5′–3′ exonuclease removes RNA leaders from Okazaki fragments to make way for polymerization. Any mismatches formed by the polymerase result in the elongating strand being displaced and hydrolyzed at the 3′–5′ exonuclease site. (Source: This drawing is primarily based on the structural studies of L. S. Beese, V. Derbyshire, and T. A. Steitz, Structure of DNA polymerase I Klenow fragment bound to duplex DNA, *Science* 260:352–355, 1993.)

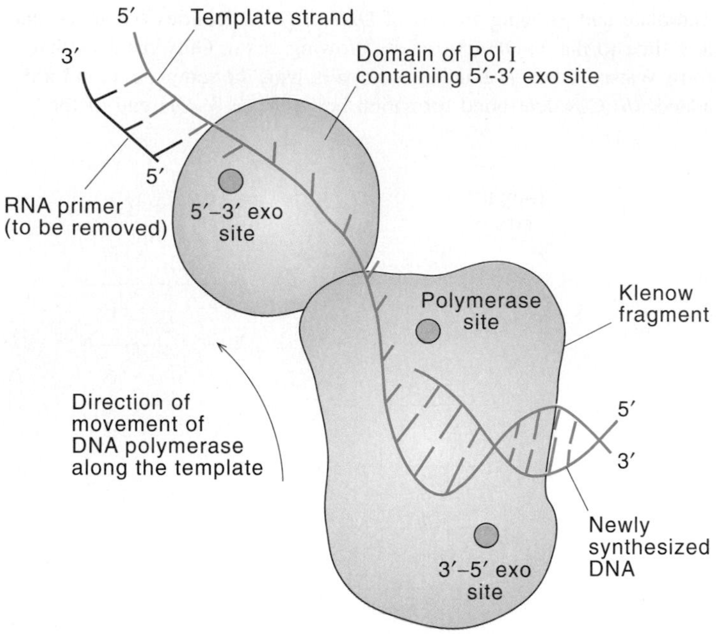

Establishing the Normal Roles of DNA Polymerases I and III

All of Kornberg's early work was done on DNA polymerase in cell-free systems. The fact that the enzyme had so many of the properties expected for a DNA-replicating enzyme led most observers to believe that it was the DNA-replicating enzyme. Final proof, however, that an enzyme functions in the same capacity *in vivo* requires the isolation of mutants that affect its behavior. Our current understanding of the role of PolI exemplifies the importance of correlating a given biochemical behavior with knowledge gained by studying mutants.

From the time of its initial discovery it took about 20 years to reach our current understanding of the physiological role of PolI. A major step in this direction was taken by Cairns and DeLucia, who laboriously scanned several thousand strains of mutagenized *E. coli* to find one that contained almost no PolI polymerizing activity (1%–2% of normal). This mutant grew well under normal conditions, suggesting that PolI was not involved in replication. However, the mutant in question was hypersensitive to ultraviolet irradiation. Since ultraviolet irradiation was known to damage DNA, the hypersensitivity of the mutant suggested that PolI might be involved in repairing chromosome damage. The discovery of the polymerase mutant had a profound effect on thought and experimental design in further studies on DNA biosynthesis. An important general principle was underscored by this unexpected finding—that a function should not be assigned to an enzyme on the basis of its *in vitro* properties alone. Only genetic mutants make meaningful *in vivo* correlates possible. In cell-free extracts from a mutant *E. coli* strain that did not contain PolI polymerizing activity, it was subsequently possible to detect two additional DNA polymerizing enzymes. These were named DNA polymerases II (PolII) and III (PolIII). The behavior of conditional lethal mutants of PolIII led to the conclusion that this is the main replication enzyme.

For a few years following the observations of DeLucia and Cairns it was assumed that PolI was not important in replication but only in repair. However, further genetic and biochemical studies provided convincing evidence that PolI is a multifunctional enzyme possessing, in addition to $5' \rightarrow 3'$ polymerizing activity, the $3' \rightarrow 5'$ and $5' \rightarrow 3'$ degradation activities described earlier. The original mutant of PolI isolated by Cairns and DeLucia was inactivated only in its polymerizing function. Subsequently, mutants in PolI that affect the $5' \rightarrow 3'$ degradation function were found to be conditionally lethal and did not permit elongation of DNA synthesis under nonpermissive conditions; thereby, it was demonstrated that this activity of the PolI enzyme is indispensable for chromosome replication. In this regard it should be mentioned that of the three known *E. coli* DNA polymerases, only PolI has a $5' \rightarrow 3'$ exonuclease activity.

Figure 26.9

Steps in the sealing of a DNA nick, catalyzed by DNA ligase. The bacterial ligase uses NAD^+ to make an enzyme–AMP intermediate. Mammalian DNA ligases and bacteriophage T4 ligase use ATP for the same purpose.

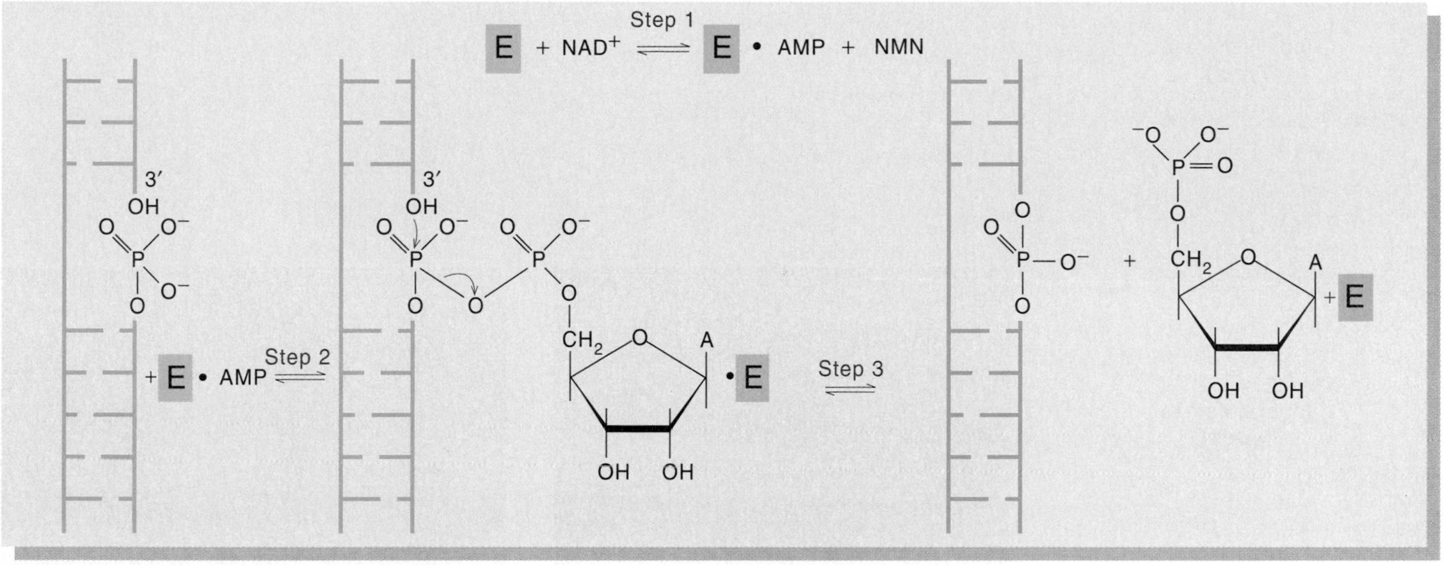

Figure 26.10

Type I and type II topoisomerases relax negatively supercoiled DNA in steps of one and steps of two, respectively. Type II topoisomerases can also add additional negative supercoils (as indicated by the double arrow). The latter reaction requires energy input, which is encoded by ATP cleavage.

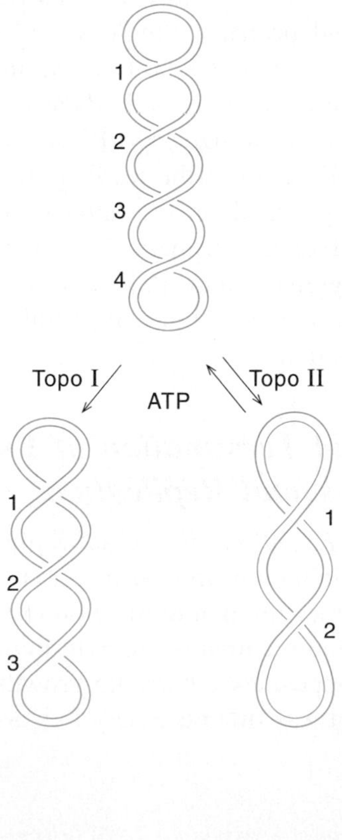

Other Proteins Required for DNA Synthesis in Escherichia coli

As mentioned earlier, discontinuous DNA synthesis necessitates the existence of an enzyme for joining the newly synthesized segments (see fig. 26.6). Such an enzyme has been found in a variety of cell types and is called polynucleotide ligase (fig. 26.9).

Topoisomerases can introduce negative supercoils into DNA or relax negatively supercoiled DNA. They can also catalyze the linking together (catenation) of double-stranded circular DNAs or the decatenation of linked circular molecules (figs. 26.10 and 26.11). Topoisomerases must serve vital functions because mutants carrying defective topoisomerases of one sort or another show severely impaired growth properties. It seems likely that decatenation and modulation of supercoiling could both be very important to cell survival.

At the growth fork it is necessary that the parental double helix be unwound to present further single-stranded regions to serve as templates for continued replication. The dnaB protein of *E. coli* is believed to be a helicase that directly catalyzes this process.

Single-strand binding protein (SSB) is found in abundance in *E. coli,* and it is believed to be bound in mass at the replication fork. This fact and other evidence described later on indicate that it plays an important role at the growth fork.

Figure 26.11

Catenation by topoisomerases. (*a*) Two circular DNAs can be catenated by type I topoisomerase only if one of the DNAs is nicked. This is not necessary when using a type II topoisomerase. (*b*) Electron micrographs of catenated DNA before (i) and after (ii) incubation with DNA gyrase. The catenate contains one large circular DNA and one small circular pBNP66 plasmid DNA. (Source: Adapted from M. Gellert, L. M. Fisher, H. Ohmori, M. H. O'Dea, and K. Mizuchi, DNA gyrase: Site-specific interactions and transient double-strand breakage of DNA, *Cold Spring Harbor Symp. Quant. Biol.* 45:301, 1981.)

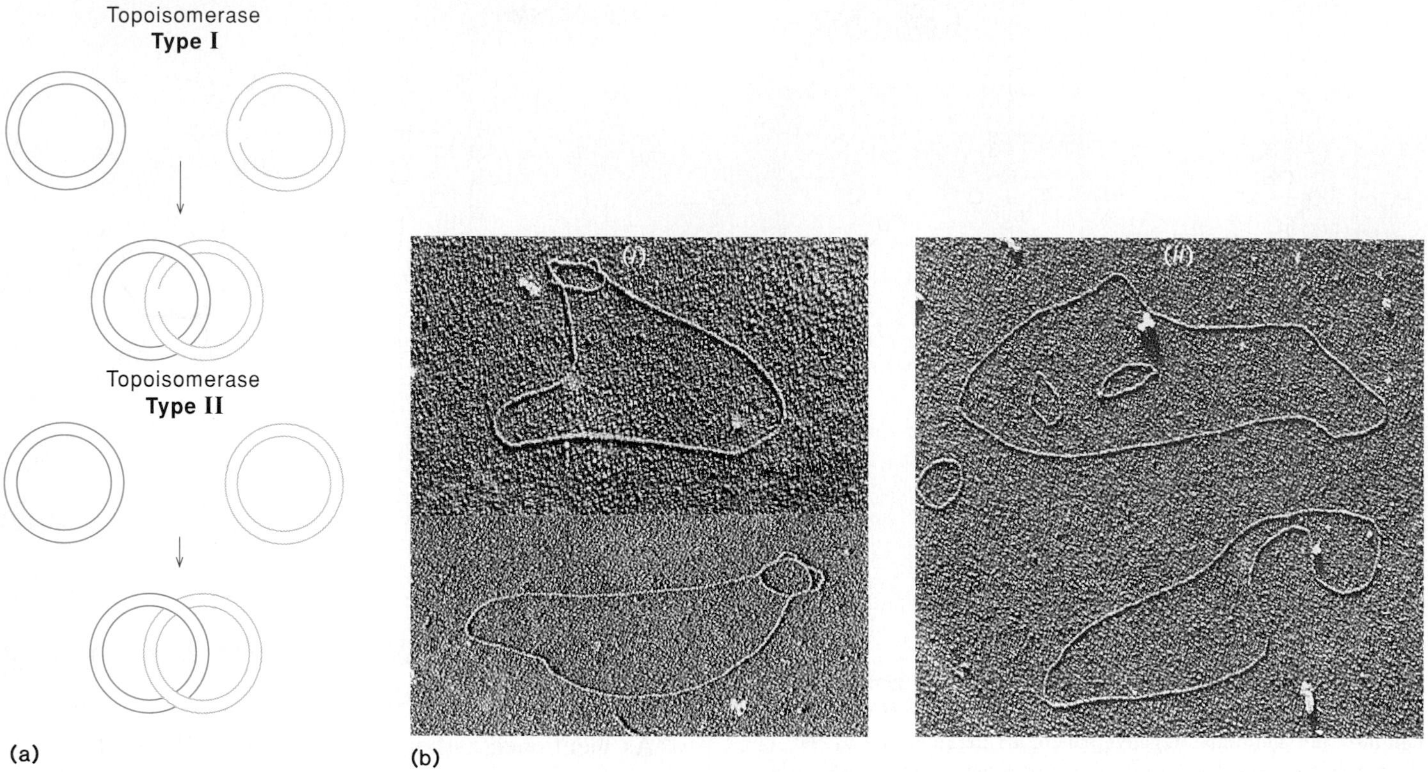

Two functions can be suggested, both inferred from its preference for binding to single-stranded DNA. It could protect single-stranded DNA from nucleases, or it could facilitate unwinding by inhibiting rewinding.

There are two enzymes that can catalyze the synthesis of RNA with the help of a DNA template. One of these, called primase is encoded by the *dnaG* gene. Primase catalyzes the synthesis of small primer RNAs that are required for DNA synthesis. The other enzyme called RNA polymerase catalyzes the synthesis of all the other RNAs found in *E. coli.* For some time it was thought that RNA polymerase catalyzed the synthesis of the primers that initiate replication of the chromosome. Presently, it is believed that all primer synthesis is catalyzed by primase.

Replication of the Escherichia coli *Chromosome*

The elongation step in *E. coli* chromosomal DNA synthesis is depicted in figure 26.12. As the DNA gradually unwinds new synthesis takes place on the two strands. At the heart of the unwinding process we find a DnaB helicase, and just behind it we find primase. Single strand binding protein (SSB) coats the newly unwound DNA strands. The primase travels in the same direction as the DnaB helicase, making periodic pauses to synthesize short RNA primers on the lagging strand. PolIII extends the leading strand and also extends the lagging strand. In the latter case the first DNA bases are added to the primers. Following the action of PolIII on the lagging strand, PolI replaces the RNA bases with DNA bases, and DNA ligase then links the short DNAs on the lagging strand.

Initiation and Termination of Escherichia coli *Chromosomal Replication*

Two aspects of *E. coli* chromosomal replication still to be considered are initiation and termination. From what has been said we conjecture that replication initiates at a unique site, proceeds bidirectionally, and terminates at a point where the two oppositely advancing growth forks meet. To study initiation it was first necessary to isolate that segment

Figure 26.12

Model for the replication fork in the *E. coli* chromosome. The DnaB protein unwinds the duplex while the primase directly behind it synthesizes RNA primer on the lagging strand. SSB protein binds to single-stranded regions wherever they are. DNA polymerase III progressively adds nucleotides to the leading strand and also adds nucleotides to the lagging strand starting at the RNA primers. After this, DNA polymerase I replaces the RNA primer with the DNA equivalent. Finally DNA ligase knits the short DNA segments together on the lagging strand. (Source: Adapted from K. Arai, R. L. Low, and A. Kornberg, movement and site selection for priming by the primosome in phage ϕX174 DNA replication, *Proc. Natl. Acad. Sci. USA* 78:711, 1981.)

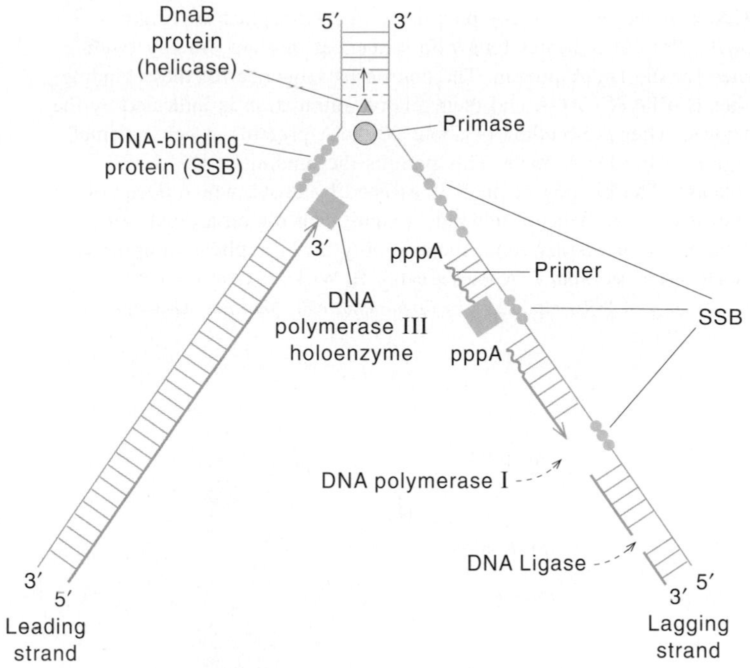

of the chromosome that carries the unique origin of replication. This was done by chopping the bacterial chromosome into small pieces, circularizing the small pieces, and then transfecting them into cells to see which pieces could sustain autonomous replication. In this way a unique segment of chromosome was discovered which contains 245 bp required to initiate replication.

With circular minichromosomes containing this unique segment termed *oriC*, it has been possible to study the initiation and replication process *in vitro*. First, in crude extracts it was possible to show that replication initiates within or near the *oriC* sequence and proceeds bidirectionally. About 13 different proteins participate in the *oriC*-directed DNA replication as judged by their activity in a reconstituted replication reaction.

Initial complex formation entails binding of the dnaA protein to several sites (probably four) of the *oriC* region (fig. 26.13). This is followed by the cooperative binding of more dnaA proteins. In all, 20–40 dnaA proteins bind to a region of about 200 bp. This binding is climaxed by a structural change in the dnaA–*oriC* complex, which results in localized melting of the DNA at one end of the *oriC* region. The melting reaction, which reproduces the local opening of the duplex requires that the DNA be negatively supercoiled. As we have already discussed, negative supercoiling favors unwinding of the duplex (see chapter 25).

The dnaB protein is transferred to this complex from a dnaB–dnaC complex. The dnaC dissociates, leaving the dnaB protein bound to the template. Melting of the DNA duplex proceeds bidirectionally from *oriC* as the dnaB helicase migrates from the dnaA–*oriC* complex to provide a template for the priming and replication enzymes. Addition of the priming and replication enzymes results in immediate initiation and elongation. Continued elongation requires gyrase and SSB. Gyrase provides a swivel to permit continued unwinding of parental strands while SSB stabilizes the transiently single-stranded regions. DNA gyrase proves useful in the late stages of replication as well, because many circular dimers are formed that require decatenation and negative supercoils so they can function again as active templates.

DNA Replication in Eukaryotic Cells

Mutant studies have not been that helpful in finding proteins with known functions in vertebrate DNA replication. The major approach in vertebrate investigations has been isolation of proteins having activities logically related to DNA replication and isolation of replicative intermediates. Different forms of DNA polymerases, DNA ligases, topoisomerases, single-strand binding proteins and unwinding enzymes have all been found.

Figure 26.13

Model of the initiation complex at the *E. coli* replication origin (*oriC*). R1-R4 indicates four 9 bp sequences that are primary binding sites for the DnaA protein. The consensus sequence for these binding sites is TTATCCACA and their relative orientation is indicated by the arrows. After the binding of about 20 dnaA proteins at *oriC,* a small region of the DNA melts. This permits the binding of the dnaB helicase. The binding of dnaB is assisted by dnaC, which does not bind itself. The helicase unwinds a significant region around *oriC,* creating the necessary room for assembly of the replication apparatus. (Source: Adapted from T. A. Baker and S. H. Wickner, Genetics and enzymology of DNA replication in *Escherichia coli, Ann. Rev. Genetics* 26:447, 1992).

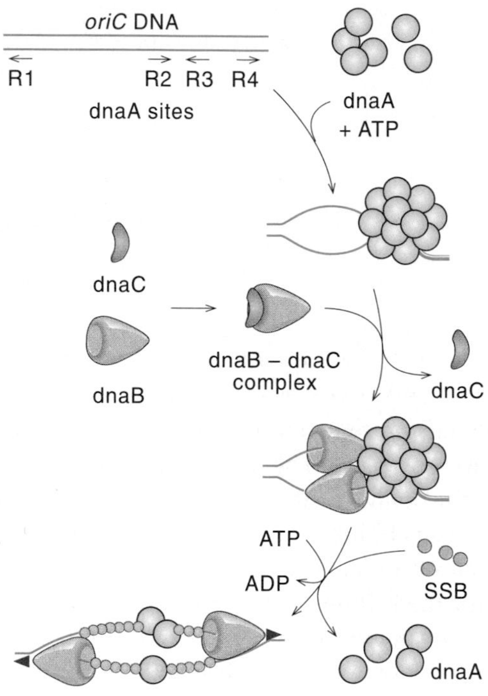

Figure 26.14

Multiple-origin model for eukaryotic chromosomal DNA replication. (*a*) Autoradiograph of short-term labeling of a eukaryotic chromosome during replication and its interpretation. (*b*) Overall replication scheme for a eukaryotic chromosome. Only a short region of the chromosome is shown. It is believed that replication origins are relatively free of proteins.

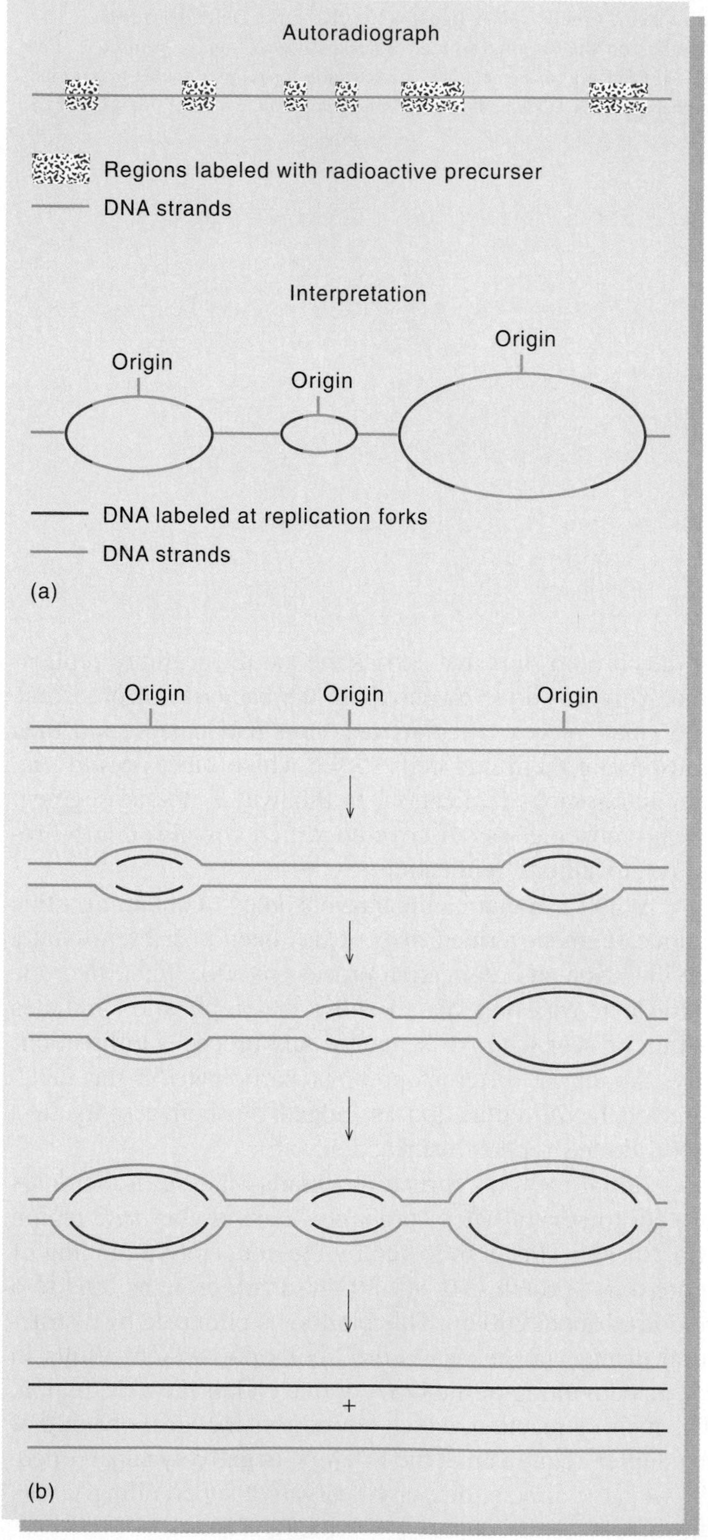

Eukaryotic Chromosomal DNA

Although general features of DNA replication in eukaryotes are thought to be similar to those of prokaryotes, there are some interesting differences. The chromosomes of higher eukaryotic organisms are quite large, in some cases 1,000 times larger than their bacterial counterparts. For these larger DNA molecules to replicate in a reasonable time, they have multiple origins of replication. The simultaneous synthesis of DNA at several points along the chromosome has been demonstrated by incorporating radioactive nucleotides into the replicating chromosomes for a short period and then observing the distribution of radioactive DNA by autoradiography (fig. 26.14). Multiple regions of incorporated label are observed, and replication proceeds bidirectionally from these regions. Termination of replication occurs at the point where the growth forks from two adjacent replication units meet (see fig. 26.14). DNA on the lagging strand of a fork is made discontinuously. The Okazaki fragments are much shorter than those found in prokaryotes, averaging only between 100 and 200 nucleotides in length. Synthesis of these DNA fragments is initiated on primers that are found covalently attached to the 5′ ends of newly synthesized fragments.

Remember that DNA in eukaryotic chromosomes is associated with histones in complexes called nucleosomes (see fig. 25.18). The disassembly of the DNA–histone complex presumably occurs directly in front of the replication fork. Two observations suggest that nucleosome disassembly may be a rate-limiting step in the migration of the replication fork in chromatin: (1) The rate of migration of replication forks in eukaryotes is slower than in prokaryotes; and (2) the length of the replication fragments on the lagging strand is similar to the length of DNA between adjacent nucleosomes (about 200 bp). Before the newly replicated DNA is reassembled into nucleosomes, the RNA primers must be removed, and the gaps must be filled by enzymes unknown at the present time. Finally, replication fragments must be linked together by DNA ligase.

SV40 Is Similar to Its Host in Its Mode of Replication

Much has been learned about the replication of chromosomal DNA in eukaryotes by studying a simple monkey virus known as SV40. The viral genome of SV40 consists of a circular duplex DNA molecule of about 5,200 bp with one origin for replication. The SV40 viral genome is complexed with histones to form a nucleoprotein structure very similar to that observed for chromatin. Since SV40 encodes only a single replication protein (T antigen), the virus must make extensive use of the host's cellular replication machinery. As a result many similarities exist between viral and cellular DNA replication. In both cases initiation of DNA synthesis involves two nascent strands. The leading strand grows continuously while the lagging strand grows discontinuously by joining together small (about 200 bp) segments of DNA that are independently initiated with RNA primers. Completion of replication occurs when two oppositely moving forks meet. In linear cellular chromosomes the two merging forks originate from adjacent origins, but in circular SV40 chromosomes they have a single origin.

The development of an efficient cell-free replication system has greatly accelerated progress in understanding the molecular mechanisms involved in SV40 DNA replication. The origin of replication is recognized by the viral T antigen. In addition to its specific binding activity, the T antigen has helicase activity. Once it is bound to the origin, T antigen enters the duplex and catalyzes the unwinding of the two DNA strands. Unwinding appears to be a critical step that establishes the replication forks and generates the single-stranded DNA regions required for the priming and elongation of nascent strands.

In addition to specific nucleotide sequence elements, the T-antigen-mediated unwinding reaction requires accessory proteins contributed by the host cell. For example, a single-stranded DNA-binding protein is required to prevent reassociation of the single strands exposed during unwinding. Such a protein has been found, and it binds specifically to single-stranded DNA.

Of four distinguishable DNA polymerase activities α, β, γ, and δ, it appears that α and δ are required for SV40 DNA replication. DNA polymerase α has long been considered the major replicative polymerase in animal cells. The enzyme is composed of four subunits. The largest subunit contains the polymerase active site and a $3' \rightarrow 5'$ exonuclease that serves a proofreading function during polymerization. The smallest subunit of DNA polymerase α contains a primase capable of synthesizing short RNA transcripts that can serve as primers for subsequent DNA chain elongation by the polymerase-carrying subunit. Each time a polymerase binds to the template primer it adds a certain number of nucleotides to the growing chain before it dissociates. A highly processive enzyme adds a great number of residues before it dissociates. DNA polymerase α is not a highly processive enzyme because fewer than 100 nucleotides are polymerized per binding event.

The properties of DNA polymerase δ, which is also required for SV40 DNA replication, contrast with those of DNA polymerase α. First δ has no primase activity, and second it is a highly processive enzyme capable of catalyzing the polymerization of more than 1,000 nucleotides per binding event.

These differences between the two polymerases support the suggestion that DNA polymerase α might be best suited to serve as the lagging-strand polymerase and that DNA polymerase δ would be best suited to serve as the leading-strand polymerase. This is because the leading-strand polymerase would be expected to be highly processive and would derive little benefit from an associated primase activity. By contrast, the lagging-strand polymerase would require only moderate processivity, and it would benefit from a tightly associated primase activity (fig. 26.15). Most recent evidence (cited in Waga and Stillman reference) suggests that DNA polymerase δ is involved in the replication of both strands while DNA polymerase α and primase are involved in the synthesis of RNA-DNA primers.

Figure 26.15

Hypothetical scheme for the concurrent replication of leading and lagging strands at the replication fork on the SV40 chromosome. The scheme follows the same general notion as one proposed by Kornberg for *E. coli* chromosomal replication. In this scheme polymerase δ functions on the leading strand, and polymerase α functions in conjunction with primase on the lagging strand. The 37-kdα protein, PCNA, greatly augments the activity of polymerase δ but has no effect on polymerase α. (Source: Adapted from B. Stillman, Initiation of eukaryotic DNA replication *in vitro*, *BioEssays* 9:56, 1988.)

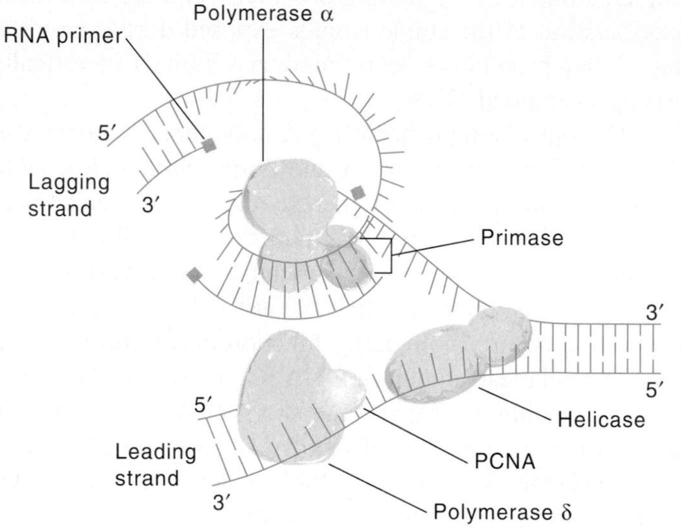

Figure 26.16

Structure of a thymine dimer formed in DNA by exposure to short-wavelength ultraviolet light.

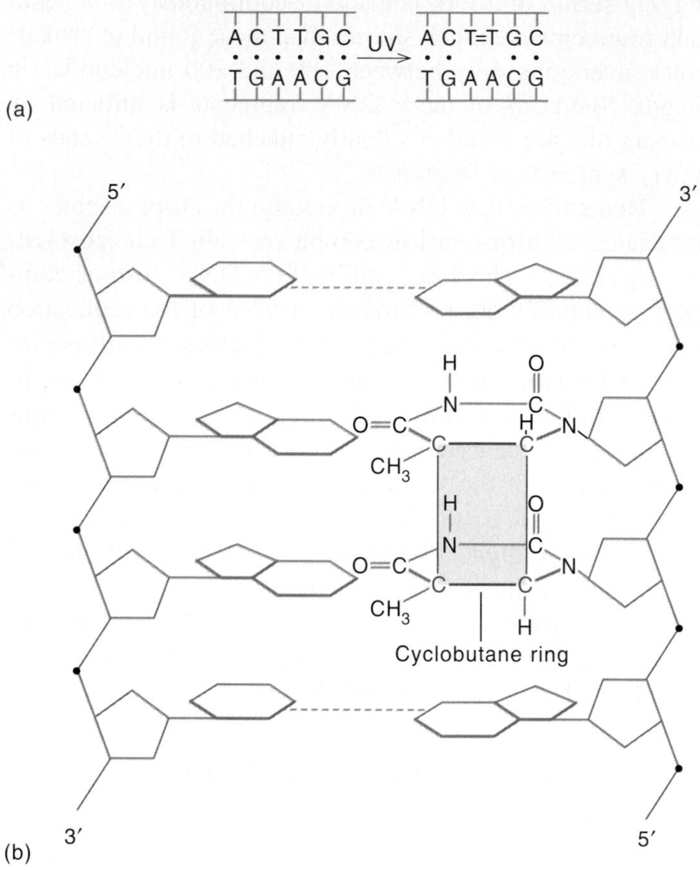

Several Systems Exist for DNA Repair

We have seen that chromosomes are usually formed by a single DNA molecule regardless of their size. Such a large molecule makes an easy target to attack and damage. In fact, limited damage occurs quite frequently. A single lesion in a DNA molecule left unrepaired could interfere with the replication process and probably would cause cell death. Clearly it is highly advantageous to have a repair system and indeed most cells have more than one.

DNA damage is caused by a variety of physical, chemical, and biological agents. Physical agents include ultraviolet light and ionizing radiation. Damage can be in the form of a missing, incorrect, or modified base or an alteration in the structural integrity of the DNA strands by breaks, cross-links, or dimerization of bases, usually pyrimidines.

Adjacent pyrimidine bases in a DNA strand form dimers with high efficiency after absorbing ultraviolet light (fig. 26.16). By contrast, purines are quite resistant to damage by ultraviolet. Pyrimidine dimers formed within an otherwise intact DNA duplex have provided a useful substrate to assay for DNA repair. These dimers can be repaired directly by enzymatic photoreactivation (fig. 26.17). The

Figure 26.17

Thymine dimers may be monomerized from DNA by enzymatic photoreactivation. In this case no nucleotides are removed in the repair reaction.

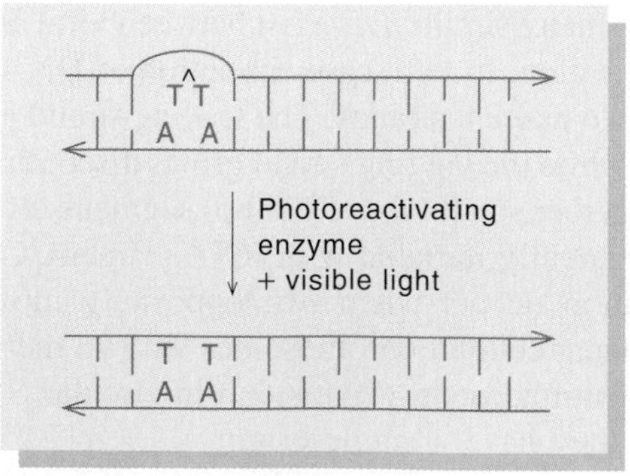

Figure 26.18

Pyrimidine dimers and other forms of DNA damage can be removed by a general excision repair mechanism. The first reaction in this form of repair involves forming nicks about the damaged region of the DNA. In (*a*) we see the mode of incision of UV-irradiated DNA by the pyrimidine-dimer-specific glycosylase and AP endonuclease activities of *Micrococcus luteus* and bacteriophage T4. In (*b*) we see the mode of incision of the uvrABC endonuclease of *E. coli.* (Source: Adapted from G. Walker, Inducible DNA repair systems, *Ann. Rev. Biochem.* 54:425, 1985.)

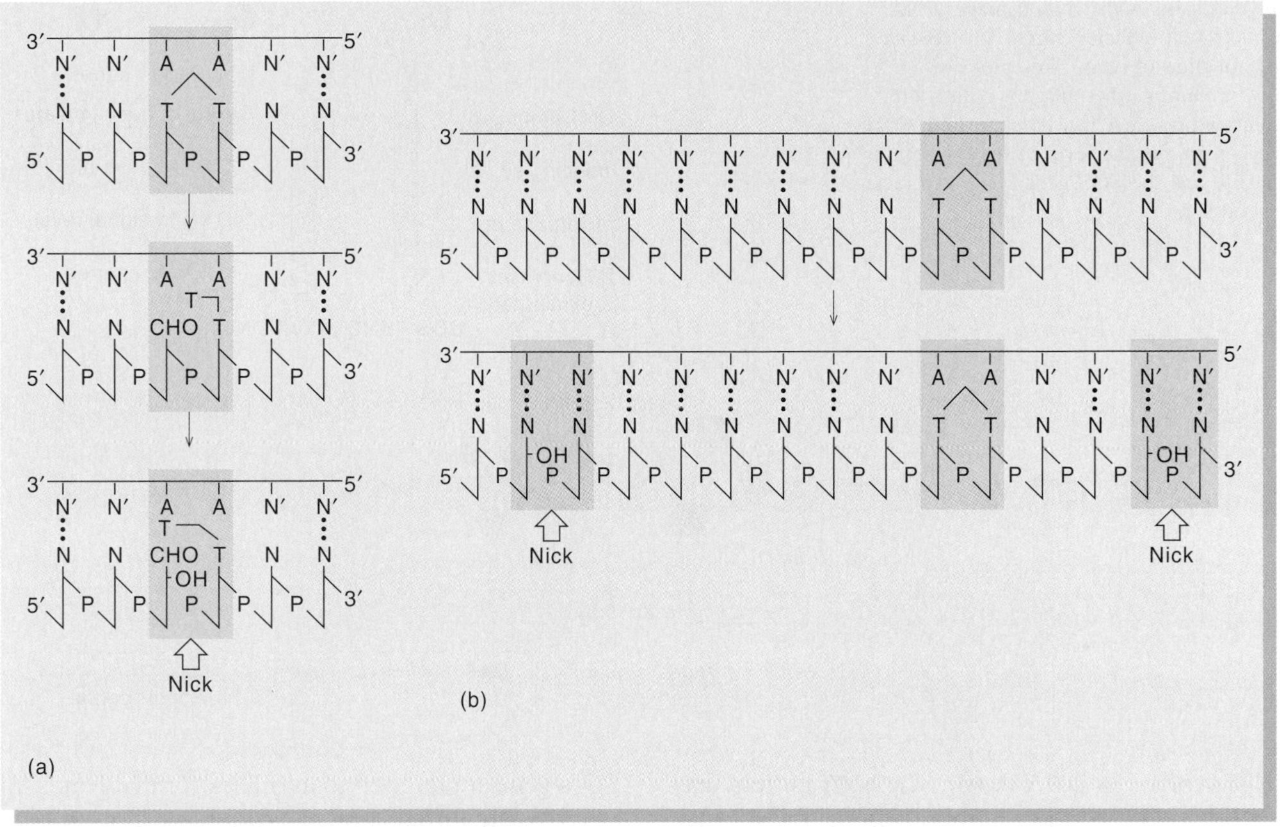

photoreactivation enzyme binds to the DNA containing the pyrimidine dimer and uses visible light to cleave the dimer without breaking any phosphodiester bonds.

Systems that function with the help of visible light are quite common. This relates to the fact that sunlight is a mixture of ultraviolet and visible light. The potentially harmful effects of moderate doses of ultraviolet are overcome by the repair processes triggered by the visible light. Ultraviolet tanning lamps usually eliminate most of the visible light and consequently could be quite harmful if used excessively or even in moderation.

Pyrimidine dimers and other forms of DNA damage can also be removed by other mechanisms that do not rely on light (fig. 26.18). This type of repair entails removal of the damaged region and resynthesis of the excised region. Special glycosylases that recognize abnormal or incorrectly paired bases can cleave N-glycosidic bonds to generate an apurinic or apyrimidinic site (see fig. 26.18*a*). Alternatively, a double incision is made in the strand that carries the lesion by a complex of three proteins encoded jointly by the *uvrA, uvrB,* and *uvrC* genes (see fig. 26.18*b*). In the case of pyrimidine dimers, an incision is made seven nucleotides 5′ to the pyrimidine dimer, followed by a second incision of three or four nucleotides 3′ to the same dimer. The *uvrD* gene product, a helicase, together with PolI and possibly a single-strand binding protein, releases the 12–13 nucleotide oligomer generated by the incision of the uvrABC enzyme complex. After release of this oligomer, PolI and DNA ligase resynthesize the excised region and ligate the nicks.

Synthesis of Repair Proteins Is Regulated

In *E. coli* the synthesis of many enzymes involved in repair is regulated by the so-called SOS system. Two proteins, lexA and recA, form the working machinery of this regulatory system (fig. 26.19). Under normal conditions the lexA protein inhibits the expression of about 17 genes (the *din* genes), the encoded proteins of which are

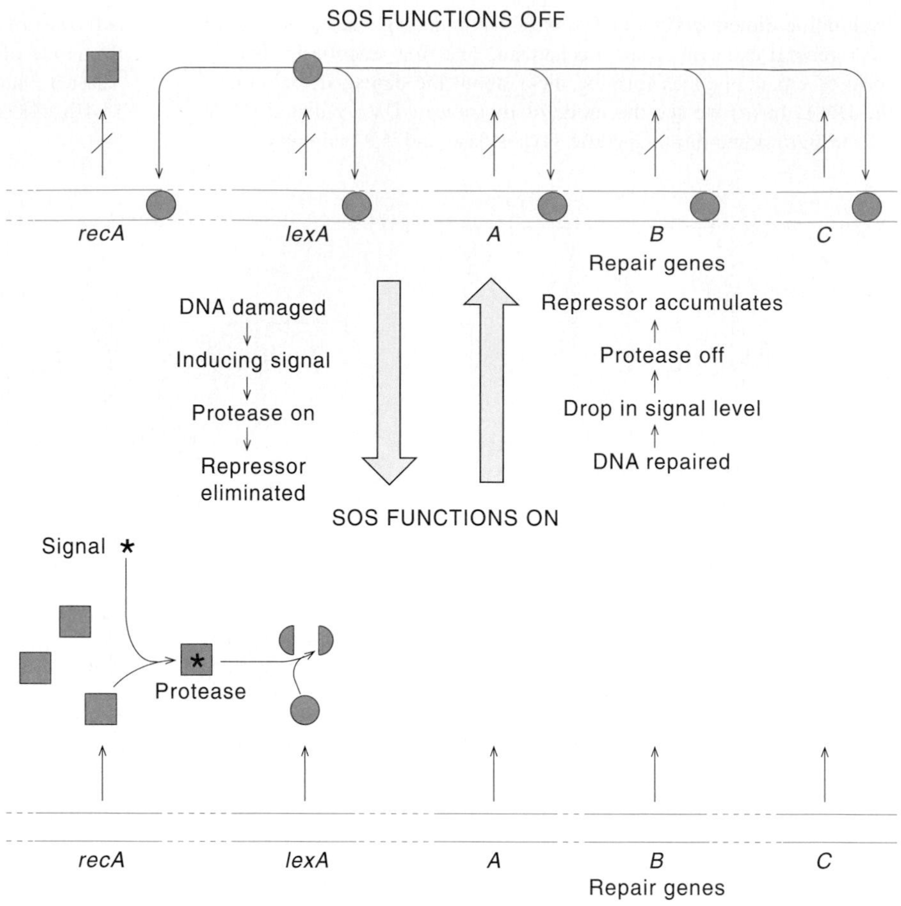

Figure 26.19

Model for the SOS regulatory system. In normally growing cells the SOS functions associated with DNA repair are not expressed. This is because lexA repressor inhibits their transcription. LexA repressor also inhibits its own expression and that of *recA*. SOS functions are turned on by a series of reactions that starts with DNA damage. DNA damage results in an inducing signal that activates the protease function of recA. This protease cleaves lexA protein, so that all genes that were formerly inhibited by lexA can be expressed. Once the damage is repaired, the level of lexA repressor builds up again, and the SOS genes return to their usual repressed state.

exclusively involved in DNA repair. The lexA protein does this by binding tightly to the control regions of these genes.

The recA protein has two functions: One is to catalyze DNA recombination (see next section), the other is to inactivate the lexA protein. It does the latter by mediating the cleavage of the lexA protein. Damage to the chromosome usually results in DNA fragments. A DNA fragment that binds the recA protein activates a highly specific protease function that cleaves the lexA protein at a specific -AlaGly-bond. The fragmented lexA protein falls off the DNA, thereby permitting the *din* genes to express at a high rate. As soon as the DNA damage has been corrected, the recA protease returns to its dormant form, and newly synthesized lexA protein is free to bind to the *din* genes again.

DNA Recombination

Individuals belonging to the same species carry approximately the same number of genes positioned in the same relative locations on their homologous chromosomes; each gene is represented by more than one variant which frequently results in different visible characteristics (phenotypes) in different individuals. The different representations of a gene are referred to as alleles for that gene. Alleles that give rise to different phenotypes serve as useful markers for detecting recombination between homologous chromosomes. Most cells (diploid cells) of an organism contain pairs of homologous chromosomes, one from each parent. During formation of the sex cells (haploid cells) a reduction takes place so that each sex cell only contains one chromosome of each type. The process of chromosome segregation, which takes place during formation of the sex cells, is called meiosis (fig. 26.20). Homologous chromosomes segregate at random during the first phase of meiosis (see fig. 26.20*a*). This results in the production of sex cells with a very large number of possible combinations of chromosomes. Some combinations of alleles undoubtedly have selective advantages over others. Indeed the primary object of mating seems to be to produce new combinations of genes. But what of the alleles on the same chromosomes? Nature has provided another mechanism for reshuffling alleles on the same chromosome, a process called recombination. Although recombination can take place in most cells of the organism, meaningful genetic recombination that can be passed from one generation to the next only takes place during the first phase of meiosis (see fig. 26.20*c*).

Figure 26.20

Meiosis. The process of meiosis involves two cell divisions with two segregation cycles, meiosis I and meiosis II. These cycles are pictured for a hypothetical cell containing four chromosomes. (*a*) During the first cycle homologous chromosomes pair and then segregate. (*b*) During the second cycle the sister chromatids from each chromosome segregate. The second cycle is very much like mitosis (see fig. 1.21). Each diploid cell that enters meiosis ultimately yields four haploid cells. These haploid cells are the sex cells for the next round of mating. (*c*) Recombination between homologous chromosomes during meiosis I leads to different combinations of alleles. In this illustration alleles for the same gene are represented by corresponding capital or lower-case letters.

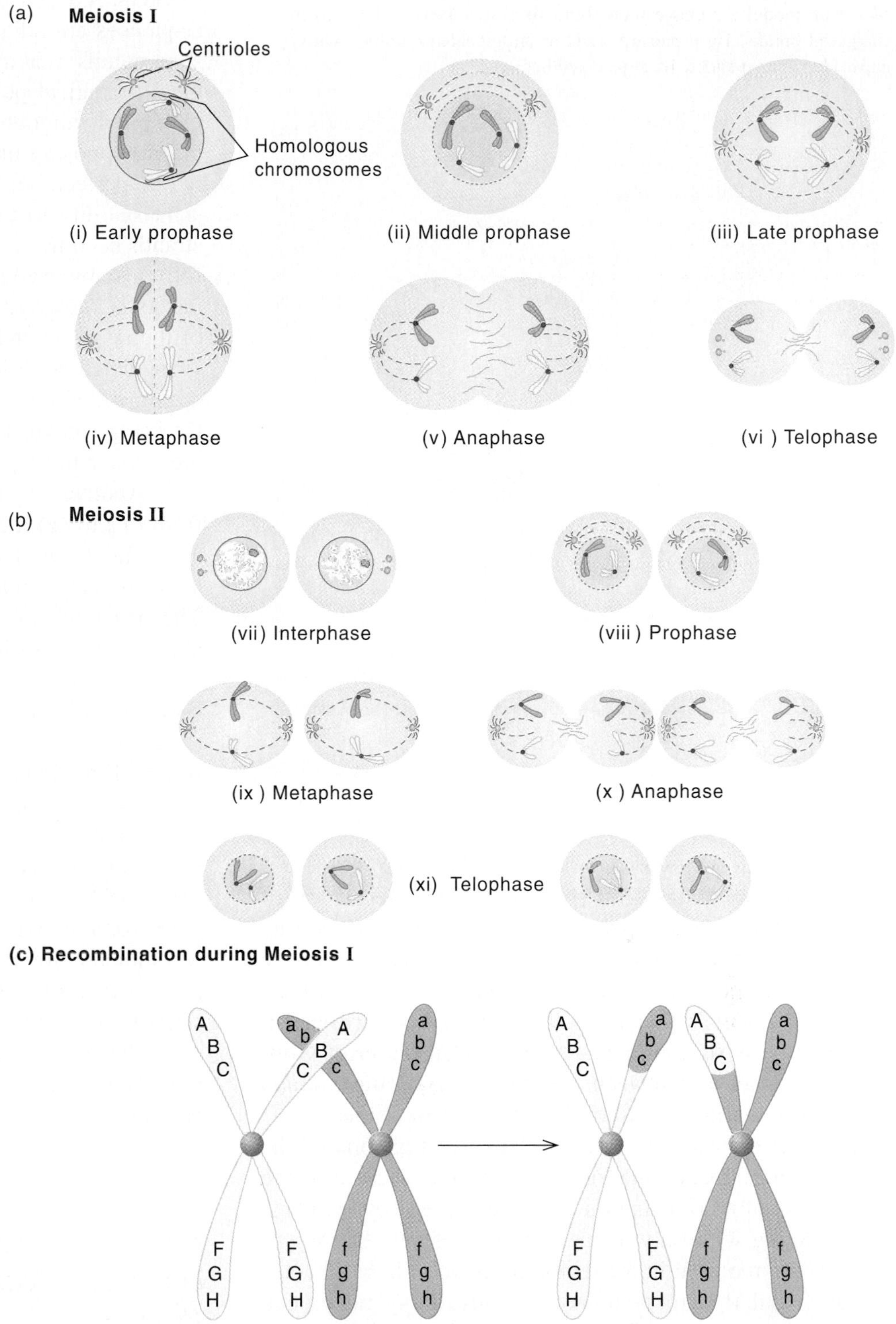

In the 1960s Matthew Meselson demonstrated that homologous recombination involves breakage and rejoining between existing chromosomes. He made this demonstration on bacteriophages carrying different alleles that infect the same cell. With the help of isotopic labeling he showed that recombinant chromosomes can form without DNA replication, thus demonstrating the breakage–rejoining mechanism. Only a small fraction of new DNA appears in the recombinant chromosome, and this can be accounted for by the repair processes that accompany breakage and rejoining.

Figure 26.21

Meselson model for phage recombination (circa 1964). Phage form staggered breaks. Base pairing leads to an annealed complex with gaps. Gaps are mended by repair synthesis.

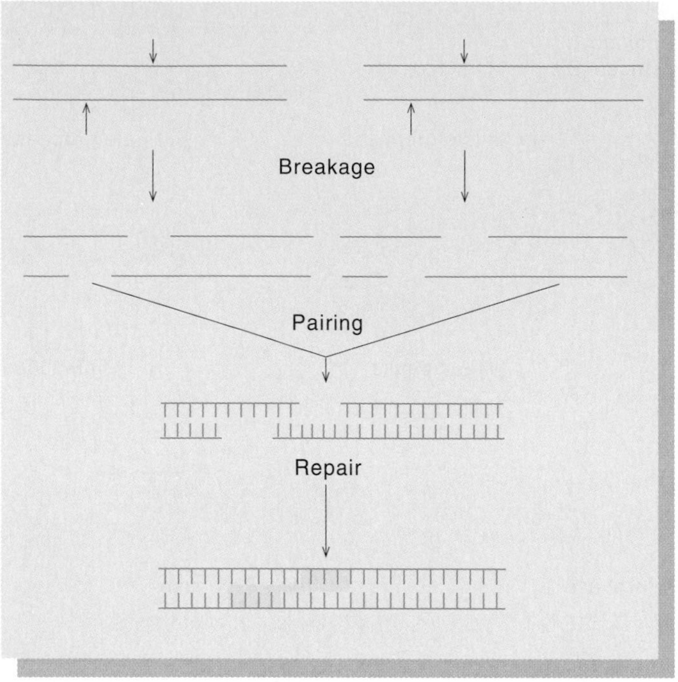

From his studies Meselson proposed a three-step model for homologous recombination (fig. 26.21): (1) Formation of staggered breaks in two parental DNAs; (2) base pairing between single-stranded regions of the two parental types; and (3) repair synthesis. The most important aspect of the Meselson model is that it explained why homologous recombination is so common—a key step in the process involves interaction between complementary single-strand regions originating from homologous chromosomes.

Meselson's results spearheaded investigations of the mechanism of recombination in eukaryotes. Here also the mechanism of recombination can be explained by breakage and rejoining with only a small amount of new DNA synthesis being required in the region of the join. In eukaryotes it was found that recombination is invariably reciprocal. Thus, if chromosome AB recombines with chromosome a,b so that Ab forms as a recombinant product, then aB also forms. R. Holliday proposed a model to explain how recombination occurs in eukaryotes (fig. 26.22). Depending on how the recombination complex finally breaks up, the genes on either side of the region of interaction may or may not show recombination.

The Holliday model is depicted in figure 26.22. Initially a pair of homologous chromosomes containing two markers, *A,a* and *B,b,* is aligned so that homologous DNA sequences are adjacent to one another (frame *a*). The first step entails making single-strand breaks at identical or nearly identical points in the two duplexes (frame *b*). The cut strands migrate and base-pair with homologous regions on the opposing unbroken DNA strands. The pairing need not be perfect, but it should be regular enough to give transient stability to the complex. Subsequently the crossover strands become covalently linked to the opposing strands to form a stable bridge structure. This bridge between the duplexes is known as the Holliday junction. Further exchange of the single strands between the interacting chromosomes is mediated by bridge migration (frame *f*), which can be accompanied by further heteroduplex formation. The more the bridge moves from the initial position of crossing over, the greater the region of heteroduplex formation.

Another way of displaying frame *f* is shown in *g*; to go from (*f*) to (*g*) the ends of the duplexes are merely pulled apart. In (*h*) we have the same structure as in (*g*), but an arrow is shown indicating a twisting of one of the duplexes. The structure in (*h*) leads to the structure in (*i*), called the chi form because it is shaped like the Greek letter chi (χ). The chi form divides by two single-strand cuts in either the horizontal direction (*i*) or the vertical direction (*j*), leading to quite different products, (*k*) or (*n*). In both structures some DNA exchange has occurred in the region of heteroduplex formation. In (*k*), the outside markers (i.e., the markers outside the region of heteroduplex formation) *A* and *B* and *a* and *b* have not recombined; that is, they are on the same DNA molecules (chromosomes) as they were before the recombination event. The structure in (*n*), however, is recombinant for the parental genetic markers. The repair process that follows exchange closes the gaps in the severed strands, (*l*) and (*o*).

Genetic investigations have supported the Holliday model and at the same time indicate that different organisms can only be explained by variations of the model.

Enzymes Have Been Found in Escherichia coli *That Mediate the Recombination Process*

Cursory inspection of both the Meselson and Holliday models suggests the need for special enzymes to mediate the recombination process. First an endonuclease is needed to make the initial strand breaks that are required for strand invasion. Second an enzyme is required to mediate formation of the recombination complex. Finally an enzyme is probably required to resolve the recombinant duplexes.

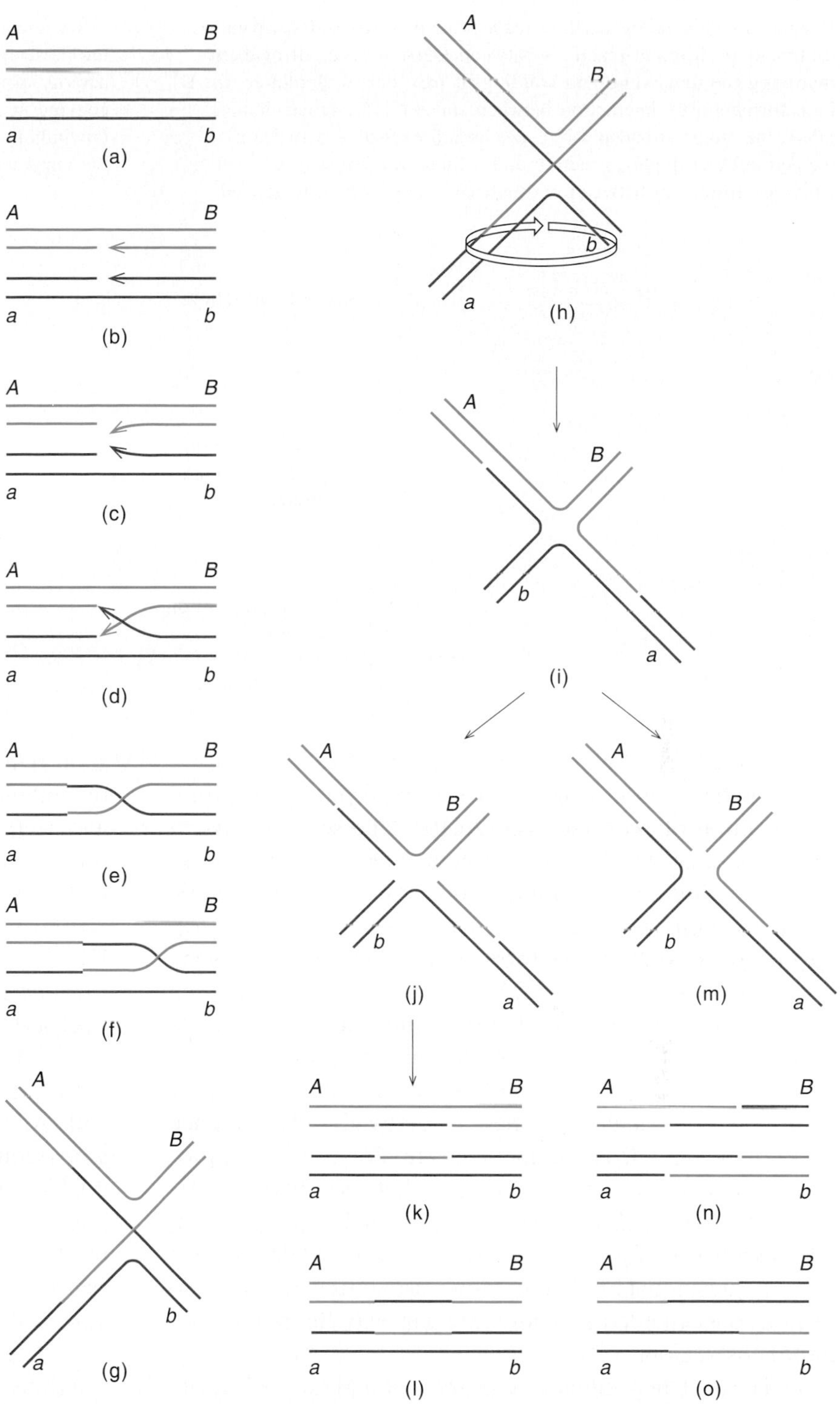

Figure 26.22

Holliday model of genetic recombination. In (*a*) we see two paired homologous chromosomes with different marker alleles, *A, B,* and *a, b.* In (*b*), single-strand cuts have been made at identical loci on the two chromosomes. In (*c*) and (*d*) we see these strands beginning to invade the duplexes of the homologous strands. The invading strands base-pair at homologous sites. In (*e*) the 3′ ends of the invading strand (indicated by the arrowheads in (*d*) become covalently linked to the 5′ end of the cut strand in place. This produces a bridge commonly called a *Holliday junction.* The bridge can migrate to the right as shown (or to the left, not shown) in (*f*) to form more heteroduplex structures. In (*g*) we see another way of displaying (*f*), and (*h*) is the same as (*g*), except for the arrow indicating the way in which the lower half of the structure is twisted to form (*i*). Then (*i*) may be cut in one of two ways to yield either (*j*) or (*m*). Parts (*k*) and (*n*) are alternative ways of displaying (*j*) and (*m*), respectively. Finally, in (*l*) and (*o*) the duplexes resulting from (*k*) and (*n*) are mended.

Genetic and biochemical studies on *E. coli* have led to the discovery of these three types of enzymes.

The search for enzyme activity associated with recombination has been based on analysis of organisms carrying mutations that affect general recombination. Such mutants, known as *rec* mutants, were first discovered more than 20 years ago, and new ones are still being found. *E. coli rec* mutations, *recA, recB, recC,* and *recD* reduce recombination efficiency to different extents. The *recA* mutants are totally deficient in homologous recombination, whereas *recB* or *recC* mutants can recombine but only about 10^{-4} times as well as wild-type cells.

Figure 26.23

Reactions catalyzed by purified recA protein *in vitro.* RecA catalyzes a number of different reactions between DNA strands, all of them involving the unwinding and winding of base-paired structures. (*a*) D-loop formation by interaction between supercoiled circular duplex DNA and single-stranded DNA. (*b*) Strand exchange between a gapped circular duplex structure and a linear duplex structure. (*c*) Complex formation between two helices, one of which is gapped.

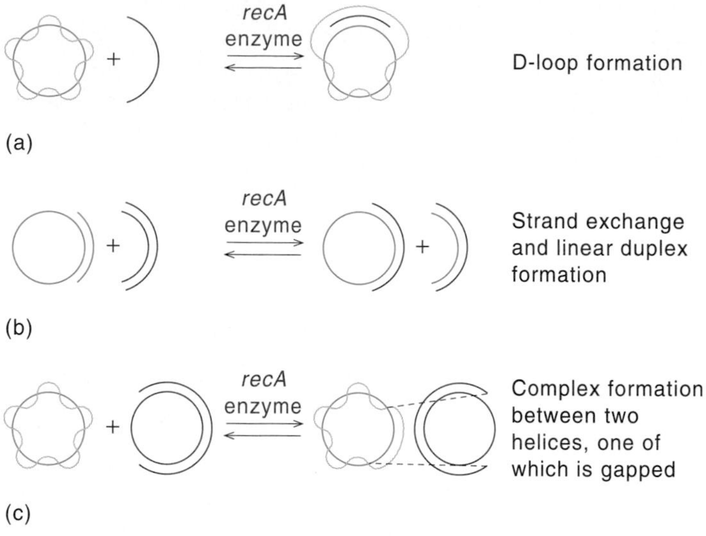

Figure 26.24

DNA unwinding by exoV enzymes. This is best observed *in vitro* in the presence of ATP and Ca^{2+}. The Ca^{2+} inhibits the exonucleolytic activity of the enzyme. The ATP provides the energy that drives the enzyme in a concerted manner into the duplex structure. The duplex unwinds in front of the path of the enzyme and rewinds in back of the enzyme.

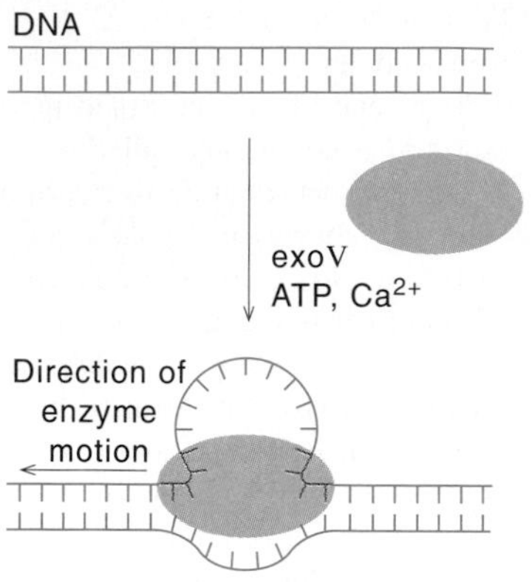

We have discussed the repair role played by the protease function of *recA* (see fig. 26.19). The second class of reactions catalyzed by recA protein indicates a prominent role in the initiation of homologous pairing during genetic recombination. Thus, the purified recA protein catalyzes various forms of complex formation and exchange between duplex and single-stranded DNAs in reactions that entail the cleavage of ATP (see fig. 26.23). D-loop formation can result when the recipient is a circular duplex and the donor is a single strand (see fig. 26.23*a*). Strand exchange can occur when a single-stranded fragment is removed in the presence of a double-stranded fragment (see fig. 26.23*b*). RecA protein also causes a complex to form between two circular helices, provided that at least one input helix is gapped on one strand (see fig. 26.23*c*). Interestingly, while the two helices must contain homologous sequences in order to form a four-stranded structure, the gap may lie in a nonhomologous region.

The fact that mutations in *recA* completely eliminate recombination, as well as the finding that the recA-related protein catalyzes strand exchange reactions like those required in the recombination models we have been discussing, makes it highly likely that *recA* plays a key role in these processes *in vivo.* If this is so, then recA protein is clearly situated at the hub of activities in the recombination complex.

The recA enzyme does not catalyze any recombination event unless at least one free end of a single strand or a DNA duplex is available. Thus, it seems unlikely that the recA protein could be involved in making the initial incision(s) in the duplex structure that is(are) required to initiate recombination. There are reasons for believing that the *recB* and *recC* genes are involved in this capacity. First, a mutation in either of these genes reduces recombination by a factor of about 10^{-4}. Second, these two genes together with *recD* encode the subunits of a nuclease known as exoV. ExoV is both a nuclease and a helicase. The helicase activity is best demonstrated *in vitro* under conditions where the nuclease activity is blocked. This can be done by adding Ca^{2+}. When linear duplex DNA is incubated with exoV in the presence of ATP and Ca^{2+}, it begins to unwind at one end. As the unwinding progresses, the single-stranded regions collapse back on each other to reform a duplex. A double-loop structure is maintained in the vicinity of the migrating enzyme (fig. 26.24). What is the point of this migrating behavior? Further experiments indicate that exoV scans the duplex looking for preferred sequences where it makes single-strand incisions.

The recombination process triggered by the initial scissions made by exoV and mediated by the recA enzyme are not sufficient for recombination between interacting chromosomes. An additional enzyme, resolvase, is required for efficient separation of the recombining chromosomes in the recA–chromosome complex. This protein is encoded by the *ruvC* gene in *E. coli* and is absolutely required for homologous recombination in the bacterium. Its effectiveness has also been demonstrated in a cell-free *in vitro* system.

Incidentally, the fact that the recA enzyme carries two enzyme activities, one for homologous recombination and one for regulating the concentration of DNA repair genes, including itself (the recA protease, which cleaves the lexA repressor), leads to the suspicion that recA is also more directly involved in DNA repair. Recombination between two homologous chromosomes badly damaged in different regions could lead to one normal chromosome. Although, technically a haploid organism, *E. coli* carries two copies of the bacterial chromosome for a considerable time prior to cell division. Thus, the necessary identical pairs of chromosomes are present for homologous recombination.

Other Types of Recombination

There are two other types of recombination, on which we will not elaborate. Site-specific recombination, is limited to highly select regions of the genome and to very specific functions. One of the best understood examples of site-specific recombination is the integration of bacteriophage λ DNA into the *E. coli* host chromosome (see chapter 30). Finally nonspecific recombination occurs between nonhomologous regions of chromosomes with little or no sequence homology. In the next section we see that at one stage in its life cycle a retrovirus recombines in this way.

RNA-Directed DNA Polymerases

It is a commonly held belief that RNA preceded DNA in the early evolution of living systems. If this is the case then the first DNA polymerases must have been capable of transferring sequence information from RNA to DNA. Enzymes of this sort are called reverse transcriptases because they do the reverse of common transcriptases (see chapter 28). Reverse transcriptases no longer play the central role in genetic information transfer, but they are still found in all species and function in a number of capacities in both cellular and viral metabolism.

Retroviruses Are RNA Viruses That Replicate through a DNA Intermediate

RNA viruses that cause tumors (oncogenic RNA viruses) are called retroviruses because their life cycle involves a DNA intermediate. The ability of retroviruses to use such a route for replication hinges on a viral-encoded enzyme called reverse transcriptase, an enzyme with three discrete activities: (1) It catalyzes the synthesis of DNA from the viral plus-strand; (2) it catalyzes the synthesis of DNA plus-strand from the viral minus-strand DNA; and (3) it catalyzes the degradation of the viral RNA from an RNA–DNA heteroduplex.

Following viral infection the virus-plus strand first functions as an mRNA for the synthesis of the viral proteins. Once the viral reverse transcriptase has been synthesized, it is used in conjunction with the viral RNA template and a tRNA primer for the synthesis of minus-strand DNA. Next the RNase H activity of the viral enzyme removes extensive sections of RNA from the DNA–RNA heteroduplex so that the newly synthesized minus-strand DNA can be used as a template for the synthesis of viral plus-strand DNA. The complete process of duplex DNA synthesis is depicted in figure 26.25. The duplex DNA is subsequently integrated more or less indiscriminately at many sites in the host genome. There it can remain indefinitely or until the cell dies. The integrated viral DNA serves as a template for the synthesis of viral plus strand, completing the nucleic acid cycle that began with virus infection. Additional properties of retroviruses, including their association with malignant transformation, are taken up in Supplement 4: Carcinogenesis.

Hepatitis B Virus Is a DNA Virus That Replicates through an RNA Intermediate

Hepatitis B is an animal virus with a small circular DNA genome. A gap in one strand is bridged by an incomplete complementary strand. The longer strand has a protein bound at its 5′ end. A DNA polymerase present in the mature virus particle can elongate the 3′ ends of the incomplete strands. An unusual feature of this virus is that replication involves an RNA intermediate that must be reverse transcribed.

Some Transposable Genetic Elements Encode a Reverse Transcriptase That Is Crucial to the Transposition Process

A widely distributed group of genetic elements exist that are capable of integrating at random sites in the host chromosome. These transposable genetic elements usually carry genes to facilitate their own transposition in addition to conferring specific properties on the host cells in which they are located. Many of these transposable elements encode a reverse transcriptase that functions in the transposition process. The first step in transposition of elements of this type involves transcription of the genetic element. The resulting RNA, like the retroviral RNA, has two functions: One as a messenger RNA for the synthesis of element-encoded proteins and one as a template for the synthesis of the transposable element DNA. The transposable element integrates at other sites on the host chromosome without any apparent regard for sequence homology at the sites of integration. Transposable genetic elements that use a reverse transcriptase in this way have been found in yeast, *Drosophila,* and mice, and they are presumed to be present in a very wide range of organisms.

Figure 26.25

A model for the generation of double-stranded DNA carrying two copies of LTR. The sequence X is a marker for the plus strand; the complementary sequence X′ occurs on the minus strand. (*a-c*) Synthesis of minus-strand DNA from the genomic RNA template using a tRNA primer (represented by inverted J in step *a*). (*d*) Degradation of the RNA portion of the resulting DNA:RNA hybrid by RNAseH. (*e*) Bridge between the newly synthesized segment of minus-strand and repeated sequences at the 3′ end of genomic RNA. (*f*) Extension of minus-strand DNA (*leftward*) and initiation of plus-strand DNA (*heavy arrow moving rightward*). (*g*) Completion of 300-nucleotide fragment of plus-strand DNA. (*h*) Bridge between sequences repeated in minus-strand DNA and in the 300-nucleotide fragment. (*i*) Completion of synthesis and tidying up: = LTR, with sequences from the 5′ and 3′ ends of the viral genome separated by the terminal repeat. In this figure red stands for RNA and blue stands for DNA.

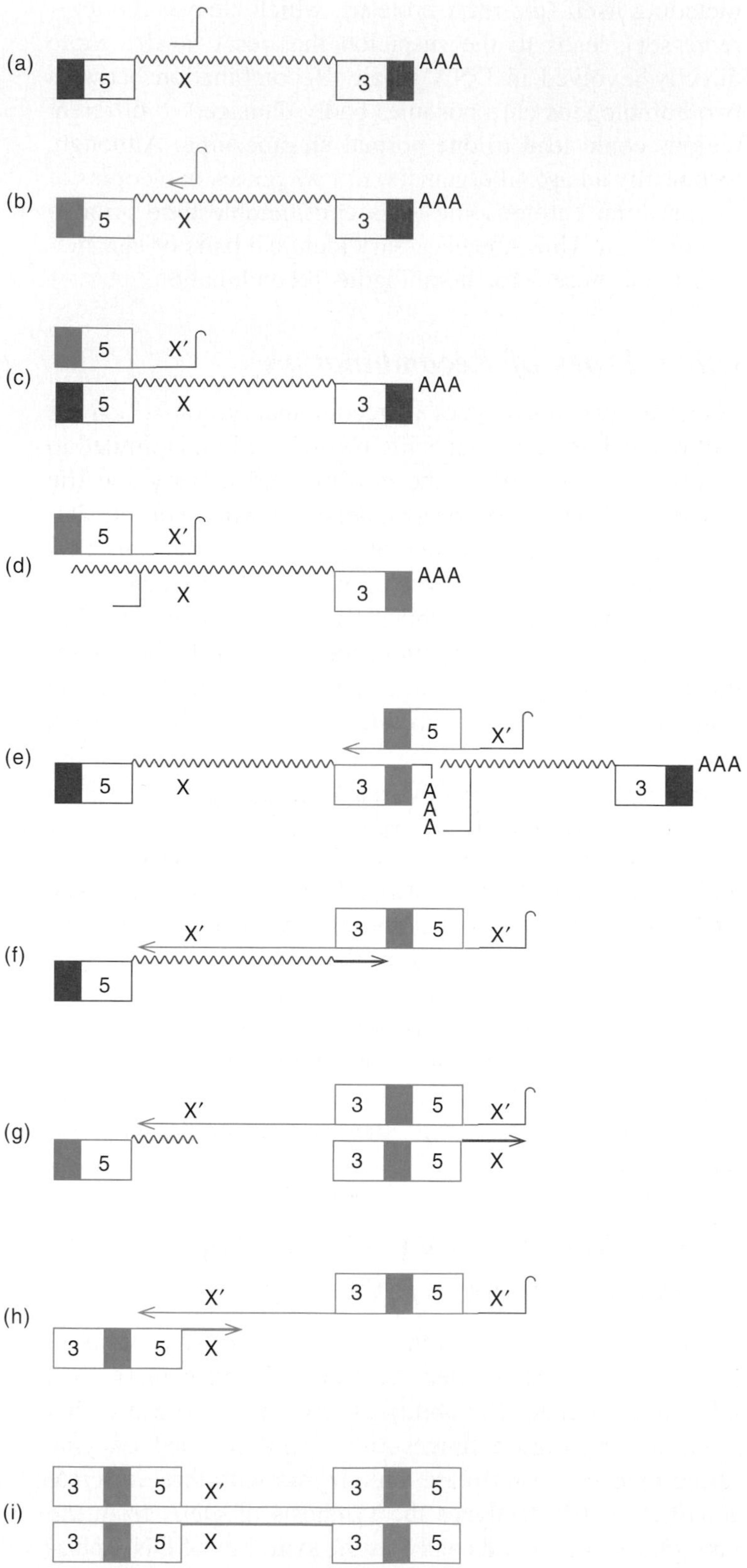

Bacterial Reverse Transcriptase Catalyzes Synthesis of a DNA–RNA Molecule

Except as intermediates, molecules in which DNA and RNA are covalently linked to each other have only been found in certain bacteria. For example, some strains of *E. coli* carry a genetic element that encodes a reverse transcriptase. This genetic element produces one long transcript that folds into a complex three-dimensional structure. Part of this molecule serves as a template for the synthesis of a reverse transcriptase, which recognizes a specific G residue on the folded RNA as a primer start site for DNA synthesis. Synthesis starts from the 2′-OH group of this G residue using another part of the RNA as a template. DNA synthesis arrests at a specific point on the RNA template. An RNase H removes that region of the RNA in which DNA–RNA heteroduplex has been formed, leaving a covalently linked DNA–RNA molecule. No function has been found for this unusual nucleic acid (see Linn and Maas reference).

Telomerase Facilitates Replication at the Ends of Eukaryotic Chromosomes

Circular bacterial chromosomes are initiated by RNA primers. At some stage the RNA primers must be eliminated and replaced by DNA. Due to the circular nature of the chromosome an upstream DNA molecule can always serve as a primer for regions from which RNA primers are eventually removed. This guarantees that the primer requirement does not interfere with complete replication of the chromosome.

Since eukaryotic chromosomes are linear, the ends of these chromosomes require a special solution to ensure complete replication. This can be seen in figure 26.26. At the very end of a linear duplex a primer is necessary to initiate DNA replication. After RNA primer removal there is bound to be a gap at the 5′ end of the newly synthesized DNA chains. Since DNA synthesis always requires a primer the usual way of filling this gap is not going to solve the problem. This dilemma is overcome by a special structure at the ends (telomeres) of eukaryotic chromosomes and a special type of reverse transcriptase (telomerase) that synthesizes telomeric DNA. In many eukaryotes the telomeres contain short sequences (frequently hexamers) that are tandemly repeated many times. Telomerase contains an RNA that binds to the 3′ ends and also serves as a template for the extension of these ends. Prior to replication, the 3′ ends of the chromosome are extended with additional tandemly repeated hexamers. The 3′ ends are extended sufficiently so that there is room to accommodate an RNA primer. In this way there is no net loss of DNA from the 5′ ends as a result of replication. After replication the 3′ end is somewhat longer than it was before because of the extra added tandem repeats. An additional mechanism is necessary to keep the telomeres from growing indefinitely by this effect. Possibly related to this requirement, it has been observed that the telomeres become shorter in somatic tissues with aging. This is believed to be due to a deficiency of telomerase in most somatic tissues.

Figure 26.26

Synthesis at the ends of a eukaryotic chromosome. One end of the linear DNA of a eukaryotic chromosome is diagrammed. A flush-ended DNA duplex presents a problem for completing synthesis at the 5′ end (*a*). This is because of the RNA primer requirement for DNA synthesis. When the primer at the 5′ end is removed there is no conventional way to fill the gap. A solution to this problem is shown in (*b*). The ends of eukaryotic chromosomal DNAs consist of highly repetitious tandem repeats (telomeres). These repeats on the 3′ end serve as both primer and template for extending the 3′ end. The extended 3′ end can accommodate a primer RNA, so after chromosomal DNA replication no loss occurs from the 5′ end of the DNA. Another process is needed to remove the extension from the 3′ end. New synthesis is indicated in red. The zigzag represents primer.

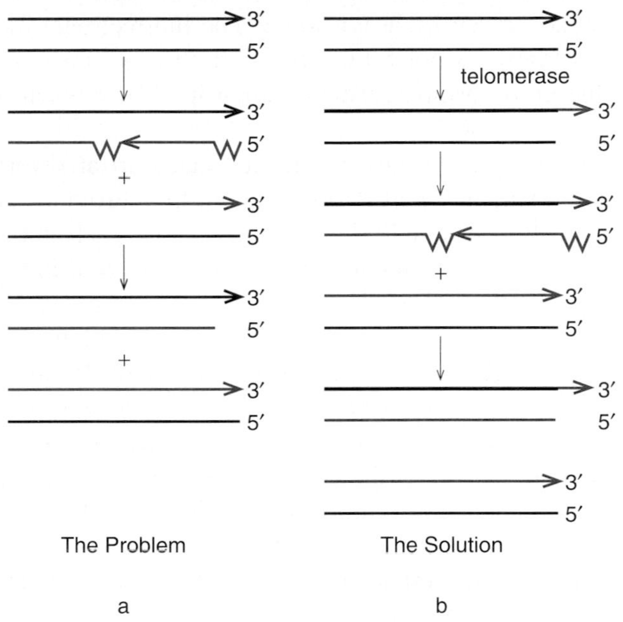

Other Enzymes That Act on DNA

Many other enzymes modify DNA and degrade it in various ways. In the next chapter we discuss some of those enzymes that have been particularly useful in DNA manipulations. In Supplement 3 (Immunobiology) we discuss DNA splicing operations that are essential for assembling complex antibody genes prior to their expression.

Summary

This chapter deals with reactions involved in DNA synthesis, degradation, repair, and recombination. The chief points to remember are as follows.

1. DNA replication proceeds by the synthesis of one new strand on each of the parental strands. This mode of replication is called semiconservative, and it appears to be universal. DNA synthesis initiates from a primer at a unique point on a prokaryotic template such as the *E. coli* chromosome. From the initiation point, DNA synthesis proceeds bidirectionally on the circular bacterial chromosome. The bidirectional mode of synthesis is not followed by all chromosomes. For some chromosomes, usually small in size, replication is unidirectional.
2. In eukaryotic systems, replication can start at several points (still not well defined) along the chromosome. Replication is usually bidirectional about each initiation site. The termination points of replication are interspersed between initiation sites. In most cases of unidirectional or bidirectional replication, synthesis occurs nearly (but not exactly) simultaneously on both strands of the parent DNA template. Because synthesis can occur only in the $5' \rightarrow 3'$ direction on the growing chain, and the two strands in the parent duplex are oriented in opposite directions, synthesis can occur continuously on only the leading strand. On the other (lagging) strand it must pause for the template to unwind. Synthesis on the lagging strand does not occur continuously but rather in small discontinuous spurts, generating Okazaki fragments.
3. Many proteins are required for DNA synthesis and chromosomal replication. These include polymerases; helicases, which unwind the parental duplex; enzymes that fill in the gaps and join the ends in the case of lagging-strand synthesis; enzymes that synthesize RNA primers at various points along the DNA template; topoisomerases, which permit rotation and supercoiling; and single-strand DNA-binding proteins, which stabilize single-stranded regions that are transiently formed during replication. Most of these proteins have been isolated from whole cells and studied in cell-free systems.
4. In *E. coli,* mutations have been isolated in the genes encoding a number of these enzymes. Many of these mutations are conditional because the functional enzymes involved are required for DNA synthesis and cell viability. Mutants carrying mutationally altered proteins have been important in confirming their roles predicted from cell-free studies.
5. In eukaryotes, considerable progress has been made in studying the *in vitro* replication of animal viruses, such as SV40. The importance ascribed to the enzymes that have been characterized is largely based on a comparison of their properties with similar prokaryotic enzymes whose functions are better understood.
6. Many enzymes that act on DNA are involved in processes other than DNA synthesis. They include DNA repair enzymes, DNA degradation enzymes, and DNA recombination enzymes.
7. Enzymes that catalyze the synthesis of DNA using an RNA template are known as reverse transcriptases. The first reverse transcriptase discovered was encoded by an RNA retrovirus. This enzyme is needed in the virus replication cycle. Some animal viruses pass through an RNA intermediate and also require a reverse transcriptase to replicate the viral DNA. Similarly, a number of transposable elements found in cellular chromosomes replicate through RNA intermediates; they usually encode a reverse transcriptase. A unique reverse transcriptase called telomerase is used to synthesize the DNA at the ends of linear eukaryotic chromosomes.

Selected Readings

Baker, T. A., and S. H. Wickner, Genetics and enzymology of DNA replication in *Escherichia coli. Ann. Rev. Genet.* 26:447–469, 1992.

Bauer, W. R., F. H. C. Crick, and J. H. White, Supercoiled DNA. *Sci. Am.* 243(4):118–133, 1980.

Beese, L. S., V. Derbyshire, and T. A. Steitz, Structure of DNA polymerase I Klenow fragment bound to duplex DNA. *Science* 260:352–355, 1993.

Beese, L. S., and T. A. Steitz, Structural basis for the $3' \rightarrow 5'$-exonuclease activity of *E. coli* DNA polymerase I: A two metal ion mechanism. *EMBO J.* 10:25–33, 1991.

Bell, S. P., and B. Stillman, ATP-dependent recognition of eukaryotic origins of DNA replication by a multiprotein complex. *Nature* 357:128–134, 1992.

Blackburn, E. H., Telomerases. *Ann. Rev. Biochem.* 61:113–129, 1992.

Bohr, V. A., and K. Wasserman, DNA repair at the level of the gene. *Trends Biochem. Sci.* 13:429–432, 1988.

Bramhill, D., and A. Kornberg, A model for initiation at origins of DNA replication. *Cell* 54:915–918, 1988.

Campbell, J. L., Eukaryotic DNA replication. *Ann. Rev. Biochem.* 55:733, 1986.

Challberg, M. D., and T. J. Kelly, Animal viruses and DNA replication. *Ann. Rev. Biochem.* 58:671–717, 1989.

Cox, M. M., and I. R. Lehman, Enzymes of general recombination. *Ann. Rev. Biochem.* 56:229–262, 1987.

Demple, B., and P. Karran, Death of an enzyme: Suicide repair of DNA. *Trends Biochem. Sci.* 8:137–139, 1983.

Dunderdale, H. J., F. E. Benson, C. A. Parsons, G. J. Sharples, R. G. Hoyd, and S. C. West, Formation and resolution of recombination intermediates by *E. coli* RecA and RunC protein. *Nature* 354:506–510, 1991.

Fangman, W. F., and B. J. Brewer, Activation of replication origins with yeast chromosomes. *Ann. Rev. Cell. Biol.* 7:375–402, 1991.

Fink, G. R., J. D. Boeke, and D. J. Garfinkel, The mechanisms and consequences of retrotransposition. *Trends Genet.* 2:118–123, 1986.

Holliday, R., A different kind of inheritance. *Sci. Am.* 260(4):60–73, 1989. On DNA methylation.

Itoh, T., and J. Tomizawa, Antisense RNA. *Ann. Rev. Biochem.* 60:631–652, 1991. Includes an excellent description of the initiation of DNA synthesis for Col E1 plasmid.

Kohlstaedt, L. A., J. Wang, J. M. Friedman, P. A. Rice, and T. A. Steitz, Crystal structure at 3.5 Å resolution of HIV-1 reverse transcriptase complexed with an inhibitor. *Science* 256:1783–1790, 1992.

Kong, X-P, R. Onrust, M. O'Donnell, and J. Kuriyan, Three-dimensional structure of the β subunit of *E. coli* DNA polymerase III holoenzyme: A sliding clamp. *Cell* 69:425–437, 1992.

Kornberg, A., and T. A. Baker, *DNA replication,* 2d ed. New York: Freeman, 1991. A magnificant up-to-date, clearly written and thorough treatment.

Landy, A., Dynamic, structural and regulatory aspects of site-specific recombination. *Ann. Rev. Biochem.* 58:913–950, 1989.

Linn, D., and W. Maas, Reverse transcriptase-dependent synthesis of a covalently linked branched DNA–RNA compound in *E. coli* B. *Cell* 56:891–904, 1989.

Lohman, T. M., W. Bujalowski, and L. B. Overman, *E. coli* single strand binding protein. *Trends Biochem. Sci.* 13:250–255, 1988.

Maxwell, A., and M. Gellert, Mechanistic aspects of DNA topoisomerases. *Adv. Prot. Chem.* 38:69–107, 1986.

McHenry, C. S., DNA polymerase III holoenzyme of *E. coli. Ann. Rev. Biochem.* 57:519–550, 1988.

Meselson, M., and F. W. Stahl, The replication of DNA in *Escherichia coli. Proc. Natl. Acad. Sci. USA* 44:671–682, 1958. A classic paper.

Messer, W., and W. Noyer-Weidner, Timing and targeting: The biological function of Dam methylation in *E. coli. Cell* 54:734–737, 1988.

Modrich, P., DNA mismatch correction. *Ann. Rev. Biochem.* 56:435–466, 1987.

Newlin, C. S., Yeast chromosome replication and segregation. *Microbiol. Rev.* 52:568–601, 1988.

Ogawa, T., and T. Okazaki, Discontinuous DNA replication. *Ann. Rev. Biochem.* 57:519–550, 1988.

Radman, M. and R. Wagner, The high fidelity of DNA replication. *Sci. Am.* 259(2):40–46, 1988.

Sancar, A., and G. B. Sancar, DNA repair enzymes. *Ann. Rev. Biochem.* 57:29–67, 1988.

Sancar, G. B., and A. Sancar, Structure and function of DNA photolyases. *Trends Biochem. Sci.* 12:259–261, 1987.

Shapiro, J. A., *Mobile Genetic Elements.* New York: Academic Press, 1983.

Smith, G. R., Homologous recombination in *E. coli:* Multiple pathways for multiple reasons. *Cell* 58:807–809, 1989.

Stahl, F. W., Genetic recombination. *Sci. Am.* 256(2):90–101, 1987.

Varmus, H., Reverse transcription. *Sci. Am.* 257(3):56–64, 1987.

Waga, S., and Stillman, B., Anatomy of a DNA replication fork revealed by reconstitution of SV40 DNA replication in vitro. *Nature.* 369:207–212, 1994.

Wang, J. C., DNA topoisomerases. *Ann. Rev. Biochem.* 54:665–697, 1985.

Wang, T. S. F., Eukaryotic DNA polymerases. *Ann. Rev. Biochem.* 60:513–553, 1991.

Problems

1. In the Meselson-Stahl experiment illustrated in figure 26.2, a sample of the DNA shown in tube 4 (labeled with ^{15}N followed by one generation in ^{14}N) was heat-denatured prior to being subjected to centrifugation in a CsCl density gradient. This gradient showed two peaks of single-stranded DNAs of different densities. How did this experiment further support the idea of semiconservative replication?
2. Explain why Cairns and coworkers used [^{3}H]-thymidine to label replicating *E. coli* DNA in the experiments shown in figure 26.3.
3. Draw the chemical reaction mechanism for the formation of a phosphodiester linkage during DNA synthesis. Discuss the significance of the pyrophosphate product that is formed. What is the significance of the Mg^{2+} requirement?
4. Cairns and De Lucia isolated a mutant strain of *E. coli* that had only about 1% of the DNA PolI activity found in wild-type cells, yet the strain replicated its DNA at a normal rate. Explain how this discovery was important in understanding the role of the different DNA polymerases in replication and repair.
5. All enzymes that make DNA in a template-dependent fashion require a primer. How does the use of a primer increase the fidelity of DNA synthesis, and why is this primer usually RNA?
6. Even with its proofreading activity, *E. coli* DNA polymerase III still exhibits a measurable rate of nucleotide misincorporation (about one mistake per 10^{10} nucleotides incorporated). Mutants of *E. coli* DNA polymerase III can be isolated that have a lower than normal rate of misincorporation. Why might such mutants, which can be said to have hyperaccurate DNA replication, be evolutionarily unfavorable?
7. *E. coli* has a genomic complexity of about 4×10^6 bp, and each replication fork can move at a rate of about 10^3 bp/s (base pairs per second). How long does it take to replicate the *E. coli* chromosome? With an ample carbon source and ideal growth conditions, cells of *E. coli* can divide in about 20 min. How can this shorter division time occur if the rate of fork migration remains constant at 10^3 bp/s?
8. Draw the chemical reaction mechanism for DNA ligase in *E. coli* (uses NAD^+ as the source of energy and forms a covalent intermediate with an ϵ-amino group of lysine). Why are ligation reactions that require ATP more thermodynamically favorable?
9. Humans have about 2.9×10^9 bp of DNA in their genome, and the replication fork migrates at the rate of approximately 30 bp/s. How long would it take to replicate the entire genome if it was a single continuous piece of DNA? How many replication origins are required to replicate this DNA in one hour?
10. Why does the reaction catalyzed by the *dnaB* gene product require ATP hydrolysis? Does this protein alter the linking number of DNA? Explain.
11. Eukaryotic DNA is replicated at a slower rate than prokaryotic DNA. One reason may be the requirement for the deposition of histone proteins on DNA (histone synthesis and DNA replication are coupled). Describe a model for the replication of eukaryotic DNA and nucleosome formation.
12. The graph shows *E. coli* labeled with radioactive thymidine for a short pulse (10 s) followed by a chase with an excess of nonradioactive thymidine. The DNA is then extracted and centrifuged in alkaline sucrose gradients (under high pH conditions the DNA denatures). Explain what these data imply, and interpret these results in light of our current model for DNA replication.

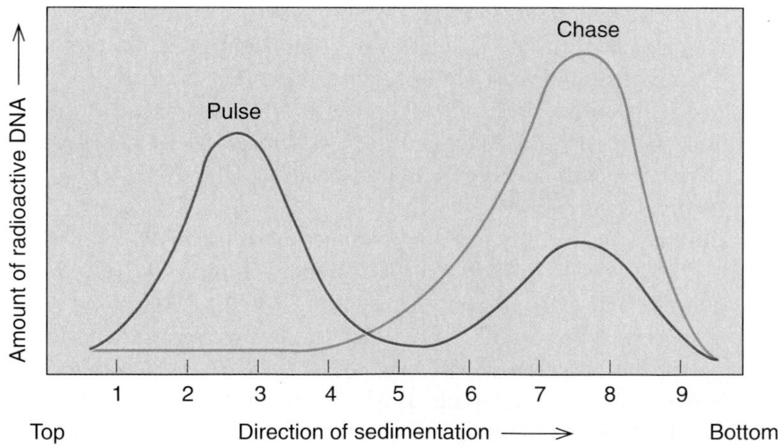

13. How is the SOS response reversed following repair of DNA?
14. Compare and contrast the excision repair and photoreactivation mechanisms for correction of ultraviolet-induced thymine dimers.
15. What is the role of the recA protein in *E. coli* DNA repair and in recombination? What use are *recA* mutants in biochemical and genetic research?
16. Normal human fibroblasts are grown in culture and then exposed to UV light. A short time later the DNA is extracted and applied to an alkaline sucrose gradient, and the data in graph (*a*) are observed. Another sample of cells is also exposed to UV light, but about 12 h are allowed to pass before the DNA is extracted and applied to an alkaline sucrose gradient (graph *b*). Explain these data from what you know about DNA repair (*Hint:* see fig. 26.18.). Another sample of fibroblast cells, isolated from a patient with xeroderma pigmentosum (a disease resulting from the inability to repair DNA damage caused by UV light) was exposed to UV light and then applied to a gradient after a short time. Would you expect the data to resemble those in graph (*a*) or (*b*)? Why?

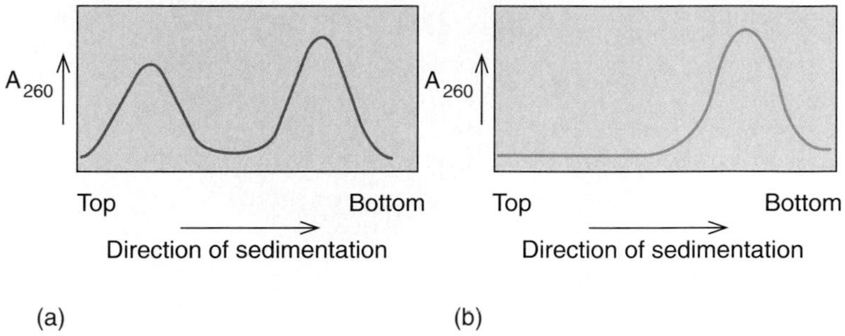

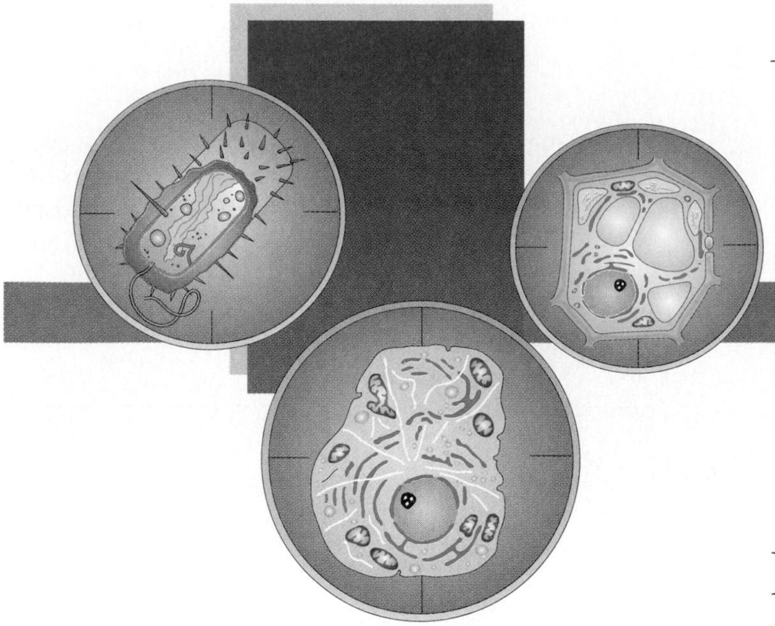

CHAPTER

27

DNA Manipulation and Its Applications

Chapter Outline

DNAs from different sources can be knitted together to produce new genetic alignments.

A broad array of techniques has made it possible to investigate the fine structure of DNA as well as to redesign existing genes. The impact of these techniques cannot be overestimated. It is possible to map and isolate genes from complex eukaryotes—feats that could never have been accomplished by classical genetic methods. Genes can be redesigned to achieve a variety of practical goals. Newly designed genes may be inserted into the same species from which the original unmodified genes came or into other species.

In this chapter we first consider DNA sequencing. Then we explore the different approaches for amplifying and isolating specific genes or gene segments. Following this, we examine the methods currently available for restructuring existing DNA sequences. Finally, we look at some of the major advances in our understanding of gene structure and location that have resulted from the new technology. This part of the discussion focuses on two examples: The mapping of the human globin gene family and the mapping of the gene responsible for the genetically inherited disease, cystic fibrosis.

Sequencing DNA

If we are going to manipulate a segment of DNA, it is most useful to know something about its primary structure. The most complete information comes from total sequence analysis.

Initial efforts at sequencing nucleic acids were confined to RNA molecules that could be readily isolated in pure form. The first sequence to be determined was that for tyrosine tRNA from yeast. From roughly 100 pounds of yeast Robert Holley was able to isolate enough of the tyrosine tRNA to carry out a sequence analysis. This historically significant effort required a number of enzymes and chromatographic techniques. We shall not elaborate on this accomplishment because sequencing of RNA is no longer done directly. In fact, the sequencing of RNA has been replaced by the sequencing of "cDNA," which results from the reverse transcription of RNA into DNA.

Two quite different methods have been developed for sequencing DNA. One method, developed by Walter Gilbert and Alan Maxam and involving cleavage of preexisting DNA, uses a chemical approach. A second method, involving premature termination of newly synthesized DNA, uses an enzymatic approach that was developed by Fred Sanger.

For pure sequencing, Sanger's is the method of choice. It employs chain-terminating dideoxynucleoside triphosphates to produce a continuous series of fragments in reactions catalyzed by DNA polymerase. Dideoxynucleoside triphosphates (ddXTPs) resemble deoxynucleoside triphosphates except that they lack a 3′ —OH group. They can add to a growing chain during polymerization, but they cannot be added onto, and as a result they act as chain terminators.

The DNA being sequenced is mixed with a suitable primer, radioactive dXTPs, DNA polymerase I (PolI), and a small amount of one ddXTP. The primers determine where DNA synthesis starts, and the ddXTP determines the base type where elongation stops. The products of four separate reaction mixtures, each differing only by the ddXTP it contains, are analyzed in figure 27.1. As depicted, reaction 1, using ddATP, contains all fragments with an A terminus; reaction 2, using ddCTP, contains all C terminations; and so on. After the newly synthesized oligonucleotides are separated from the template by denaturation they are fractionated by electrophoresis for a limited time on polyacrylamide gels. The positions of the fragments on the gel are detected by autoradiography. The sequence is read directly from the composite autoradiogram, starting with the fastest moving (smallest) band at the bottom of the gel and moving up. If the first band is in reaction 3, it is a G residue; the next band up, appearing from reaction 4, would be T; and so on. Up to 800 residues can be read from a single gel.

Methods for Amplification of Select Segments of DNA

Cellular genomes are very large; even *E. coli* contains more than a million base pairs, and eukaryotic genomes frequently contain a billion or more base pairs in one complete genome. Because of their large size it is impractical to fractionate the cellular genome and expect to obtain enough of a particular DNA segment for sequencing or other investigations. Two methods of amplifying defined segments of DNA have been developed. The first of these takes the desired segment and amplifies it *in vitro* with DNA PolI using DNA primers that bind to the ends of the region of interest. This method involves repeated cycles of synthesis and is appropriately named the polymerase chain reaction (PCR) method. The second method inserts the DNA segment of interest into a plasmid or virus that can be amplified *in vivo*. We discuss both of these methods because they are both useful in many ways.

Amplification by the Polymerase Chain Reaction

PCR entails enzymatic amplification of specific DNA sequences using two oligonucleotide primers that flank the DNA segment to be amplified (fig. 27.2). The primers must complement opposite strands so that after annealing, their 3′ ends in effect face each other (see fig. 27.2*b*).

The PCR procedure has three steps, which are usually repeated many times in a cyclical manner:

1. Denaturation of the original double-stranded DNA sample at high temperature
2. Annealing of the oligonucleotide primers to the DNA template at low temperature (37°C)
3. Extension of the primers using DNA polymerase

Figure 27.1

The Sanger dideoxynucleoside method of sequencing DNA. (*a*) A suitable template is chosen, and the primer is chosen so that DNA synthesis begins at the point of interest. The primer is radioactively labeled. In addition to the template–primer complex the reaction mixture contains all four radioactive deoxyribonucleoside triphosphates and small amounts of a single dideoxynucleoside triphosphate. The dideoxy compound serves as a chain terminator. (*b*) After synthesis in the presence of DNA polymerase I, the products of the reaction mixture are separated by gel electrophoresis and analyzed by autoradiography. The gel is run under denaturing conditions in warm urea so that single-stranded fragments separate strictly according to size. For a given dideoxy compound all fragments terminating with that particular base should give rise to bands on the gel. The interpretation of the gel pattern is given in (*c*). The smallest labeled fragment moves the fastest and appears at the bottom of the gel. (*d*) A typical sequencing film. The sequence begins CAAAAAACGG. (Courtesy of GIBCO-BRL, Life Technologies, Inc., Gaithersburg, Md.)

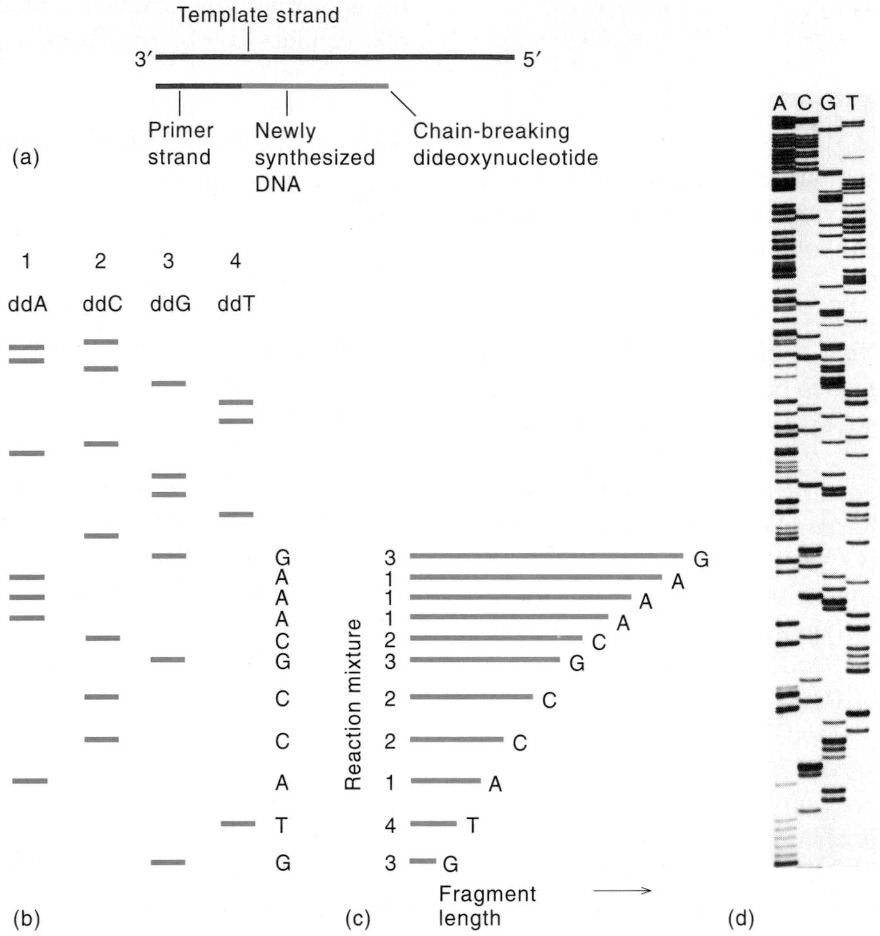

These steps are illustrated in figure 27.2. Each set of three steps comprises a cycle. The extension products of one primer provide a template for the other primer in a subsequent cycle so that each successive cycle essentially doubles the amount of DNA. This results in the exponential accumulation of the specific target fragment by approximately 2^n, where n is the number of cycles. The specific target fragment is also referred to as the "short product" and is defined as the region between the 5′ ends of the extension primers. Each primer is physically incorporated into one strand of the short product.

Other products are also synthesized during the succession of cycles, such as the "long product" of indefinite length, which is derived from the template molecules. However, the amount of long product only increases arithmetically during each cycle of the amplification process because the quantity of original template remains constant.

At the end of the PCR process the short product is so overwhelmingly abundant compared with the long product that its purification is not required for most purposes.

Figure 27.2

Steps in the polymerase chain reaction (PCR). The DNA to be amplified is denatured and annealed with two oligonucleotides that flank the region of interest. These oligonucleotides (or primers) are extended. Extension continues to the ends of the DNA strands. The products are again denatured and annealed to primers for a second round of extension. This process of denaturation, annealing, and primer extension is repeated many times. The primary product of the reaction is duplex DNA, bounded by the sequences of the primers. (From J. L. Marx, Multiplying genes by leaps and bounds, *Science* 240:1408–1410, June 10, 1988. Copyright 1988 by the AAAS. Reprinted by permission.)

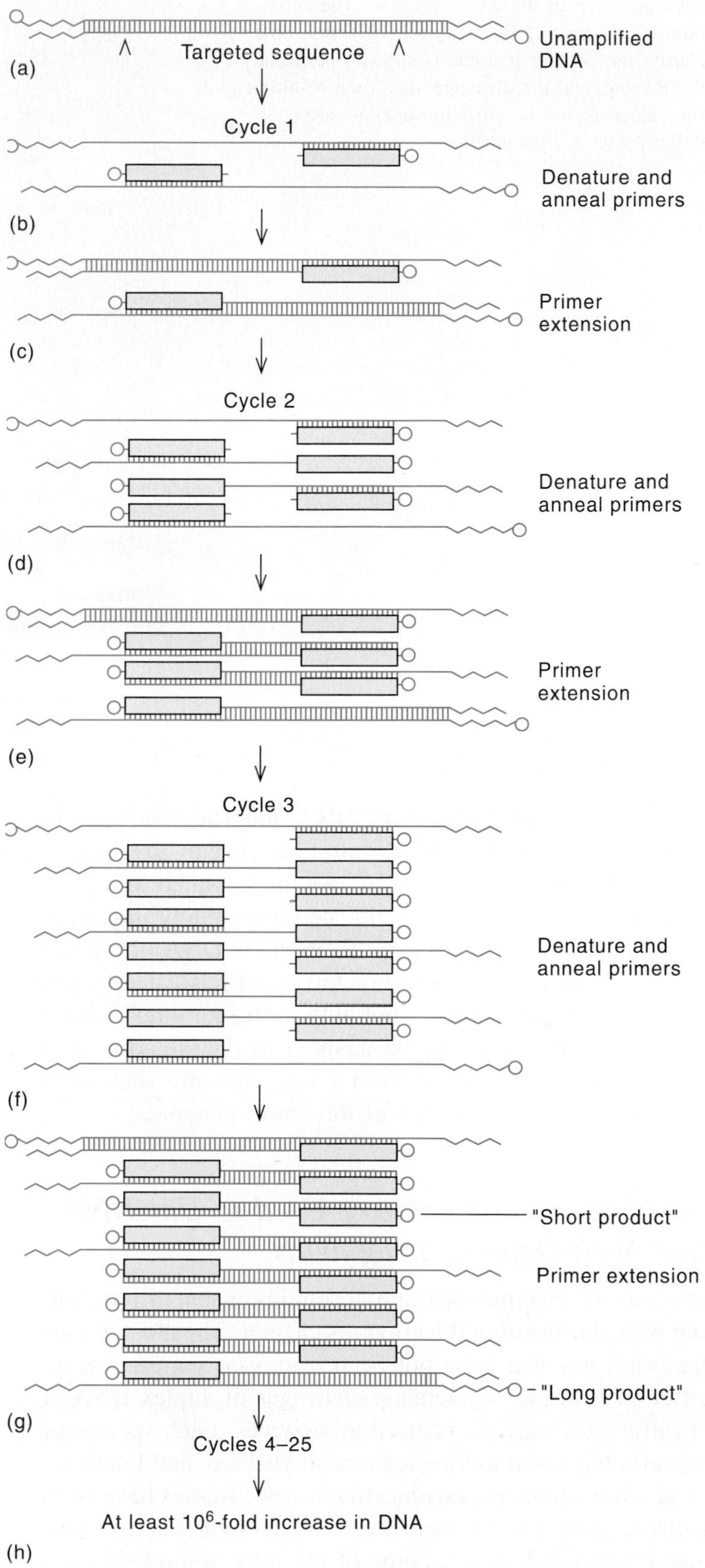

Figure 27.3

Cleavage map of the SV40 genome. The zero point of the map is the unique *Eco*R1 site. For clarity, the circular genome is shown opened at the R1 site, and the cleavage sites (and resulting fragments) for each restriction enzyme are indicated on a separate line.

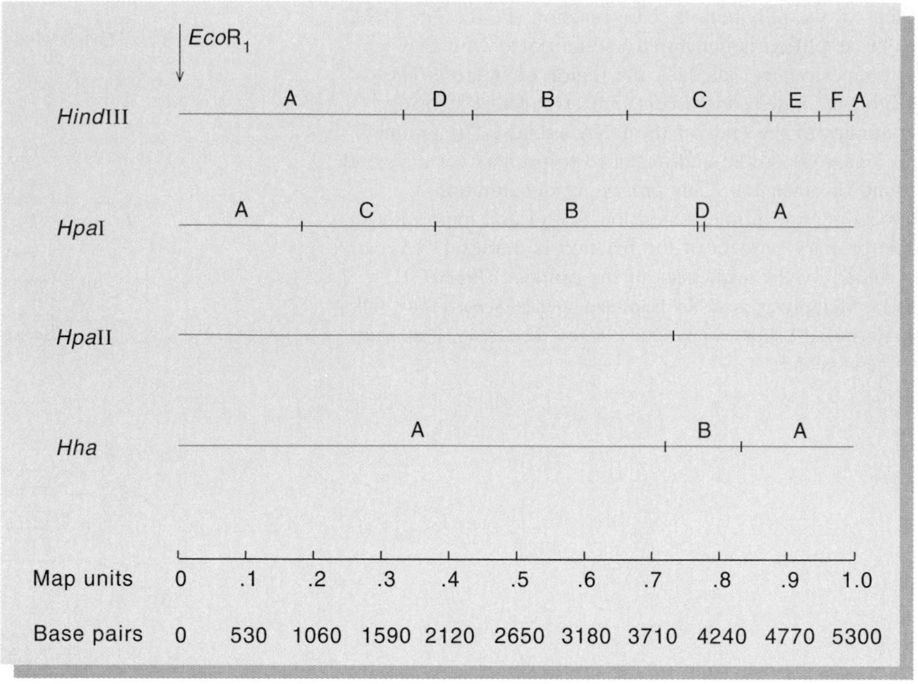

DNA Cloning

The second method for DNA amplification is more complicated than PCR, but it has several advantages. DNA to be amplified by cloning is linked to a plasmid or a virus that can be replicated indefinitely in the appropriate host cell. After amplification the DNA of interest can be cut from the plasmid or virus and reisolated by gel electrophoresis. Cloning is not only useful for amplifying a segment of DNA, it can be adapted to the isolation of a DNA segment of interest from a large mixture such as is obtained from the isolation of the entire genome.

Restriction Enzymes Are Used to Cut DNA into Well-Defined Fragments

Most of the enzymes that are absolutely essential for cloning were discussed in the previous chapter. The most important enzymes that have not been discussed yet are the restriction enzymes. Systematic cleavages of duplex DNA at specific sites requires restriction enzymes. Each species of bacteria harbors a unique restriction enzyme, and hundreds of restriction enzymes with different specificities have been isolated, giving researchers a great deal of choice as to how and where DNA is cut. Some of the most commonly used restriction enzymes and their recognition sites are indicated in table 27.1. Most of these enzymes recognize a sequence of either four or six contiguous base pairs. The cleavage sites are situated so that a blunt-ended or staggered-ended DNA results from the cleavage reaction. As a rule the recognition sites are located on an axis of symmetry so that the freshly cleaved segments have identical structures at their ends.

A viral genome cleaved exhaustively with a particular restriction enzyme usually yields several fragments. Some restriction enzyme cleavage sites for the 5,300 bp (5.3 kb) SV40 virus genome are shown in figure 27.3. The duplex fragments obtained after cleavage can be separated according to size by gel electrophoresis. Nondenaturing conditions are used so that the duplex strands stay together. The larger a fragment is, the slower it migrates on the gel. After electrophoresis for a time sufficient to separate the fragments, the gel is stained with a fluorescent dye such as ethidium bromide and viewed under long-wavelength ultraviolet light (long-wavelength UV is used because it does not damage the DNA). Individual fragments may be extracted from the gel for sequencing, PCR amplification, or cloning (described later on).

The problem of determining how a set of restriction fragments are normally connected is resolved by determining the sequences by a second set of fragments cut with a different restriction enzyme. The overlapping information obtained from the two sets of fragments permits a determination of the complete sequence of the intact genome. The

Table 27.1

Recognition Sequences and Cutting Sites of Selected Restriction Enzymes

Enzyme	Recognition Sequences	Enzyme	Recognition Sequences
AluI	AG↓CT TC↑GA	HpaII	C↓CGG GGC↑C
BamHI	G↓GATCC CCTAG↑G	KpaI	GGTAC↓C C↑CATGG
BglII	A↓GATCT TCTAG↑A	MboI	↓GATC CTAG↑
ClaI	AT↓CGAT TAGC↑TA	PstI	CTGCA↓G G↑ACGTC
EcoRI	G↓AATTC CTTAA↑G	PvuI	CGAT↓CG GC↑TAGC
HaeII	GG↓CC CC↑GG	SalI	G↓TCGAC CAGCT↑G
HindII	GTPy↓PuAC CAPu↑PyTG	SmaI	CCC↓GGG GGG↑CCC
HindIII	A↓AGCTT TTCGA↑A	XmaI	C↓CCGGG GGGCC↑C

strategy of sequencing overlapping fragments is identical to that used in primary structure determination of proteins (see chapter 3).

Plasmids Are Used to Clone Small Pieces of DNA

In a simple procedure for "DNA cloning," an autonomously replicating plasmid and insert DNA are cut with a restriction enzyme and then the pieces are annealed and covalently joined by the action of DNA ligase. The resulting recombinant molecules are then transfected into *E. coli,* where they replicate. When plasmid vectors are used, a population of permeabilized cells is bathed in the plasmid DNA containing the inserted DNA. Because only a small number of cells become transfected by this procedure, a way to select cells that carry the desired hybrid plasmids is needed.

A particularly useful plasmid vector for selecting transfected cells called pBR322 is itself a hybrid plasmid (fig. 27.4). This plasmid contains two genes, *amp*^{r} and *tet*^{r}, which confer resistance to penicillin and tetracycline, respectively. *Pst*I restriction fragments of foreign DNA may be inserted into the unique *Pst*I restriction site on pBR322 (see fig. 27.4). This is done by digesting pBR322 with *Pst*I, mixing with the restriction fragments to be cloned at low temperatures to permit annealing to take place between the two DNAs, and finally ligating the annealed fragments with DNA ligase. The product contains some of the original pBR322 and some pBR322 with inserted foreign DNA. When this mixture is used in transfection, most cells are not transfected, some are transfected with pBR322, and some are transfected with the desired hybrid plasmid. The three types of cells may be readily distinguished by their drug-

Figure 27.4

Structure of the pBR322 plasmid (*a*) and construction of a hybrid plasmid containing the pBR322 vector and a segment of foreign DNA (*b*). For pBR322 the unique sites for various restriction enzymes are indicated. Also indicated are the locations of the tetracycline (*tet*^{r}) and the ampicillin (*amp*^{r}) resistance genes and the origin for DNA replication. The hybrid plasmid is constructed by treating the plasmid and the foreign DNA with the *Pst*I restriction enzyme and mixing the two DNAs together in the presence of DNA ligase.

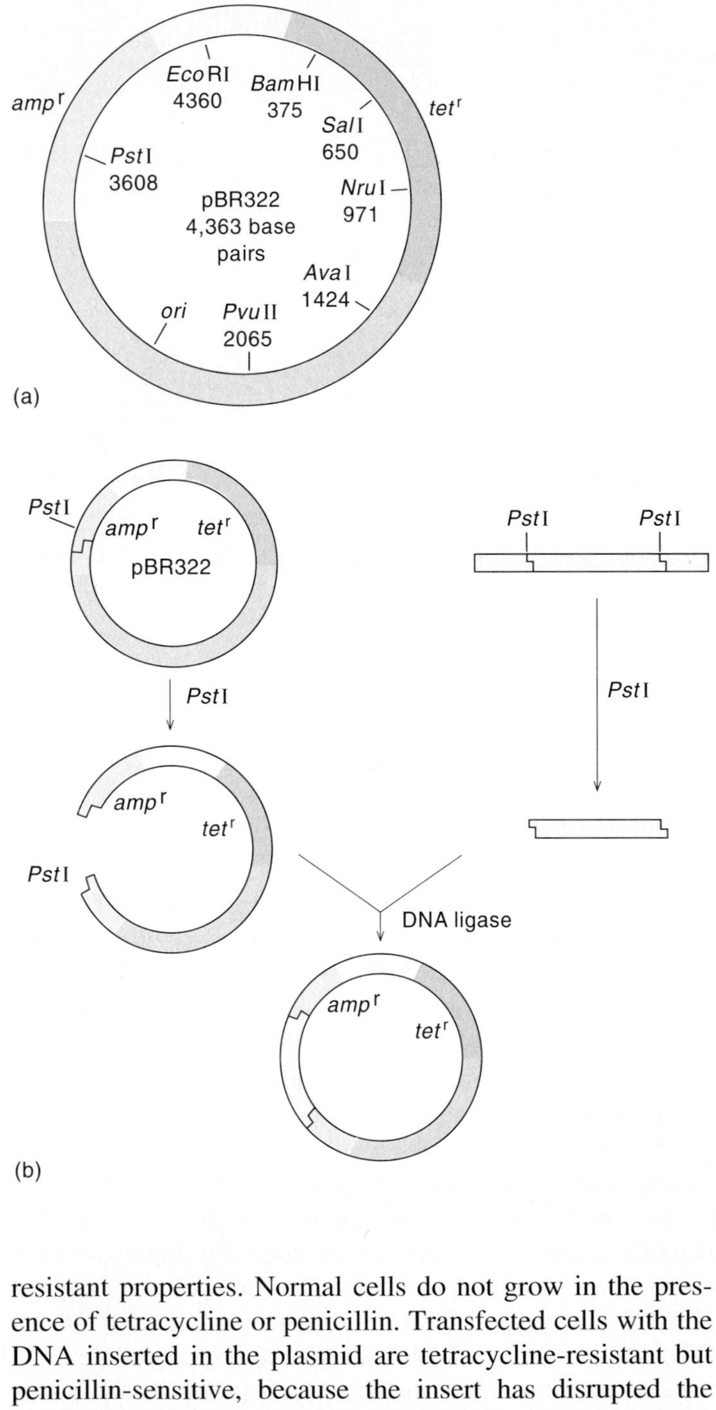

Figure 27.5

Application of the replica plating technique to the detection of hybrid plasmid-containing cells. About 10^7 bacteria are spread on a plate. After overnight growth the plate (master plate) appears as a uniform "lawn" of bacteria, but in reality it consists of very small colonies that have merged to give a uniform appearance. A piece of velvet is lightly pressed against the surface of this lawn, and some cells stick to the velvet. Several essentially identical impressions of the lawn are transferred to fresh plates, which contain normal medium or normal medium supplemented with antibiotics. The results on the replica plates after overnight growth are indicated. The plate in normal medium again gives rise to a lawn of cells as virtually all of the cells transferred grow into colonies. The plates containing antibiotics only give rise to a few colonies, each of which is derived from a single cell that carries the plasmid-conferred drug resistance(s).

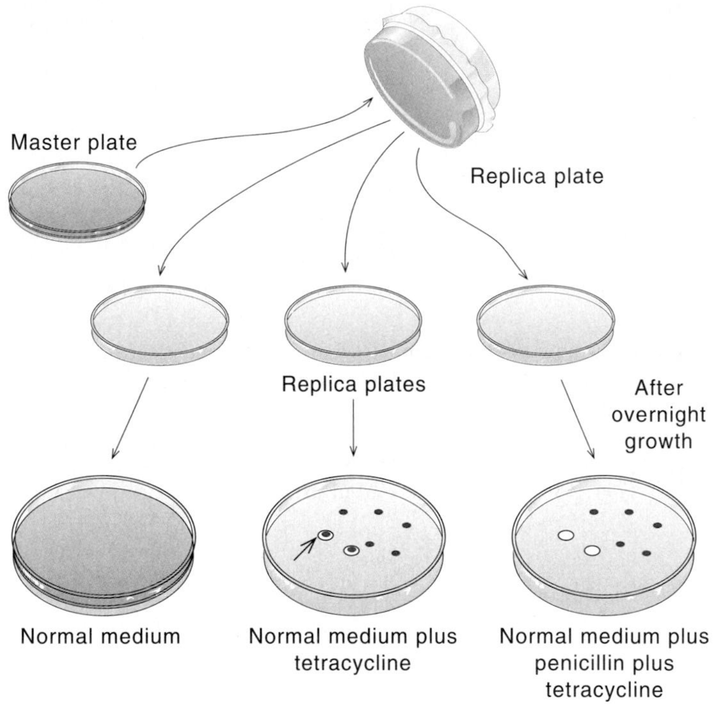

resistant properties. Normal cells do not grow in the presence of tetracycline or penicillin. Transfected cells with the DNA inserted in the plasmid are tetracycline-resistant but penicillin-sensitive, because the insert has disrupted the *amp*^{r} gene. Cells containing the desired plasmids can be distinguished from those containing pBR322 by "replica plating" (fig. 27.5). The first step when using this approach is to spread a large population of treated bacteria on an agarose plate containing growth medium. Within 12 h a seemingly homogenous "lawn" of cells develops on the surface of the agarose. Actually, the lawn results from the growth of many microcolonies to the point of confluency. At this point a piece of velvet is lightly pressed against the surface of the plate, and this impression is transferred to other agarose plates containing growth medium with tetracycline or penicillin plus tetracycline. Only the transfected cells produce colonies on the plates containing the antibiotics, and because of their small number, each of these gives rise to readily detectable clones. The clones present on the tetracycline-containing plates, which are missing on the penicillin plus tetracycline plates, most likely contain the

Figure 27.6

Electrophoretogram of restriction enzyme digests of pBR322 and pBR322 with a DNA insert at the *Pst*I site. The insert is assumed to have no internal *Bam*HI restriction sites. In channels A and B the pBR322 is predigested with *Pst*I and *Bam*HI, respectively. The resulting DNA migrates with the same mobility because the plasmid has one site for each of these enzymes and therefore has the same molecular weight. In C and D the hybrid plasmid containing a DNA insert is treated with *Bam*HI and *Pst*I, respectively. In C the hybrid plasmid has been linearized by one cut at the *Bam*HI site in the *tet*^{r} gene. It runs more slowly than the pBR322 because it is larger due to the DNA insert. In D the plasmid cuts at two *Pst*I sites located between the pBR322 sequences and the insert sequences. Consequently, one segment migrates at the rate of a linearized pBR322 plasmid. The other segment, also linearized, migrates at a rate characteristic of the size of the DNA insert. The electrophoresis is run from left to right; fragments are stained with ethidium bromide and photographed with UV light.

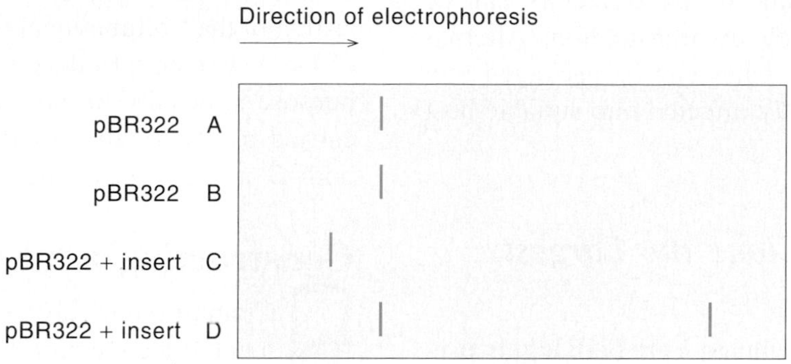

Figure 27.7

The nutrient agar plate contains a continuous lawn of *E. coli* bacteria except for circular clearings that represent phage plaques. Each plaque was originally derived from a single phage infecting a single *E. coli* bacterium. After infection the phage multiplies, ultimately producing about 100 mature viruses. The phages also produce an enzyme that causes the harboring cell to lyse. When this happens the phages are released, and each of them infects a neighboring cell and goes through the same infectious cycle. The process continues. Each cycle takes about 30 min. Eventually a visible clearing can be seen on the plate.

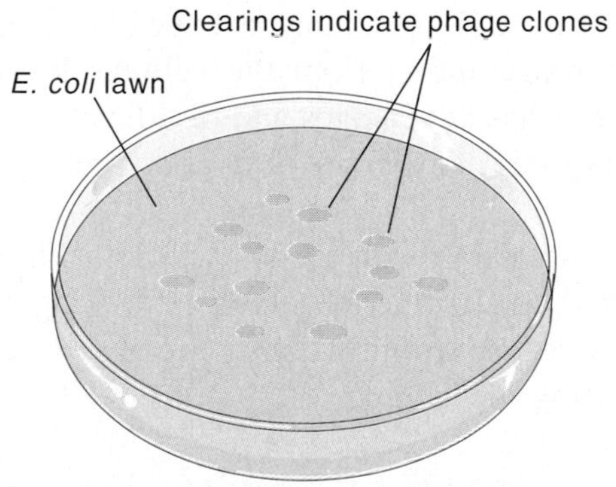

desired hybrid plasmids (see fig. 27.5). These clones are usually plucked from the tetracycline plates and retested to eliminate any uncertainty about the original drug testing. Once this is confirmed, the appropriate plasmid-containing cells are grown in liquid culture. After a moderate density of growth is achieved, the plasmid DNA is selectively amplified by overnight growth. Plasmid DNA replication continues for several hours, until each cell contains 1,000 to 2,000 copies of the small circular plasmid DNA. This DNA is readily separable from the host DNA and can be characterized by its rate of migration on gel electrophoresis (fig. 27.6) or other more specific tests to determine if it contains the inserted DNA sequence. If desired, the inserted sequence may be removed from the plasmid vector by digestion with *Pst*I, the restriction enzyme used in the initial construction of the hybrid plasmid. The cleaved fragments can be separated readily by gel electrophoresis. Many plasmid vectors, other than pBR322, have specific advantages for other purposes.

Bacteriophage λ Vectors Are Useful for Cloning Larger DNA Segments

Bacteriophage λ possesses a number of advantages as a cloning vector. DNA fragments as large as 24 kb can be propagated using such vectors. The primary pool of clones can be amplified by limited phage growth as plaques, and the entire collection of phage clones (recognized as clear plaques) can be stored for long periods in a small volume (fig. 27.7).

Because it does not accommodate molecules of DNA that are much longer than the viral genome, the use of λ as a vector for cloning substantial DNA fragments requires the removal of a significant portion of the viral DNA beforehand. Fortunately, the central third of the genome contains genes that are not essential for phage production and can therefore be deleted.

Bacteriophage λ vectors that accommodate foreign DNA fragments generated by a variety of restriction endonucleases have been constructed. The recombinant DNA molecules that incorporate some of these vectors can be introduced directly into *E. coli* by transfection. Alternatively, recombinant DNA molecules can be packaged into phage particles and subsequently infected into suitable host cells.

Cosmids Are Used to Clone the Largest Segments of DNA[a]

Although plasmids and bacteriophage λ are both highly useful vectors, the size of the DNA fragments that can be cloned in them is limited. With plasmids, the larger the fragment of foreign DNA inserted, the lower the efficiency of ligation and transfection, making the cloning of DNA fragments larger than 15 kp experimentally difficult. In λ vectors, the length of the nonessential region of λ DNA limits fragment size to 24 kb or less. Also, the original λ vectors do not allow propagation of viable bacterial cells that carry the inserted DNA fragment; the insert is propagated as part of a virus that lyses the cell.

Cosmids were developed as vectors for cloning large DNA fragments. The first part of their name, ''cos,'' comes from the fact that cosmids contain the cohesive ends, or *cos,* sites of normal λ. These ends are essential for packaging the DNA into λ phage heads. The last part of their name, ''mid,'' indicates that cosmids carry a plasmid origin of replication like the one found in the pBR322 plasmid. Such cosmids can be used for cloning in the same way as any other plasmid vector. However, because cosmids also contain the *cos* sites, cosmid DNA along with an inserted DNA fragment can be packaged as a λ phage. The result after packaging is a defective but nevertheless infectious phage particle. Once the cosmid and the inserted DNA fragment are introduced by infection into a λ-sensitive cell, the plasmid replicates. Since cosmids lack the entire bacteriophage genome except for the region adjacent to the *cos* sites, these vectors can propagate exogenously derived DNA fragments of up to 40–50 kb in length.

[a] Yeast artificial chromosomes have been developed that can clone hundreds of kilobases of DNA (see Watson, Gilman, Wilkowski, and Zoller reference).

Shuttle Vectors Can Be Cloned into Cells of Different Species

Vectors that include replication systems derived from more than one host species are known as shuttle vectors. Such vectors commonly include a replication system able to function in *E. coli* and one that works in a second host, which may be bacterial or eukaryotic. Initial cloning and amplification of the DNA segment to be studied is often carried out in *E. coli* because it is easier to make large quantities when culturing in *E. coli.* The recombinant DNA molecule, consisting of the ''bifunctional vector'' plus the cloned segment of DNA, is then introduced into the second host, where the purpose is usually to measure the expression of the genes carried by the vector. Shuttle vectors that can replicate in both *E. coli* and yeast are the most common.

Constructing a ''Library''

Cloning can involve a single vector-linked DNA fragment or a collection of independently isolated vector-linked DNA fragments derived from a single organism. Such a collection is termed a ''library'' and may serve as the source of well-defined sequences from a given organism. Each clone of a library harbors a particular DNA segment from the desired organism. Within the entire library a sequence may be repeated, but other sequences may be missing. The ideal library, which can only be approached, represents all of the sequences with the smallest possible number of clones.

A library from the same cell or organism can be prepared in two ways. The genome may be fragmented and ligated to the appropriate vector to produce a genomic DNA library. An alternative approach is to construct a cDNA library in which the DNA fragments to be cloned are obtained by reverse transcription from the cellular RNA. Each of these libraries has advantages and disadvantages, and for a specific purpose, one library is usually preferred over the other.

The vast majority of DNAs within a library are uncharacterized. As a rule, the task of finding the desired genes or sequences within a library greatly exceeds the task of constructing the library.

A Genomic DNA Library Contains Clones with Different Genomic Fragments

A major concern in constructing a genomic DNA library is to maximize the probability that all segments of the genome are represented. If the genomic DNA is prepared by cutting

Table 27.2

Theoretical Number of Clones Required to Fully Represent the Entire Genome of Various Organisms

Size of Cloned DNA Fragment (bp)	Genome Size (bp)		
	2 × 10^6 (e.g., bacteria)	*2 × 10^7 (e.g., fungi)*	*3 × 10^9 (e.g., mammals)*
5×10^3	400	4,000	600,000
10×10^3	200	2,000	300,000
20×10^3	100	1,000	150,000
40×10^3	50	500	75,000

with a restriction enzyme, an added concern is that the enzyme cleaves genes of interest at one or more sites. To increase the likelihood of isolating desired genes in one piece, different restriction enzymes can be used on different parallel preparations. But even if the genes of interest are not cut by the enzyme(s) chosen, the DNA fragments produced may be inconveniently small to work with. An enzyme that recognizes a sequence of six bases (a six-cutter) gives an average fragment size of 4,096 bp[b], which is a reasonable size for making a plasmid library but much smaller than the size desirable for cloning in λ or a cosmid vector. Therefore, when large, randomly generated fragments are desired, the method of choice usually entails making an incomplete digest with a four-cutter restriction enzyme, which produces overlapping ends that can be readily cloned into the chosen vector as described earlier. The extent of digestion is controlled so that cleavage occurs at only some of the restriction enzyme recognition sites and the average size of the fragments produced is in the desired range. The conditions used thus depend on whether the product is going to be cloned in a plasmid, a λ phage, or a cosmid vector. Table 27.2 gives the minimum number of clones (i.e., the size of the library) required to fully represent the entire genome in a genomic DNA library, as a function of the average size of the cloned fragments and the size of the genome. Since DNA fragments in a population are cloned on a random basis, the chance of finding a given single-copy gene in a library of the indicated size is 50%. A clone bank should be 3 to 10 times the minimum size to give a high probability that a particular segment is represented.

[b] $(\frac{1}{4})^6 = 1/4{,}096$

A cDNA Library Contains Clones Reflecting the mRNA Sequences

A cDNA library consists of a collection of clones that contain DNA copies of the cellular or organismic RNA. If the RNA is obtained from a differentiated multicellular organism, then the library varies in composition according to the type of cell used as the RNA source and to the physiological state of the cell. This variation is a reflection of the relative abundances of particular mRNAs made by different cell types. If a cDNA species corresponding to a particular gene product is desired, it is often possible to select a cell type suspected to synthesize a large amount of the corresponding mRNA or mRNA-related protein. Thus, pituitary cells can be used if cDNA encoding growth hormone is desired, whereas liver cells can be used if a serum albumin cDNA is the goal. The mRNAs present in low amounts clearly require the screening of a larger library than the mRNAs present in medium or high abundance.

Once the crude mRNA fraction has been isolated from the chosen cells or tissue, it is converted to duplex DNA molecules with the help of reverse transcriptase. This duplex DNA does not have "sticky ends" for insertion into a vector. For this purpose DNA linkers are attached to the ends. Linkers are synthetic single-stranded oligonucleotide segments (6, 8, 10, or 12 bases in length) that self-associate to form symmetrical, blunt-ended, double-stranded molecules containing the recognition sequence for a particular restriction enzyme. Figure 27.8 shows an eight-base linker (CCTGCAGG) containing a *Pst*I recognition site. This linker self-associates to produce an eight-base, blunt-ended, duplex structure that adds to the double-stranded cDNA in the presence of T4 ligase. The resulting product is treated

Figure 27.8

Insertion of cDNA into pBR322 plasmid by the linker method. The strategy here is to open up the plasmid with a restriction enzyme that makes staggered cuts and to attach linkers that contain the same recognition site to the cDNA. After the linkers are attached to the cDNA, the duplex is treated with the same restriction enyzme (*Pst*I) to expose the overhangs. The two DNAs are mixed together and ligated. After transfection, cells containing the hybrid plasmids are recognized by tetracycline resistance and ampicillin sensitivity. Identification of the insert is discussed in the text.

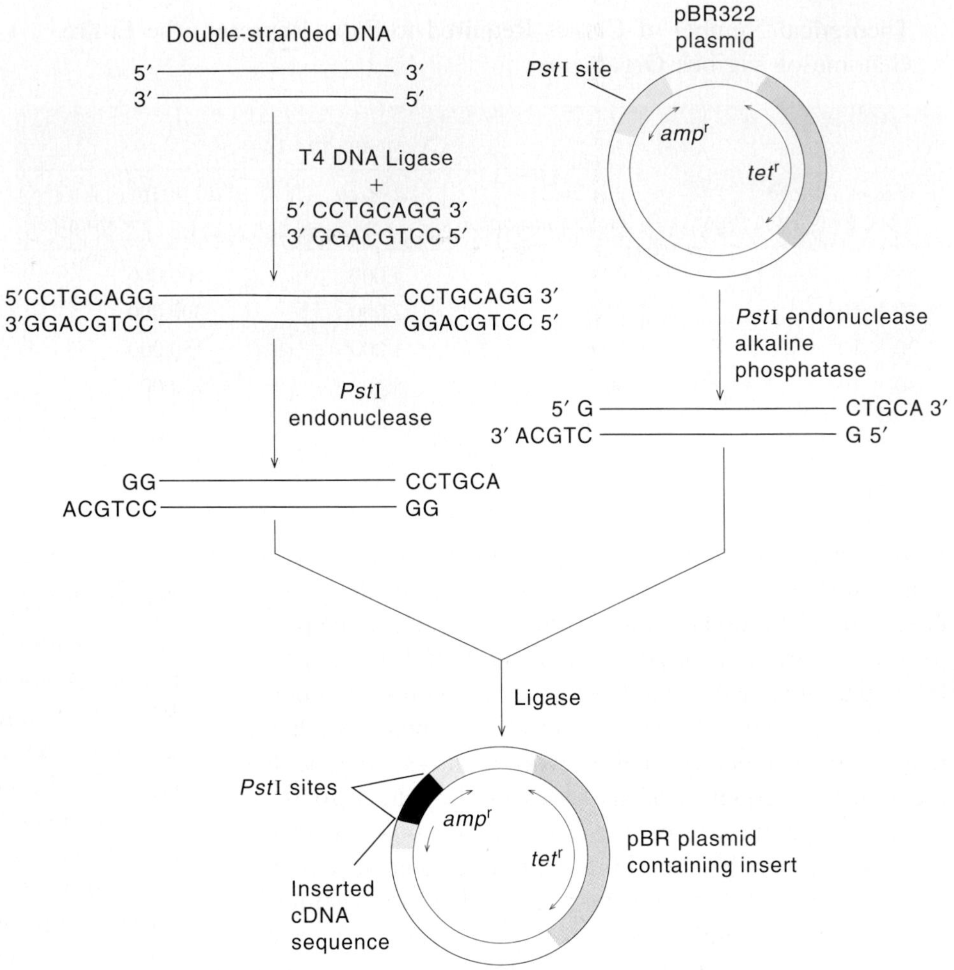

with *Pst*I to produce the characteristic 3′ overhang. The plasmid, linearized with *Pst*I, and the two DNAs are mixed and reacted with ligase to produce plasmid with the insert.

Numerous Approaches Can Be Used to Pick the Correct Clone from a Library

A library can contain thousands or even tens of thousands of different kinds of clones (see table 27.2), making it a challenging endeavor to isolate a clone with the DNA of interest. Most currently used procedures for screening large numbers of colonies for plasmids or phage that contain specific DNA inserts are variants of the colony hybridization method developed by Grunstein and Hogness. This procedure makes use of a specific radioactive probe that contains some sequences complementary to those in the DNA of interest. The colonies to be screened are first grown on agar petri plates (fig. 27.9). A replica of each plate is made on another agar plate, which is stored for reference. A replica is also made on a nitrocellulose filter. The colonies formed on the filter are lysed and the contents denatured simultaneously by treatment with sodium hydroxide. After heating, the denatured DNA is fixed on the filter at each site where a colony was located. The DNA on the filter is then hybridized with a radioactively labeled nucleic acid probe complementary to the specific DNA sequence to be selected. The presence of hybridized probe at sites occupied by DNA derived from colonies that include the DNA fragment of interest is detected by autoradiography. The colony whose DNA hybridizes with the nucleic acid probe can then be picked from the reference plate, which contains a viable bacterial colony at a corresponding location.

Cloning in Systems Other than Escherichia coli

Despite the success and broad applications of *E. coli* cloning systems, instances occur in which gene products cannot be made in this bacterium. Either they are not synthesized in

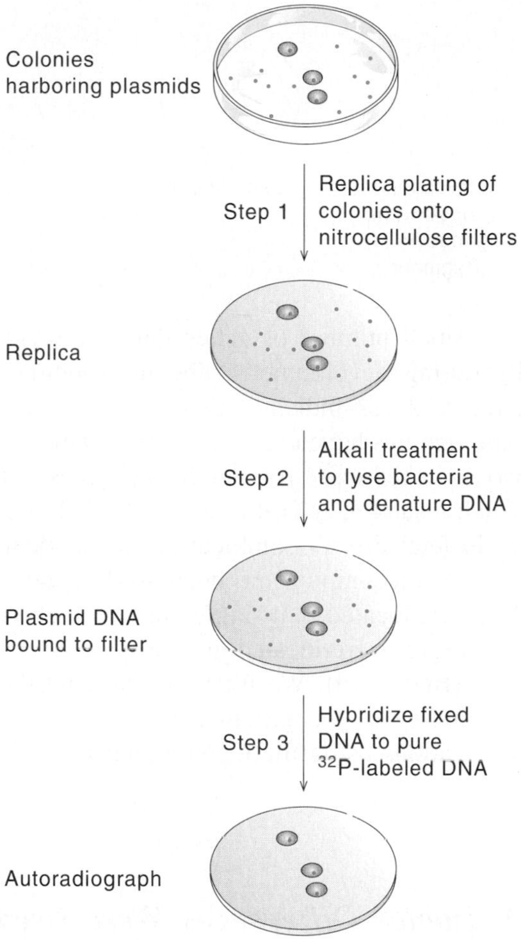

Figure 27.9

Colony hybridization procedure used to identify bacterial clones harboring a plasmid containing a specific DNA. Step 1: Replica-plate the colonies containing plasmids onto nitrocellulose paper. Step 2: Lyse cells with NaOH and fix denatured DNA to paper. Step 3: Hybridize to ^{32}P-labeled DNA carrying the desired sequence and autoradiograph the product. Locations of desired DNA should be emphasized in the autoradiograph. Clones carrying desired plasmids (circled) may then be isolated from a corresponding agar replica plate carrying untreated colonies.

their entirety, or they are rapidly broken down after synthesis. In addition, the study of certain processes indigenous to other species (e.g., photosynthesis, antibiotic production) often requires the use of a host bacterial species that naturally carries out the process, which may exclude *E. coli.*

Effective cloning systems are available for a variety of bacterial hosts, including *Bacillus subtilis, Streptomyces spp.,* and *Agrobacter tumefaciens.* Cloning systems have also been developed for eukaryotic hosts such as the yeast *Saccharomyces cerevisiae,* mammalian cells in tissue culture, and plant cells.

Site-Directed Mutagenesis Permits the Restructuring of Existing Genes

By combining different procedures of manipulation, it is now possible to make discrete changes in genes. This technique, called site-directed mutagenesis, is one of the most important in modern genetics and biochemistry. The first site-directed mutagenesis studies were carried out by David Shortle and Daniel Nathans in 1978 with the help of the mutagen sodium bisulfite, which deaminates C residues so that they become converted into U residues.

Directed mutagenesis as it is practiced today is based on the chemical synthesis of a deoxyoligonucleotide that contains discrete changes in its sequence from that normally observed in the genome under investigation. These changes may be single-base or multibase; they may involve base changes, base deletions, or base additions.

Many variations of site-directed mutagenesis exist. One can start out with a circular, single-stranded DNA and anneal it to a synthetic primer DNA carrying the desired changes (fig. 27.10). This primer can be extended, and the resulting product can be transfected. Finally, one selects clones of cells containing the plasmid with the desired changes.

The polymerase chain reaction (PCR) will probably be the method of choice in the future for carrying out site-directed mutagenesis. In its most general form the use of PCR for this purpose requires a piece of duplex starting DNA, two outside flanking primers that are perfect complements to segments of opposing strands, and two complementary primers with the desired changes in their sequence as diagrammed in figure 27.11. The steps followed parallel the steps of PCR amplification.

PCR amplification can be coupled with classical cloning methods using cloning vectors. For this purpose the PCR product should contain restriction sites at its ends that are suitable for cloning. Thus, PCR amplification and cloning need not be thought of as alternatives for particular purposes but as complementing each other to give a greater variety of approaches.

Recombinant DNA Techniques Were Used to Characterize the Globin Gene Family

The human globin family is a paradigm for studying differential gene activity during development and the molecular basis of genetic disorders in gene expression. Hemoglobin is a tetramer containing two α-like and two β-like subunits (see chapter 5). These proteins are en-

Figure 27.10

Scheme for oligonucleotide-directed mutagenesis of double-stranded circular plasmid DNA. Supercoiled plasmid circles are nicked in one strand and rendered partially single-stranded by treatment with exonuclease. The gapped circles are hybridized with a homologous oligodeoxynucleotide carrying, by design, some mismatches. *In vitro* DNA synthesis, primed in part by the oligodeoxynucleotide, leads to heteroduplex plasmid circles. (Source: After G. Dalbadie-McFarland, L. W. Cohen, A. D. Riggs, C. Morin, K. Itakura, and J. H. Richards, Oligonucleotide-directed mutagenesis as a general and powerful method for studies of protein function, *Proc. Natl. Acad. Sci. USA* 79:6408–6412, 1982.)

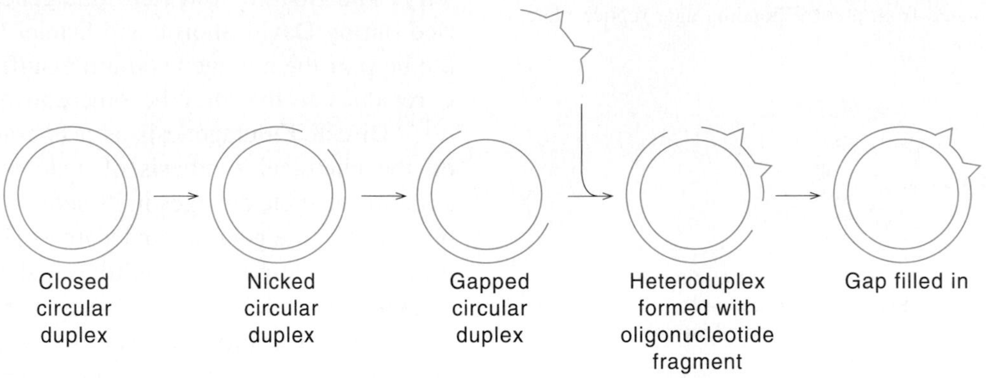

Closed circular duplex
Nicked circular duplex
Gapped circular duplex
Heteroduplex formed with oligonucleotide fragment
Gap filled in

Figure 27.11

Illustration of a general method of mutagenesis using PCR. Primers are represented as short lines with arrowheads pointing in the 3′ direction. The bump in primers 2 and 3 and their products represent a mismatched base, a deliberate alteration in base sequence from that present in the starting DNA. Of the four major products resulting from step 3 only D is extendable by DNA polymerase.

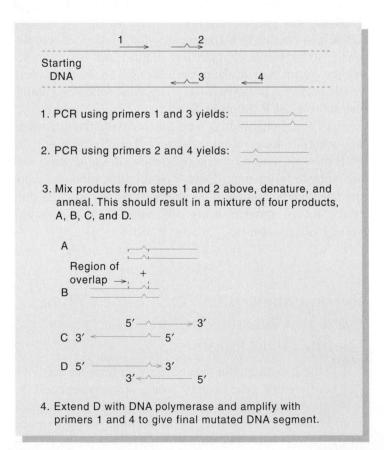

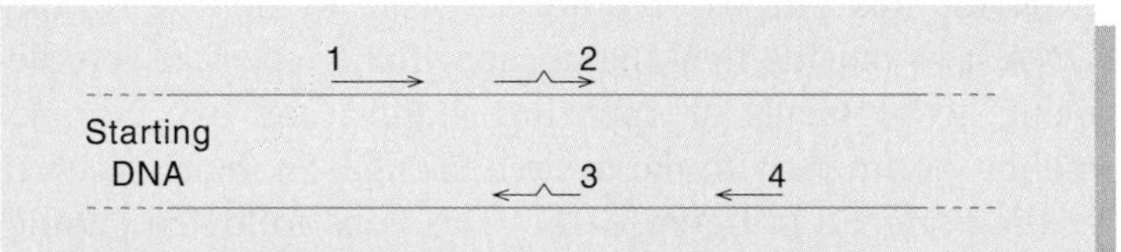

5′
4. Extend D with DNA polymerase and amplify with primers 1 and 4 to give final mutated DNA segment.

coded by a small number of genes that are expressed sequentially during development. The information summarized in figure 27.12 indicates that the α-like and β-like globin gene families have coordinated programs for expression: Two switches exist for the β-like genes, whereas a single switch results in activation of adult α-globin production early in fetal life. A combination of classical and recombinant DNA techniques has been used to show that the α-like genes are located in a single cluster on chromosome 16, and the β-genes are located in a single cluster on chromosome 11 (fig. 27.13). We focus on the contributions to our understanding of the globin genes that have resulted from investigations using the recombinant DNA approach.

DNA Sequence Differences Were Used to Detect Defective Hemoglobin Genes

1. PCR using primers 1 and 3 yields:
2. PCR using primers 2 and 4 yields:
3. Mix products from steps 1 and 2 above, denature, and anneal. This should result in a mixture of four products, A, B, C, and D.

Region of overlap

5′

C 3′
5′
D 5′

All of the hemoglobin genes are represented by two or more alleles within the human population; the genes are said to be polymorphic. This polymorphism frequently shows up in readily detectable phenotypes when the differences occur in vital areas of the polypeptide chains. Polymorphisms show up in the DNA even more frequently because the DNA contains sequence differences in both the coding and the noncoding regions of a gene, and these differences can be detected even when no visible effect is apparent in the organism. Because of their frequency and ease of detection by recombinant DNA methods, DNA polymorphisms have become extremely useful in mapping the human genome.

Figure 27.12

Changes in types of hemoglobin observed in early development. A single switch in gene expression is observed for α-like chains. Two switches in gene expression are observed for β-like chains. The corresponding tetrameric hemoglobin molecules observed at different stages in development are also indicated.

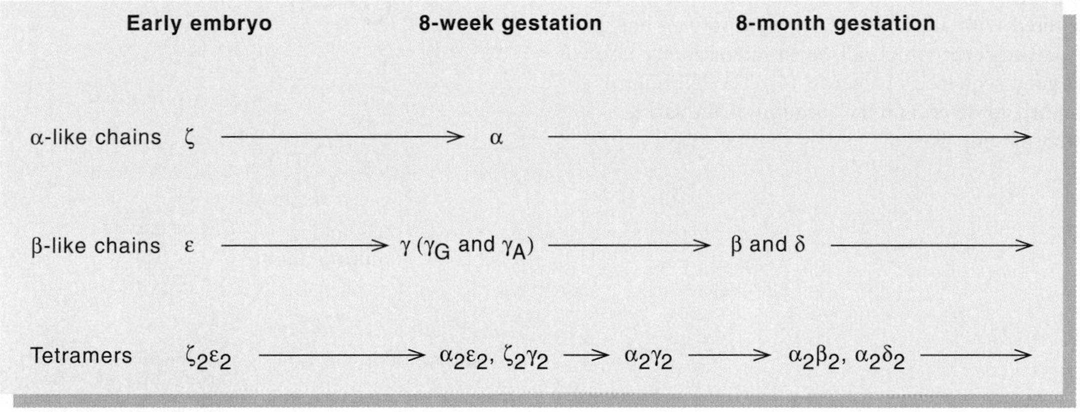

Figure 27.13

The chromosomal localization and genomic organization of the human globin genes. The α- and β-globin gene complexes are positioned on chromosomes 16 and 11, respectively. For each complex, the arrangement of genes on the chromosome is depicted above, and the general structure of the major gene is shown below, together with the location of the intervening sequences, or introns (IVS), and codon numbers. Coding regions are shown by solid boxes and IVS regions by open boxes. Genes with the ψ symbol in front are called pseudogenes because they are sequence-related but not expressed.

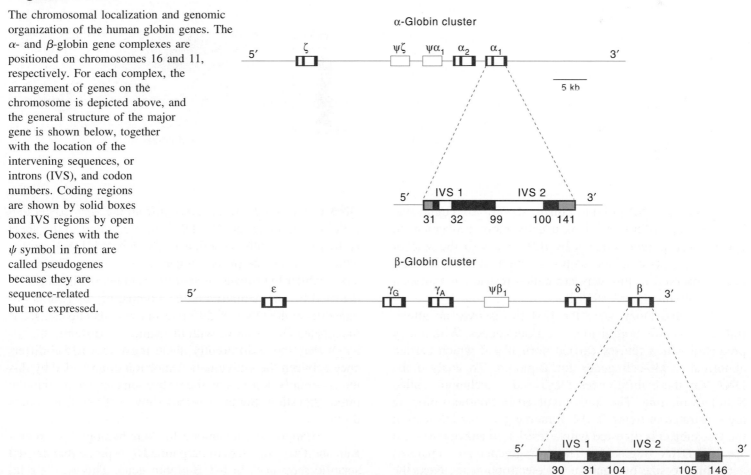

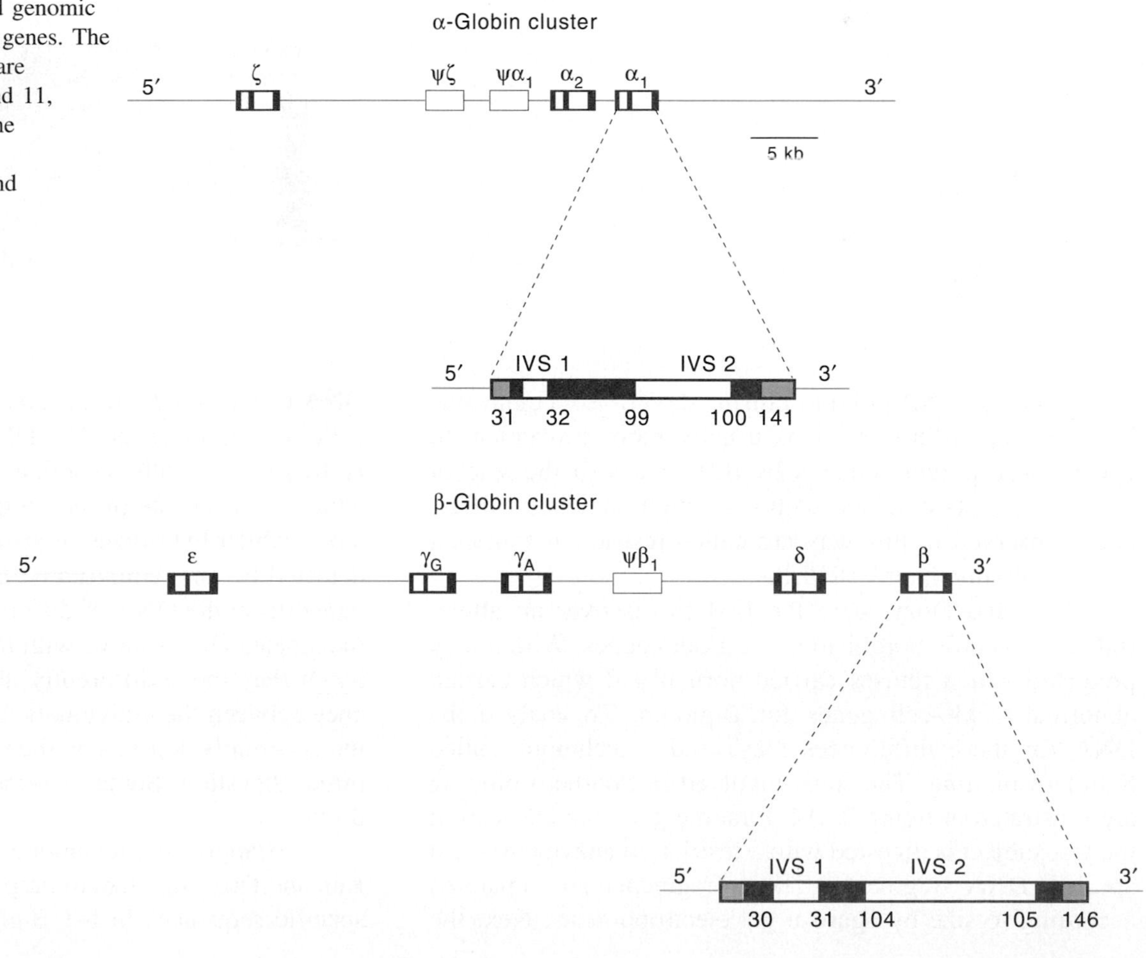

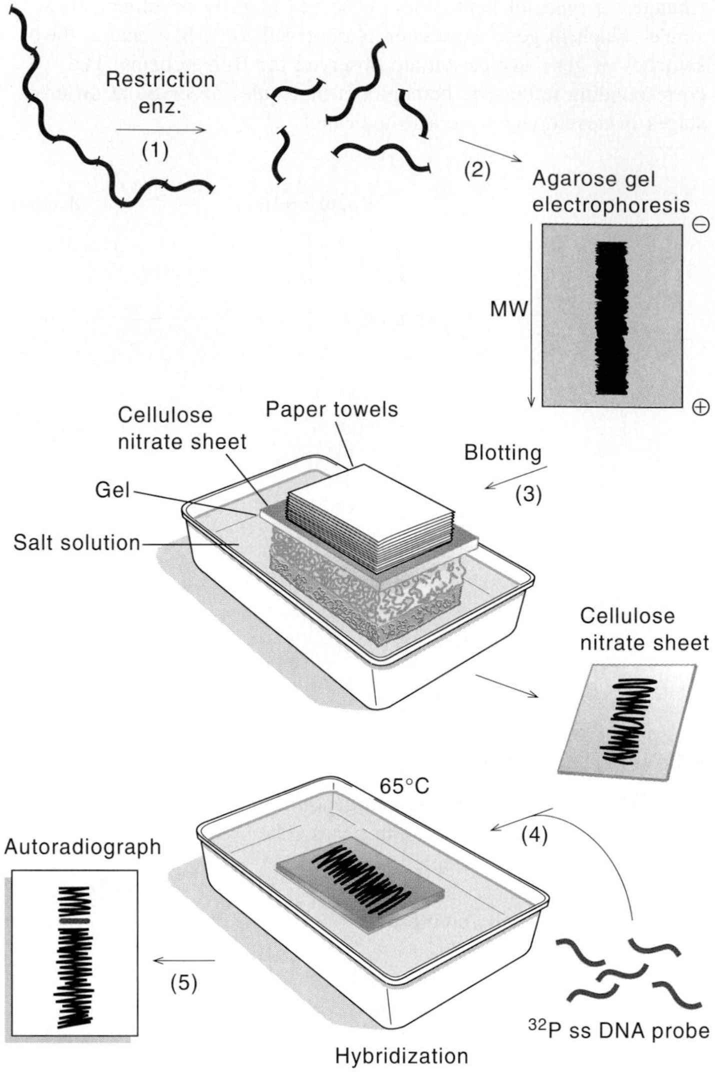

Figure 27.14

The steps involved in assaying by Southern blotting. The DNA to be analyzed is digested with a restriction enzyme (1). The resulting fragments are electrophoresed on an agarose gel (2). The DNA fragments on the gel are transferred to a cellulose nitrate sheet by placing the cellulose nitrate sheet next to the gel and passing solvent through the gel into the sheet. Flow of the solvent is maintained by blotting the far side of the cellulose nitrate sheet with paper towels. The DNA, first denatured with alkali, flows with the solvent but gets stuck in the sheet (3). The sheet is hybridized to radioactively labeled DNA containing the gene sequence of interest (4). The hybridized sheet is autoradiographed to determine the location of the labeled restriction fragment on the gel (5).

Whereas DNA polymorphisms should be recognizable by sequence differences, it is usually more convenient to detect these polymorphisms by differences in the size of DNA fragments obtained with restriction enzymes. Differences observed in this way are called restriction fragment length polymorphisms (RFLPs).

Kan and Dozy were the first to discover an allele-linked DNA polymorphism in the globin genes. With it they predicted which fetuses carried normal and which carried abnormal sickle-cell genes for β-globin. To analyze the DNA for these differences they used a technique called Southern blotting. The steps involved in Southern blotting are illustrated in figure 27.14. First the genomic DNA from the test subject is digested with a restriction enzyme to yield specific DNA fragments. These fragments are separated according to size by agarose gel electrophoresis. Next the DNA is denatured and transferred from the agarose gel to a cellulose nitrate sheet. The DNA firmly bound to the sheet is hybridized with a radioactively labeled DNA probe, which carries some of the sequences of interest. The radiolabel, which hybridizes to specific regions of the sheet, is detected by autoradiography. By comparing the results obtained from the DNA of different individuals one can see if the labeled DNAs move with the same or a different mobility. If they move differently, there must be a RFLP difference between the individuals. Detection of an RFLP by this means usually depends on the restriction enzyme used in the initial digestion. Some enzymes show a difference; others do not.

To apply this technique to their hemoglobin studies Kan and Dozy first had to prepare a DNA probe that carried specific sequences in the β-globin gene. This was a rela-

Figure 27.15

Inheritance pattern of an RFLP associated with sickle-cell disease. Humans carry two alleles for the same gene, and each offspring inherits one allele from each of its parents in an entirely random fashion. Normal individuals are homozygous for normal *Hb* alleles; individuals with sickle-cell trait are heterozygous, with the one normal *Hb* allele and one Hb^s allele; and individuals with sickle-cell disease are homozygous for the Hb^s allele. At the top (*a*) we see a three-generation pedigree analysis for a family that carries both the normal and the sickle-cell gene for β-globin. Males are represented by squares and females by circles. A purple circle or square indicates an individual who is homozygous normal. A half-filled circle or square (red/purple) indicates a heterozygous individual with sickle-cell trait. A filled circle or square (red) indicates a homozygous individual with sickle-cell disease. In (*a*) both sets of grandparents produce a heterozygous individual with sickle-cell trait. Because one of the grandparents is homozygous normal and the other is heterozygous, there is a 50% chance that the grandparent mating will give rise to a heterozygous offspring as shown and also a 50% chance that they will have normal offspring (not shown). The two heterozygous parents have an increased chance of having abnormal offspring because in this mating each parent carries one abnormal gene or allele. There is a 25% chance of a homozygous sickle-cell anemic offspring, a 50% chance of an abnormal offspring with sickle-cell trait, and a 25% chance of a normal offspring. Below the pedigree chart is the electrophoretic pattern of a *Hpa*I digest probed with β-globin cDNA by the Southern blotting technique (*b*). At the bottom we see an interpretation of the normal and abnormal DNAs (*c*).

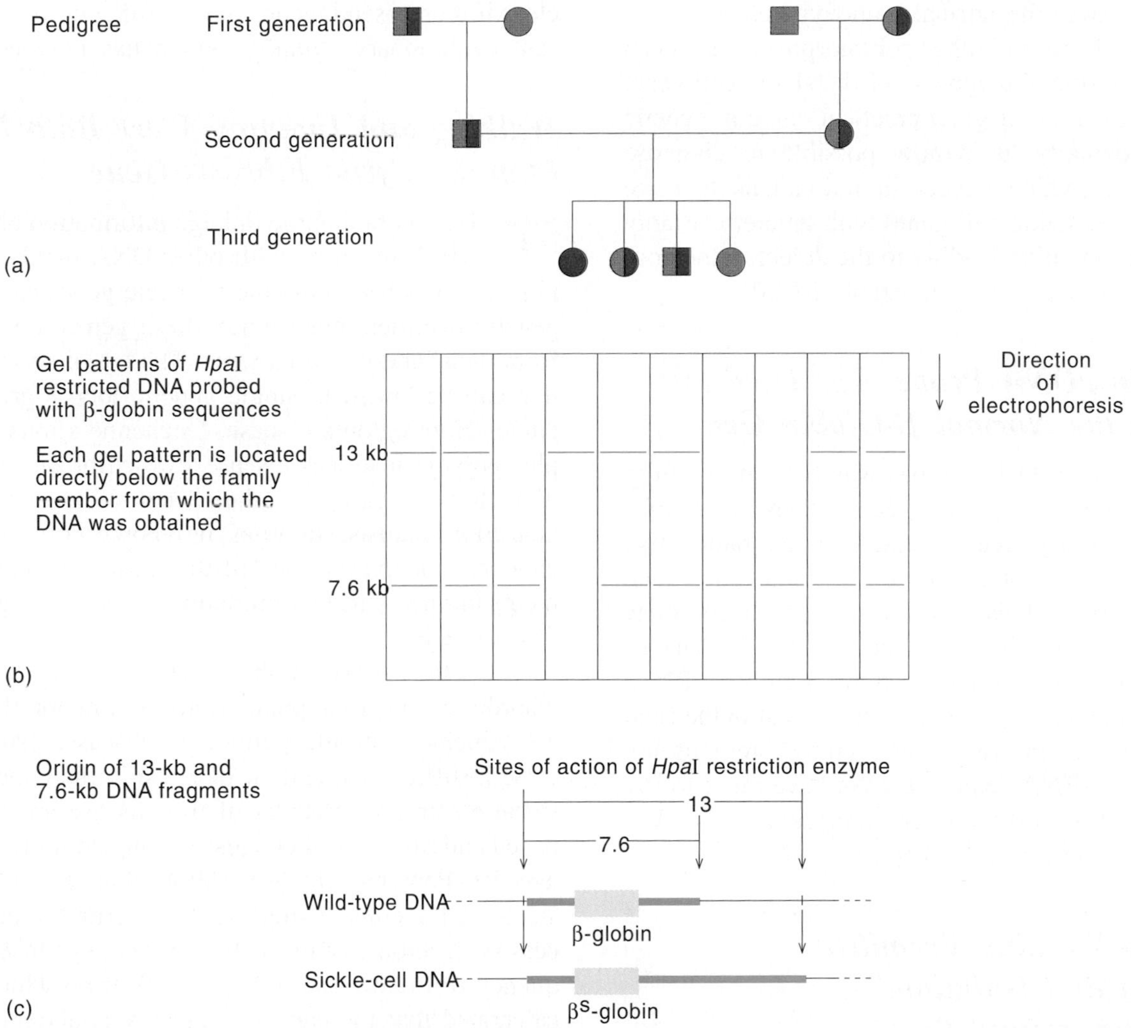

tively simple task because reticulocytes, which contain vast quantities of hemoglobin, are also greatly enriched in the β-globin mRNA. From the purified β-globin mRNA isolated from reticulocytes, a cDNA probe was made with the help of reverse transcriptase. Using a *Hpa*I restriction enzyme digest of the total genomic DNA in conjunction with Southern blotting, Kan and Dozy showed that the normal β-globin gene was contained within a 7.6-kb *Hpa*I restriction fragment, whereas the β-globin gene of sickle-cell anemia, Hb^s, was contained within a 13-kb fragment (fig. 27.15). Further analysis showed that the RFLP resulted from a *Hpa*I restriction site 5 kb to the 3′ side of the β-

globin gene, that was present in the normal case and absent in Hb^s. Subsequent analysis showed sequence differences within the coding regions. One may wonder why the RFLP outside the coding region was so commonly associated with the abnormal gene. A possible explanation for this is that in the distant past a mutation occurred which resulted in the 7.6-kb type and a few 13-kb types before introduction of the sickle-cell gene mutation. After the Hb^s mutation was introduced into the 13-kb type, it became greatly expanded because of the selective advantages of this gene in heterozygotes. In this connection it should be noted that heterozygotes carrying one normal gene and one sickle-cell gene fare far better when infected by malaria. As a result in areas where malaria is prevalent the heterozygote has a selective advantage over the normal homozygote.

Knowledge of this and other polymorphisms has been used for pre- or postnatal diagnosis of the sickle-cell gene. Such information can be of great practical value in genetic counseling. Incidentally, it is now possible to diagnose sickle-cell disease (which occurs in individuals that are homozygous for the sickle-cell gene) with greater certainty because the point mutation leading to the defect in the coding region itself produces a recognizable RFLP.

The β-Globin cDNA Probe Was Used to Characterize the Normal β-Globin Gene

Detailed mapping with DNA probes was first successfully executed on the human β-globin gene. All members of a human genomic library that annealed to the radioactive cDNA probe for the β-globin gene were isolated, and each of these was sequenced. This analysis resulted in a complete description of the β-globin gene (see fig. 27.13). The β-globin gene is appreciably longer than the β-globin mRNA; in addition to containing regions that are present in the final mRNA the gene contains two ''intervening'' regions not represented in the mRNA sequences. We have more to say about the significance of these intervening sequences (IVS, also called introns) in the next chapter.

Chromosome Walking Permitted Identification and Isolation of the Regions around the Adult β-Globin Genes

The original cDNA probe carrying the β-globin mRNA sequences could only detect members in the genomic library that contained sequences homologous to those present in the probe. To explore the region flanking the β-globin gene the genomic library was probed further. For this purpose members of the genomic library that hybridized with the original cDNA were themselves converted into radioactive probes, and these were used to locate additional members of the library that contained sequences flanking the β-globin gene. A cyclic repetition of this process resulted in a gradual extension of sequence information in and around the β-globin gene. Using the library in this manner to extend the map is known as chromosome walking (fig. 27.16). A parallel approach was used to extend the map around the adult α-globin gene (see fig. 27.13). It can be seen that in both cases several genes occur in a cluster for each of the protein types. Most of these can be correlated with the genes that are expressed at different times during development. In addition, genes, called pseudogenes, occur that have strong sequence similarities to known genes but are never expressed. It is not clear if these pseudogenes have a function or simply represent evolutionary ''junk,'' which has not been removed.

Walking and Jumping Were Both Used to Map the Cystic Fibrosis Gene

By combining the linkage information obtained from RFLP mapping with other DNA manipulation techniques, it has been possible to locate genes causing serious genetic disorders even when these genes are only known from their inheritance patterns. The list of serious disorders that can be linked to single genetic loci is growing. It includes Huntington's disease, Duchenne's muscular dystrophy, polycystic kidney disease, cystic fibrosis, chronic granulomatous disease, peripheral neurofibromatosis, central neurofibromatosis, familial polyposis coli, and multiple endocrine neoplasia. One of the most spectacular achievements has been the determination of the gene causing cystic fibrosis (CF).

Cystic fibrosis is the most common serious genetic disorder in caucasian populations. The major clinical symptoms include chronic pulmonary disease, pancreatic exocrine insufficiency, and an increase in the concentration of sweat electrolyte. Bearers of this disease are readily diagnosed and often die of congestive lung complications before age 30. Pedigree analysis shows that, as in the case of sickle-cell anemia, a single gene inherited in autosomal recessive fashion results in the disease syndrome. The frequency of the disease is 1 in 2,000, from which it may be calculated that the carrier frequency is about 1 in 20 (the frequency of the heterozygote for a rare allele is twice the square root of the frequency of the homozygote). By classical genetic analysis the CF locus has been assigned to the long arm of chromosome 7 near the *met* locus. A map of the region containing the *met* locus and the CF gene is shown in figure 27.17.

In many genetic disorders, cystic fibrosis included, the analysis begins before the responsible gene and its protein

Figure 27.16

The linkage map of the human β-globin gene locus as shown by the structural analysis of overlapping λ genomic clones. Both $\lambda H\beta G1$ and $\lambda H\beta G3$ clones contained the entire β-globin gene. Other clones detected by "walking" led to the discovery of other β-globinlike genes. These included four genes that are expressed and two pseudogenes that are not expressed. The genomic segments of the clones isolated are shown together with the cleavage sites for the enzyme *Eco*R1. The numbers on the top line indicate the size of the fragments in kilobase pairs.

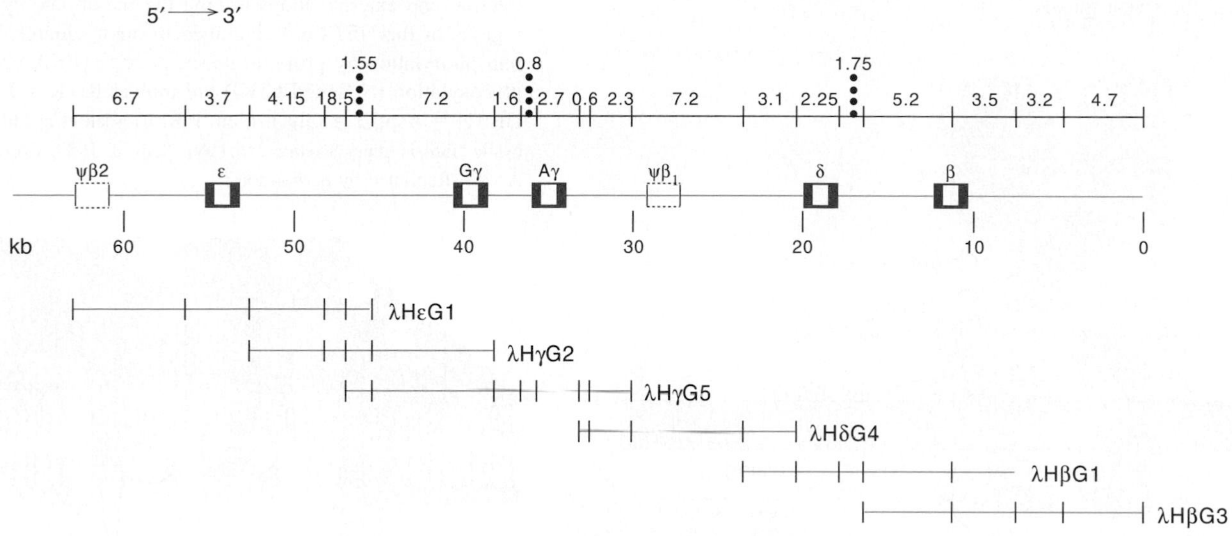

Figure 27.17

Map of restriction fragment length polymorphisms (RFLPs) closely linked to the cystic fibrosis (CF) gene. The inverted triangle near the right-hand end indicates the location of the ΔF_{508} mutation characteristic of most persons with cystic fibrosis disease. (Source: Adapted from B-S. Kerem, J. M. Rommens, J. A. Buchanan, D. Markiewicz, T. K. Cox, A. Chakravarti, M. Buchwald, and L-C. Tsui, Identification of the cystic fibrosis gene: Genetic analysis, *Science* 245:1075, 1989.)

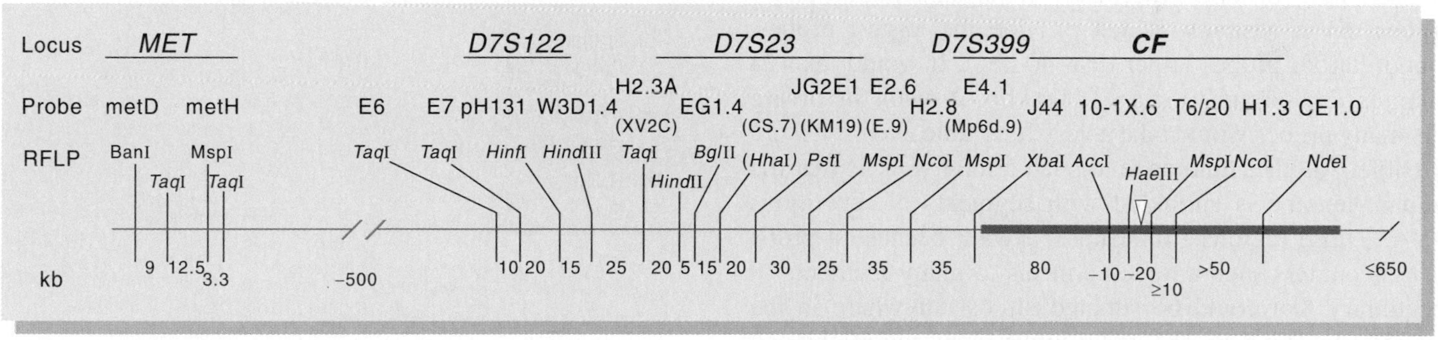

product are known. Thus, one cannot locate the gene directly, as in the case of the β-globin gene and then determine its approximate location in subsequent analysis. For genes such as those responsible for cystic fibrosis the approximate location is determined by conventional genetic analysis, and then one attempts to close in on the gene by a process called "reverse genetics." Conventional genetic mapping by recombination frequency is not practical with human genes below map distances of 1 centimorgan (cM)[c] because of the small number of test recombinant crosses that are ordinarily available for observation. Unfortunately, a centimorgan on the human genome is equivalent to a

[c] Genetic loci 1 cM apart recombine 1% of the time at meiosis.

Figure 27.18

Jumping and walking to find the cystic fibrosis gene. Following each jump, the locus defined by the jump was used as a starting point for a chromosome walk. Each DNA segment so found was hybridized to a sweat gland cDNA library until a match was found. It seemed likely that the sweat gland cDNA library would have a good representation of the cystic fibrosis gene transcript because the disease involves the sweat glands.

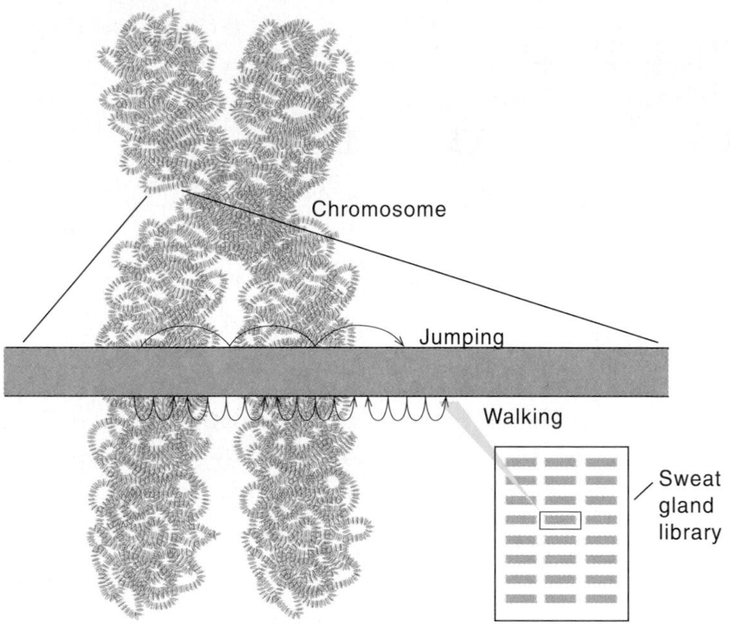

Figure 27.19

Predicted structure of the CFTR protein. Cylinders represent membrane-spanning helical segments. The cytoplasmically oriented NBFs are shown as blue spheres with slots to indicate the points of entry of nucleotides. R represents the large polar domain, which is linked to two halves of the protein molecule. Charged amino acids are shown as small circles with the charge sign. Net charges on the internal and external loops joining the membrane cylinders and on regions of the NBFs are contained in open squares. Potential sites for phosphorylation by protein kinases A or C (PKA or PKC) and N-glycosylation (N-linked CHO) are indicated. (K = Lys; R = Arg; H = His; D = Asp; E = Glu.) (From J. R. Riordan et al., Identification of the cystic fibrosis gene, *Science* 245:1066, Setp. 8, 1989. Copyright 1989 by the AAAS. Reprinted by permission.)

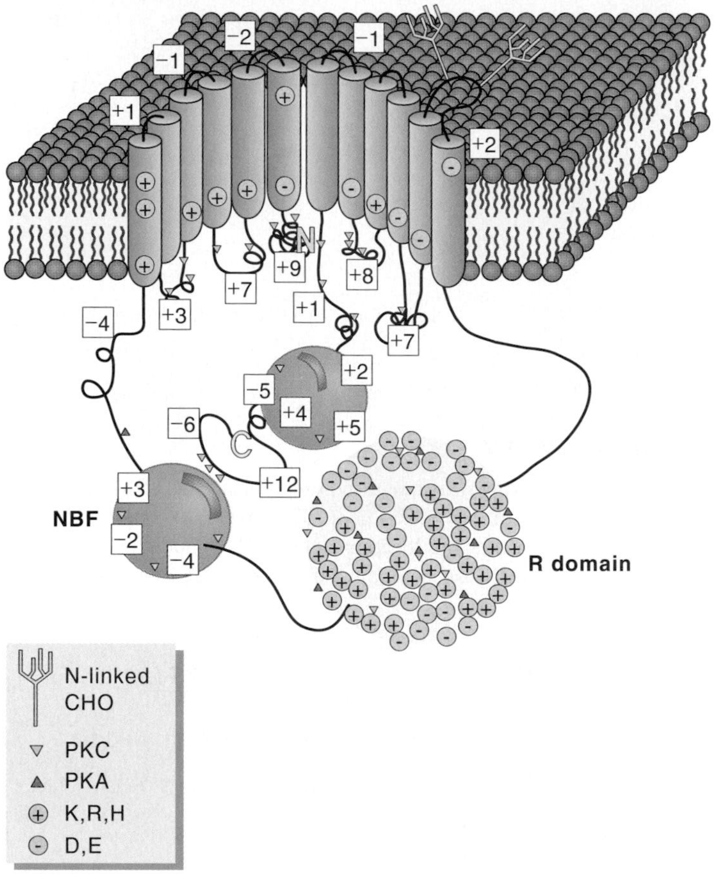

physical distance of about 1,000 kbp, so this presents the problem of ''walking along'' a chromosome for a million base pairs. Chromosome walking procedures are not suited to such long distances because of the size limitation of probes. Even cosmids, which provide the largest probes, cannot harbor probes larger than 40 kbp′. It would take 25 cosmids end-to-end to span 1,000 kbp. A walk involving this many probes would take a very long time even if it were possible. Another hindrance to such a long walk is that the human genome is sprinkled with segments of repetitious DNA. Those regions ''interrupt'' a walk because a probe that encounters such a region anneals to many members of the library that could be situated almost anywhere in the genome. Walking breaks down under such conditions and would benefit by an alternative procedure. To traverse long distances of the genome and to skip troublesome regions of repetitious sequences, a procedure called chromosome ''jumping'' was devised. Probes for jumping carry small segments from the same chromosome, which are ordinarily about 500 kb apart. The precise distance is variable and a function of how the jumping probes are made. We do not go into the complexities here of constructing such a library except to say that there are several ways in which this can be done.

Usually a jumping library and a walking library are prepared from the same genome and used in conjunction with each other. The region around each locus recognized by the jumping library is scrutinized by the walking library as shown in figure 27.18.

This complex analysis would be in vain if one did not have some criterion for knowing when the goal of finding the cystic fibrosis gene was reached. This is where clues from the physiological nature of the condition became useful. In a brilliant strategy a cDNA library was prepared from the mRNA fraction of sweat gland tissue. Recall that in cystic fibrosis sweat glands malfunction; therefore, the mRNA for the CF gene might be well represented in the mRNA fraction of the sweat gland cells.

While the walking and jumping process was in progress each new segment mapping in the general region of interest was tested against the sweat gland cDNA library. Finally, a member of the walking library was found that annealed with a member of the sweat gland cDNA library. Was this match fortuitous or did it mean the CF gene had been found? To answer this question the cDNA discovered in this way was used to probe the genomic library. By this means a gene was mapped that extended over a region of about 250 kb with 23 introns. A unique transcript, approximately 6,500 nucleotides in length was detected in extracts of sweat gland tissue which matched the transcript size expected from this gene. The protein predicted from a sequence analysis of this transcript consists of two similar motifs, each with (1) a domain having properties consistent with membrane association and (2) a domain believed to be involved in ATP binding (fig. 27.19). Finally it was discovered that many CF patients carry a three base deletion in this transcript which should result in the loss of a phenylalanine residue from the protein. This defect correlates with the notion that CF patients have a faulty membrane protein that leads to the secretion problems characteristic of the disease. The fact that the abnormal gene is located as close as one can tell by classical genetics to the CF locus adds additional support to the notion that the CF gene has been found.

Finding the disease gene does not, of course, mean that a cure is in the offing. However, the characterization of the disease gene will be a tremendous aid in diagnosing carriers and fetuses that are homozygous for the disease gene. It also should be a help in focusing approaches to finding a cure for the disease.

Summary

1. Sequencing DNA uses chemical methods to cleave specific bases in preexisting DNA or carries out the synthesis under conditions where synthesis is interrupted at specific bases.
2. A specific segment of DNA can be synthesized *in vitro* by the polymerase chain reaction. Short segments of DNA bordering the segment of interest are added to a mixture containing the segment of interest, a DNA polymerase, and the deoxyribotriphosphate substrates. The DNAs are first denatured, then annealed, and then synthesized. This cycle is repeated 20 or more times by raising the temperature to stop synthesis and lowering the temperature for annealing and synthesis. The outcome is a mixture in which the vast majority of the DNA is newly synthesized DNA bounded by the sequences of the added primers.
3. Another procedure for amplification cuts DNA containing the segment of interest into small pieces with a restriction enzyme. The cut pieces are incorporated into a plasmid or virus "vector" to be amplified in a suitable host. After growth, the mixture is plated to produce a mixture of bacterial or viral clones. The clone or clones of interest are identified often by hybridization of the clones after replica plating with a radioactive probe, followed by autoradiography to find the clone of interest.
4. Most cloning has been done in *E. coli.* Yeast is the most used eukaryotic host. Cloning is also possible in a number of plant and animal cells.
5. Mapping with recombinant DNA probes was first applied to the human globin genes. Starting probes were obtained by isolating the globin messenger from reticulocytes and converting it into a cDNA, which was used to scan a human genomic library for cross hybridizing members. Once detected and purified, these cross hybridizing members carrying globin messenger sequences were themselves converted to radioactive probes and used to further scan the genomic library for nearby sequences. By repeating this cycle several times, a process known as chromosome walking revealed a region around the adult hemoglobin gene that contained several closely related genes associated with hemoglobin.
6. Frequently, alleles of the same gene can be distinguished by restriction site differences in the genes themselves or in nearby locations. Alleles identified in this way are said to show restriction fragment length polymorphism. The allele responsible for sickle-cell disease was identified in this way.
7. The cystic fibrosis gene has been mapped by chromosome walking and jumping, a newer approach in which the relevant probes contain segments of the

genome that are normally located about 500 kbp from one another. A cDNA library was made from normal sweat gland tissue, chosen because of the disease's association with abnormal release of sweat salt suggested that the sweat gland would contain an abundance of the messenger associated with the gene. By hybridizing the genomic DNA probes with the cDNA sweat gland library, a segment of genome was identified as a candidate for the cystic fibrosis gene. This gene was characterized in detail and found to encode a complex transmembrane protein that carries a specific amino acid change in over half of the persons with cystic fibrosis. This correlation is overwhelming support that the gene responsible for cystic fibrosis has been mapped and characterized.

Selected Readings

Caruthers, M. H., A. D. Barone, S. L. Beaucage, D. R. Dodds, E. F. Fisher, L. J. McBride, M. Matteucci, Z. Stabinsky, and J. Y. Tang, Chemical synthesis of deoxyoligonucleotides. *Methods Enzymol.* 154:287–313, 1987.

Cohen, S. N., A. Change, H. Boyer, and R. Helling, Construction of biologically functional bacterial plasmids *in vitro*. *Proc. Natl. Acad. Sci. USA* 70:3240–3244, 1973.

Gusella, J. F., DNA polymorphism and human disease. *Ann. Rev. Biochem.* 55:831–854, 1986.

Hunkapiller, T., R. J. Kaiser, B. F. Koop, and L. Hood, Large-scale and automated DNA sequence determination. *Science* 254:59–67, 1991. State of the art on the mammoth project to sequence the human genome.

Jackson, D. A., R. H. Symons, and P. Berg, Biochemical method for inserting new genetic information into DNA of Simian Virus 40: Circular SV40 DNA molecules containing lambda phage genes and the galactose operon of *E. coli. Proc. Natl. Acad. Sci. USA* 69:2904–2909, 1972.

Kerem, B., J. M. Rommens, J. A. Buchanan, D. Markiewicz, T. K. Cox, A. Chakravarti, M. Buchwald, and L. C. Tsui, Identification of the cystic fibrosis gene: Genetic analysis. *Science* 245:1073–1079, 1989.

Mansour, S. L., K. R. Thomas, and M. R. Capecchi, Disruption of the proto-oncogene *int-2* in mouse embryo-derived stem cells: A general strategy for targeting mutations to non-selectable genes. *Nature* 336:348–352, 1988.

Maxam, A. M., and W. Gilbert, A new method of sequencing DNA. *Proc. Natl. Acad. Sci. USA* 74:560–564, 1977.

Mullis, K. B., The unusual origin of the polymerase chain reaction. *Sci. Am.* 262:56–65, 1990.

Riordan, J. R., J. M. Rommens, B. S. Kerem, N. Alon, R. Rozmahel, Z. Grezelczak, J. Zielenski, S. Lok, N. Plasvsic, J.-L. Chou, M. T. Drumm, M. C. Iannuzzi, F. S. Collins, and L-C. Tsui, Identification of the cystic fibrosis gene: Cloning and characterization of complementary DNA. *Science* 245:1066–1073, 1989.

Sambrook, J., E. F. Fritsch, and T. Maniatis, *Molecular Cloning: A Laboratory Manual,* 2d ed., Cold Spring Harbor Laboratory Press, Cold Spring Harbor, N.Y, 1989. A three-volume collection that is thorough.

Sanger, F., Sequences, sequences, and sequences. *Ann. Rev. Biochem.* 57:1–28, 1988. A scientific memoir.

Sanger, F., and A. R. Coulson, A rapid method for determining sequences in DNA by primed synthesis with DNA polymerase. *J. Mol. Biol.* 94:444–448, 1975.

Watson, J. D., M. Gilman, J. Witkowski, and M. Zoller, *Recombinant DNA,* 2d ed. Scientific American Books. New York: W. H. Freeman Company, 1992. This text is an excellent elementary text on the subject of recombinant DNA. It contains many exciting chapters on specific applications and is extremely well referenced.

Problems

1. Read the rest of the sequence in the autoradiogram in figure 27.1d as far as possible.
2. What are the major advantages of the polymerase chain reaction (PCR) method for amplifying defined segments of DNA as opposed to the use of conventional cloning methods? How might the PCR method be used to test for infection with the AIDS virus and how would this be an improvement over the antibody test currently used? (The current ELISA test is an indirect test for the presence of antibodies against the HIV proteins.)
3. Calculate the frequency of occurrence of restriction sites for PstI and HindIII in the DNA from a thermophile (80% G + C) and from *E. coli* (52% G + C).

4. You just isolated a novel recombinant clone and purified the desired insert (a 10,000 bp linear duplex DNA) from the vector. Now you wish to map the recognition sequences for restriction endonucleases A and B. You cleave the DNA with these enzymes and fractionate the digestion products according to size by agarose gel electrophoresis. Comparison of the pattern of DNA fragments with marker DNAs of known sizes yields the following results:
   (a) Digestion with A alone gives two fragments, of lengths 3,000 and 7,000 bp.
   (b) Digestion with B alone generates three fragments, of lengths 500, 1,000, and 8,500 bp.
   (c) Digestion with A and B together gives four fragments, of lengths 500, 1,000, 2,000 and 6,500 bp.

   Draw a restriction map of the insert, showing the relative positions of the cleavage sites with respect to one another.
5. Draw the ends of a DNA fragment digested with the restriction endonuclease *Bam*HI. How do these ends differ from those generated by *Mbo*I? If *Mbo*I and *Bam*HI ends were to be ligated together, would the resulting junction be cleavable by *Bam*HI or *Mbo*I?
6. Describe a procedure for cloning a DNA fragment into the *Bam*HI site of pBR322.
7. How large a genomic library should you construct in order to detect and isolate a 15-kb gene out of a genome containing 3×10^9 bp?
8. If you were interested in isolating a cDNA for human serum albumin, why would you use a cDNA library established from mRNA isolated from liver? If you wanted to isolate the gene for albumin, why would you use a genomic library established from any human tissue?
9. Which of the *E. coli* vectors on the left (a, b, c) would be used to achieve the cloning objectives on the right (1–5)?

(a) Plasmid	(1) Genomic library
(b) Cosmid	(2) DNA sequencing
(c) Lambda	(3) cDNA library
	(4) Small inserts
	(5) Genomic walking

10. Site-directed mutagenesis is one of the most powerful tools available to the biochemist. What are some of the applications of this technique? How can the PCR method be used to do site-directed mutagenesis, and what is the advantage of this method?
11. The Southern blot technique is often used to compare genes from different organisms. For example, one could use the human globin gene probe described in the text to determine the extent of homology between globin genes from different primates. How could one reduce the stringency of the hybridization conditions (step 4 of fig. 27.14) to permit such a ''heterologous hybridization''?
12. An unusual feature of the sickle-cell variant of the β-globin gene is that it directly alters a cleavage site for restriction endonuclease *Mst*II. *Mst*II recognizes the sequence CCTGAGG, which is mutated to CCTGTGG in the sickle-cell gene. How would you use this information and the Southern blot method to analyze fetal cells in amniotic fluid to determine whether the fetus carries sickle-cell anemia? What problems might you encounter in using this method?
13. Describe the procedure called ''chromosome jumping.'' How was this procedure used to map the cystic fibrosis gene?
14. Describe a procedure using the PCR technique that could be used to determine whether a normal individual is a carrier of the cystic fibrosis ΔF_{508} mutation. What problems could you anticipate with this method?

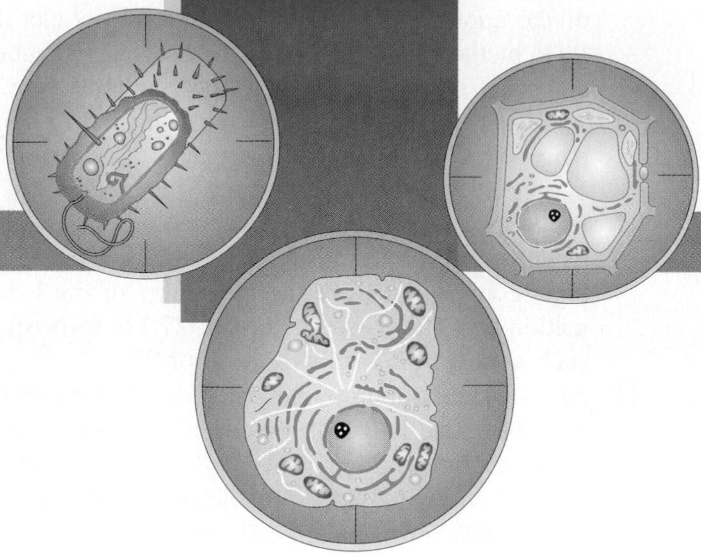

CHAPTER

28

RNA Synthesis and Processing

Chapter Outline

Single stranded RNA molecules are synthesized by a template mechanism from select regions of the DNA genome.

In prokaryotes DNA, RNA, and protein synthesis all take place in the same cellular compartment. In eukaryotes the DNA is compartmentalized in the cell nucleus, and it became clear long before the biochemistry of these three processes was understood that DNA synthesis takes place in the nucleus, whereas the bulk of protein synthesis takes place in the cytoplasm. From these observations on eukaryotes it was self-evident that DNA cannot be directly involved in the synthesis of protein but must somehow transmit its genetic information for protein synthesis to the cytoplasm. Careful experiments with radioactive labels were used to demonstrate that RNA synthesis takes place in the nucleus; much of this RNA is degraded rather quickly, but the portion that survives is mostly transferred to the cytoplasm (fig. 28.1). From observations of this kind it became clear that RNA was the prime candidate for the carrier of genetic information for the synthesis of proteins.

In this chapter we focus on the structure and metabolism of the major classes of cellular RNA.

Figure 28.1

In eukaryotic cells the nuclear membrane separates the processes of RNA and protein synthesis. This can be demonstrated with radioactive substrates that are precursors of RNA and protein. Immediately after exposure of cells to labeled precursors, the RNA label becomes fixed in the nucleus, and the protein label becomes fixed in the cytoplasm. Eventually most of the labeled RNA becomes transferred to the cytoplasm, and a fraction of the labeled protein becomes transferred to the nucleus.

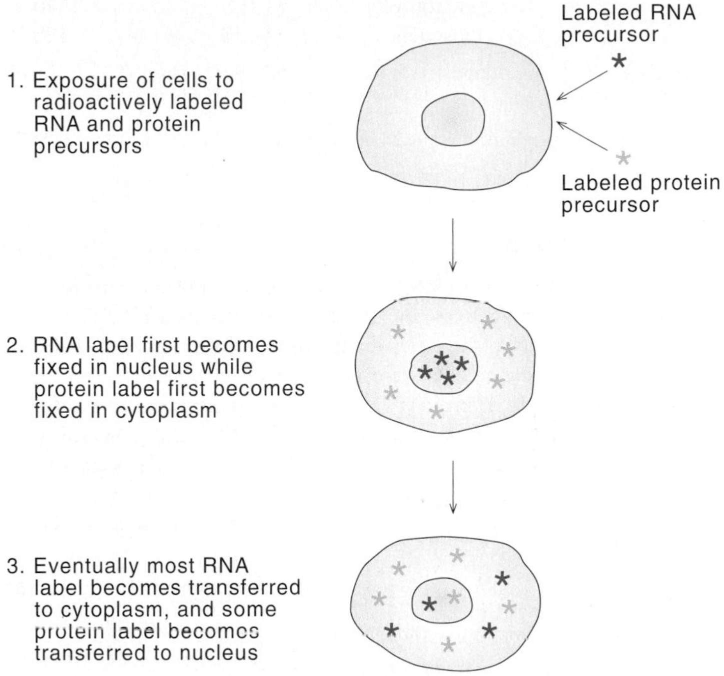
Labeled RNA precursor
1. Exposure of cells to radioactively labeled RNA and protein precursors
Labeled protein precursor
2. RNA label first becomes fixed in nucleus while protein label first becomes fixed in cytoplasm
3. Eventually most RNA label becomes transferred to cytoplasm, and some protein label becomes transferred to nucleus

The First RNA Polymerase to Be Discovered Did Not Require a DNA Template

The first enzyme discovered that could catalyze polynucleotide synthesis was a bacterial enzyme called polynucleotide phosphorylase. This enzyme, isolated by Severo Ochoa and Marianne Grunberg-Manago in 1955, could make long chains of 5′-3′-linked polyribonucleotides starting from nucleoside diphosphates. However, there was no template requirement for this synthesis, and the sequence was uncontrollable except in a crude way by adjusting the relative concentrations of different nucleotides in the starting materials.

Shortly after Ochoa's studies, Sam Weiss began a search for a DNA-directed RNA polymerase. His experimental design was influenced by the theory that RNA must be made on a DNA template if it is to carry the genetic message. He was also influenced by Kornberg's discovery that DNA synthesis required nucleoside triphosphates for substrates rather than nucleoside diphosphates. With crude liver extracts, Weiss was able to demonstrate a capacity for RNA synthesis that was severely inhibited by DNase. Weiss's results touched off systematic investigations of RNA metabolism in many laboratories. A continuous expansion of research effort from that time on has yielded a wealth of understanding about the transcription process and related aspects of RNA metabolism.

DNA–RNA Hybrid Duplexes Suggest That RNA Carries the DNA Sequences

Sol Spiegelman reasoned that if RNA was made on a DNA template, it should be complementary to one of the DNA chains. In this case it should be possible to make a DNA–RNA duplex by annealing, as was done for complementary DNA chains (see chapter 26). Using ^{3}H-labeled T2 bacteriophage DNA and ^{32}P-labeled RNA that was synthesized after T2 infection of *E. coli* cells, he was able to show that the newly synthesized RNA forms a specific complex with the viral DNA (fig. 28.2). It was presumed that the specific complex must involve Watson-Crick-like base pairing between an RNA and a DNA chain, with uracil playing the role of thymine in the RNA. Complementary interaction was a strong indication that the RNA was a product of DNA-directed synthesis. Although most RNAs are synthesized as a result of transcription from a DNA template, different strategies are used in their synthesis and in their posttranscriptional modification. The strategy used in a given case is strongly related to the function of

Table 28.1

Types of RNA in *Escherichia coli*

Type	Function	Number of Different Kinds	Number of Nucleotides	Percent of Synthesis	Percent of Total RNA in Cell	Stability
mRNA	Messenger	Thousands	500–6000	40–50	3	Unstable ($t_{1/2}$ = 1–3 min)
rRNA	Structure and function of ribosomes	3 {23S, 16S, 5S}	2800 1540 120	50	90	Stable
tRNA	Adapter	50–60	75–90	3	7	Stable

(Source: G. Zubay, *Biochemistry,* 2d ed., Macmillan, New York, 1988, p. 893.)

Figure 28.2

Demonstration that phage RNA has sequences complementary to phage DNA. S. Spiegelman and B. Hall used ^{3}H-labeled T2 DNA and ^{32}P-labeled RNA, the latter made after T2 infection of *E. coli* cells. The two nucleic acids were mixed together, annealed, and then centrifuged to equilibrium in a CsCl density gradient. In (*a*) the DNA was first heat-denatured and then mixed with RNA and annealed at 65°C prior to centrifugation. In (*b*) the DNA denaturation step was left out. Most of the RNA goes to the bottom of the centrifuge tube because of its high density. The DNA bands about one-third of the way from the top. In (*a*) some RNA also bands at approximately the same location as the DNA, but in (*b*) this is not the case. The comigration of a fraction of the RNA with the DNA is believed to be due to the formation of a specific DNA–RNA hybrid duplex. The RNA is much smaller than the DNA, so the RNA in the hybrid duplex migrates at the density of the DNA. In (*b*) no hybrid duplex forms because the DNA was not denatured before carrying out the annealing process.

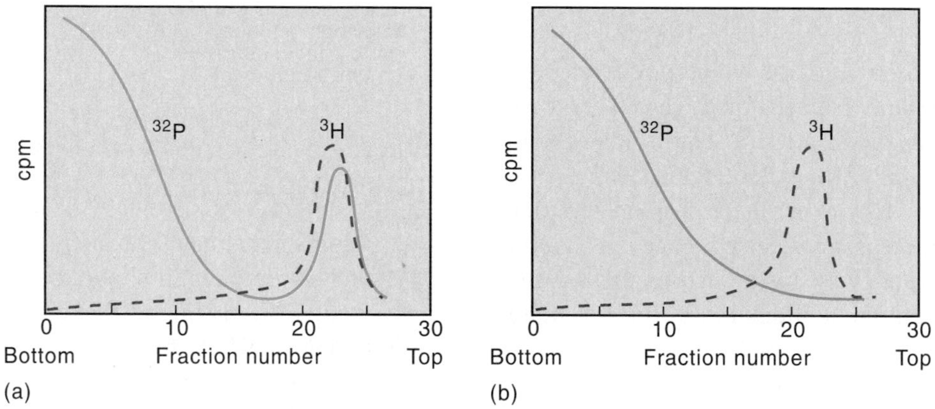

the RNA. For this reason, it is important that we discuss the different classes of RNA found in the cell before we consider their mode of synthesis or postsynthesis modification.

There Are Three Major Classes of RNA

There are three major types of RNA that are transcribed from the DNA template: Messenger RNA (mRNA), ribosomal RNA (rRNA), and transfer RNA (tRNA). These three RNAs work together in protein synthesis. Some of the properties of the three RNAs as they are found in *E. coli* are summarized in table 28.1.

Messenger RNA Carries the Information for Polypeptide Synthesis

Messenger RNA carries the message from the DNA for the synthesis of a polypeptide chain with a specific sequence of amino acids. Each protein is unique, resulting in an enormous variety of mRNAs. It should be noted (see table 28.1) that although 40%–50% of the RNA synthesized in *E. coli*

Figure 28.3

The tertiary structure of yeast phenylalanine tRNA. (*a*) The full tertiary structure. Purines are shown as rectangular slabs, pyrimidines as square slabs, and hydrogen bonds as lines between slabs. (Source: From G. J. Quigley and A. Rich, Structural domains of transfer RNA molecules, *Science,* 194:796, 1976.) (*b*) Nucleotide sequence. Residues that appear in most of the yeast tRNAs and residues that appear to be constantly a purine or a pyrimidine are indicated. Residues involved in tertiary base pairing are shown connected by solid lines. Several of the nucleotides are methylated. These are indicated by a small m. (Illustration copyright by Irving Geis. Reprinted by permission.)

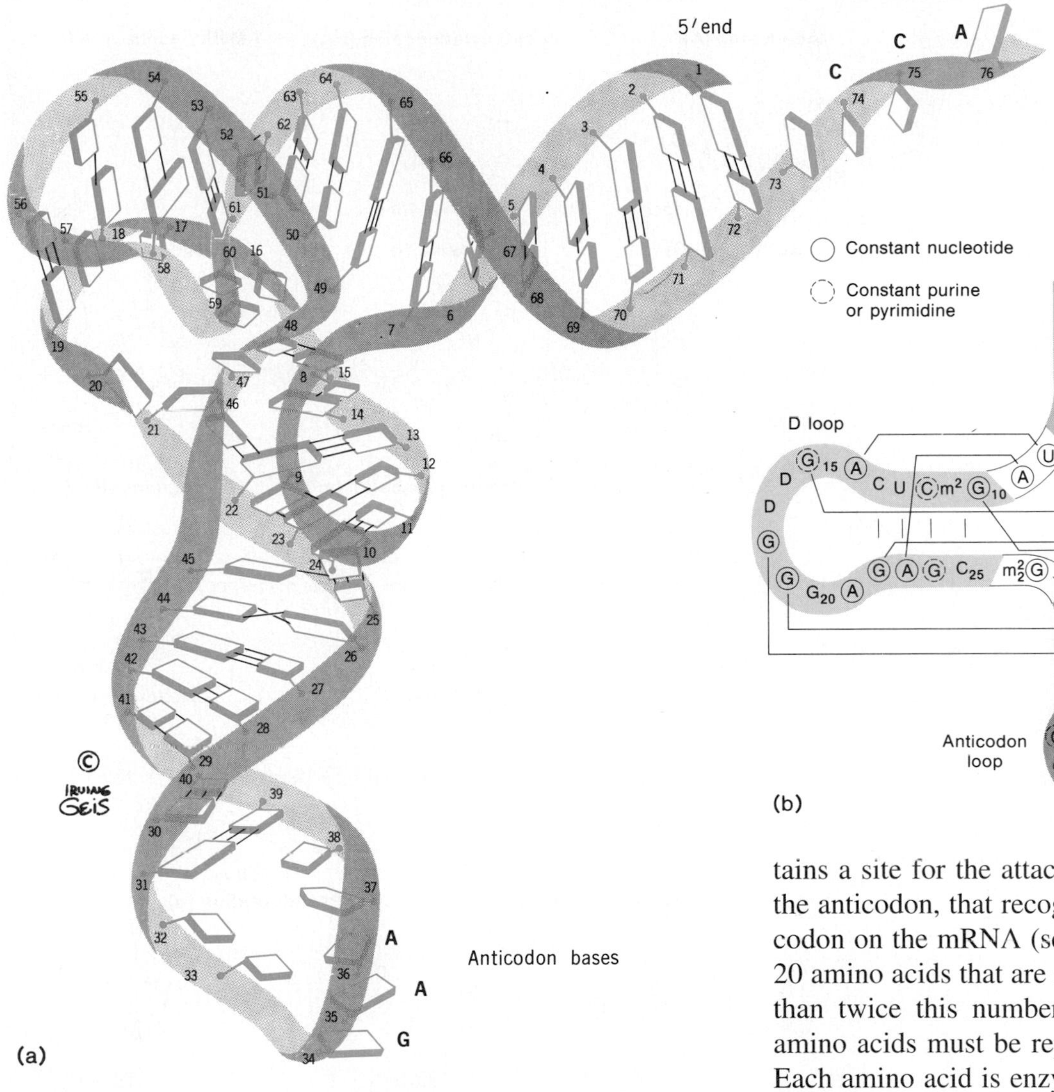

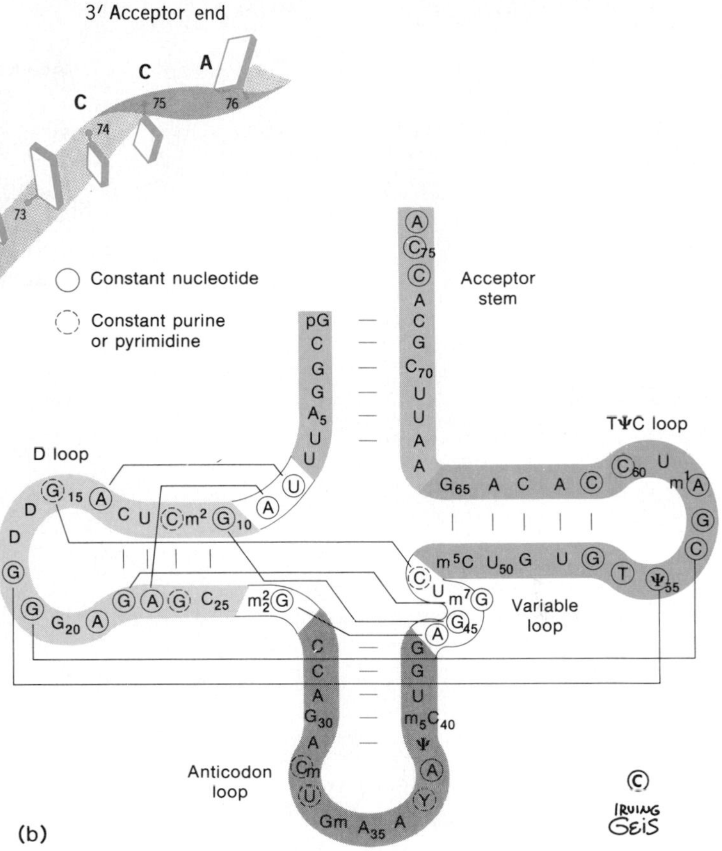

is mRNA, at any given time mRNA only accounts for about 3% of the cellular RNA. This is because bacterial mRNA is very unstable, with an average half-life of only 1–3 min. Messenger RNAs are heterogenous in both size and sequence, reflecting the fact that each mRNA encodes the information for the synthesis of a different protein(s).

Transfer RNA Carries Amino Acids to the Template for Protein Synthesis

The function of transfer RNA is to carry amino acids to the mRNA template, where the amino acids are linked into a specific order. For this purpose each tRNA molecule contains a site for the attachment of an amino acid and a site, the anticodon, that recognizes the corresponding three-base codon on the mRNA (see chapter 29). Since there are only 20 amino acids that are incorporated into proteins and more than twice this number of tRNAs in most cells, several amino acids must be represented by more than one tRNA. Each amino acid is enzymatically attached to the 3′ end of one or more tRNA by a specific aminoacyl-tRNA synthase that recognizes both the amino acid and the tRNA.

The primary structure is known for all *E. coli* tRNAs, and the three-dimensional structures of some of them have been determined by x-ray crystallography. The three-dimensional structure of phenylalanine tRNA of yeast is shown in figure 28.3. All tRNAs show similarities in their folded structures with four loops. Except for the variable loop, the loops are usually the same size in different tRNAs. The $tRNA^{Phe}$ molecule contains 20 bp that are hydrogen-bonded in Watson-Crick fashion and an additional 40 or so hydrogen bonds formed in other ways. These additional hydrogen bonds and the accompanying base stacking stabi-

Figure 28.4

The structures of some modified nucleosides found in tRNA. The parent ribonucleosides are shown on the left in yellow screens. The other bases found in RNA result from post-transcriptional modification.

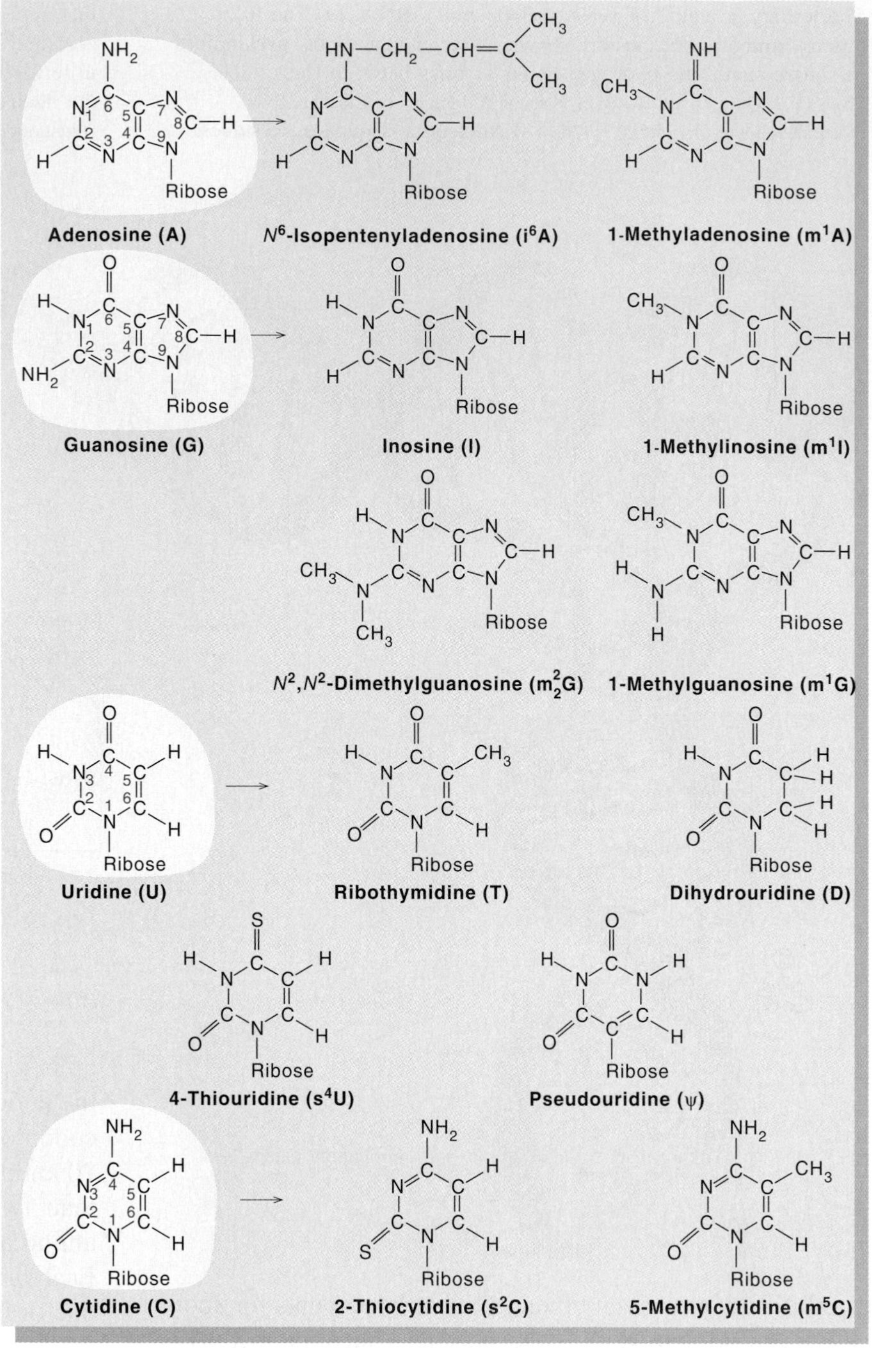

lize the $tRNA^{Phe}$ in the complex folded structure shown in figure 28.3*a*.

All tRNAs contain several ribonucleotides that differ from the usual four (12 in the case of $tRNA^{Phe}$). The structures for some of these are shown in figure 28.4. Only four ribonucleotides are incorporated into RNA in the transcription process. All of the rare bases found in the mature tRNA result from posttranscriptional modification.

The complex folded structures adopted by tRNAs illustrates the fact that nucleic acids with a properly adjusted primary sequence can adopt complex secondary and tertiary structures. Apropos of this, Francis Crick once said that transfer RNA is an RNA molecule trying to look like a protein.

Figure 28.5

Composition of the *E. coli* ribosomes. The 70S ribosome can dissociate into a 50S and a 30S subunit. *In vitro* this can be done by lowering the Mg ion concentration. The individual subunits can be dissociated into their constituent RNAs and proteins by exposure to urea denaturant. Molecular weights are given for the subunits and the proteins, and the numbers of nucleotides are given for the RNAs.

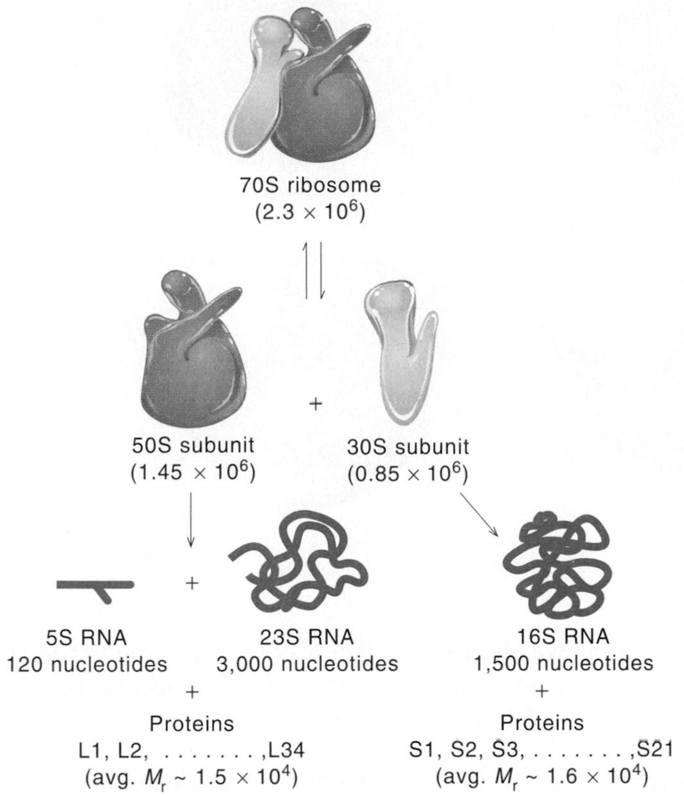

Ribosomal RNA Is an Integral Part of the Ribosome

The bulk of the cellular RNA is ribosomal RNA. Although seven genes exist in *E. coli* for rRNA, they all lead to essentially the same three ribosomal RNA molecules (see table 28.1) which differ substantially in size. The three rRNAs are always found in a complex with proteins in a functional component known as the ribosome. The ribosome is the site where mRNA and tRNAs meet to engage in protein synthesis. In *E. coli,* ribosomes are referred to as 70S particles, a measure of their rate of sedimentation and hence their size (S refers to Svedberg units, which are defined in chapter 6). A 70S ribosome consists of two dissociable subunits: A 50S subunit and a 30S subunit. Each of these contains both RNA and protein. The 50S subunit contains 23S and 5S rRNAs. The 30S subunit contains a single 16S rRNA (fig. 28.5). Eukaryotic ribosomes are similar in structure, although they are somewhat larger (80S) and contain mostly larger rRNAs (25–28S, 18S, 5S, and an additional 5.5–5.8S; this additional rRNA corresponds in sequence to the first 150 nucleotides of the prokaryotic large rRNA subunit). Chloroplasts and mitochondria have ribosomes and rRNA that are distinctly different from those present in the cytoplasm and strongly resemble those of prokaryotes, testifying to their evolutionary origin from bacteria.

The Fine Structure of the Ribosome Is Beginning to Emerge

Results from many different experimental approaches are beginning to coalesce to produce a three-dimensional picture of the ribosome that includes the location of its individual structural components and functional sites. The development of this picture has been especially challenging because the ribosome is large, fragile, and structurally complex and has resisted efforts to produce crystals that are capable of giving high-resolution structural information.

The current view of the overall morphology of ribosomes is based largely on electron-micrographic studies of the subunits of *E. coli.* From this analysis it appears that both subunits are asymmetrical (fig. 28.6).

The relative location of individual ribosomal proteins within the two subunits has been examined in two ways. One method involves determining which ribosomal proteins can be chemically cross-linked to each other and has yielded an elaborate grid of spatial relationships based on the frequency of cross-linking. The other method relies on neutron diffraction whereby the individually deuterated ribosomal proteins are located within the ribosomal subunit. The two methods of determining the location of ribosomal proteins have yielded a consistent spatial picture.

Recently, information concerning protein locations within the 30S subunit has been combined with a secondary structure model of 16S rRNA and the location of protein binding sites in rRNA to generate a partial three-dimensional picture of where the rRNA and proteins are situated in this ribosomal subunit (fig. 28.7).

Overview of the Transcription Process

All DNA-dependent RNA polymerases carry out the following reaction:

$$\text{NTP} + (\text{NMP})_n \xrightarrow[\text{DNA}]{\text{Mg}^{2+}} (\text{NMP})_{n+1} + \text{PP}_i$$

The subsequent breakdown of PP_i ensures the irreversibility of this reaction, as is the case in DNA polymerization reactions; this helps explain why both DNA and RNA polymer-

Figure 28.6

Gross shapes of the *E. coli* ribosomal subunits and the ribosome.

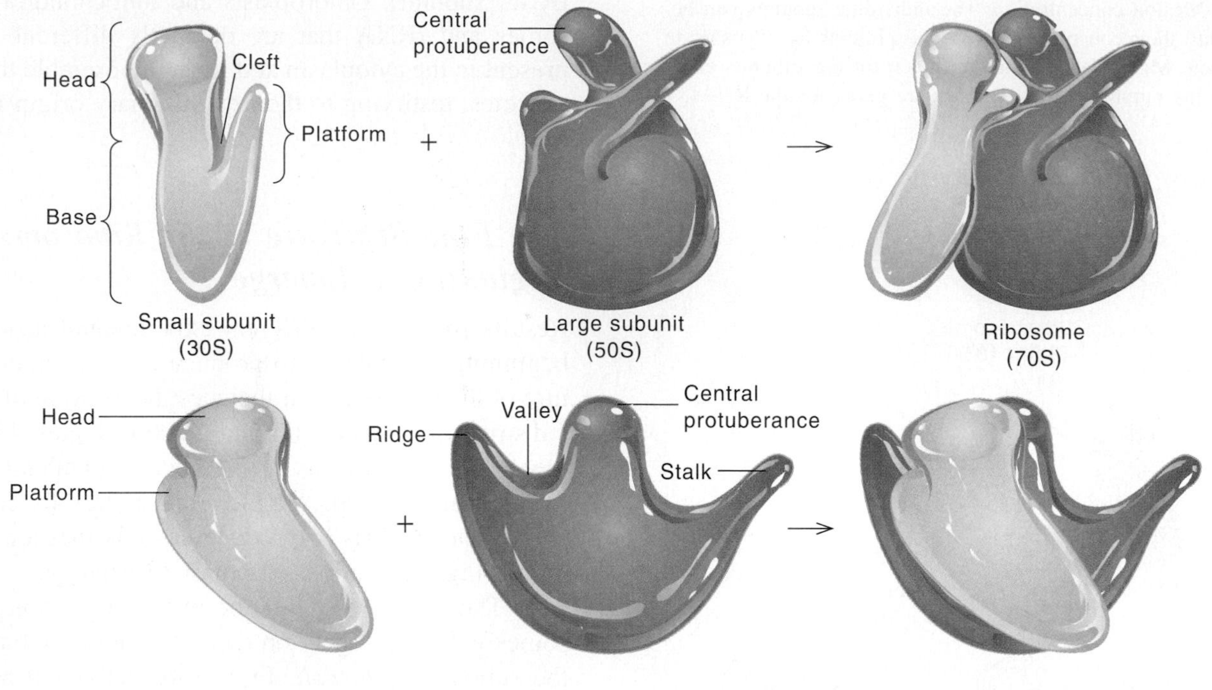

ase utilize NTPs rather than NDPs. The DNA template strand determines which base is added to the growing RNA molecule. For example, a cytosine in the template strand of DNA means that a complementary guanine is incorporated at the corresponding location of the RNA. Synthesis proceeds in a $5' \rightarrow 3'$ direction, with each new nucleotide being added onto the 3′—OH end of the growing RNA chain.

The overall process for RNA synthesis on a duplex DNA template can be conceptually divided into initiation, elongation, and termination (fig. 28.8). In the initiation phase of the reaction, the RNA polymerase binds at a specific site on the DNA called the promoter. Here it unwinds and unpairs a small region of the DNA. Elongation begins by the base pairing of ribonucleotide triphosphates to one strand of the DNA, followed by the stepwise formation of covalent bonds from one base to the next. As elongation proceeds the DNA unwinds progressively in the direction of synthesis. The short stretch of DNA–RNA hybrid formed during synthesis is prevented from becoming longer than 10–20 bp by a rewinding of the DNA and the simultaneous displacement of the newly formed RNA. Termination occurs at a sequence recognized by the RNA polymerase as a stop signal. At this point the ternary complex of DNA, RNA, and polymerase breaks up.

First we discuss the process of transcription in bacteria and then in eukaryotes. In both cases we consider the properties of the RNA polymerase(s) first.

Bacterial RNA Polymerase Contains Five Subunits

Escherichia coli contains one RNA polymerase, which transcribes all three major types of RNA. The active enzyme is a pentamer containing four different polypeptide chains with a total molecular weight of about 500,000. The subunits of the enzyme can be separated by electrophoresis on polyacrylamide gels. The four different polypeptide chains, termed β', β, σ^{70}, and α, have molecular weights of 155,000, 151,000, 70,000, and 36,500, respectively. Additional σ-like proteins have been identified, which we discuss later.

A complex with the subunit structure $\alpha_2\beta\beta'\sigma^{70}$ can carry out the functions necessary for synthesis of RNA and is referred to as the holoenzyme. Holoenzyme can be reversibly separated into two components by chromatography on a phosphocellulose column.

$$\underset{\text{Holoenzyme}}{\alpha_2\beta\beta'\sigma^{70}} \rightleftharpoons \underset{\text{Core polymerase}}{\alpha_2\beta\beta'} + \underset{\text{Sigma-70}}{\sigma^{70}}$$

Figure 28.7

Arrangement of components in the *E. coli* 30S ribosomal particle. In (*a*) the relative locations of the ribosomal proteins, numbered 1–21, are shown. (Illustration prepared by Dr. Malcolm Capel from data described in M. S. Capel, M. Kjeldguard, D. M. Engelman, *J. Mol. Biol.* 200:66–87, 1988.) In (*b*) the conformation of the rRNA is shown. The location of the proteins is indicated by numbers given in the figure. (Illustration prepared by S. Stern, B. Weiser, and H. F. Noller from data described in *J. Mol. Biol.* 204:447–481, 1988.)

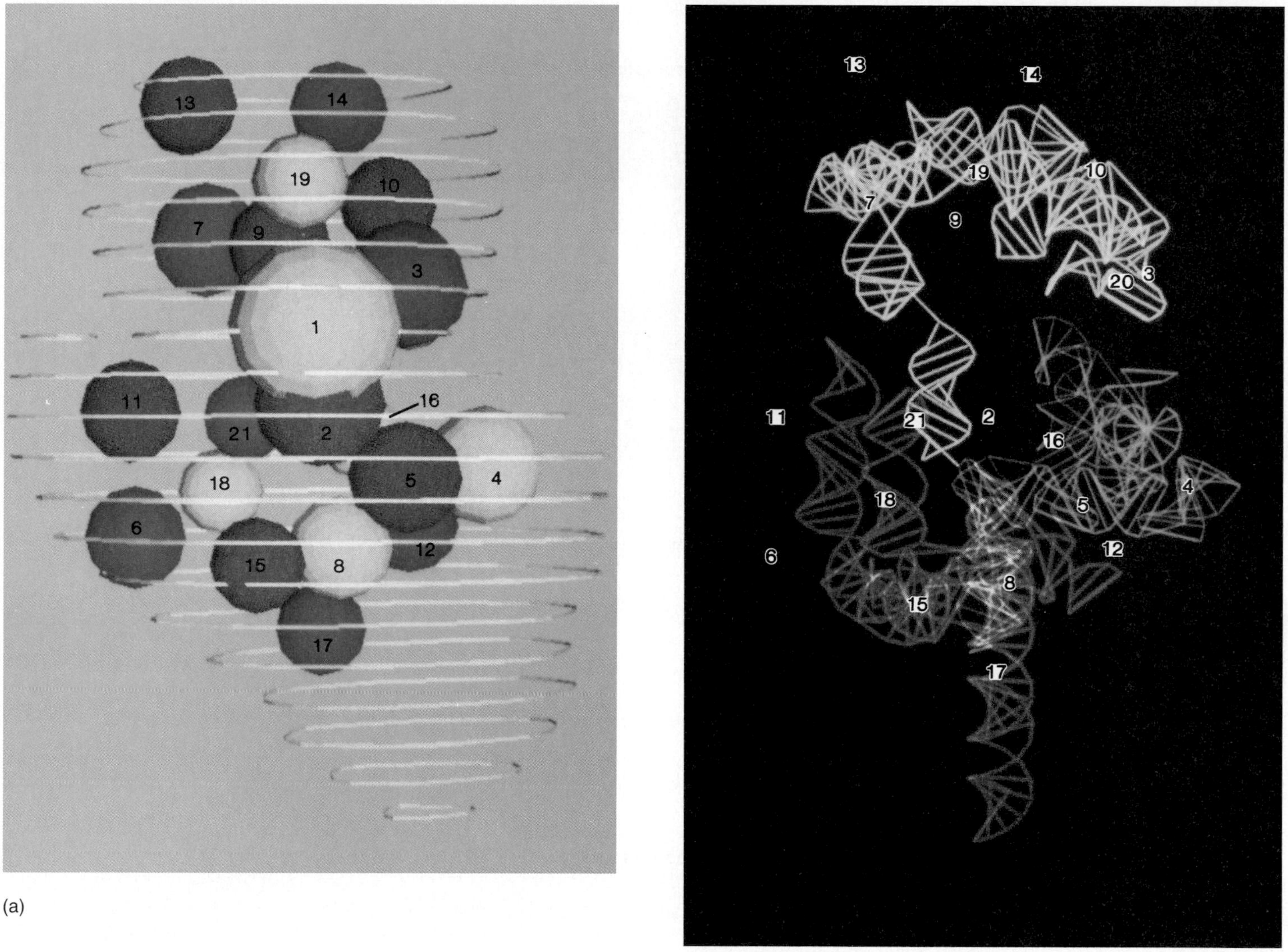

(a)

(b)

The enzyme without the sigma factor, called core polymerase, retains the capability to synthesize RNA, but it is defective in the ability to bind and initiate transcription at true initiation sites on the DNA. In fact when RNA polymerase was first purified from crude extracts it was missing the σ factor. The assay for polymerase involved the use of DNA with single-strand nicks. When a DNA template was used that did not have single-strand nicks, this enzyme was not active. This led to a search for a missing factor. When this factor (σ^{70}) was added back to the purified core enzyme and the uncut DNA template, the enzyme was able to bind tightly and selectively initiate RNA chains. Because of its role in binding and initiation, σ^{70} is often referred to as an initiation factor.

The precise functions of the subunits of the core enzyme are not known. β' is a basic (positively charged) polypeptide thought to be involved in DNA binding. The β subunit is the site of binding of several inhibitors of transcription and is thought to contain most or all of the active sites for phosphodiester bond formation. The α subunit is necessary for reconstituting active enzyme from separated subunits.

Figure 28.8

Overview of RNA synthesis.

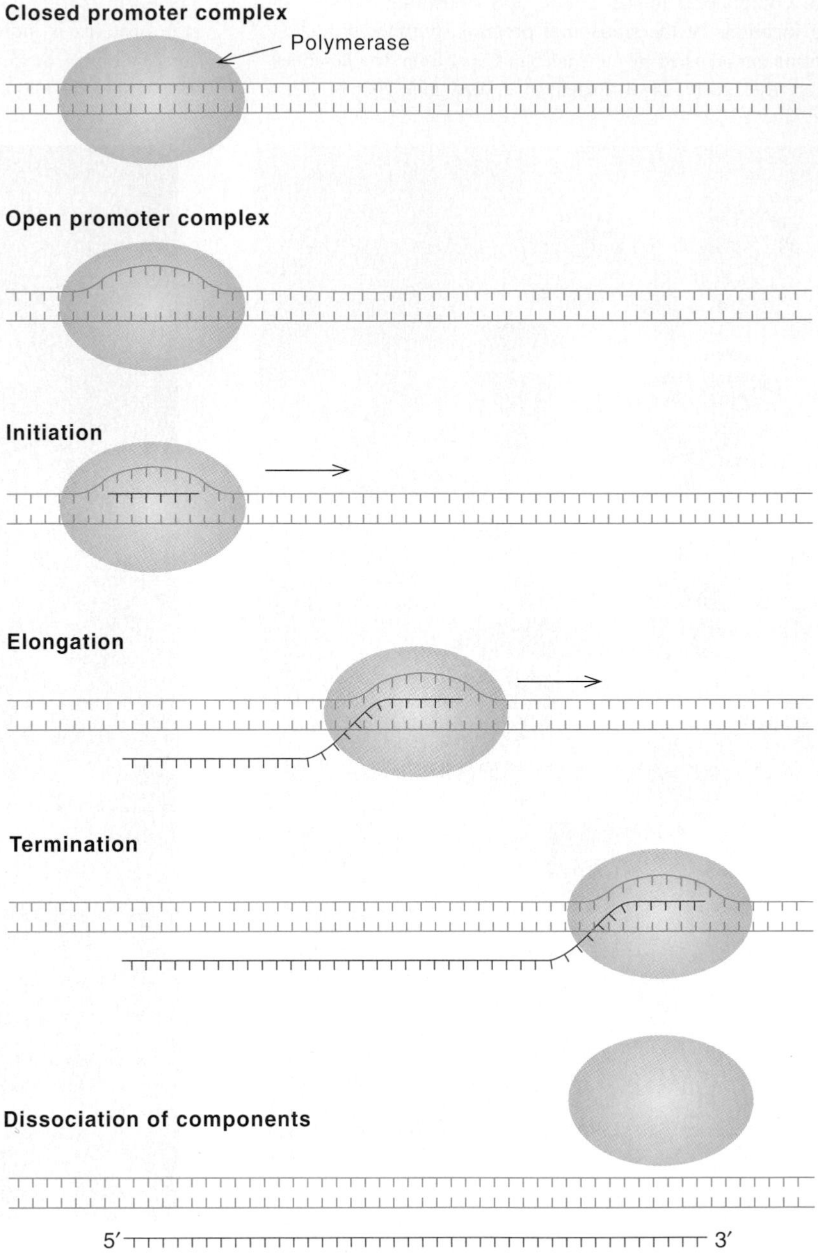

Binding at Promoters

In the cell, RNA polymerase transcribes only select regions of the DNA; in these regions it synthesizes RNA that is complementary to one of the DNA strands. This selective action is possible because the holoenzyme is able to recognize and form a stable complex with DNA at specific promoters. RNA polymerase is able to form unstable nonspecific complexes at any place on the template, mainly by an interaction with the DNA phosphates, but it either rapidly dissociates and rebinds or slides along the DNA until it reaches a promoter. It then forms a moderately stable complex with the promoter, most likely interacting with particular nucleotides in the −35 and −10 regions of the promoter. Initially the complex contains DNA in the double-helical or unmelted state; in this state they are referred to as closed promoter complexes (fig. 28.9). The next step involves breaking the H bonds (melting) over a stretch of about 10 bp of DNA, from positions −9 to 2, and a conformational change in the polymerase; this complex is termed the open promoter complex.

Common features have been identified in the sequence of over 250 different *E. coli* promoters at two locations: One 6-bp region centered at −35 (35 bp upstream of the initia-

Figure 28.9

Various types of RNA polymerase–DNA binding complexes. A nonspecific complex is formed at any point along the DNA. The closed promoter complex is formed at a polymerase-binding site. Following the formation of the closed promoter complex, an open promoter complex is formed at the same site.

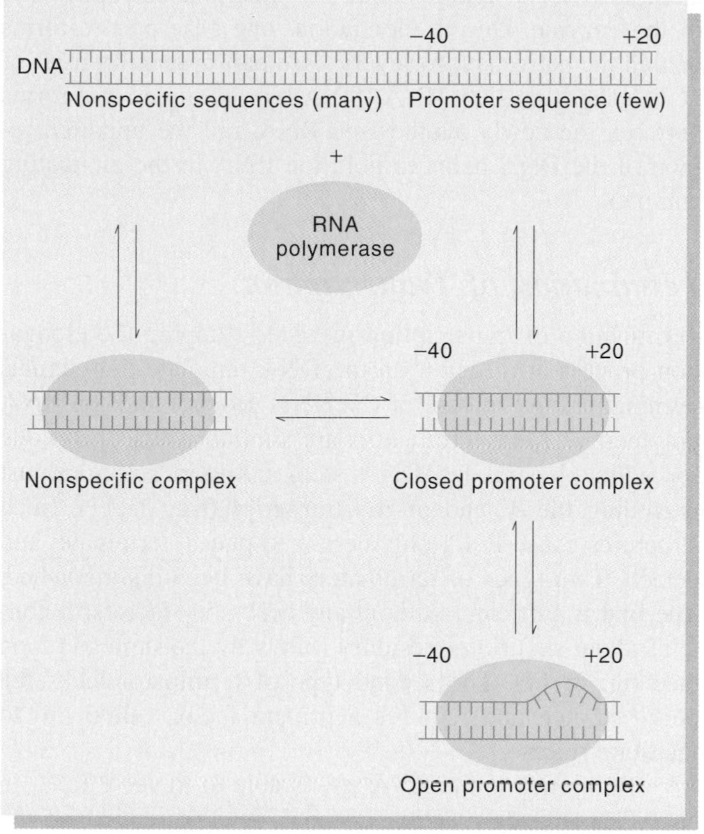

tion site) and one 6-bp region centered at −10 contain sequences that are similar but not identical in all bacterial promoters. The average sequences in both of these regions (TTGACA at −35 and TATAAT at −10) are termed the consensus sequences. A promoter with the consensus sequences in both regions leads to a very high level of transcription. Remarkably no naturally occurring promoter has the consensus sequence. It seems likely that this is part of a strategy which ensures that regulatory proteins can stimulate promoters and thereby exert an influence over their activity. This issue is taken up in chapters 30 and 31. Promoter strength in the absence of regulatory proteins decreases the more the actual sequences deviate from the consensus sequences, with some bases being more important than others. Naturally occurring promoters differ greatly in strength, which is best defined in terms of the frequency of RNA initiation. Some promoters are very weak, such as that associated with the *lac* repressor gene (discussed in chapter 30); as a result the *lac* repressor gene is transcribed only once in 20–40 min. At the other extreme one finds very strong promoters for rRNA genes, which are transcribed at the rate of one per second. This 2,000-fold difference in transcription rate is primarily a function of the base sequence of the promoter.

Initiation at Promoters

Once the polymerase binds to the promoter and strand separation occurs, initiation usually proceeds rapidly (1–2 s). The first, or initiating, NTP, which is usually ATP or GTP, binds to the enzyme. The binding is directed by the complementary base in the DNA template strand at the start site. A second NTP binds, and initiation occurs on formation of the first phosphodiester bond by a reaction involving the 3′-hydroxyl group of the initiating NTP with the inner phosphorus atom of the second NTP. Inorganic pyrophosphate derived from the second NTP is a product of the reaction. This process is illustrated in figure 28.10.

Alternative Sigma Factors Trigger Initiation of Transcription at Different Promoters

Although most promoters in *E. coli* appear to utilize σ^{70} as an initiation factor, several additional sigma-like initiation factors have been discovered that bind to core polymerase to form holoenzymes that recognize different species of promoters. One such factor is the product of the *rpoH* gene, which is involved in heat shock regulation. This factor is called sigma-32 (σ^{32}) because of its molecular weight of 32,000. It becomes bound to core polymerase in *E. coli* cells that have been subjected to heat shock and directs the polymerase to bind to and initiate at a class of promoters (the heat shock promoters) responsible for high-level expression of a dozen or so heat shock genes. Another such factor, σ^{54}, (M_r = 54,000) encoded by the *rpoN* gene is required for expression of certain genes involved in nitrogen metabolism. The consensus sequences associated with different classes of promoters are given in table 28.2.

Elongation of the Transcript

After initiation has occurred, chain elongation proceeds by the successive binding of the nucleoside triphosphate complementary to the base at the growth point in the template strand, bond formation with pyrophosphate release, and translocation of the polymerase one base farther along on the template strand. Transcription proceeds in the 5′ → 3′ direction, antiparallel to the 3′ → 5′ strand of the templating DNA strand. Once elongation has produced an RNA chain about 10 bases long, the σ subunit dissociates from the holoenzyme, leaving core polymerase to continue the

Figure 28.10

Details of phosphodiester bond formation. The α, β, and γ phosphates are indicated on the initiating NTP, which in this case is ATP. The colored ovals represent NTP-binding sites on the RNA polymerase. The biochemistry of bond formation in RNA synthesis is very similar to that in DNA synthesis.

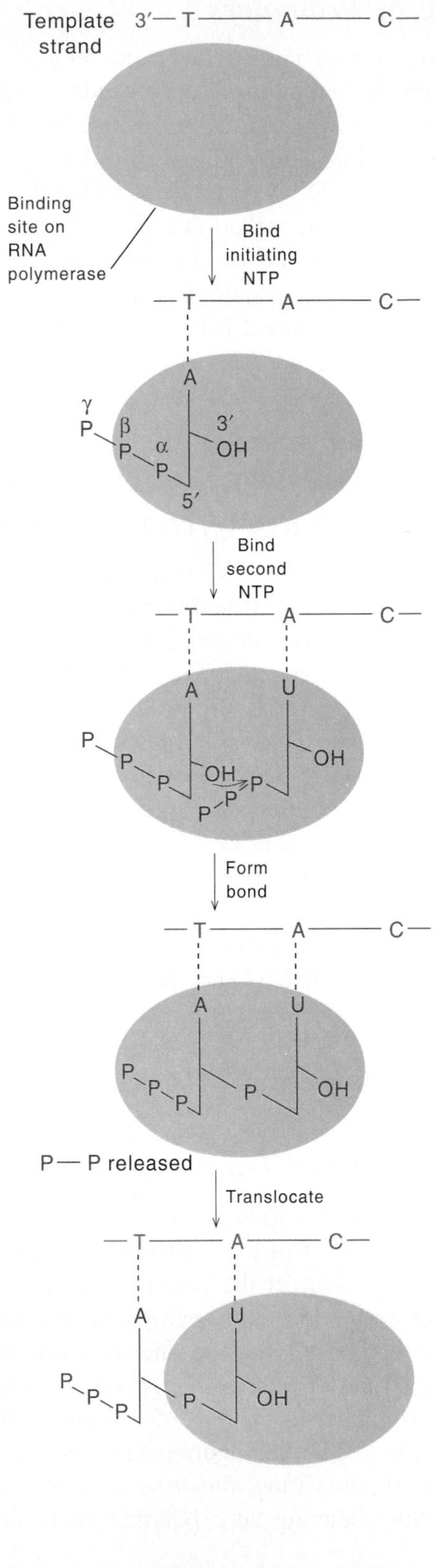

elongation reaction until a terminator signal is reached. The released σ is available to bind to a free core polymerase and re-form a holoenzyme capable of binding at the same or other promoters where it initiates additional RNA chains.

As the polymerase traverses the DNA, it must continually cause a melting or strand separation of the DNA so that a single DNA template strand is available at the active site of the enzyme. During elongation, one base pair re-forms behind the active site for every base pair opened in front of it. The short transient RNA–DNA hybrid duplex that forms between the newly synthesized RNA and the unpaired region of the DNA helps to hold the RNA to the elongating complex.

Termination of Transcription

Termination of transcription involves stopping the elongation process at a region on the DNA template that signals termination and release of the RNA product and the RNA polymerase. Most terminators are similar in that they code for a double-stranded RNA stem-and-loop structure just preceding the 3′ end of the transcript (fig. 28.11). Such structures cause RNA polymerase to pause, terminate, and detach. Two types of terminators have been distinguished. The first is sufficient without any accessory factors; it contains about six uridine residues following the stem and loop (see fig. 28.11). The second type of terminator lacks the polyU stretch and requires a protein factor called rho to facilitate release.

Rho binds to the RNA and is able to traverse RNA in an ATP-requiring reaction in the $5' \rightarrow 3'$ direction on the RNA. C residues on the RNA are absolutely essential for this to happen. A helicase activity in rho is essential for its termination activity. If rho catches up with the RNA polymerase at a pause site or a termination site, it is highly likely to catalyze termination. This general picture of rho action is supported by the finding that interjection of a translational block that inhibits ribosome movement encourages premature release of the transcript. In *E. coli,* ribosomes normally traverse the nascent transcript before transcription is completed. These ribosomes make it difficult for rho to bind and reach the elongating polymerase. A translational block facilitates rho binding and migration to the elongating RNA polymerase.

Comparison of Escherichia coli *RNA Polymerase with DNA PolI and PolIII*

In table 28.3 the bacterial RNA polymerase and the bacterial DNA polymerase are compared. Both types of enzymes are DNA-template-directed and require 4 NTPs and a divalent cation. While RNA polymerase makes single-stranded

Table 28.2

Summary of *Escherichia coli* Sigma Factors

Sigma Factor	Gene	Consensus Sequence -35 Region	-10 Region	Genes Recognized
σ^{70}	*rpoD*	TTGACA	TATAAT	most genes
σ^{32}	*rpoH (htpR)*	CTTGAA	CCCCAT-TA	heat-shock regulated
σ^{54}	*rpoN (ntrA)*	$CTGGCACN_5TTGCA$		nitrogen regulated
σ^{E}	not identified	GAACTT	TCTGA	*rpoH, htrA*
σ^{F}	not identified	TAAA	GCCGATAA	flagellar, chemotaxis
σ^{K}	*rpoK (katF)*	Unknown		*KatE*

Figure 28.11

Important features of a typical transcription unit. DNA is shown with promoter and terminator regions expanded below. RNA is transcribed starting in the promoter region at +1 and ending after the stem and loop of the terminator. The protein resulting from translation of this RNA is shown above with its N and C termini indicated.

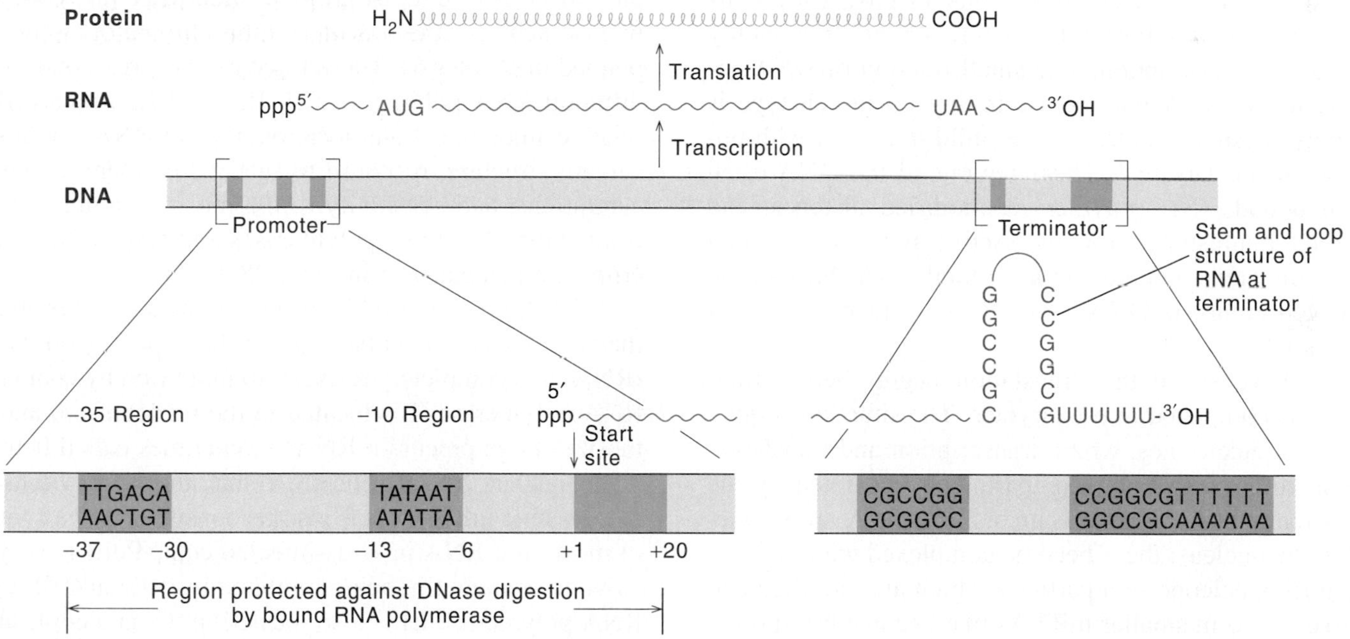

chains involved directly or indirectly in protein synthesis, the DNA polymerases replicate and repair the duplex DNA. Only RNA polymerase recognizes start-and-stop sequences in the duplex DNA. Thusfar only the DNA polymerases have been shown to have a proofreading function, but the possibility of a proofreading function for RNA polymerase has not been outruled.

Important Differences Exist between Eukaryotic and Prokaryotic Transcription

Although the basic mechanism by which RNA is synthesized is quite similar in prokaryotes and eukaryotes, several important differences occur. Most parts of the bacterial

Table 28.3

Comparison of *Escherichia coli* RNA Polymerase with DNA Polymerases I and III

	RNA Polymerase	DNA Polymerases I and III
Similarities		
DNA-template-directed	Yes	Yes
Requires 4 NTPs	Yes (rNTPs)	Yes (dNTPs)
Requires divalent cation	Yes	Yes
Differences		
Function	Transcription	Replication and repair
Initiates chains	Yes	No
Terminates chains	Yes	No
Recognizes sequences	Yes	No
Uses intact duplex template	Yes	No
Product	Single-strand RNAs	Duplex DNA strands
Proofreading	?	Yes

DNA are readily accessible to RNA polymerase binding and transcription. By contrast, most DNA in eukaryotic cells exists in a condensed form (chromatin), which is not readily accessible to transcription. The small fraction of DNA accessible to the RNA polymerase in any given cell type is especially sensitive to cleavage by mild treatment with bovine pancreatic DNase I. These regions of the DNA often contain bound RNA polymerase, modified histones, and additional nonhistone proteins. Active regions are often undermethylated compared with the total DNA. Most of the methylated groups in DNA are on the C residues in the CG sequence.

We have stated that translation begins before transcription is completed in prokaryotes. The situation is quite different in eukaryotes, where transcription and translation occur in different cellular compartments separated by the nuclear membrane. Large precursors of mRNA are synthesized in the nucleus; these become complexed with proteins to form ribonucleoprotein particles which are modified and processed to form smaller mRNAs that become transported across the nuclear membrane to the cytoplasm.

Eukaryotes Have Three Nuclear RNA Polymerases

Unlike prokaryotes, in which all major types of RNA are synthesized by one RNA polymerase, eukaryotic cells contain three nuclear DNA-dependent RNA polymerases, each responsible for synthesizing a different class of RNAs.

Nuclear extracts can be fractionated by chromatography on DEAE-cellulose to give three peaks of RNA polymerase activity (the use of column chromatography is explained in chapter 6). These three peaks correspond to three different RNA polymerases (I, II, and III), which differ in relative amount, cellular location, type of RNA synthesized, subunit structure, response to salt and divalent cation concentrations, and sensitivity to the mushroom-derived toxin α-amanitin. The three polymerases and some of their properties are summarized in table 28.4.

RNA polymerase I is located in the nucleolus and synthesizes a large precursor that is later processed to form rRNA. It is completely resistant to inhibition by α-amanitin. RNA polymerase II is located in the nucleoplasm and synthesizes large precursor RNAs (sometimes called heterogeneous nuclear RNA, or hnRNA) that are processed to form cytoplasmic mRNAs. It is also responsible for the synthesis of most viral RNA in virus-infected cells. PolII is very sensitive to α-amanitin, being inhibited by 50% at 0.05 μg/ml. RNA polymerase III is also located in the nucleoplasm and synthesizes small RNAs, such as 5S RNA and the precursors to tRNAs. This enzyme is somewhat resistant to α-amanitin, requiring about 5 μg/ml to reach 50% inhibition.

Eukaryotic RNA Polymerases Are Not Fully Functional by Themselves

Crude enzyme preparations of RNA polymerases I, II, and III have been shown to be capable of selective transcription

Table 28.4

Comparison of Eukaryotic DNA-Dependent RNA Polymerases

Type	Location	RNAs Synthesized	Sensitivity to α-Amanitin
RNA polymerase I	Nucleolus	Pre-rRNA	Resistant
RNA polymerase II	Nucleoplasm	hnRNA, mRNA	Sensitive
RNA polymerase III	Nucleoplasm	Pre-tRNA, 5S RNA	Sensitive to very high levels
Mitochondrial	Mitochondria	Mitochondrial	Resistant
Chloroplast	Chloroplasts	Chloroplast	Resistant

of defined DNA templates, initiating at sites known in some cases to be utilized *in vivo*. Fractionation of these polymerase-containing extracts has revealed many additional protein factors that stimulate *in vitro* transcription. These factors are divided into two categories. Basal transcription factors are required for the transcription of virtually all genes of a particular class. In addition to basal transcription factors there are an array of factors required for activated transcription. Transcription studies are presently in a stage of rapid development, and the distinction between factors required for basal transcription and activated transcription is not always clear. The situation is much more complex than for prokaryotes.

First we discuss basal transcription factors (fig. 28.12). With all three polymerases, two or more basal transcription factors are required, and some or most of these factors must bind to the promoter before the polymerase can bind.

Because many eukaryotic genes have been cloned during the last few years, it has become possible to compare the DNA sequences preceding genes that may act as promoter-like signals for RNA polymerase II. One feature that stands out is a common sequence, TATAAA, called the TATA box, found usually 25–30 bp before the transcription start site in many but not all PolII-transcribed genes.

Phil Sharp and Leonard Guarente showed that at least four transcription factors are required in addition to polymerase II for initiation from the major late promoter of adenovirus. *In vitro* studies indicate that these factors assemble in an orderly fashion (see fig. 28.12*b*). First the TFIID complex binds to the TATA box. Sequential binding of TFIIA, TFIIB, RNA polymerase II, and TFIIE follow. It is believed that this multifactor complex functions for a large number of eukaryotic promoters that contain TATA boxes.

The promoter region for RNA polymerase I also involves the cooperative binding of additional basal transcription factors. In this case two transcription factors bind upstream of the transcription start site (see fig. 28.12*a*).

In contrast to the initiation complex for RNA polymerases I and II, the region necessary for selective transcription of 5S RNA by RNA polymerase III is located in a region 40–80 bp downstream of the transcription start site (see fig. 28.12*c*). One of the additional protein complexes needed for selective transcription of the 5S gene (TFIIIA) binds to this site, and in conjunction with two other protein factors (TFIIIB and TFIIIC), directs RNA polymerase III to bind and initiate transcription. The binding site for factor TFIIIA was determined by the DNA "footprinting" technique (Methods of Biochemical Analysis 28A), a technique commonly used to locate binding sites for specific DNA-binding proteins.

PolIII polymerase also transcribes tRNA genes. In this case TFIIIA does not participate in the transcription process (see fig. 28.12*c*).

Recent Evidence Suggests That the TATA-Binding Protein May Be Required by All Three Polymerases

Sometimes it is said that the basal transcription factor TFIID is the TATA-binding protein. Actually TFIID is a multiprotein complex of which the TATA-binding protein is but one component. Initially it was thought that the TATA-binding protein was exclusively involved with PolII promoters that contain a TATA element. However, closer examination of the different transcription systems indicates that the TATA-binding protein is part of the transcription initiation complex for PolII promoters that do not contain a TATA element. The TATA-binding protein is also required for transcription of ribosomal RNA genes. PolIII polymerase also may use the TATA-binding protein. In all of those instances where the TATA sequence is missing from the promoter, the exact function of the TATA-binding protein is not clear since it does not bind directly to the DNA in such

Figure 28.12

Formation of the initiation complex for transcription for the three major classes of eukaryotic RNA polymerase: PolI, PolII, and PolIII. Each polymerase consists of many subunits (not shown). In addition to the firmly bound subunits, a number of protein factors, called transcription factors (TFs), only associate with the polymerases at the initiation site for transcription. For all three polymerases some of these transcription factors must bind to the promoter before the polymerase can bind. In each case we see an orderly progression of binding of factors and polymerase. In the case of PolIII, three transcription factors are involved for 5S rRNA genes, and only two are involved for tRNA genes.

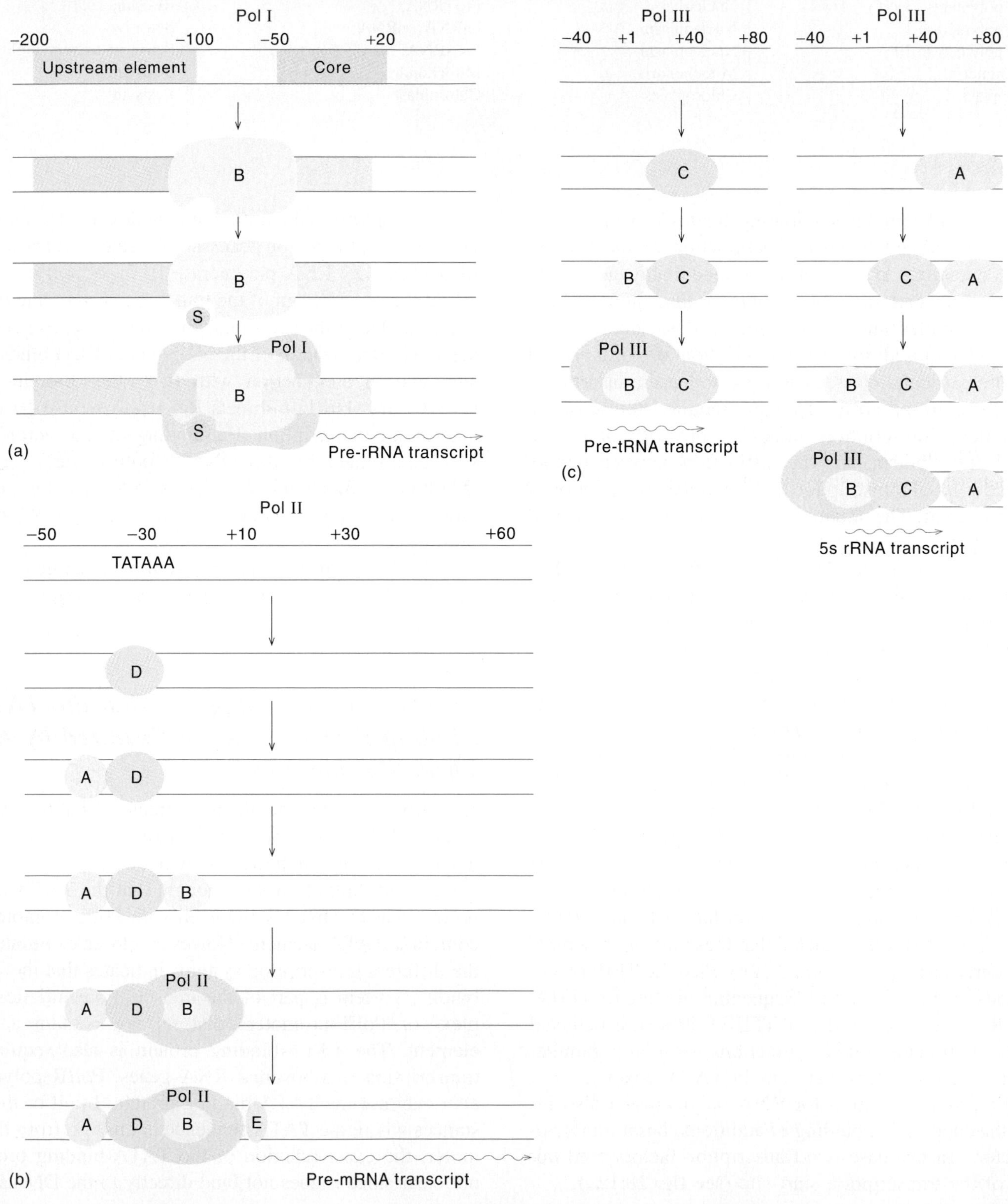

Figure 28.13

Cis elements involved in transcription in yeast (*a*) and in vertebrates (*b*). Upstream activator sequences (UAS) in yeast are similar in function to upstream enhancers in vertebrates. Yeast has no parallel to downstream enhancers found in vertebrates.

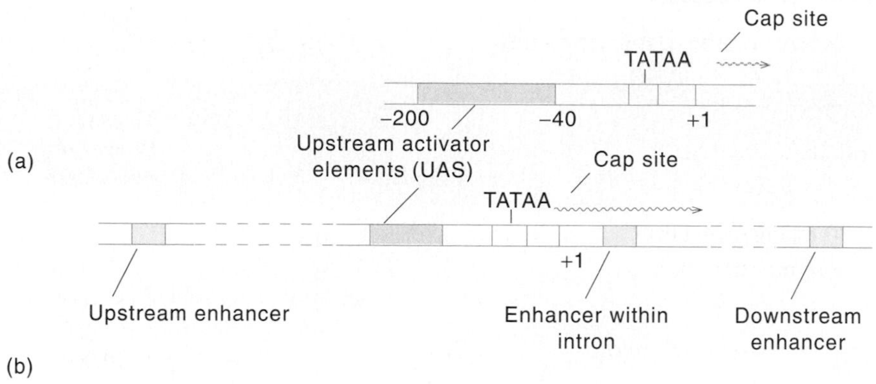

complexes. Perhaps it helps to link other transcription factors to the polymerase.

In Eukaryotes Promoter Elements Are Located at a Considerable Distance from the Polymerase Binding Site

Eukaryotic promoters often contain binding sites adjacent to the polymerase-binding site where additional protein factors may bind. Promoter regions further removed from the polymerase-binding site are also quite common. The transcription factors that bind to these regions tend to be less general and more gene-specific than the factors that bind next to the polymerase. Considerable differences occur between the additional promoter elements found in yeast and vertebrates. In the case of yeast, additional promoter elements called upstream activator sequences (UAS) are usually located 40–200 bp upstream from the transcription start site (see fig. 28.13). The UAS elements bind additional transcription activation factors, which interact with those that bind at the TATA box. In vertebrates activator elements called enhancers are found upstream, downstream, and even in the middle of genes. Remarkably these elements may be located as far as 10 kb from the gene they influence. Like UAS elements, enhancers serve as binding sites for additional transcription activators. We have more to say about transcription activators in chapter 31 when we discuss regulation of gene expression in eukaryotes.

Many Viruses Encode Their Own RNA Polymerases

Three strategies are used by different DNA viruses to accomplish transcription of viral DNA. The first type utilizes the host RNA polymerase, in some cases modifying it or synthesizing new promoter-specific factors to direct it to read the viral promoters. Examples of such viruses are bacteriophage ϕX174, and T4 (of *E. coli*).

The second type of virus utilizes the host RNA polymerase to transcribe ''early'' viral genes including a gene for a new RNA polymerase that transcribes exclusively the remaining ''late'' viral genes. *E. coli* bacteriophages T7 and T3 are the best known examples of this type. T7 RNA polymerase recognizes specifically the T7 late promoters, all of which contain a nearly identical sequence of 18–22 nucleotides immediately upstream of the 5′-triphosphate terminal GTP start site. T7 RNA polymerase also recognizes specific termination points on the template and ignores the ones normally recognized by *E. coli* RNA polymerase.

A third type of virus, exemplified by bacteriophage N4, carries a virus specific RNA polymerase in its virion. This polymerase enters the cell together with the viral DNA and transcribes some early viral genes. Some of these genes code for specificity factors that direct the host RNA polymerase to transcribe late genes. Vaccinia virus is another example of a virus that contains a virion-encapsulated RNA polymerase.

RNA-Dependent RNA Polymerases of RNA Viruses

The RNA genomes of single-stranded RNA bacterial viruses, such as Qβ, MS2, R17, and f2, are themselves mRNAs. Bacteriophage Qβ codes for a polypeptide that combines with three host proteins to form an RNA-dependent RNA polymerase (replicase). The three host proteins are ribosomal protein S1 and two elongation factors for protein synthesis: EF-Tu and EF-Ts (see table 28.5). The Qβ replicase functions exclusively with the Qβ RNA plus strand template. It first makes a complementary RNA transcript (minus strand) and ultimately uses the minus strand as

Table 28.5

RNA-Synthesizing Enzymes

	Template	Primer	Molecular Weight of Subunit(s)	Gene Name	Substrate	Inhibition by Rifampicin
Template-dependent						
Enzymes from bacteria						
Holoenzyme *(E. coli)*	DNA	—	155,000 151,000 70,000 36,500	*rpoC* *rpoB* *rpoD* *rpoA*	4 NTPs	Yes
DNA primase	DNA	—	65,000	*dnaG*	4NTP, 4 dNTP	No
Enzymes from phage or phage-infected bacteria						
T7 RNA polymerase	T7 DNA	—	99,000	T7 *gene1*	4 NTPs	No
N4 RNA polymerase	N4 DNA	?	350,000	Viral	4 NTPs	No
Qβ replicase	Qβ RNA	—	65,000 55,000 43,000 35,000	*rpsA* Viral *tuf* *tsf*	4 NTPs	No
Template-independent						
CCA enzyme	—	3′ end tRNA	45,000	*cca*	CTP ATP	No
Poly(A) polymerase (eukaryotic)	—	3′ end mRNA			ATP	No
Polynucleotide phosphorylase	—	3′ end RNA	86,000 48,000	*pnp*	4 NDPs	No

(Source: G. Zubay, *Biochemistry,* 2d ed., Macmillan, New York, 1988, p. 813)

a template to synthesize multiple copies of viral RNA plus strands. Like the DNA-dependent RNA polymerases, the replicase utilizes rNTPs and transcribes in the $5' \rightarrow 3'$ direction. The phage RNA must first act as an mRNA to direct the synthesis of the aforementioned component of the replicase, since uninfected cells do not have an RNA-dependent RNA polymerase or replicase.

RNA tumor viruses (retroviruses) that infect animal cells exhibit a different replication strategy. In their virions they carry an enzyme that uses the viral RNA as a template to synthesize a DNA copy (see chapters 26 and 27). This DNA becomes integrated into the host genome. Subsequently the viral RNA is transcribed from the integrated viral DNA using host cell RNA polymerase.

Other Types of RNA Synthesis

In addition to the cellular enzyme(s) that catalyzes DNA-directed RNA synthesis, cellular enzymes are involved in polyribonucleotide synthesis that do not use a template. Some of the properties of these enzymes are summarized in table 28.5. We have already mentioned polynucleotide phosphorylase in this chapter, and in chapter 26 we discussed the importance of DNA primase to DNA synthesis.

Two enzymes are known to add ribonucleotides post-transcriptionally to the 3′ hydroxyl end of specific RNAs. One adds the CCA sequence found in all tRNAs at their 3′ ends. The 3′ terminal adenine in this sequence serves as the amino acid attachment site. The 3′-CCA is relatively unstable and is continually being rejuvenated by this enzyme,

Figure 28.14

Comparative processing of major transcripts in prokaryotes and eukaryotes.

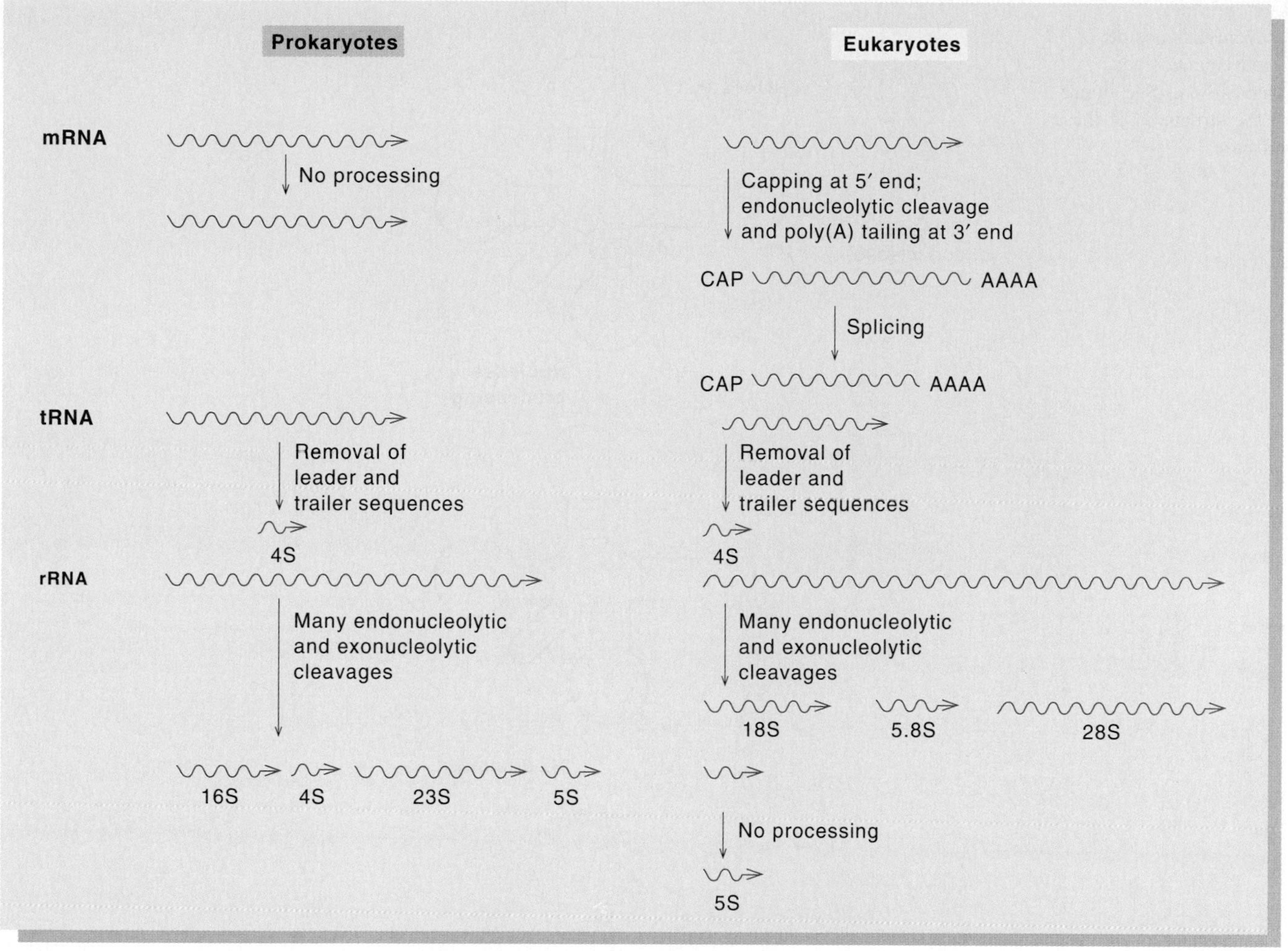

called the CCA enzyme, or tRNA nucleotidyltransferase.

In eukaryotes, 100–200 adenosine residues are added to the 3′ ends of most mRNAs by a poly(A) polymerase. This addition occurs in the nucleus before the mRNA is fully processed and transported to the cytoplasm.

Posttranscriptional Alterations of Transcripts

Most RNAs are not made in their final functional forms as they peel off the DNA template (fig. 28.14). They must undergo backbone phosphodiester bond cleavages into smaller molecules (processing) and individual base changes (modification). The types of alterations that pre-tRNA and pre-rRNA transcripts undergo are very similar in prokaryotes and eukaryotes. We focus on the situation in *E. coli* because it is the best understood.

Processing and Modification of tRNA Requires Several Enzymes

Transfer RNAs are processed from larger precursors in both prokaryotic and eukaryotic cells. This processing involves two types of nucleases: Endoribonucleases, that cleave at internal sites in the RNA, and exonucleases, that remove nucleotides from the ends of the chains.

Figure 28.15

Processing and modification of *E. coli* tyrosine tRNA. (T = ribothymidine; ψ = pseudouridine; i^6A = isopentyladenosine; mG = methylguanosine; s^4U = thiouridine.) See figure 28.4 for the structures of these modified bases.

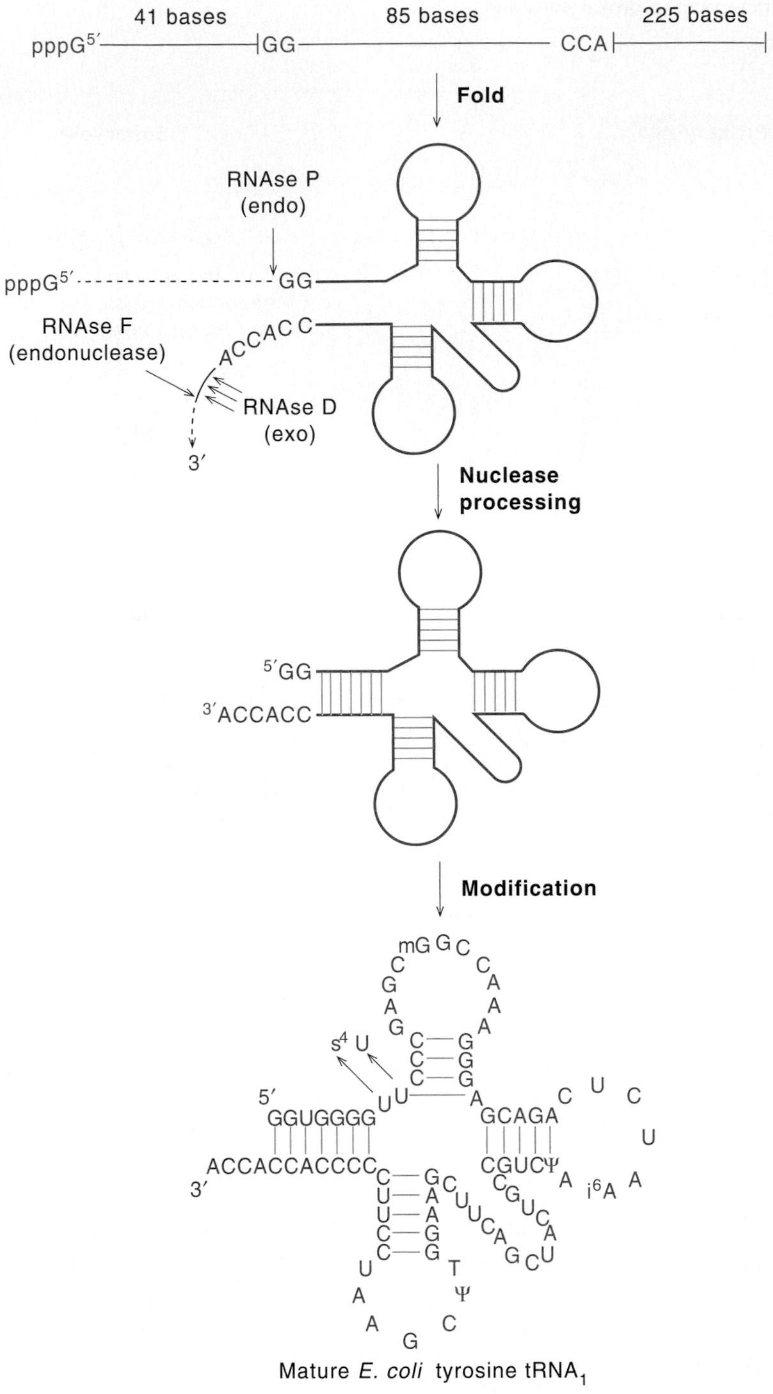

The processing steps of *E. coli* tyrosine $tRNA^{Tyr}$ are diagrammed in figure 28.15. The initial transcript has, in addition to the 85 nucleotide residues of the final product, 41 residues at the 5′ end and 225 residues at the 3′ end; it probably folds to form the typical cloverleaf structure of the mature tRNA prior to processing. Processing begins when a specific endonuclease called RNaseF cleaves the precursor at a site three nucleotides beyond what will be the 3′ end of the mature tRNA. Another endonuclease, RNaseP, then cleaves the remaining RNA to produce the mature 5′ end. At the 3′ end, exonuclease RNaseD sequentially removes additional nucleotides usually until it reaches the 3′ terminal CCA sequence. Some tRNAs encode the CCA terminal sequence; some do not. Following processing individual bases

Figure 28.16

Processing of *E. coli* ribosomal RNA. The ribosomal RNA is transcribed as one long RNA molecule, which contains the sequences for the three ribosomal RNAs and one or two tRNA molecules. Many processing sites occur and many different enzymes are involved in the processing, as indicated by the vertical arrows and the symbols associated with these arrows. The various nucleases are described in table 28.7. (Source: Adapted from D. Apirion and P. Gegenheimer, Processing of bacterial RNA, *FEBS Lett.* 125:1, 1981.)

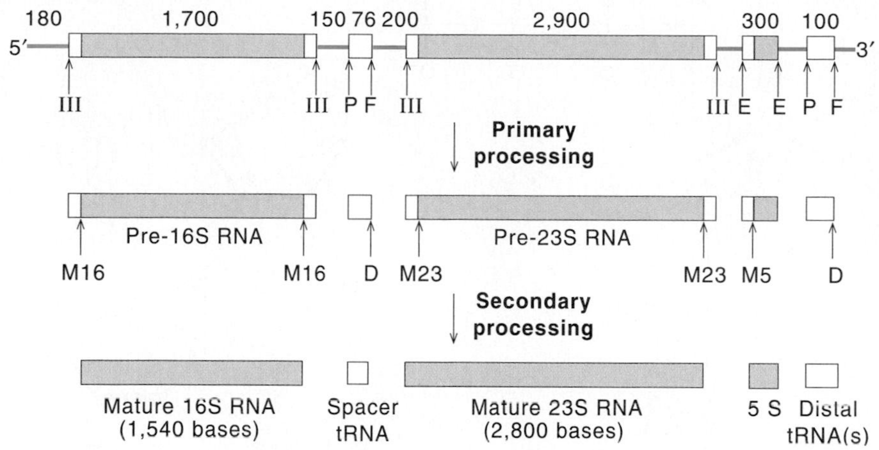

on the tRNA molecule are modified by a variety of enzymes, including methylases, deaminases, thiolases, pseudourydylating enzymes, and transglycosylases. Some of the modified bases that result from the action of these enzymes are illustrated in figure 28.4.

Processing of Ribosomal Precursor Leads to Three RNAs

Both eukaryotic and prokaryotic cells synthesize large precursors to rRNA that are processed to produce the mature rRNAs. The scheme for processing rRNA in *E. coli* is summarized in figure 28.16.

The initial transcript is over 5,500 nucleotides long and includes, from the 5′ end of the RNA: A 16S rRNA, a spacer region with one or two tRNAs, a 23S rRNA, and a 5S rRNA, and in some cases one or two additional tRNAs. Extra bases are found preceding and following each of these RNAs. Primary processing events include the endonucleolytic action by RNaseIII to produce pre-16S and pre-23S RNAs. This is followed by the action of specific ribonucleases to produce the tRNAs and pre-5S rRNA. Secondary processing by endonucleases M16, M23, and M5 results in mature 16S, 23S, and 5S RNAs, respectively. Extra bases on the 3′ end of the tRNAs are removed by exonuclease RNaseD. This processing scheme has been deduced by observing the accumulation of intermediates in mutant strains defective in one or more of the nucleases and by cleaving the intermediates *in vitro* with purified or partially purified nucleases.

Eukaryotic Pre-mRNA Undergoes Extensive Processing

In prokaryotes, most mRNAs function in translation with no prior alterations. Indeed little opportunity arises for any alterations because translation starts on the nascent transcript before transcription has been completed. By contrast, in eukaryotes transcription occurs in the nucleus, and a transcript undergoes extensive changes before being transported to the cytoplasm. First, the 5′ end of the message is modified, a process called capping (fig. 28.17). This usually occurs before transcription has been completed. Changes at the 3′ end of the transcript are part of the termination mechanism because the PolII enzyme does not appear to recognize any termination signal. The RNA polymerase transcribes well beyond the useful part of the message. These extended transcripts are cleaved about 12 nucleotides downstream of an AAUAA sequence, after which poly(A) polymerase adds about 200 adenylate residues to the 3′ end. In many cases, especially in higher eukaryotes, noncoding regions called introns are removed from interior locations, and the remaining message is reunited. This process is called splicing because it is similar to the process of splicing in film editing. We discuss splicing in some detail.

To assist in the modification and processing of mRNAs, eukaryotic cells contain in their nuclei small nuclear RNAs (snRNAs), which are complexed with specific proteins to form small nuclear ribonucleoprotein particles (snRNPs). These RNAs have been named U1, U2, U3, . . . , U13 and range in size from 100 to 220 bases. One, U3, is

Figure 28.17

Structure of the 5′ methylated cap of eukaryotic mRNA. A 7-methylguanosine (in red) is attached through a triphosphate linkage formed between its 5′-OH and the 5′-OH of the terminal residue in the initial transcript. Note that the 2′-OH groups on the last two bases of the initial transcript have also been modified by methylation (in red). N_1, N_2, and N_3, can be any purine or pyrimidine bases.

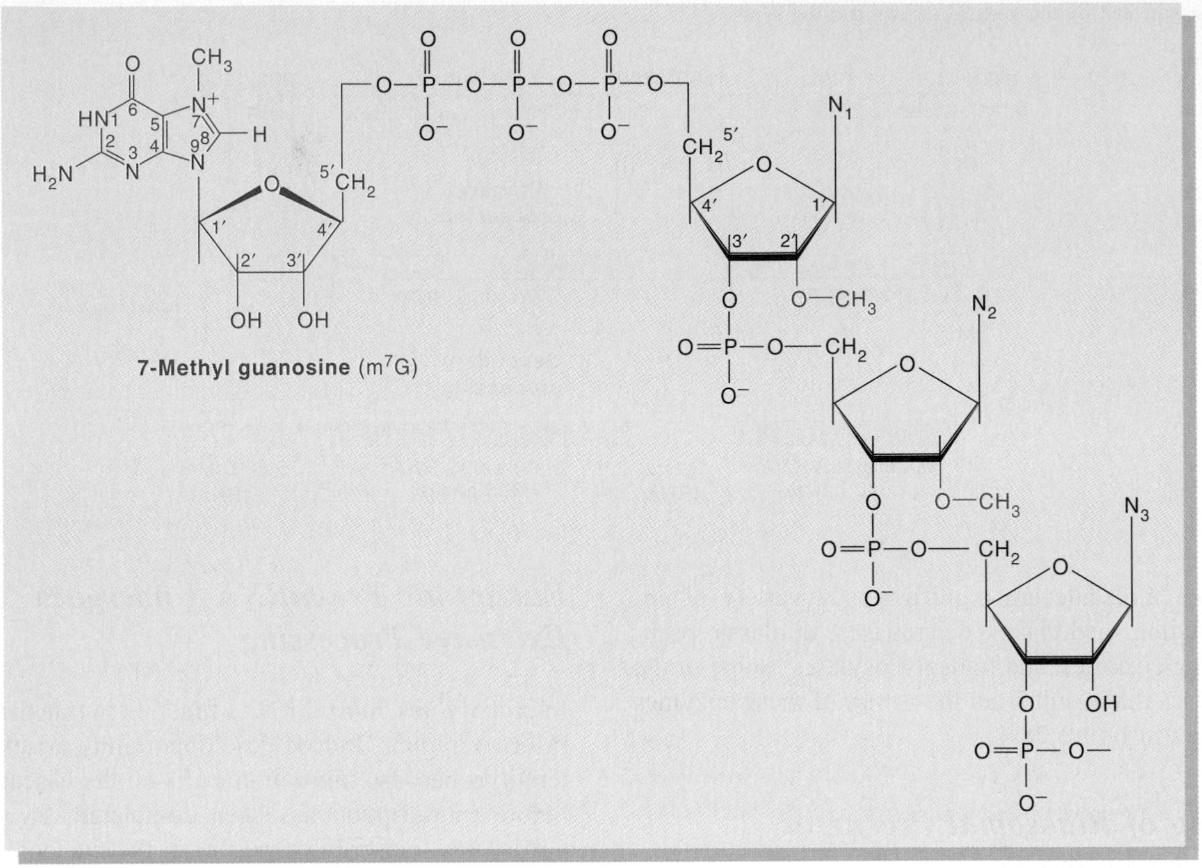

found in the nucleolus, the site of rRNA synthesis. All are very abundant, and as many as 1 million copies of most of them may occur per nucleus. Their base sequences are highly conserved among organisms, and all seem to contain unusual trimethylguanosine structures at their 5′ ends. Each snRNP contains an snRNA and 6–12 proteins, some of which are common to all snRNPs. The snRNPs appear to be involved in the processing and modification of RNAs, including splicing (U1, U2, U5, U6), polyadenylation of pre-mRNA (U11), formation of 3′ ends of histone mRNAs (U7), and maturation of rRNA (U3).

Splicing Entails the Removal of Internal Sequences. It was established in the early 1970s that a great deal of RNA (hnRNA) turns over in the nucleus without ever reaching the cytoplasm. What could this RNA be? Was it a specific type of RNA that never made it to the cytoplasm, or was it evidence for processing of mRNA precursors? The big surprise came in 1977 when Ric Robert's and Phil Sharp's laboratories simultaneously discovered that the mRNAs of adenovirus undergo extensive processing in which internal segments are removed from the mRNA precursors. Very soon thereafter it was found that this phenomenon was general and widespread, especially in higher eukaryotes.

One of the early demonstrations of sequence removal resulted from the finding that mouse β-globin precursor mRNA did not form a perfect hybrid with DNA complementary to mature mRNA. When the hybrid was observed with an electron microscope, a loop in the DNA of the heteroduplex appeared, which suggested that the precursor contained internal sequences not present in the mature mRNA (fig. 28.18). These noncoding intervening sequences (introns) are interspersed with coding sequences (exons).

The presence of introns implies a function, but in

Figure 28.18

Electron micrograph showing mouse β-globin precursor mRNA (nascent transcript) hybridized with DNA (cDNA) complementary to mature mRNA (upper photo). A control experiment (lower photo) shows that mature mRNA forms a perfect hybrid with DNA complementary to mature mRNA (cDNA) as expected. The intron region in the nascent transcript is indicated by a loop in the upper figure. Schematics indicating the RNA and cDNA are shown on the right. A second small intron is present in the precursor mRNA near the 5′ end and is the reason that the 5′ end of the RNA is not hybridized to the cDNA in the upper photo. (From A. Kinniburgh, J. Mertz, and J. Ross, The precursor of mouse β-globin contains two intervening sequences, *Cell* 14:681, 1978. © Cell Press.)

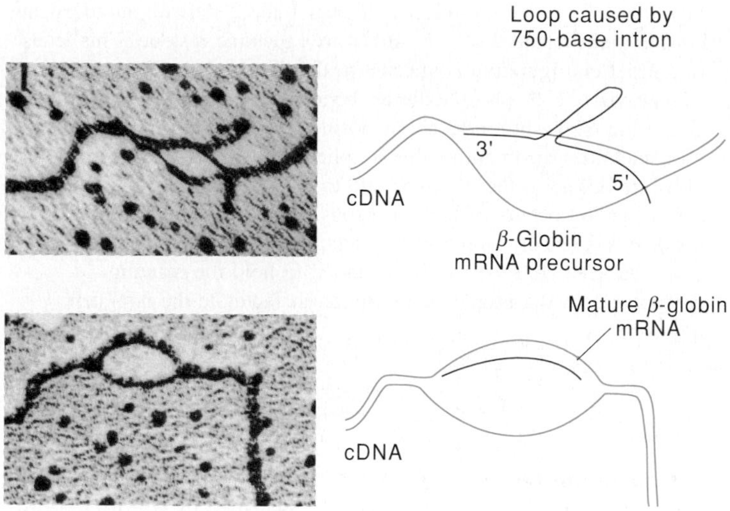

many cases the presence of introns may represent no more than a stage in the evolution of a gene. This argument is supported by the finding that introns are far less common in unicellular eukaryotes such as the yeast *Saccharomyces,* and they are very rare in prokaryotes such as *E. coli.* Frequently, splice points are correlated with "domains" that define protein structural units (see chapter 4). Similar domains are often seen in different proteins. For example, the exons of hemoglobin encode three structural domains of different types, whereas the heavy-chain immunoglobulin exons encode four domains that are quite similar in structure.

Splicing Is a Two-Step Process. By attaching the promoter for the bacteriophage SP6 RNA polymerase to the β-globin gene, it has become possible to transcribe the gene *in vitro* with SP6 RNA polymerase to produce abundant amounts of β-globin pre-mRNA. This source of precursor mRNA was used to determine the steps and factors involved in splicing. The process involves cleavage of the pre-mRNA at the 5′ splice site to generate the 5′-proximal exon and an RNA species containing the intron in a "lariat" configuration connected to the distal exon (fig. 28.19). The lariat is formed via a 2′-5′ phosphodiester bond, which joins the 5′ terminal guanosine of the intron to an adenosine residue within the intron at a spot 18–40 nucleotides upstream of the 3′ splice site. These two RNA species are probably held together in a noncovalent complex until the next step (see fig. 28.19, step 2) in the reaction, which entails cleavage at the 3′ splice site to generate the free intron RNA and ligation of the two exons via a 3′-5′ phosphodiester bond. U1 and U2 snRNAs, as well as several additional protein factors, are necessary for the reactions to occur *in vitro.* These various factors are assembled in a large 40S–60S ribonucleoprotein called a spliceosome.

Removal of internal sequences in eukaryotes is not restricted to mRNA processing. It also occurs in the processing of rRNA and some tRNAs. In tRNAs the mechanism appears to be different in that the signal for splicing originates not from the primary sequence but from the secondary or tertiary structure of the pre-tRNA.

RNA Editing Involves Changing Some of the Primary Sequence of a Nascent Transcript. On some occasions, the nascent transcript requires alterations in its sequence to convert it into a translatable mRNA. For example, in the mitochondrial message for cytochrome c oxidase from the protozoan *Leishmania tarentolae,* several U residues are added at different points to the nascent transcript (fig. 28.20). To make the editing changes a so-called guide RNA must form a complex with the nascent transcript. The changes in sequence are dictated by the complementary sequence in the guide RNA in the region where editing occurs. In addition to the guide RNA a special enzyme system is required, which is capable of inserting and removing bases at various points in the nascent transcript.

Editing seems like a roundabout way to get a functional message. Why not just make the transcript so that it can be translated in the first place. Possibly, editing serves a regulatory function. Alternatively it has been proposed that editing may be a device used to accelerate the evolutionary process for a gene.

Figure 28.19

Splicing scheme for pre-mRNA. In step 1 the 2′-OH on an adenosine attacks a phosphate that is 5′-linked to a guanine residue. This leads to a lariat configuration connected to the distal exon. The lariat is formed via a 2′-5′ phosphodiester bond, which joins the 5′ terminal guanosine of the intron to an adenosine residue within the intron, 18–40 nucleotides upstream of the 3′ splice site. In the next step a cleavage occurs at the 3′ splice site to generate the free intron RNA, and a ligation occurs of the two exons via a 3′-5′ phosphodiester bond. Bases usually found in the region of the splice sites are indicated. The spliceosome is presumed to hold the reacting components in the proper juxtaposition to facilitate the two-step reaction.

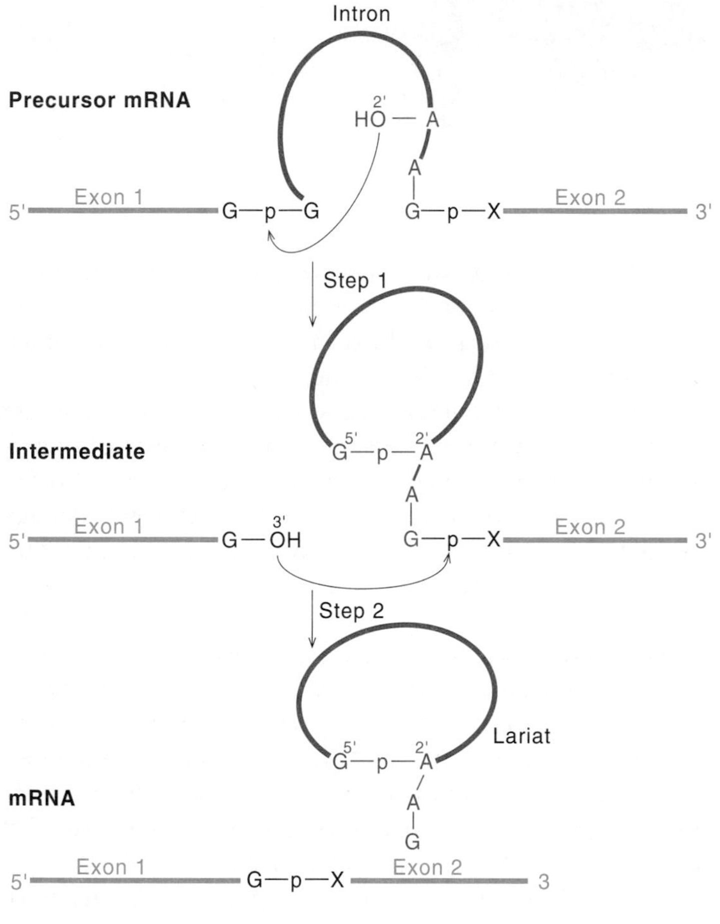

Figure 28.20

Editing of the cytochrome b message in the flagellated protozoan *Leishmania tarentolae*. Editing, which involves the insertion of 11 U residues, creates a translatable RNA where no translation was possible before. Structure shown for mRNA is after editing.

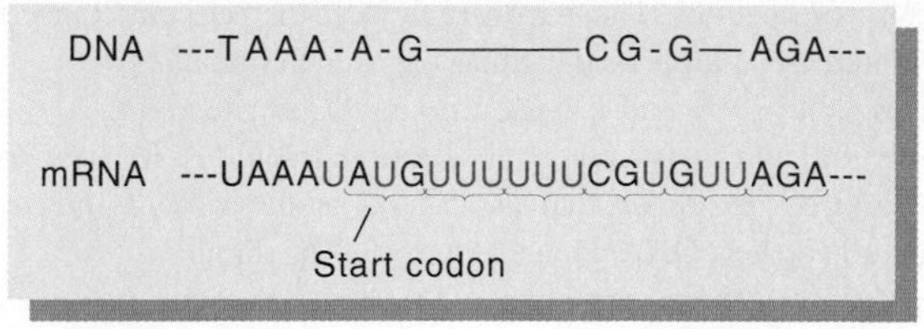

Some RNAs Function Like Enzymes

For the more than 50 years since Sumner's discovery of urease, evidence has grown for all enzymes being proteins. This isn't quite true, as coenzymes frequently show similar activities to enzymes to which they are normally associated. Moreover, if we accept the broader definition of an enzyme to include the ability to bring reactive groups in close proximity so that they are encouraged to react, then we must regard all nucleic acid templates as enzymes. Nevertheless, it came as a giant surprise in the early 1980s when it was found that a number of biochemical activities normally associated with protein enzymes were shared by RNAs. In this section we describe some of these results, the new approaches they have led to, and the far-reaching implications of these findings.

Some RNAs Are Self-Splicing

In 1982, Tom Cech discovered that when the pre-rRNA of the protozoan *Tetrahymena* was incubated with Mg^{2+} and guanosine monophosphate, splicing of the RNA occurred without the involvement of any protein. This ability of RNA to carry out its own splicing has given rise to the term "ribozyme."

The mechanism of self-splicing in this case is somewhat different from that observed in the spliceosome reaction (fig. 28.21). First the 3′ hydroxyl group of the guanosine cofactor attacks the phosphodiester bond at the 5′ splice site. This is followed by another transesterification reaction in which the 3′ hydroxyl group of the upstream RNA attacks the phosphodiester bond at the 3′ splice site, thereby completing the splicing reaction. The final reaction products include the spliced rRNA and the excised oligonucleotide.

Since its initial discovery, self-splicing has been found to occur for RNAs from a wide variety of organisms. Certain precursor RNAs that exhibit self-splicing produce lariats, just like those seen in the commonly observed splicing reactions that are catalyzed by spliceosomes (see fig. 28.19). These findings suggest that at one time all splicing reactions were RNA-catalyzed.

Some Ribonucleases Are RNAs

Ribozyme activities are not confined to splicing reactions. Among the large number of RNases, the most sophisticated enzymes are probably those involved in processing because they must attack preRNAs at specific sites. In his studies on

Figure 28.21

Self-splicing of pre-rRNA from the protozoan *Tetrahymena.* The first step is a transesterification reaction in which the 3′ hydroxyl group of a guanosine attacks the phosphodiester bond at the 5′ splice site. The second step involves another transesterification reaction in which the 3′ hydroxyl group of the upstream exon attacks the phosphodiester bond at the 3′ splice site and displaces the 3′ hydroxyl group of the intron.

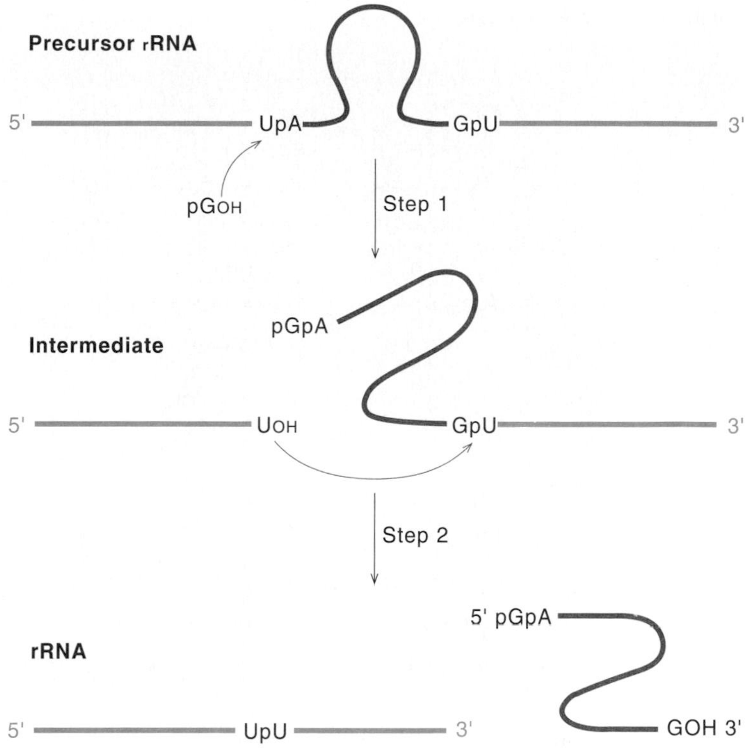

the RNaseP-processing enzyme, Sid Altman discovered that this enzyme is a ribonucleoprotein composed of both a 20 kilodalton protein and a 377-nucleotide RNA molecule. Stimulated by Cech's results, Altman tested the RNA component of RNaseP alone and found that it still had catalytic activity.

Since then other RNA phosphodiesterases have been discovered. Of special interest it was found that single-stranded plant RNAs, called viroids, cleave the viroid RNA at specific linkages. This activity is essential to the life cycle of the viroid.

Ribosomal RNA Catalyzes Peptide Bond Formation

Recently Harry Noller has extended the range of ribozymes to include peptide bond catalysis. Noller and his colleagues found that the removal of all protein from the ribosomes of certain thermophilic bacteria left the ribosomes with the ability to catalyze peptide bond formation.

Catalytic RNA May Have Evolutionary Significance

It seems likely that the range of catalytic activities for RNA will be expanded by future discoveries. For many years prior to the discovery of RNA catalysis, molecular evolutionists struggled with the question of which came first, RNA or protein. Proteins make excellent enzymes, but they lack the template activity required for transmission of information and evolution. On the other hand, nucleic acids (RNA) have excellent template activity, like DNA, but they did not have any known catalytic activities. For a period of time it was impossible to imagine how either of these molecules could have evolved first since proteins cannot exist without templates for ordering amino acids in a polypeptide chain and RNAs cannot exist without enzymes. This dilemma appears to have been resolved by the recent discovery of ribozymes. Now a reasonable argument can be made that the first enzymes were RNAs and the first proteins were made under the direction of ribozymes.

Although this argument gives us a theoretically satisfying resolution to a central dilemma in molecular evolution, it still remains to demonstrate that RNAs have the versatility to function as enzymes for a much wider range of reactions. In this regard it seems unlikely that the full range of ribozymes that probably existed in early evolution are still present. More than likely most primitive ribozymes have long since been replaced by more efficient protein enzymes.

This creates a dilemma of a different sort. How are we to demonstrate ribozyme activities that no longer exist? Perhaps we must design our own ribozymes to demonstrate the full potential for RNA to act as an enzyme. This approach is being used by Jack Szostak and Gerald Joyce, who indepen-

Figure 28.22

Inhibitors of RNA synthesis.

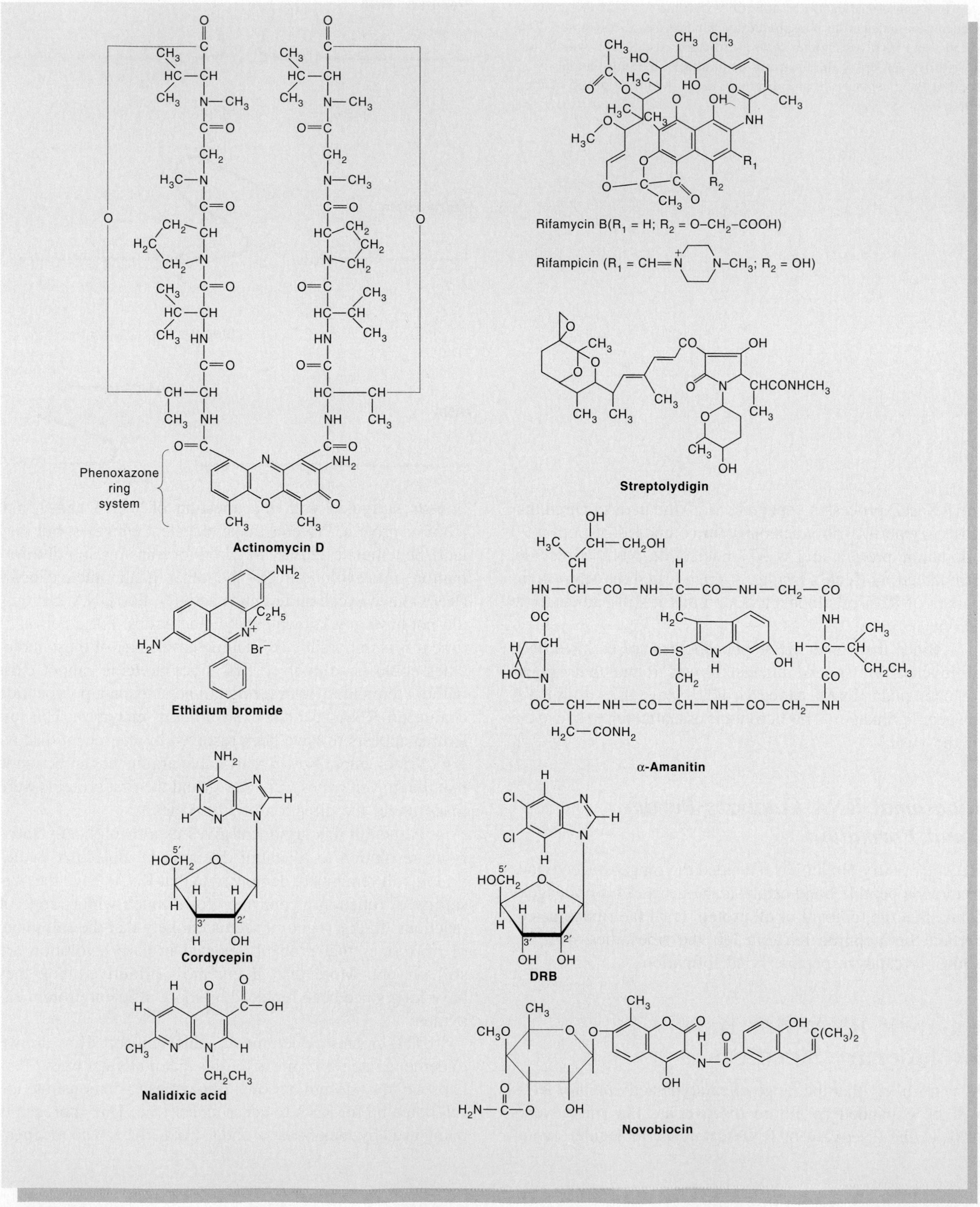

dently have developed *in vitro* systems to select for RNA molecules with predesignated activities. So far, Joyce has selected for a *de novo* synthesized RNA that can cleave DNA, while Szostak has selected for an RNA that catalayzes primer extension by trinucleotides. Rapid progress using this approach to assess the potential of RNA to function as an enzyme is expected.

Inhibitors of RNA Metabolism

A large variety of inhibitors of RNA synthesis have been identified. Some of these inhibitors have proved useful in elucidating transcription mechanisms, and some have facilitated selection for superior mutant strains with enzymes that are resistant to their inhibition. The inhibitors fall into three classes (fig. 28.22).

Some Inhibitors Act by Binding to DNA

The best known example of inhibitors that bind to DNA is actinomycin D, an antibiotic produced by *Streptomyces antibioticus.* The inhibition of RNA synthesis is caused by the insertion (intercalation) of its phenoxazone ring between two G-C base pairs, with the side chains projecting into the minor groove of the double helix, hydrogen-bonded to guanine residues. RNA polymerase binding to DNA that contains actinomycin D is only slightly impaired, but RNA chain elongation in both eukaryotes and prokaryotes is blocked. Ethidium bromide also intercalates into DNA and at low concentrations preferentially binds to negatively supercoiled DNA (see chapter 25). It has been used to selectively inhibit transcription in mitochondria, which contains supercoiled DNA.

Some Inhibitors Bind to RNA Polymerase

Rifampicin is a synthetic derivative of a naturally occurring antibiotic, rifamycin, that inhibits bacterial DNA-dependent RNA polymerase but not T7 RNA polymerase or eukaryotic RNA polymerases. It binds tightly to the β subunit. Although it does not prevent promoter binding or formation of the first phosphodiester bond, it effectively prevents synthesis of longer RNA chains. It does not inhibit elongation when added after initiation has occurred. Another antibiotic, streptolydigin, also binds to the β subunit; it inhibits all bond formation.

The most useful inhibitor of eukaryotic transcription has been α-amanitin, a major toxic substance in the poisonous mushroom *Amanita phalloides.* The toxin preferentially binds to and inhibits RNA polymerase II (see table 28.4). At high concentrations it also can inhibit RNA polymerase III but not RNA polymerase I or bacterial, mitochondrial, or chloroplast RNA polymerases.

Some Inhibitors Are Incorporated into the Growing RNA Chain

Cordycepin in its 5′-triphosphorylated form is a substrate analog that is incorporated into growing RNA chains by most RNA polymerases. It causes chain termination after incorporation, since it does not contain the 3′ hydroxyl group necessary for the formation of the next phosphodiester bond.

Summary

In this chapter we described the synthesis, transcription, and posttranscriptional reactions undergone by the three major classes of RNA. The main points we covered are as follows:

1. RNA is synthesized in the 5′ → 3′ direction by the formation of 3′-5′-phosphodiester linkages between four ribonucleoside triphosphate substrates, analogous to the process of DNA synthesis. The sequence of bases in RNA transcripts catalyzed by DNA-dependent RNA polymerases is specified by the complementary sequences of the DNA template strand.
2. Some newly synthesized RNA transcripts are the functional species, whereas others must be modified or processed into the mature functional species. Modifying enzymes add nucleotides to the 5′ or 3′ ends or alter bases within the RNA, such as by methylation of specific residues. Specific processing enzymes cleave RNA internally, splice together noncontiguous regions of a transcript, or remove nucleotides from the 5′ or 3′ ends.
3. The major classes of RNA in both prokaryotes and eukaryotes are messenger RNA, ribosomal RNA, and

transfer RNA. These distinct classes of RNA play specific functional or structural roles in the translation of genetic information into proteins.
4. DNA-dependent synthesis of RNA in *E. coli* is catalyzed by one enzyme, consisting of five polypeptide subunits. The complete holoenzyme is composed of four polypeptides (the core enzyme) and an additional polypeptide that confers specificity for initiation at promoter sequences in the DNA template.
5. The steps involved in transcription include binding of polymerase at the initiation site, initiation, elongation, and termination.
6. In eukaryotes most transcription takes place in the nucleus. Three nuclear RNA polymerases, I, II and III, are responsible for the synthesis of rRNA, mRNA, and small RNA transcripts, respectively. The polymerases contain more subunits than in *E. coli,* and other proteins must bind at the initiation sites or near the initiation sites before the polymerases can begin transcription.
7. Viruses sometimes use the host RNA polymerase in a modified form and sometimes synthesize their own RNA polymerase.
8. One of the more interesting processing reactions of nascent transcripts involves the removal of internal sequences. This type of processing is referred to as splicing. Most splicing reactions appear to require host proteins; however, some only require RNA, a fact demonstrating that RNA is capable of functioning like an enzyme in making and breaking of phosphodiester linkages in polyribonucleotides. Catalytic RNAs may have evolutionary significance.
9. Inhibitors of RNA synthesis may be classified according to their mechanism of action. Some bind to DNA, some bind to RNA, and some are incorporated into the growing RNA chain during transcription.

Selected Readings

Ahsen, U., and H. F. Noller, Footprinting the sites of interaction of antibiotics with catalytic group I intron RNA. *Science* 260:1500–1503, 1993.

Altman, S., M. Baer, C. Guerrier-Takada, and A. Vioque, Enzymatic cleavage of RNA by RNA. *Trends Biochem. Sci.* 11:515–518, 1986.

Barinaga, M., Ribozymes: Killing the messenger. *Science* 262:1512–1514, 1993.

Bass, B. L., Splicing: The new edition. *Nature* 352:283–284, 1991.

Bear, D. G., and D. W. Peabody, The *E. coli* rho protein: An ATPase that terminates transcription. *Trends Biochem. Sci.* 13:343–348, 1988.

Beaudry, A. A., and G. R. Joyce, Directed evolution of an RNA enzyme, *Science* 257:635–641, 1992.

Bjork, G. R., J. U. Ericson, C. E. D. Gustafsson, T. G. Hdagervall, Y. H. Josson, and P. M. Wikstrom, Transfer RNA modification. *Ann. Rev. Biochem.* 56:263–287, 1987.

Buratowski, S., S. Hahn, L. Guarente, and P. A. Sharp, Five intermediate complexes in transcription initiation by RNA polymerase II. *Cell* 56:549–561, 1989.

Burtis, K. C., and B. X. Baker, *Drosophila* double sex gene controls somatic sexual differentiation by producing alternatively spliced mRNAs encoding related sex-specific polypeptides. *Cell* 56:997–1010, 1989.

Cattaneo, R., RNA editing: In chloroplast and brain. *Trends Biochem. Sci.* 17:4–6, 1992. A recent review dealing with RNA editing.

Cech, T., RNA editing: World's smallest introns? *Cell* 64:667–669, 1991.

Cech, T. R., RNA as an enzyme. *Sci. Am.* 225(5):64–75, 1986.

Cech, T. R., and B. L. Bass, Biological catalysis by RNA. *Ann. Rev. Biochem.* 55:599–630, 1986.

Chambon, P., Split genes. *Sci. Am.* 244:60–66, 1981.

Chowrira, B. M., A. Berzal-Herranz, and J. M. Burke, Novel guanosine requirement for catalysis by the hairpin ribozyme. *Nature* 354:320–323, 1991. In some cases the guanine amino group serves a catalytic role in self-splicing.

Crick, F., Central dogma of molecular biology. *Nature* 227:561–563, 1970.

Dahlberg, A. E., The functional role of ribosomal RNA in protein synthesis. *Cell* 57:525–529, 1989.

Darnell, J. E., Jr., RNA. *Sci. Am.* 253(4):68–78, 1985.

Deutscher, M. P., The metabolic role of RNases. *Trends Biochem. Sci.* 13:136–139, 1988.

Dorit, R. L., L. Schoenbach, and W. Gilbert, How big is the universe of exons? *Science* 250:1377–1382, 1990.

Forster, A. C., A. C. Jeffries, C. C. Sheldon, and R. H. Symons, Structural and ionic requirements for self-cleavage of virusoid RNAs and trans self-cleavage of viroid RNA. *Cold Spring Harb. Symp. Quant. Biol.* 52:249–259, 1987.

Futcher, B., Supercoiling and transcription, or vice versa? *Trends Genet.* 4:271–272, 1988.

Geiduschek, E. P., and G. P. Tocchini-Valentini, Transcription by RNA polymerase III. *Ann. Rev. Biochem.* 57:873–914, 1988.

Haas, E. S., D. P. Morse, J. W. Brown, F. J. Schmidt, and N. R. Pace, Long-range structure in ribonuclease P RNA. *Science* 254:853–856, 1991.

Hall, B. D., and S. Spiegelman, Sequence complementarity of T2-DNA and T2 specific RNA. *Proc. Natl. Acad. Sci. USA* 47:137–146, 1964. The first use of RNA–DNA hybridization.

Helmann, J. D., and M. J. Chamberlin, Structure and function of bacterial sigma factors. *Ann. Rev. Biochem.* 57:839–872, 1988.

Hou, Y.-M., and P. Schimmel, A simple structural feature is a major determinant of the identity of a transfer RNA. *Nature* 333:144–145, 1988.

Khoury, G., and P. Gruss, Enhancer elements. *Cell* 33:313–314, 1983. The first report on enhancers.

Koleske, A. J., and R. A. Young, An RNA polymerase II holoenzyme responsive to activation. *Nature* 368:466–473, 1994.

Landweber, L. R., and W. Gilbert, RNA editing as a source of genetic variation. *Nature* 363:179–182, 1993.

Leff, S. D., M. G. Rosenfeld, and R. M. Evans, Complex transcriptional units: Diversity in gene expression by alternative RNA processing. *Ann. Rev. Biochem.* 55:1091–1117, 1986.

Lorsch, J. R., and J. W. Szostak, In vitro evolution of new ribozymes with polynucleotide kinase activity. *Nature* 371:31–36, 1994.

Lührmann, R., B. Kastner, and M. Bach, Structure of spliceosomal snRNPs and their role in pre-mRNA splicing. *Biochem. Biophys. Acta* 1087:265–292, 1990.

Nikoliu, D. B., S-H. Hu, J. Lin, A. Gasch, A. Hoffman, M. Horikoshi, N-H. Chua, R. G. Roeder, and S. K. Burly, Crystal structure of TF IID TATA-box binding protein. *Nature* 360:40–45, 1992.

Padgett, R. A., P. J. Grabowski, M. M. Komarska, S. Seller, and P. A. Sharp, Splicing of messenger RNA precursors. *Ann. Rev. Biochem.* 55:1119–1150, 1988.

Patzelt, E., K. L. Perry, and N. Agabian, Mapping of branch sites in trans-spliced pre-mRNAs of *Trypanosoma brucei. Mol. Cell. Biol.* 9:4291–4297, 1989. Cases where the splicing event involves transcripts from different chromosomes.

Petska, S., J. A. Langer, K. C. Zoon, and C. E. Samuel, Interferons and their actions. *Ann. Rev. Biochem.* 56:757–777, 1987.

Peterson, M. G., J. Inostroza, M. E. Maxon, F. Osvaldo, A. Admon, D. Reinberg, and R. Tjian, Structure and functional properties of human general transcription factor IIt. *Nature* 354:369–373, 1991.

Proudfoot, N. J., How RNA polymerase II terminates transcription in higher eukaryotes. *Trends Biochem. Sci.* 114:105–110, 1989.

Steitz, J. A., ''Snurps.'' *Sci. Am.* 258(6):56–63, 1988.

Stuart, K., RNA editing in mitochondrial mRNA of trypanosomatids. *Trends Biochem. Sci.* 16:68–72, 1991. RNA editing is processing that involves the removal, addition, and modification of nucleotides in the coding regions of nascent transcripts.

Thompson, C. C., S. L. McKnight, Anatomy of an enhancer. *Trends in Genetics* 8:232–236, 1992.

Weiner, A. M., mRNA splicing and autocatalytic introns: Distant cousins or the products of chemical determination? *Cell* 72:161–164, 1993.

Xing, Y., C. V. Johnson, P. R. Dobner, and J. B. Lawerence, Higher level organization of individual gene transcription and RNA splicing, *Science* 259:1330, 1993.

Problems

1. Compare the reactions catalyzed by RNA and DNA polymerases. What are the similarities and differences?
2. One strand of DNA is completely transcribed into RNA by RNA polymerase. The base composition of the DNA template strand is: G = 20%, C = 25%, A = 15%, T = 40%. What would you expect the base composition of the newly synthesized RNA to be?
3. Although 40 to 50% of the RNA being synthesized in *E. coli* at any given time is mRNA, only about 3% of the total RNA in the cell is mRNA. Explain.
4. In *E. coli,* the mRNA fraction is heterogeneous in size, ranging from 500 to 6,000 nucleotides. The largest mRNAs have many more nucleotides than needed to make the largest proteins. Why are some of these mRNAs so large in bacteria?
5. Base-stacking interactions are an important stabilizing force in nucleic acid structures. Describe how base stacking contributes to the tertiary structure of the tRNA molecule.
6. Why is a single-strand binding protein or a DNA helicase not required for transcription as they are for replication?
7. When yeast phenylalanine tRNA is digested with a small amount of RNase (partial digest) such as T2 RNase, almost all of the cleavage sites are found in the anticodon loop. Explain this observation.
8. Speculate on the advantage of having three rRNAs (16S, 23S, and 5S) as part of the same RNA precursor (such as is found in *E. coli*)?
9. In *E. coli* the precise spacing between the −35 and −10 conserved promoter elements has been found to be a critical determinant of promoter strength. What

does this suggest about the interaction between RNA polymerase and these elements? What sorts of evidence could you obtain about this interaction by doing ''footprint'' experiments? (Also explain how you would do these experiments.)

10. Human RNA polymerase II generates RNA at a rate of approximately 3,000 nucleotides per minute at 37°C. One of the largest mammalian genes known is the 2,000-kbp (1 kbp = one kilo base-pair = 1,000 base-pairs) gene encoding the muscle protein, dystrophin. How long would it take one RNA polymerase II molecule to completely transcribe this gene? How long would it take to transcribe the adult β-globin gene (1.6 kbp)?
11. In early research with intact eukaryotic mRNA, the RNA appeared to have two 3′-ends and no 5′-terminus. Explain these observations.
12. What is unique about the promoter for RNA polymerase III? Diagram how this promoter works.
13. It has recently been demonstrated that the TATA-binding protein, in addition to flattening and widening the DNA by interacting (atypically) with the minor groove, also introduces a sharp bend ($>100°$) into its recognition sequence on binding. How might these changes in the structure of the DNA facilitate assembly of the RNA polymerase II transcription initiation complex?
14. Propose a sequence for the guide RNA that directs the editing of precursor RNA to yield the mRNA shown in figure 28.20.
15. A particular eukaryotic DNA virus is found to code for two mRNA transcripts, one shorter than the other, from the same region on the DNA. Analysis of the translation products reveals that the two polypeptides share the same amino acid sequence at their amino-terminal ends but are different at their carboxyl-terminal ends. The longer polypeptide is coded by the shorter mRNA! Suggest an explanation.
16. Although a phosphodiester bond must be formed between the upstream and downstream exons, there is no direct ATP requirement for splicing of mRNA introns or for splicing of the *Tetrahymena* pre-rRNA intron. Explain this observation.

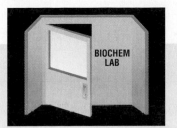

Use of the Footprinting Technique to Determine the Binding Site of a DNA-binding Protein

To determine the site of action of the transcription factor TFIIIA, the approximate location of its binding site on a 5S gene of *Xenopus borealis* was determined by the so-called ''footprinting'' technique. DNA fragments containing the 5S gene were labeled at the 5′ end with ^{32}P mixed with TFIIIA and then digested with DNAseI. The resulting DNA fragments were electrophoresed on a polyacrylamide gel that was subsequently autoradiographed (see fig. 1). Regions of the DNA that were protected from DNase attack by TFIIIA binding appear as a blank spot (footprint) on the autoradiogram. The footprint shows that the protected region is situated between the 45th and the 90th base pair.

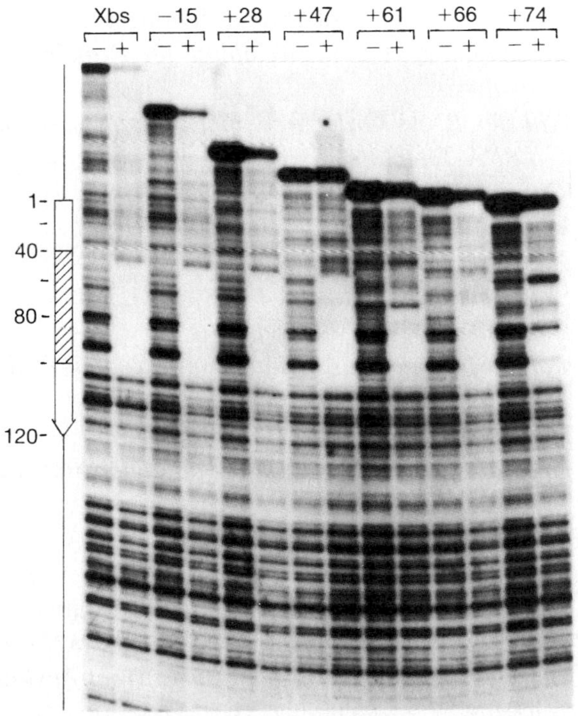

Figure 1

DNaseI protection (footprinting) experiment on 5S DNA of *Xenopus*. The diagram on the left indicates the region on the gel that corresponds to the 5S RNA gene. Arrow points in the direction of transcription. Cross-hatched area indicates the region that binds transcription factor protein. Column labeled Xbs refers to an intact gene containing 160 bp of the 5′ flanking sequence, the 5S RNA gene (120 bp), and the 3′ flanking sequence (138 bp). The various deleted 5S DNAs are preceded by 74 bp of the plasmid pBR322 sequence. Numbers in other columns refer to portions of the 5S gene that have been deleted. All samples were subjected to partial digestion with DNaseI, then were electrophoresed and autoradiographed. In + columns, transcription factor protein was added before DNaseI treatment. (Courtesy Donald D. Brown of the Carnegie Institute of Washington.)

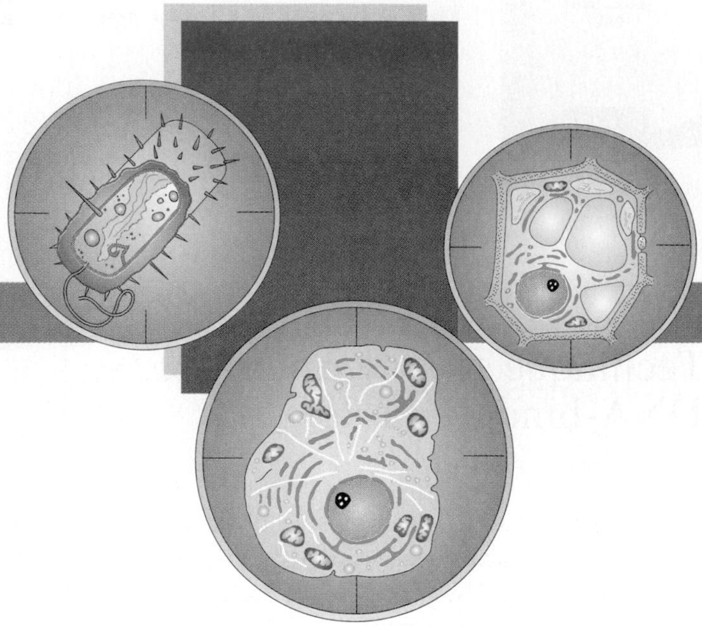

CHAPTER

29

Protein Synthesis, Targeting, and Turnover

Chapter Outline

The arrangement of amino acids in polypeptide chains is determined by the arrangement of codons in messenger RNA molecules.

Proteins are informational macromolecules, the ultimate heirs of the genetic information encoded in the sequence of nucleotide bases within the chromosomes. Each protein is composed of one or more polypeptide chains, and each peptide chain is a linear polymer of amino acids. The order of the amino acids commonly found in the polypeptide chain is determined by the order of nucleotides in the corresponding messenger RNA template. In this chapter we examine four aspects of protein metabolism (fig. 29.1): (1) The process whereby amino acids are ordered and polymerized into polypeptide chains; (2) posttranslational alterations in polypeptides, which occur after they are assembled on the ribosome; (3) the targeting process whereby proteins move from their site of synthesis to their sites of function; and (4) the proteolytic reactions that result in the return of proteins to their starting material, amino acids.

The Cellular Machinery of Protein Synthesis

Amino acids are assembled into polypeptides on ribosomes. Prior to their interaction with messenger RNA (mRNA), amino acids are covalently attached to transfer RNA (tRNA) to form aminoacyl-tRNAs. The aminoacyl-tRNAs attach to specific sites on the mRNA. Messenger RNA contains the instructions for translation in the form of the genetic code that specifies the amino acid sequence of the polypeptide to be synthesized. Each ribosome binds to and moves along the messenger RNA while producing a single polypeptide. The direction or polarity of translation is from the 5′ to the 3′ terminus on the message, whereas the polypeptide is synthesized from the amino to the carboxyl terminus. As a rule, the information in a single mRNA molecule is translated simultaneously by a number of ribosomes that form a structure called a polysome. In prokaryotic cells, translation of mRNA begins while it is still being transcribed so that assemblies of nascent polypeptides and mRNAs can be seen on chromosomal DNA (fig. 29.2). In eukaryotic cells, transcription and translation are separate, and polysomes occur either free in the cytosol or bound to membranes in the endoplasmic reticulum (fig. 29.3).

Messenger RNA Is the Template for Protein Synthesis

The mRNA molecule carries the genetic message in the form of a sequence of nucleotides that determines the order of amino acids in the polypeptide chain. Each amino acid is represented in the mRNA by a sequence of three nucleotides called codons. Codons are arranged in a contiguous reading frame, which is flanked on either side by bases that are not translated. These untranslated regions frequently have roles in regulating the processing and expression of the message. The 5′ end of the reading frame begins with a start codon, usually consisting of the nucleotides AUG, a sequence that codes for the amino acid methionine. Methionine is always used to initiate translation. The 3′ end of the reading frame contains one or more of three stop codons: UAA, UAG, or UGA. Stop codons serve as signals to terminate the polypeptide chain. The 3′ end of the message in eukaryotes usually contains a posttranscriptionally added poly(A) tail. This poly(A) tail has no known role in translation except that it may increase the lifetime of the message.

The 5′ end of the mRNA plays a special role in the selection of the start codon. This selection process occurs in fundamentally different ways in prokaryotes and eukaryotes. In prokaryotes (fig. 29.4) the mRNA contains a specific ribosome-binding site upstream of the initiating start codon. This binding site allows the ribosome to identify the correct initiating AUG. Prokaryotic mRNAs are frequently polycistronic, meaning that they encode more than one polypeptide chain. Internal ribosome-binding sites allow for independent initiation at internal reading frames. Eukaryotic mRNAs, on the other hand (see fig. 29.4), are usually monocistronic, encoding only one polypeptide chain. In keeping with this architecture, eukaryotic ribosomes interact with a ribosome entry site at the 5′ terminus of mRNA and move along it by a scanning mechanism to find the start codon. Identification of the ribosome entry site may be partly facilitated by the cap structure (see fig. 28.17) at the 5′ end of the mRNA. Initiation in eukaryotes usually occurs at the first AUG sequence downstream from the ribosome entry site. In multicellular eukaryotes the initiating AUG sequence is selected by a preferred initiation context of adjacent bases.

Transfer RNAs Order Activated Amino Acids on the mRNA Template

Transfer RNAs contain two crucial functional sites: A site for attachment of the amino acids and a site that interacts with the mRNA during translation on the ribosome. At least one type of tRNA corresponds to each of the 20 types of amino acids that are incorporated into proteins. For accurate translation to occur, tRNAs must be distinguishable from one another by the molecules that recognize these specific types while still being recognizable to the molecules that interact with all tRNAs. A major challenge has been to understand how the similarities and differences in the structures of different tRNAs are related to the functions that

Figure 29.1

Overview of reactions in protein synthesis. (aa_1, aa_2, aa_3 = amino acids 1, 2, 3.) Protein synthesis requires transfer RNAs for each amino acid, ribosomes, messenger RNA, and a number of dissociable protein factors in addition to ATP, GTP, and divalent cations. First the transfer RNAs become charged with amino acids, then the initiation complex is formed. Peptide synthesis does not start until the second aminoacyl tRNA becomes bound to the ribosome. Elongation reactions involve peptide bond formation, dissociation of the discharged tRNA, and translocation. The elongation process is repeated many times until the termination codon is reached. Termination is marked by the dissociation of the messenger RNA from the ribosome and the dissociation of the two ribosomal subunits. The polypeptide chain sometimes folds into its final form without further modifications. Frequently the folded polypeptide chain is modified by removal of part of the polypeptide chain or by addition of various groups to specific amino acid side chains. The completed polypeptide chain migrates to different locations according to its structure. It may remain in the cytosol or it may be transported into one of the cellular organelles or across the plasma membrane into extracellular space. Proteins have different lifetimes. Eventually a protein is degraded, usually down to its component amino acids.

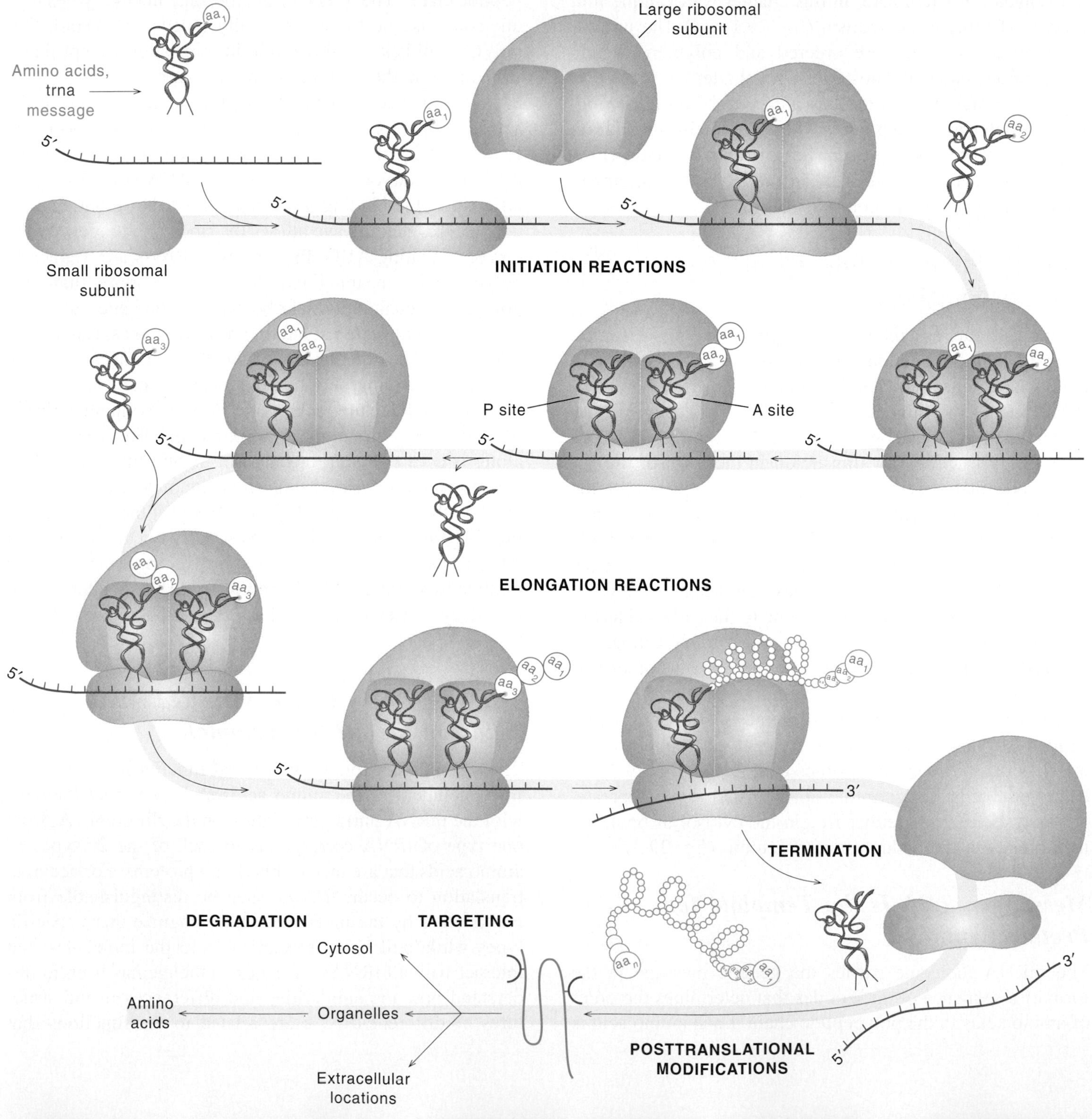

Figure 29.2

(*a*) Electron micrograph of *E. coli* polysomes. Ribosomes are the dark structures connected by the faintly visible mRNA strand. A DNA strand connecting the polysomes from which the mRNA is being transcribed is visible as a horizontal line. (From O. L. Miller, B. A. Hankalo, and C. A. Thomas, "Visualization of bacterial genes in action," *Science* 169:392, 1970, © 1970 by AAAS.) (*b*) Line drawing for clarification. In bacteria, translation usually begins before transcription is completed, resulting in a DNA–mRNA–ribosome complex as shown. As the mRNA grows, the number of ribosomes associated with it increases. Here the mRNA appears to be growing from left to right. It is not possible to see the growing polypeptide chains on the ribosome because of the staining procedure used and the relatively small size of the polypeptide chains.

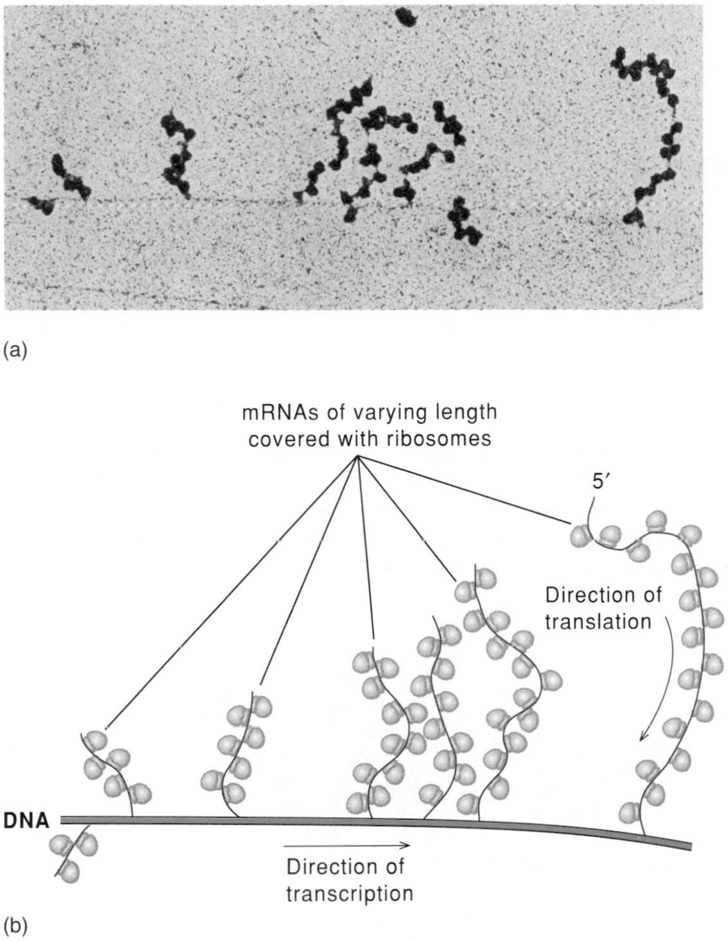

Figure 29.3

(*a*) Electron micrograph of mammalian rough endoplasmic reticulum. Continuous sheets of membrane create a compartment distinct from the surrounding cytosol. (From S. L. Wolfe, *Biology of the Cell,* 2d ed., 1981, Wadsworth Publishing Co.) (*b*) Clarifying line drawing shows expanded region of endoplasmic reticulum with ribosomes attached to the surface of the membrane facing the cytosol. The term "rough endoplasmic reticulum" arose because at low magnification the attached ribosomes give the endoplasmic reticulum a rough appearance.

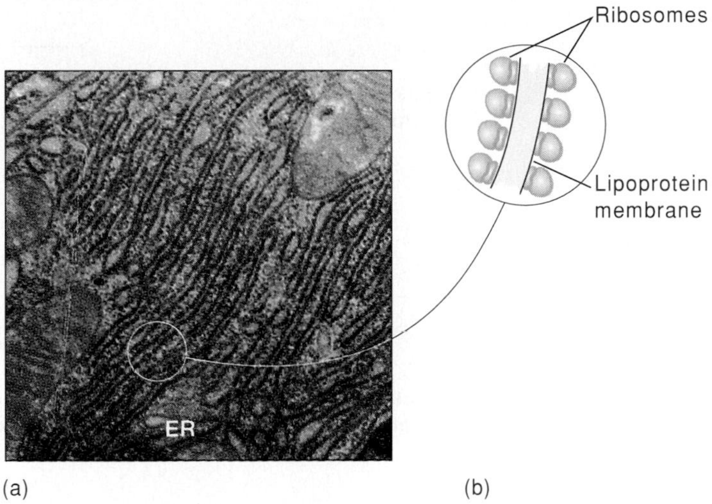

they play in protein synthesis. Here we summarize the structural features found in all tRNAs that allow them to perform their common functions.

Transfer RNAs that correspond to a single amino acid type are known as cognate tRNAs. In writing, the different cognate tRNAs are designated by a superscript. For example, $tRNA^{Phe}$ and $tRNA^{Ser}$ designate the tRNAs that correspond to the amino acids phenylalanine and serine, respectively. Some amino acids have several different cognate tRNAs, which are called isoacceptor tRNAs. Isoacceptor tRNAs are distinguished from one another by subscripts. A single cell typically contains a total of 50 or more different tRNAs.

The tRNAs are built on a structural theme so similar that a single figure can describe the common features of their primary, secondary, and tertiary structures (fig. 29.5). All tRNAs are composed of a single polynucleotide chain of from 70–95 residues. This chain folds back on itself to form a cloverleaf structure composed of four double-stranded stems and four single-stranded loops. The overall folding is such that 5′ and 3′ ends of the molecule are brought together to create a stem with seven regular Watson-Crick base pairs. This stem is known as the acceptor stem because during protein synthesis the amino acid is transiently attached to the ribose at this terminus. The 3′ end of the molecule is always terminated with the sequence CCA, which identifies the tRNA to the components that interact with all of these molecules during protein synthesis.

The unpaired loops of tRNA are named according to their unique structural features. Loop I varies in size from 7 to 11 unpaired bases and frequently contains the unusual base dihydrouracil; it is designated the D loop. Loop II contains the three bases known as the anticodon and is therefore designated the anticodon loop. This portion of tRNA plays a

Figure 29.4

Simplified diagram of mRNA structure. (*a*) Typical eukaryotic mRNA. An AUG start codon is located near the 5′ end of the mRNA. The single reading frame ends with one of the three trinucleotide sequences that represents a stop codon. Frequently, but not always, the 5′ end of the mRNA is capped, and the 3′ end contains a poly(A) tail. The cap structure is described in figure 28.17. (*b*) Typical bacterial mRNA. Some bacterial mRNAs contain more than one reading frame; some contain only one. Each reading frame contains a start and a stop codon. Recognition of the start codons is facilitated by the presence of a ribosome recognition sequence (Shine-Dalgarno). The space between any two reading frames in the same transcript varies from one transcript to the next.

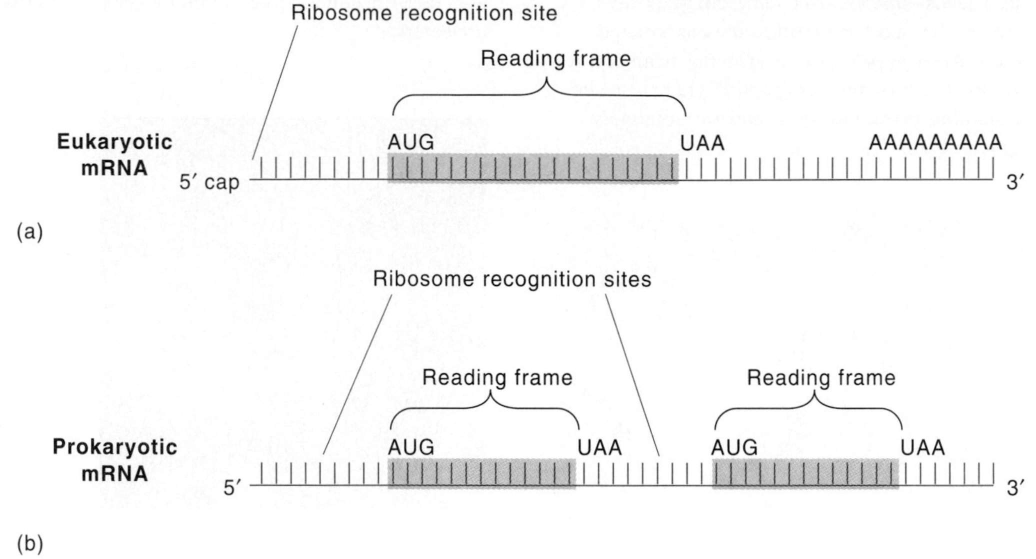

Figure 29.5

The general structure of a tRNA molecule. (*a*) Representation in the form of a cloverleaf, which is the simplest way of observing the secondary structure. (*b*) A more realistic drawing of the three-dimensional folded structure. Color coding shows how the various loops in the cloverleaf structure correspond to the parts of the folded structure. For a more detailed drawing showing the types of hydrogen bonds and the structures of the individual bases, see figure 28.3. (From A. Rich and S. H. Kim, The three-dimensional structure of transfer RNA, *Sci. Amer.* 238:52–62, January 1978. Copyright © 1978 by Scientific American, Inc. All rights reserved. Reprinted by permission.)

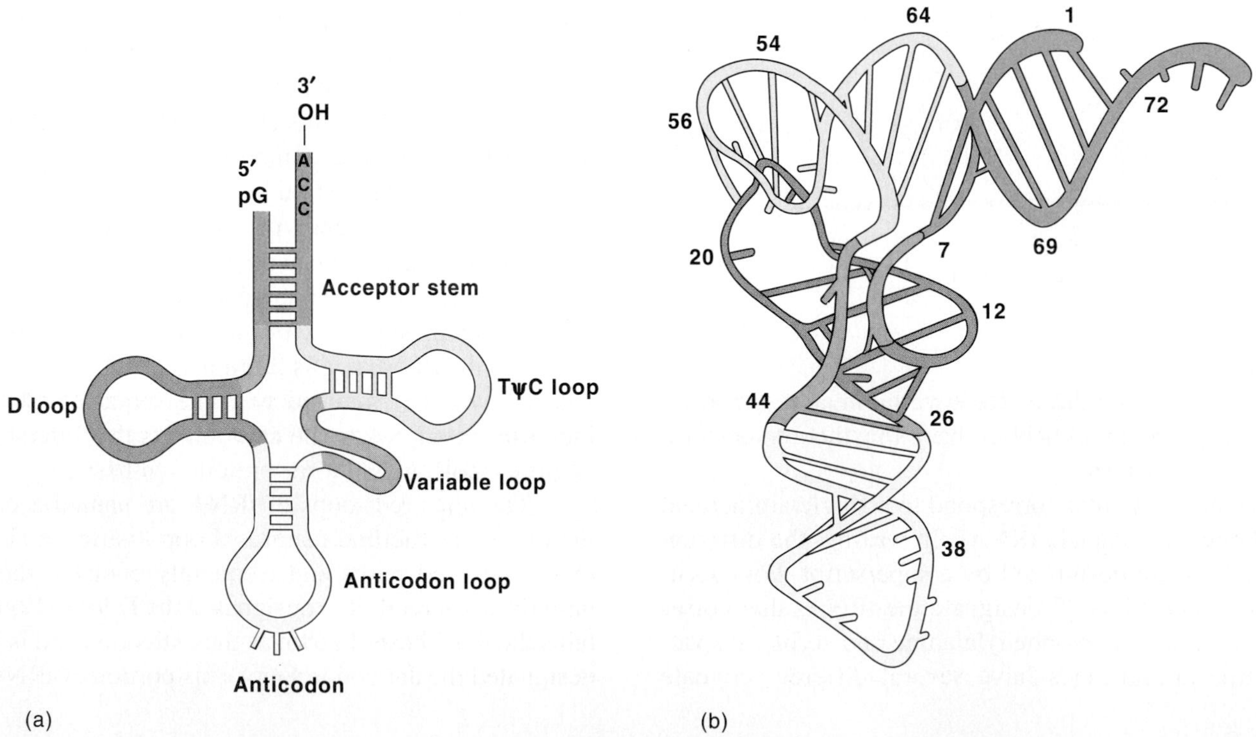

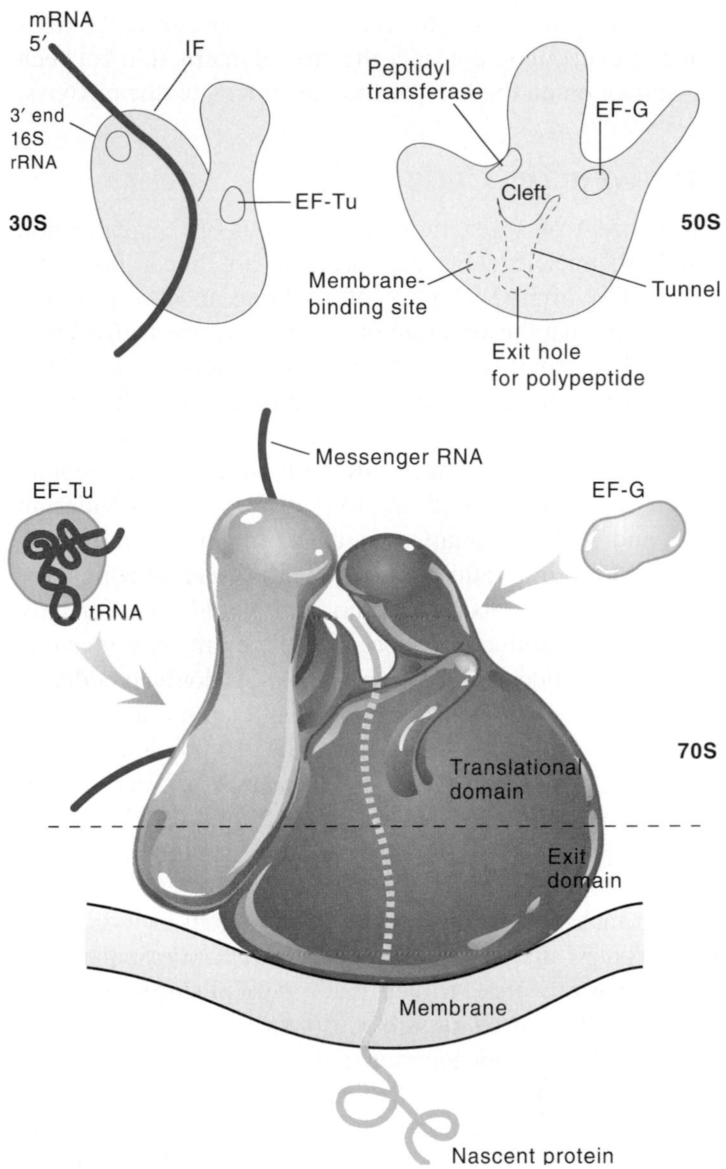

Figure 29.6

Functional sites on the prokaryotic ribosome. This figure shows a ribosome in the process of elongating a polypeptide chain. The mRNA and the EF-Tu-aminoacyl-tRNA complex are more closely associated with the 30S subunit. The peptidyl transferase and the EF-G elongation factor are associated with the 50S subunit, as is the nascent polypeptide chain. IF and EF are standard abbreviations for proteins that function as initiation factors and elongation factors, respectively. (From C. Bernabeu and J. A. Lake, Nascent polypeptide chains emerge from the exit domain of the large ribosomal subunit: Immune mapping of the nascent chain, *Proc. Nat'l. Acad. Sci. USA* 79:3111–3115, 1982. Copyright © 1982 National Academy of Sciences, Washington, D.C. Reprinted by permission.)

key role in translation by pairing with the complementary codon in mRNA and thus serves to align amino acids in their appropriate sequence. Loop III, the variable loop, may contain as few as 3 or as many as 21 bases, making it the major site of size variation in tRNA. Loop IV contains the unusual ribothymidine and pseudouridine bases as part of an invariant sequence; for this reason it is known as the TψC loop.

All tRNAs in solution fold into a three-dimensional L-shaped structure like that of $tRNA^{Phe}$ (fig. 29.5*b*). This structure is composed of two helical arms joined at right angles. The ribose moiety to which the amino acid is joined is at the end of one arm, identifying it as the acceptor arm. The anticodon is at the end of the other arm, identifying it as the anticodon arm. The 3′ terminal ribosome and anticodon are separated by a diagonal distance of about 75 Å.

Ribosomes Are the Site of Protein Synthesis

We considered the structure of the ribosome in some detail in the previous chapter without referring to those sites that are functionally important in protein synthesis. Here we can localize some of the functional sites on the two ribosomal subunits (fig. 29.6). The mRNA binds to the smaller ribosomal subunit. The peptidyl transferase is an integral part of the 50S subunit, and the elongation factor EF-G binds to the 50S subunit. The nascent polypeptide chain exits through a channel in the 50S subunit. Two functional sites occur on the 50S subunits, called the P site and the A site, where adjacent tRNAs are bound to the messenger (see fig. 29.1).

Before we discuss the reactions that occur on the ribosome, it is useful to consider the crucial interaction between the anticodon on the tRNAs and the codons on the mRNAs.

The Genetic Code

The concept of a genetic code grew out of the realization that both nucleic acids and proteins are linear polymers made of a limited number of building blocks, plus the knowledge that the structure of proteins is genetically determined. The genetic code is simply the sequence relationship between nucleotides in genes (or mRNA) and the amino acids in the proteins they encode. The coding ratio is the number of nucleotides in an mRNA required to represent an amino acid. Since there are four different nucleotides in RNA and 20 different amino acids in proteins, the minimum acceptable coding ratio is 3 : 1. A singlet code, in which one nucleotide represents one amino acid, could code for only four different amino acids because there are only four different nucleotides in DNA or RNA. A doublet code, in which two nucleotides code for a single amino acid, could code for 16 amino acids since there are 16 possible doublet sequences in RNA. A triplet code, made from four nucleotides taken three at a time, generates a total of 64 possible triplet sequences. This would be more than enough to represent the 20 amino acids found in proteins.

In fact, a triplet code creates a problem of a different sort. What is the use of the extra trinucleotide sequences? Allowing for the three triplets that represent stop signals, we are still left with 61 possible triplets to code for just 20 amino acids. A nondegenerate code, in which a unique triplet codes for each amino acid, would imply that 41 triplet sequences are never found in translation reading frames. On the other hand, a degenerate code, in which all triplets are used, requires that each amino acid be represented, on the average, by more than one triplet sequence. We return to this point later. For now, we simply note that in most cases, point mutations in which one base pair in the DNA is replaced by another result either in no amino acid change or in a change to another amino acid. This is the result that we expect if most of the triplet sequences do in fact represent amino acids.

Another question relates to the location of codons in the translation reading frame. Codons could be arranged in a close-packed side-by-side arrangement. In this event each nucleotide within the translation reading frame would represent a code letter within one, and only one, codon. Alternatively, codons could be separated by one or more ''spacer nucleotides.'' In this event some nucleotides within the reading frame would represent code letters within codons, while others would serve as spacer nucleotides. Finally, we can imagine an overlapping codon arrangement, in which all nucleotides represent code letters, and some nucleotides represent code letters in more than one codon. However, biochemical experiments that we describe later in this chapter favor a nonoverlapping triplet code without spacers. Further genetic experiments in support of this type of code (not elaborated on here) indicate that when a single base pair is added or deleted, a change occurs in the reading frame of the message downstream of the change. Such an alteration in the DNA is known as a frameshift mutation. A frameshift mutation produces a new termination point because the frameshift changes the reading of the original stop codon and produces a new stop codon.

The Code Was Deciphered with the Help of Synthetic Messengers

In 1961, Marshall Nirenberg and Heinrich Mattaei were attempting to synthesize proteins in cell-free extracts of *E. coli.* These extracts, containing the essential cellular components, were supplemented with nucleotides, salts, and radioactive amino acids, which were used so that the expected small amounts of protein could be detected. For mRNA they chose to use viral RNA from tobacco mosaic virus (TMV), because their goal was to make viral protein. As a control for these first ''incorporation'' experiments, they needed an mRNA that would not be expected to code for a protein. For this purpose, they chose poly(U), which contained only uridine residues. When Nirenberg and Mattaei added this control RNA to their cell-free system, substantial incorporation was evident. They showed that this synthetic messenger specifies the synthesis of a polypeptide containing only phenylalanine residues (polyphenylalanine).

A wave of excitement was set in motion by this discovery. Soon afterwards it was demonstrated that poly(A) promotes polylysine synthesis and poly(C) promotes polyproline synthesis. From these observations it seemed clear that the code words UUU, AAA, and CCC correspond to the amino acids phenylalanine, lysine, and proline, respectively.

Polynucleotide phosphorylase was used to produce ''RNA'' of random sequence, the composition of which reflected the mixture of nucleoside diphosphates in the reaction mixture. Mixed polynucleotides containing two bases were used in the incorporating system and shown to incorporate a pattern of amino acids consistent with a triplet code, but the observed incorporation could not define the code sequence.

The chemical synthesis of short oligonucleotides of known sequence provided a way out of the dilemma. First,

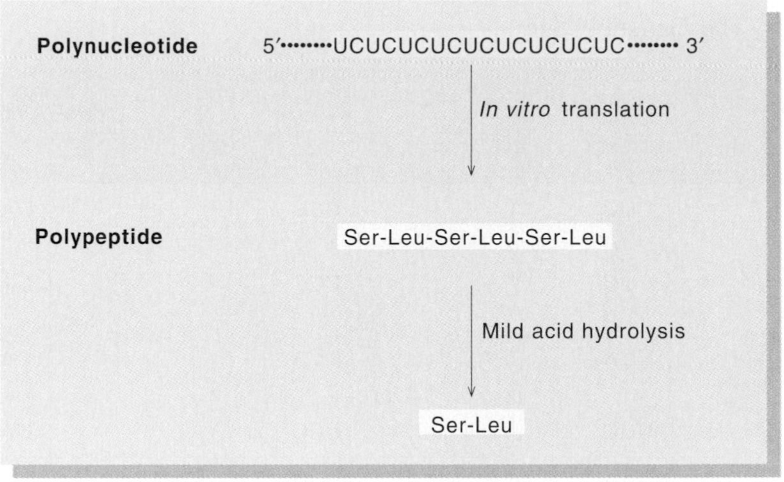

Figure 29.7

Demonstration that poly(UC) codes for a Ser-Leu repeating peptide. When poly(UC) containing a strict alternating sequence of U and C is used as an mRNA, only serine and leucine are incorporated. Analysis of the polypeptide product indicates that it contains an alternating sequence of serine and leucine.

Philip Leder and Nirenberg showed that nucleotide triplets cause the specific binding of aminoacyl tRNA to ribosomes. They cleverly observed that this binding could be readily measured by simply passing a solution containing radioactive aminoacyl-tRNA, ribosomes, and the trinucleotide through a nitrocellulose filter. Ribosomes, along with bound ligands, were adsorbed to the filter, and the coding specificity of some triplets was quickly established. Thus UUC, UCU, and CUU were shown to specify phenylalanine, serine, and leucine, respectively. However, many triplets did not give clear binding signals and there was concern that the binding reaction might not display the correct specificity.

The code was ultimately defined by the translation of polynucleotides of repeating sequences. These polynucleotides with long repeat sequences were produced by chemically synthesizing short, defined-sequence oligomers and then amplifying them enzymatically. When repeating dinucleotides were translated (fig. 29.7), they yielded repeating dipeptides, and when repeating trinucleotides (not shown) were translated, they yielded up to three individual peptides, each containing only a single amino acid. Eventually the sequence of bases from natural messages was correlated with the sequence of amino acids in the proteins they encoded.

The Code Is Highly Degenerate

A triplet code, one made from four nucleotides taken three at a time, generates a total of 64 different triplet sequences or codons. Three of these codons, as we will see, are utilized to terminate translation and are not generally used to specify amino acids. The remaining 61 codons and the 20 amino acids can be neatly summarized by grouping codons with the same first and second bases into a grid (table 29.1). The four horizontal sections are composed of codons with the same first base. The four vertical sections are composed of codons with the same second base. The boxes representing the vertical–horizontal intersections contain codon families, the members of which differ only in their 3′ terminal base. For example, the codons UCU, UCC, UCA, and UCG constitute a family encoding serine. Thus, the genetic code is degenerate, that is, one amino acid is generally specified by multiple or synonymous codons.

As we have indicated, the codon AUG is the only one generally used to specify methionine, but it serves a dual function in that it is also used to initiate translation. Occasionally, GUG and UUG are also read as an initiating codon in bacteria, but in internal positions these codons are always read as valine and leucine, respectively. In eukaryotes, initiation at codons other than AUG is much less frequent than in prokaryotes. Weak initiation occasionally occurs at GUG, CUG, and ACG codons in eukaryotic systems. The UGA triplet also serves a dual function; it is usually recognized as a stop, but on occasion it serves as a codon for selenocysteine (box 29A).

Organisms differ in the frequency with which they utilize synonymous codons. Some synonymous codons are used frequently, and some are used infrequently. For *E. coli* and yeast this frequency of use, known as codon usage, correlates with the abundance in the organism of the tRNAs that recognize particular synonymous codons. Also, for the same two species, proteins that occur in the greatest abundance employ high-usage codons more frequently and low-usage codons less frequently than the average proteins. It is not clear that this same trend is followed in multicellular eukaryotes, such as *Drosophila* and humans.

Table 29.1

The Genetic Code

	Second Position								
First Position (5′ end)	*U*		*C*		*A*		*G*		Third Position (3′ end)
U	UUU	Phe	UCU	Ser	UAU	Tyr	UGU	Cys	U
	UUC		UCC		UAC		UGC		C
	UUA	Leu	UCA		UAA	STOP	UGA	STOP	A
	UUG		UCG		UAG		UGG	Trp	G
C	CUU	Leu	CCU	Pro	CAU	His	CGU	Arg	U
	CUC		CCC		CAC		CGC		C
	CUA		CCA		CAA	Gln	CGA		A
	CUG		CCG		CAG		CGG		G
A	AUU	Ile	ACU	Thr	AAU	Asn	AGU	Ser	U
	AUC		ACC		AAC		AGC		C
	AUA		ACA		AAA	Lys	AGA	Arg	A
	AUG	Met	ACG		AAG		AGG		G
G	GUU	Val	GCU	Ala	GAU	Asp	GGU	Gly	U
	GUC		GCC		GAC		GGC		C
	GUA		GCA		GAA	Glu	GGA		A
	GUG		GCG		GAG		GGG		G

Wobble Introduces Ambiguity into Codon–Anticodon Interactions

For 61 triplets to act as codons, tRNAs must interact specifically with each triplet. Strict Watson-Crick base pairing between codon and anticodon would require 61 different anticodons and, correspondingly, 61 different tRNAs. As the characterization of tRNAs progressed it became clear that in many cases individual tRNAs could recognize more than one codon. In all cases the different codons recognized by the same tRNA were found to contain identical nucleotides in the first two positions and a different nucleotide in the third position (in the 3′ position of the codon). This relationship is the reason that the genetic code can be so neatly

BOX

29A Site-Specific Variation in Translation Elongation

The unusual amino acid selenocysteine (a derivative of cysteine in which the sulfur atom is replaced by a selenium atom) is an essential component in a small number of proteins. These proteins occur in prokaryotes and eukaryotes ranging from *E. coli* to humans. In all cases, selenocysteine is incorporated into protein during translation in response to the codon UGA. This codon usually serves as a termination codon but occasionally, in some required but unknown context of bases, is used to specify selenocysteine instead.

In *E. coli,* the products of four genes (*selA, selB, selC,* and *selD*) are required for the incorporation of selenocysteine. The product of the *selC* gene is $tRNA^{Ser}$ (a suppressor tRNA the anticodon of which is UCA). The first step in the incorporation of selenocysteine is catalyzed by seryl-tRNA synthase. The products of *selA* and *selD* function in the subsequent conversion of Ser-$tRNA^{Ser}$ to selenocysteyl-$tRNA^{Ser}$. The probable pathway is

$$tRNA^{Ser} \longrightarrow \text{seryl-}tRNA^{Ser} \longrightarrow \text{phosphoseryl-}tRNA^{Ser} \longrightarrow \text{selenocysteyl-}tRNA^{Ser}$$

Incorporation of selenocysteyl-$tRNA^{Ser}$ into protein in response to the UGA codon requires SELB (the protein product of the *selB* gene in *E. coli*). SELB is homologous in sequence to EF-Tu and probably replaces it in translation by specifically recognizing selenocysteyl-tRNA and UGA in the appropriate sequence context. Selenocysteyl-$tRNA^{Ser}$, in combination with SELB, must be capable of competing with termination factors for the translation of the termination codon when it occurs in the ''right'' context of bases. This process is known as site-specific variation in translation elongation.

Other proteins of unknown function that are homologous to EF-Tu are known to exist. Conceivably, these proteins might participate in the incorporation of other rare amino acids into unique positions in proteins.

arranged as shown in table 29.1, by codon families that differ in their third base.

The 3′ terminal redundancy of the genetic code and its mechanistic basis were first appreciated by Francis Crick in 1966. He proposed that codons and anticodons interact in an antiparallel manner on the ribosome in such a way as to require strict Watson-Crick pairing (that is, A-U and G-C) in the first two positions of the codon but to allow other pairings in its 3′ terminal position. Nonstandard base pairing between the 3′ terminal position of the codon and the 5′ terminal position of the anticodon alters the geometry between the paired bases; Crick's proposal, labeled the wobble hypothesis, is now viewed as correctly describing the codon–anticodon interactions that underlie the translation of the genetic code.

By inspecting the geometry that would result from different wobble pairings (fig. 29.8) and recognizing that inosine, the deaminated form of adenine, frequently occurs in tRNAs in the 5′ position of the anticodon, Crick grasped the relationship between codon and anticodon. According to this relationship (table 29.2), when C or A occurs in the 5′ position of an anticodon, it can pair only with G or U, respectively, in the 3′ position of a codon. Transfer RNAs containing either G or U in the 5′, or wobble, position of the anticodon can each pair with two different codons, whereas an inosine (I) in this position produces a tRNA that can pair with three codons differing in the 3′ base. Subsequent sequence analysis of many tRNAs has proved this hypothesis as a feature of the translation of the ''universal'' genetic code (but see the following discussion).

A careful comparison of the wobble rules with the genetic code indicates that the minimum number of tRNAs required to translate all 61 codons is 31. With the addition of $tRNA_i^{Met}$ the total comes to 32. Most cells contain many more than this minimum number of tRNA types.

Figure 29.8

Examples of standard (*a*) and wobble (*b* and *c*) base pairs formed between the first base in the anticodon and the third base in the codon.

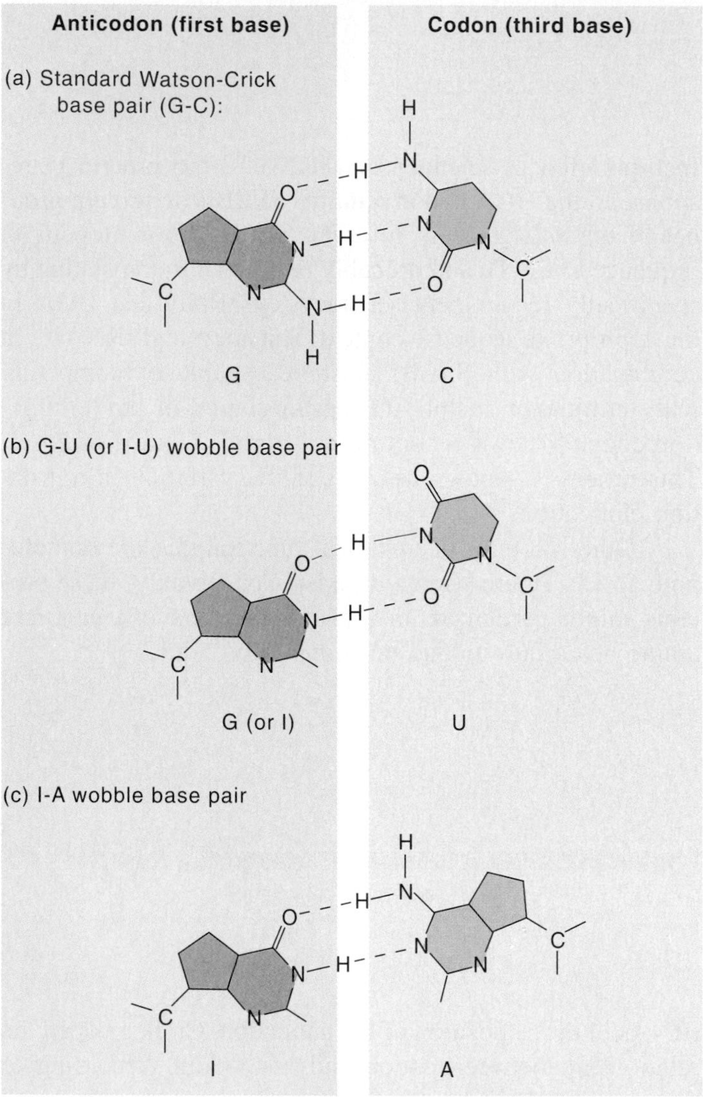

The Code Is Not Quite Universal

It was originally believed that exactly the same genetic code is utilized by all genetic systems. Initial experiments supported this conclusion, since it was found that some mammalian mRNAs could be faithfully translated in cell-free bacterial systems. We now know, however, that significant variations in the meaning of specific code words occur in many genetic systems. The exact scope and nature of these variations are still being discovered, but the current view is that the variations in genetic meaning reflect divergences from the standard, or "universal," genetic code described

Table 29.2

The Wobble Rules of Codon–Anticodon Pairing

5′ Base of Anticodon	3′ Base of Codon
C	G
A	U
U	A or G
G	C or U
I	U, C, or A

earlier, rather than independent origins of the genetic code.

Many of the known variations in the genetic code are found in genes of mitochondria and chloroplasts. It is easy to see why these genetic systems might be more plastic, since they frequently encode only 10–20 proteins. The remainder of the organellar proteins are derived by importing nuclear gene products.

The tRNAs used to translate mitochondrial mRNAs are entirely derived from mitochondrial chromosomes. The first clue that something was unusual about the mitochondrial genetic code was that only 24 types of tRNA could be found. According to Crick's rules of wobble, 32 tRNAs minimally are required for the translation of all 61 codons. One possible solution to this conundrum was that mitochondrial genes do not utilize all 61 codons. Another possibility was that the wobble rules might be different for mitochondria. In fact, the latter is the case. Crick's original wobble rules stated that at least two different tRNAs are needed to translate four codon families. In all of these cases (with the exception of the codon family specifying Arg; see the footnote to table 29.3), single tRNAs have been found responsible for specifying all four code words, and these tRNAs all contain a U in the "wobble" position of their anticodons. It appears that the mitochondrial ribosome allows these tRNAs to pair with all four members of the codon family. The six mitochondrial tRNAs that pair with the normal two codons contain an altered U in the wobble position, and this modification causes them to conform to the normal "wobble" rules.

Another peculiarity of the mitochondrial code emerged from a study of yeast codon usage. By comparing tables 29.3 and 29.1, you will see that the mitochondrial code has several differences in code word meaning. The codons beginning with CU represent Thr instead of Leu, the AUA codon represents Met instead of Ile, and the UGA codon represents Trp rather than a stop signal.

Table 29.3

The Genetic Code of Yeast Mitochondria

	Second Position												
First Position (5′ end)	U			C			A			G			Third Position (3′ end)
U	UUU	Phe	AAG	UCU	Ser	AGU	UAU	Tyr	AUG	UGU	Cys	ACG	U
	UUC			UCC			UAC			UGC			C
	UUA	Leu	AAU*	UCA			UAA	STOP		UGA	Trp	ACU*	A
	UUG			UCG			UAG			UGG			G
C	CUU	Thr	GAU	CCU	Pro	GGU	CAU	His	GUG	CGU	Arg	GCA[b]	U
	CUC			CCC			CAC			CGC			C
	CUA			CCA			CAA	Gln	GUU*	CGA			A
	CUG			CCG			CAG			CGG			G
A	AUU	Ile	UAG	ACU	Thr	UGU	AAU	Asn	UUG	AGU	Ser	UCG	U
	AUC			ACC			AAC			AGC			C
	AUA	Met	UAC[a]	ACA			AAA	Lys	UUU*	AGA	Arg	UCU*	A
	AUG			ACG			AAG			AGG			G
G	GUU	Val	CAU	GCU	Ala	CGU	GAU	Asp	CUG	GGU	Gly	CCU	U
	GUC			GCC			GAC			GGC			C
	GUA			GCA			GAA	Glu	CUU*	GGA			A
	GUG			GCG			GAG			GGG			G

Source: From S. G. Bonitz et al., Codon recognition rules in yeast mitochondria, in *Proc. Natl. Acad. Sci. USA* 77:3167, 1980.

The codons (5′ → 3′) are at the left and the anticodons (3′ → 5′) are at the right in each box. (* designates U in the 5′ position of the anticodon that carries the $—CH_2NH_2CH_2COOH$ grouping on the 5′ position of the pyrimidine.)

[a]Two tRNAs for methionine have been found. One is used in initiation and one is used for internal methionines.

[b]Although an Arg tRNA has been found in yeast mitochondria, the extent to which the CGN codons are used is not clear.

The Rules Regarding Codon–Anticodon Pairing Are Species-Specific

The genetic code differs very little between species. By contrast, considerable differences occur between species in the anticodon translation system of tRNA, as evidenced by the mitochondrial tRNA system. In all systems the bases in the anticodon–codon complex run antiparallel, as in standard double-helix pairing, and in all cases only Watson-Crick-like base pairing occurs between the first two bases in the codon and the opposing bases in the anticodon segment of the tRNA. However, for the 3′ base in the codon, the rules for pairing vary with the species and with the base in question. These rules, summarized in table 29.4, are as follows.

When the 5′ base in the anticodon is a G it can pair with either a U or a C, and this is true in all organisms. When the 5′ base in the anticodon is an A it can pair only with a U in the codon. However, it is rare that an A is found in this position of the anticodon. An A in this position in eukaryotes is usually deaminated to an inosine (I) base, which has an expanded capacity for pairing. Base A can pair only with U, but I can pair with U, C, or A. In eubacteria, deamination is limited to the conversion of the ACG sequence to an ICG anticodon. When C is the 5′ base in the anticodon it pairs with G only. The only known exception to this rule is found in eubacteria, in which the C is covalently modified in the tRNA that recognizes the AUA codon. Thus, in this one instance the modified C can pair with a 3′ A in the codon. A U base in the 5′ position of the anticodon shows the greatest variability. An unmodified U can pair with any of the four bases in the 3′ position of the codon.

Table 29.4

Anticodon–Codon Pairing

Anticodon First Base	Codon Third Base	Examples
U	U, C, A, G	Mitochondrial code in family boxes
*U	A, G	Mitochondrial code in two-codon sets
†U	A	Eukaryotes
‡U	U, A, G	Eubacteria in family boxes
C	G	All codes
*C	A	Bacteria, isoleucine codon AUA
G	U, C	All codes
A	U	Rare
I	U, C, A	Eukaryotes, ICG in eubacteria

*, †, ‡ = Various modifications of U (Yokoyama et al., 1985)

*C = modified C

(From Thomas Jukes et al., *Cold Spring Harbor Symp. Quant. Biol.* 32:775, 1987. Copyright © 1987 Cold Spring Harbor Laboratory Press, Cold Spring Harbor, N.Y. Reprinted by permission.)

This situation is reflected in the U-family box in mitochondria (table 29.3).

U can also be modified in various ways. In mitochondria one type of modification permits a U to pair with either an A or a G in the two-codon sets. In eukaryotes, another type of U modification limits U to pairing with an A in the codon, and in eubacteria, a third type of U modification permits U pairing with U, A, or G but not C in family boxes.

The Steps in Translation

Protein synthesis involves more than 100 different proteins and more than 30 kinds of RNA molecules. The process begins by the attachment of amino acids to specific tRNA molecules. Subsequent steps take place on the ribosome; amino acids are transported to the ribosome on their tRNA carriers, and they do not leave the ribosome until they have become an integral part of a polypeptide chain.

Synthases Attach Amino Acids to tRNAs

A unique class of enzymes, called aminoacyl-tRNA synthases, attach amino acids to their cognate tRNAs. This attachment serves two functions: (1) The linkage between amino acid and tRNA activates the amino acid, making the subsequent formation of a peptide bond energetically favorable. (2) The tRNA directs the amino acid to a designated location on a messenger RNA so that the amino acid is incorporated at the appropriate location in the polypeptide chain.

All synthase reactions proceed in two separate steps (fig. 29.9). In the first step the synthase recognizes its corresponding amino acid and its second substrate, ATP, and forms a mixed anhydride bond between the carboxyl group of the amino acid and the phosphate of AMP with the release of PP_i.

$$\text{Amino acid} + \text{ATP} \longrightarrow \text{aminoacyl-AMP} + PP_i \quad \textbf{(1)}$$

The equilibrium constant for this reaction is about 1, so that the energy derived from the cleavage of the phosphate anhydride of ATP is conserved in the mixed anhydride. Aminoacyl-AMP remains tightly bound to the enzyme and, as we will soon show, this fact has allowed researchers to crystallize this important complex and analyze its structure.

The second reaction catalyzed by the aminoacyl-tRNA synthases results in the attachment of the amino acid through an ester linkage to the 3′ terminal ribose of tRNA:

$$\text{Aminoacyl-AMP} + \text{tRNA} \longrightarrow \text{aminoacyl-tRNA} + \text{AMP} \quad \textbf{(2)}$$

Synthases differ with respect to their site of attachment to tRNA. Some synthases form the 2′ ester, some form the 3′ ester, and still others produce a mixture of the two. The specificity of the synthases was determined by analyzing their ability to act on tRNA derivatives lacking one or the other terminal hydroxyl group. Once esterified to the terminal ribose, the aminoacyl group can migrate between the vicinal 2′ and 3′ hydroxyl groups. Thus, in cells, aminoacyl-tRNAs are mixtures of 2′ and 3′ esters. Only the 3′ derivative is a substrate for the subsequent transpeptidation reaction catalyzed by the ribosome.

The sum of the two reactions catalyzed by aminoacyl-tRNA synthases is

$$\text{Amino acid} + \text{ATP} + \text{tRNA} \longrightarrow \text{aminoacyl-tRNA} + \text{AMP} + PP_i \quad \textbf{(3)}$$

This overall reaction is reversible but is driven to completion by the subsequent hydrolysis of PP_i to two equivalents of P_i through the action of ubiquitous pyrophosphatases. Thus, the formation of aminoacyl-tRNA consumes two equivalents of ATP. The energy that ultimately drives the formation of the peptide linkage during protein synthesis is derived from the ester linkage that joins amino acids to tRNA.

Figure 29.9

Formation of aminoacyl-tRNA. This is a two-step process involving a single enzyme that links a specific amino acid to a specific tRNA molecule. In the first step (1) the amino acid is activated by the formation of an aminoacyl-AMP complex. This complex then reacts with a tRNA molecule to form an aminoacyl-tRNA complex (2).

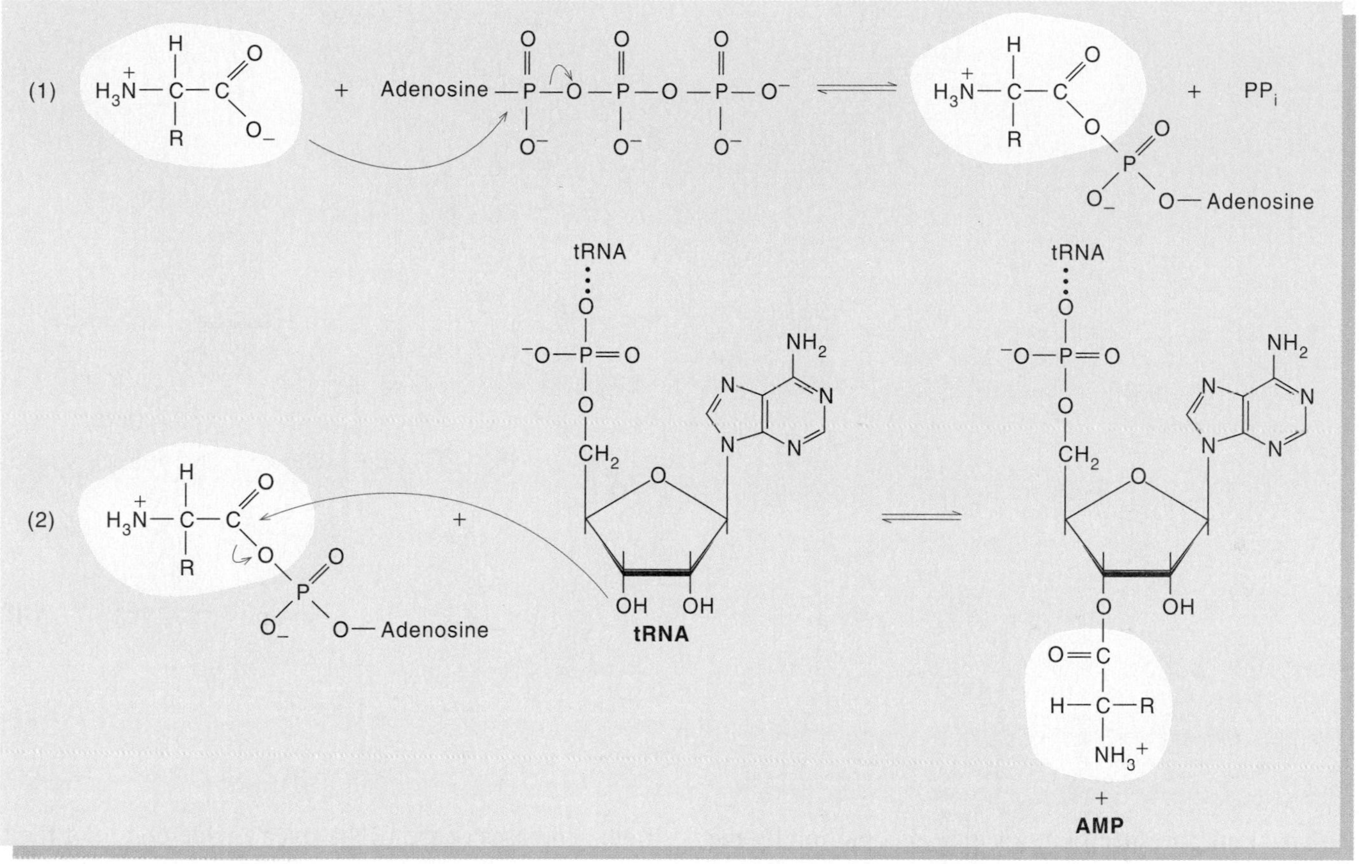

Each Synthase Recognizes a Specific Amino Acid and Specific Regions on Its Cognate tRNA

Our understanding of synthase reactions and the types of the active sites involved in these reactions was advanced substantially by the crystallization and structural solution of tyrosyl-tRNA synthase complexed with the reaction intermediate tyrosyl-adenylate (fig. 29.10). The reaction intermediate is bound in a deep cleft in the enzyme and interacts with it through 11 hydrogen bonds. Six of these bonds are with the AMP moiety, and five are with the tyrosyl moiety of the intermediate. The amino acid selectivity of tyrosyl-tRNA synthase is thus determined primarily by the formation of specific hydrogen bonds with the amino acid.

Most cells contain only one synthase for each of the 20 amino acids specified by the genetic code. Each enzyme must be capable of recognizing its unique amino acid and one or more cognate tRNAs. Solving the puzzle of how synthases recognize tRNAs has been one of the major challenges in understanding the nature of the translation of the genetic code itself. The identity relationship between the tRNA and its synthase has, in fact, been called the "second genetic code."

Surprisingly, synthases fall into two categories with respect to the importance of the anticodon as a specificity element that they recognize. Some synthases (17 in *E. coli*) recognize the anticodon and some do not (3 in *E. coli*). This distinction has been demonstrated in several ways. First, some tRNAs can be genetically altered in their anticodon

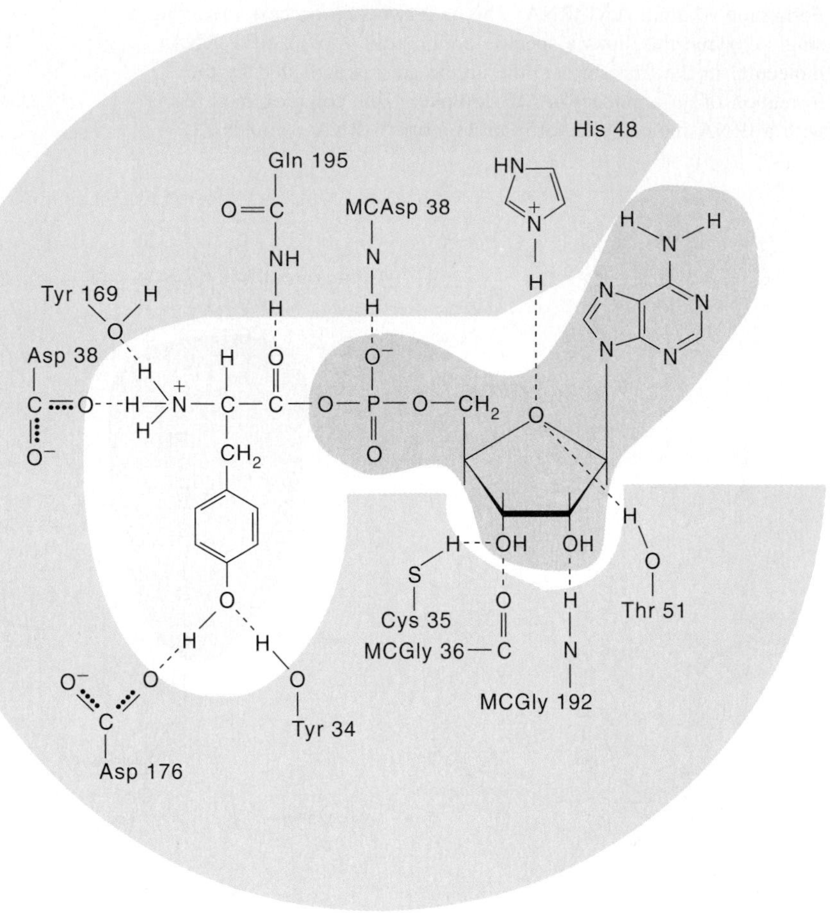

Figure 29.10

Recognition of an adenylylated amino acid by the proper synthase. Shown is adenyl tyrosine bound to the tyrosyl synthase. A network of H bonds not only stabilizes the reaction but also serves to discriminate between different amino acid residues. MC designates main chain (backbone) carbonyl or amino groups participating in hydrogen bonding.

without changing the specificity of their recognition by the cognate synthase. Other synthases apparently require an unaltered anticodon in order to recognize their cognate tRNAs. Four examples are shown in figure 29.11.

The crystal structure of glutaminyl-tRNA synthase, complexed with tRNA and ATP, has been determined by Tom Steitz and his colleagues (fig. 29.12). This accomplishment provided the first structure of a tRNA–protein complex and thus offers important insight into the general nature of the recognition of tRNAs by proteins.

Glutaminyl-tRNA synthase is known to require the integrity of the anticodon loop of tRNA for recognition and is thus presumed to interact with both "ends" of the tRNA. In keeping with this interpretation, the synthase is asymmetrical and longer ($\approx$ 100 Å) than the tRNA. Moreover, the crystal structure demonstrates the occurrence of significant contact between the protein and bases in the anticodon loop. The additional contacts between the synthase and tRNA appear to occur along the inside of the L-shaped structure. The recognition of the tRNA appears to arise from direct hydrogen bonding between amino acid residues of the protein and bases of the tRNA over a wide region of the tRNA structure.

The acceptor stem of the bound tRNA substrate plunges deeply into the active site pocket created by the dinucleotide fold, also the site of ATP binding. Surprisingly, binding in the active site induces a significant conformational change in the 3′ terminal CCA acceptor sequence. This conformational change appears to cause the melting of the A-U pair at the end of the acceptor stem in a manner that allows these bases to interact with amino acid side chains of the protein. At this point in our understanding, it seems likely that the interactions that allow synthases to identify their cognate tRNAs are both diverse and complex.

Aminoacyl-tRNA Synthases Can Correct Acylation Errors

Many aminoacyl-tRNA synthases appear to contain a second catalytic site that serves to correct errors by hydrolyzing incorrectly matched amino acids and tRNAs. These proofreading hydrolysis reactions are best understood in the case

Figure 29.11

Identity elements in four tRNAs. Each circle represents one nucleotide. Filled circles indicate nucleotides that serve as recognition elements to the appropriate aminoacyl-tRNA synthase. It is possible that other identity elements occur in these structures that are still to be discovered. (From L. H. Schulman and J. Abelson, Recent excitement in understanding transfer RNA identity, *Science* 240:1591, June 17, 1988. Copyright 1988 by the AAAS. Reprinted by permission.)

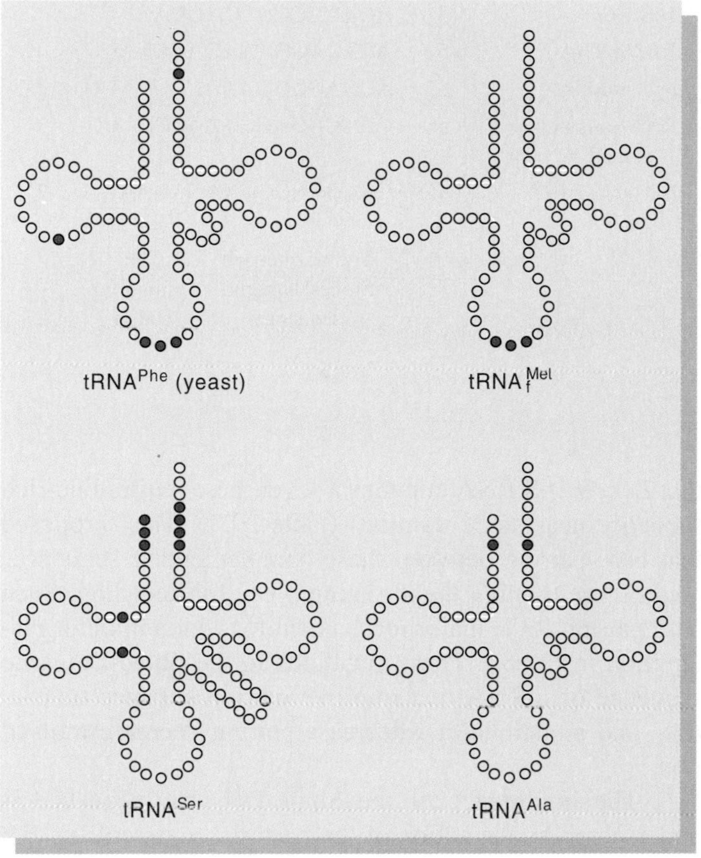

Figure 29.12

Solvent-accessible surface representation of the GlnRS enzyme complexed with tRNA and ATP. The region of contact between tRNA and protein extends across one side of the entire enzyme surface and includes interactions from all four protein domains. The acceptor end of the tRNA and the ATP are seen in the bottom of the deep cleft. Protein is inserted between the 5′ and 3′ ends of the tRNA and disrupts the expected base pair between U1 and A72. (From M. G. Rould, J. J. Persona, D. Söll, and T. Steitz, "Structure of *E. coli* glutamyl-tRNA synthetics complexed with $tRNA^{Gln}$ and ATP at 2.8-Å resolution, implications for tRNA discrimination," *Science* 246:1135–1142, 1989, © 1989 by the AAAS.)

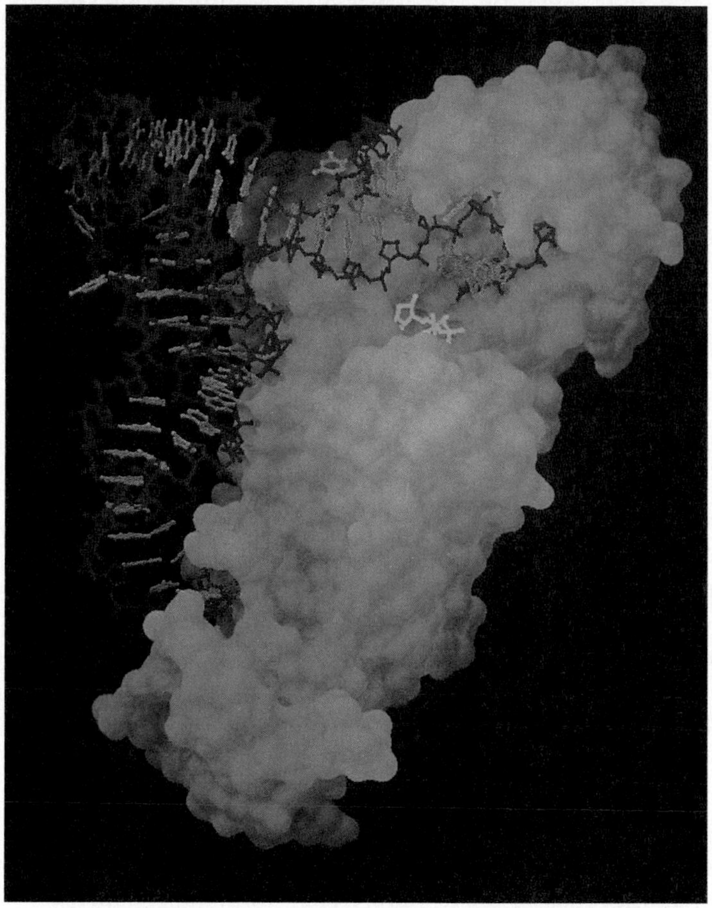

of isoleucyl-tRNA synthase. The amino acids isoleucine and valine, which differ by only a single methylene group, are among the most difficult to discriminate. Isoleucyl-tRNA synthase is capable of distinguishing these two amino acids at its aminoacylation site but occasionally and mistakenly forms valyl-$tRNA^{Ile}$. When isoleucyl-tRNA synthase encounters valyl-$tRNA^{Ile}$ the hydrolytic active site specifically degrades it:

$$\text{Valyl-tRNA}^{\text{Ile}} + H_2O \longrightarrow \text{valine} + \text{tRNA}^{\text{Ile}} \qquad (4)$$

In this way the erroneous incorporation of valine in place of isoleucine is avoided. The sequential action of aminoacylation and proofreading sites contributes to an overall error frequency of translation of less than 1 in 10,000.

A Unique tRNA Initiates Protein Synthesis

The translation of every protein begins with the incorporation of the amino acid methionine. A unique initiator tRNA, $tRNA_i^{Met}$, is responsible for the incorporation of this initiating methionine in all protein-synthesizing systems, and it also plays an important role in selecting the appropriate translation start site in mRNA. Generally, only two tRNAs occur in cells that specify methionine. We designate the one that is responsible for the incorporation of internal methionines simply as tRNA^{Met}. A single methionyl-tRNA synthase is responsible for activating both methionine isoacceptor tRNAs.

The functional discrimination between the two isoacceptor tRNAs is accomplished by protein initiation and elongation factors. Initiation factors recognize tRNA$_i^{Met}$, and elongation factors recognize tRNA^{Met}. Clearly, tRNA$_f^{Met}$ and tRNA^{Met} must possess structural features that allow them to be distinguished from all other tRNAs by the single methionyl-tRNA synthase, but they must be sufficiently different that they can be discriminated by the factors.

In prokaryotic systems, a specific formylating enzyme exists that can recognize tRNA$_f^{Met}$ and formylate, its amino terminus utilizing N^{10}-formyltetrahydrofolate as the formyl donor. This reaction serves to ensure that the initiator does not participate in the elongation reactions. For unknown reasons this recognition step is not possessed by eukaryotic systems.

Table 29.5

The Sequence of the 3′ End of *E. coli* 16S RNA and Some Shine-Dalgarno Sequences at the 5′ End of Bacterial mRNAs

	The pyrimidine-rich complement to the Shine-Dalgarno sequence
16S rRNA	3′ ··· HOAUUCCUCCA CUA ··· 5′
lacZ mRNA	5′ ··· ACACAGGAAAC AGCUAUG ··· 3′
trpA mRNA	5′ ··· ACGAGGGGAAAUCUGAUG ··· 3′
RNA polymerase β mRNA	5′ ··· GAGCUGAGGAACCCUAUG ··· 3′
r-Protein L10 mRNA	5′ ··· CCAGGAGCAAA GCUAAUG ··· 3′
	The purine-rich Shine-Dalgarno sequence; The initiation codon ↑

Translation Begins with the Binding of mRNA to the Ribosome

One of the major demands of protein synthesis is to select the appropriate initiator codon, generally AUG, for translation. This is accomplished at the level of the ribosome by the binding of the small ribosomal subunit to mRNA. The recognition of the appropriate start codon occurs in different ways in prokaryotes and eukaryotes.

In eukaryotes, the AUG sequence closest to the 5′ end of the mRNA usually serves as the start codon for the single protein encoded by each mRNA. The small ribosomal subunit binds at the 5′ end of the mRNA and moves along the strand until it encounters an AUG sequence that is recognized by base pairing with the anticodon of Met-tRNA$_i^{Met}$. In higher eukaryotes, but not in yeast, the recognition of this initiating AUG is facilitated by an initiation context of flanking bases. The preferred initiation context is GCCGCCpurCCAUGG. How this sequence is recognized is unknown, but when the initiation context of the first AUG departs from this sequence, the 40S subunit can bypass it and initiate at the next AUG downstream.

In prokaryotic systems, the initiating AUG codon may occur at any point within the mRNA, and more than one start site may occur within the same mRNA. How do prokaryotic ribosomes select the appropriate initiating codon from the much more abundant internal AUG sequences? The answer was suggested in the early 1970s when Shine and Dalgarno noticed that bacterial mRNAs contain a complementary purine-rich region (which has become known as the Shine-Dalgarno sequence) centered approximately 10 bases toward the 5′ side of the initiating AUG sequence and that *E. coli* 16S RNA contains a seven-base pyrimidine-rich sequence near its 3′ terminus (table 29.5). They proposed that base pairing between these complementary sequences could serve to align the initiating AUG for decoding. Such base pairing is the major mechanism for codon initiator recognition in *E. coli.* Thus, mutations in the Shine-Dalgarno sequence of mRNA that improve pairing enhance translation, and mutations that decrease pairing decrease translation.

The importance of the Shine-Dalgarno sequence is underscored by the action of the bacterial toxin colecin E3. This toxin inactivates the small subunit of the prokaryotic ribosome by cleavage of about 50 residues from the 3′ terminus of 16S rRNA. The cleavage disrupts the sequence that is complementary to the Shine-Dalgarno sequence and thus specifically inhibits the initiation process. Because of the fundamental differences between prokaryotic and eukaryotic initiation just described, colecin E3 does not inhibit the eukaryotic ribosome.

Dissociable Protein Factors Play Key Roles at the Different Stages in Protein Synthesis on the Ribosome

At each stage in protein synthesis on the ribosome—initiation, elongation, and termination—a different set of protein factors is engaged by the ribosome. Why do such protein factors, which are crucial to the translation, exist separate from the ribosome? Why must they cycle on and

Figure 29.13

Formation of the initiation complex for protein synthesis in prokaryotes. *E. coli* has three initiation factors bound to a pool of 30S ribosomal subunits. One of these factors, IF-3, holds the 30S and 50S subunits apart after termination of a previous round of protein synthesis. The other two factors, IF-1 and IF-2, promote the binding of both fMet-$tRNA^{fMet}$ and mRNA to the 30S subunit. The binding of mRNA occurs so that its Shine-Dalgarno sequence pairs with 16S rRNA and the initiating AUG sequence with the anticodon of the initiator tRNA. The 30S subunit and its associated factors can bind fMet-$tRNA^{fMet}$ and mRNA in either order. Once these ligands are bound, IF-3 dissociates from the 30S subunit, permitting the 50S subunit to join the complex. This releases the remaining initiation factors and hydrolyzes the GTP, which is bound to IF-2.

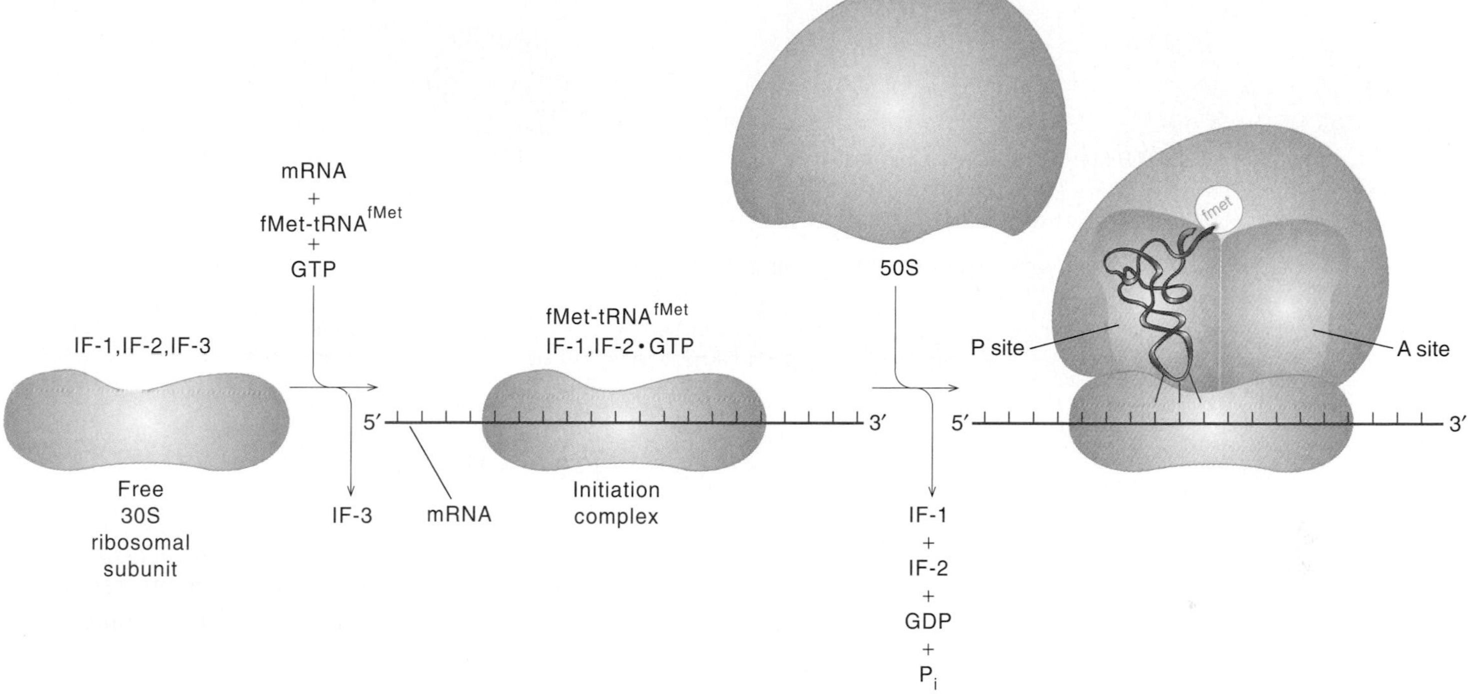

off the ribosome, and why are their functions not performed by firmly bound ribosomal proteins? The answers to these questions appear to lie in the economy of structure that is afforded by the cycling strategy. Since factors cannot interact simultaneously with the ribosome, it makes sense for the ribosome to interact with them sequentially at a single site.

Protein Factors Aid Initiation

Even though specific differences distinguish the initiation process in eukaryotes and prokaryotes, three things must be accomplished to initiate protein synthesis in all systems: (1) The small ribosomal subunit must bind the initiator tRNA; (2) the appropriate initiating codon on mRNA must be located; and (3) the large ribosomal subunit must associate with the complex of the small subunit, the initiating tRNA, and mRNA. Nonribosomal proteins, known as initiation factors (IFs), participate in each of these three processes. IFs interact transiently with a ribosome during initiation and thus differ from ribosomal proteins, which remain continuously associated with the same ribosome.

E. coli has three initiation factors (fig. 29.13) bound to a small pool of 30S ribosomal subunits. One of these factors, IF-3, serves to hold the 30S and 50S subunits apart after termination of a previous round of protein synthesis. The other two factors IF-1 and IF-2, promote the binding of both fMet-$tRNA_i^{Met}$ and mRNA to the 30S subunit. As we noted before, the binding of mRNA occurs so that its Shine-Dalgarno sequence pairs with 16S RNA and the initiating AUG sequence with the anticodon of the initiator tRNA. The 30S subunit and its associated factors can bind fMet-$tRNA_i^{Met}$ and mRNA in either order. Once these ligands are found, IF-3 dissociates from the 30S, permitting the 50S to join the complex. This releases the remaining initiation factors and hydrolyzes the GTP that is bound to IF-2. The initiation step in prokaryotes requires the hydrolysis of one equivalent of GTP to GDP and P_i.

Initiation of protein synthesis in eukaryotic cells requires a much more complex spectrum of initiation factors, abbreviated eIF. At least nine separate factors have been identified, some of which are composed of as many as 11 different peptide subunits. The exact function of only a few

Figure 29.14

Formation of the initiation complex for protein synthesis in eukaryotes. The reaction begins with the small subunit held apart from the large subunit by an antiassociation factor and ends with the hydrolysis of GTP and joining of the large subunit as in prokaryotes. The intervening reactions are different. A much more complex spectrum of initiation factors (eIFs) is involved, and the exact function of only a few of these factors is known with certainty. The Met-tRNA^{fMet} first binds to the small subunit as a ternary complex with the eIF-2 and GTP. This ternary complex then binds to the 5′ end of mRNA with the aid of several factors, one of which, eIF-4F, contains a subunit that specifically binds to the terminal cap structure of the mRNA. Binding to mRNA is followed by a scanning reaction that moves the small subunit along the mRNA, usually to the first AUG, in a reaction driven by the hydrolysis of ATP to ADP and P_i.

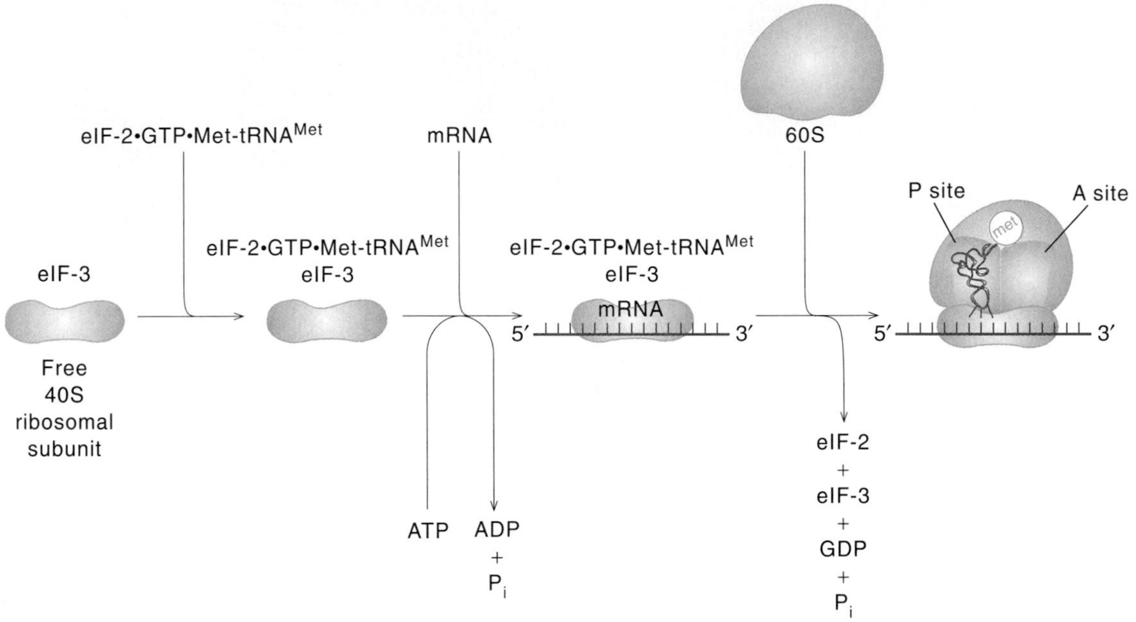

of these factors is known. The main features of the process are outlined in figure 29.14. As in the prokaryotic system, the reaction begins with the small subunit held apart from the large subunit by an antiassociation factor and ends with the hydrolysis of GTP and joining of the large subunit. The intervening reactions are different in prokaryotes and eukaryotes. The Met-tRNA$_i^{Met}$ first binds to the small subunit as a ternary complex with eIF-2 and GTP. The resulting preinitiation complex then binds to the 5′ end of mRNA with the aid of several factors, one of which, eIF-4F, contains a subunit that specifically binds to the terminal cap structure of mRNA. Binding to mRNA is followed by a scanning reaction that moves the small subunit along the mRNA, usually to the first AUG, in a reaction driven by the hydrolysis of ATP to ADP and P_i. The mRNA binding and scanning reactions, which have no counterpart in the prokaryotic system, position the ribosome on the initiating AUG sequence in the appropriate flanking nucleotide sequence.

Three Elongation Reactions Are Repeated with the Incorporation of Each Amino Acid

At the conclusion of the initiation process, the ribosome is poised to translate the reading frame associated with the initiator codon. The translation of the contiguous codons in mRNA is accomplished by the sequential repetition of three reactions with each amino acid. These three reactions of elongation are similar in both prokaryotic and eukaryotic systems; two of them require nonribosomal proteins known as elongation factors (EF). Interestingly, the actual formation of the peptide bond does not require a factor and is the only reaction of protein synthesis catalyzed by the ribosome itself.

The elongation reactions begin with the binding of the aminoacyl-tRNA specified by the codon immediately adjacent to the initiator codon. The binding of this aminoacyl-tRNA is catalyzed by an aminoacyl-tRNA binding factor,

Figure 29.15

Addition of the second aminoacyl-tRNA to the ribosome complex and the accompanying EF-Tu, EF-Ts cycle in *E. coli.* The purpose of the cycle is to regenerate another protein aminoacyl-tRNA complex suitable for transferring further aminoacyl-tRNAs to the A site on the ribosome.

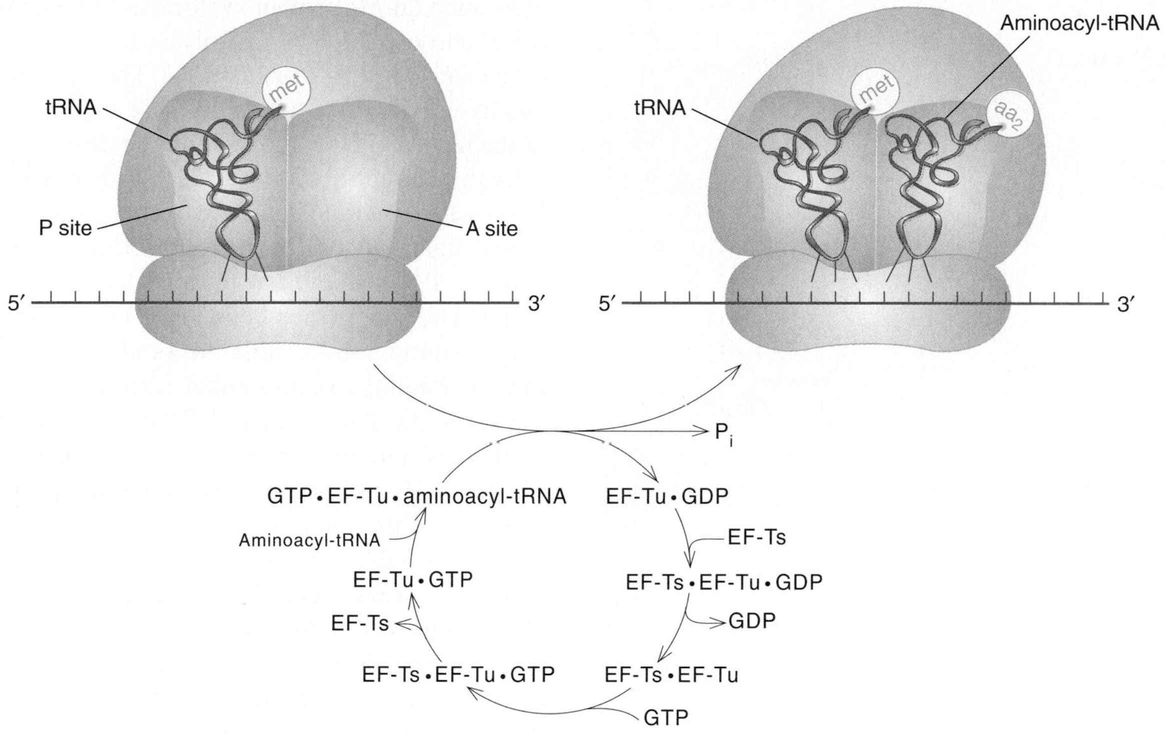

designated EF-Tu in bacteria and EF-1 in eukaryotic systems. The factor interacts with the ribosome as a ternary complex bound to both aminoacyl-tRNA and GTP. The binding of this complex to the ribosome is coupled to GTP hydrolysis. A productive complex forms if, and only if, the anticodon of the tRNA in the complex is complementary to the codon bound to the A site on the ribosome. Following the binding of aminoacyl-tRNA, EF-Tu is released from the ribosome as a complex with GDP. A second elongation factor, EF-Ts, catalyzes the regeneration of the EF-Tu–GTP complex so that it can again bind aminoacyl-tRNA. The series of reactions involved in this regeneration is depicted in figure 29.15. EF-1 is a multisubunit protein that combines the functional properties of EF-Tu and EF-Ts.

Peptide bond formation occurs immediately following the dissociation of the binding factor from the ribosome. This reaction is known as transpeptidation, and the enzymatic center that catalyzes it is known as peptidyl transferase, although it promotes the conversion of an ester to a peptide bond. Transpeptidation is generally assumed to proceed by nucleophilic attack by the amino group of the incoming aminoacyl-tRNA on the carbonyl of the ester of peptidyl-tRNA with the formation of a tetrahedral intermediate (fig. 29.16).[a]

The final step in elongation is known as translocation (fig. 29.17). This reaction, like aminoacyl-tRNA binding, is catalyzed by a factor (the translocation factor, known as EF-G in prokaryotic systems and EF-2 in eukaryotic systems) that cycles on and off the ribosome and hydrolyzes GTP in the process. The overall purpose of translocation is to move the ribosome physically along the mRNA to expose the next codon for translation.

The structural differences between bacterial and eukaryotic elongation factors are highlighted by the selective action that diphtheria has on eukaryotic systems (box 29B).

[a]H. F. Noller, V. Hossarth, and L. Zimniak. Ribosomes stripped of all their protein can still catalyze transpeptidation. *Science* 256:14–16, 1992.

Figure 29.16

Formation of the first peptide linkage. The formylmethionine group is transferred from its tRNA at the P site to the amino group of the second aminoacyl-tRNA at the A site of the ribosome. This involves nucleophilic attack by the amino group of the second amino acid on the carboxyl carbon of the methionine. The resulting bond formation attaches both amino acids to the tRNA at the A site.

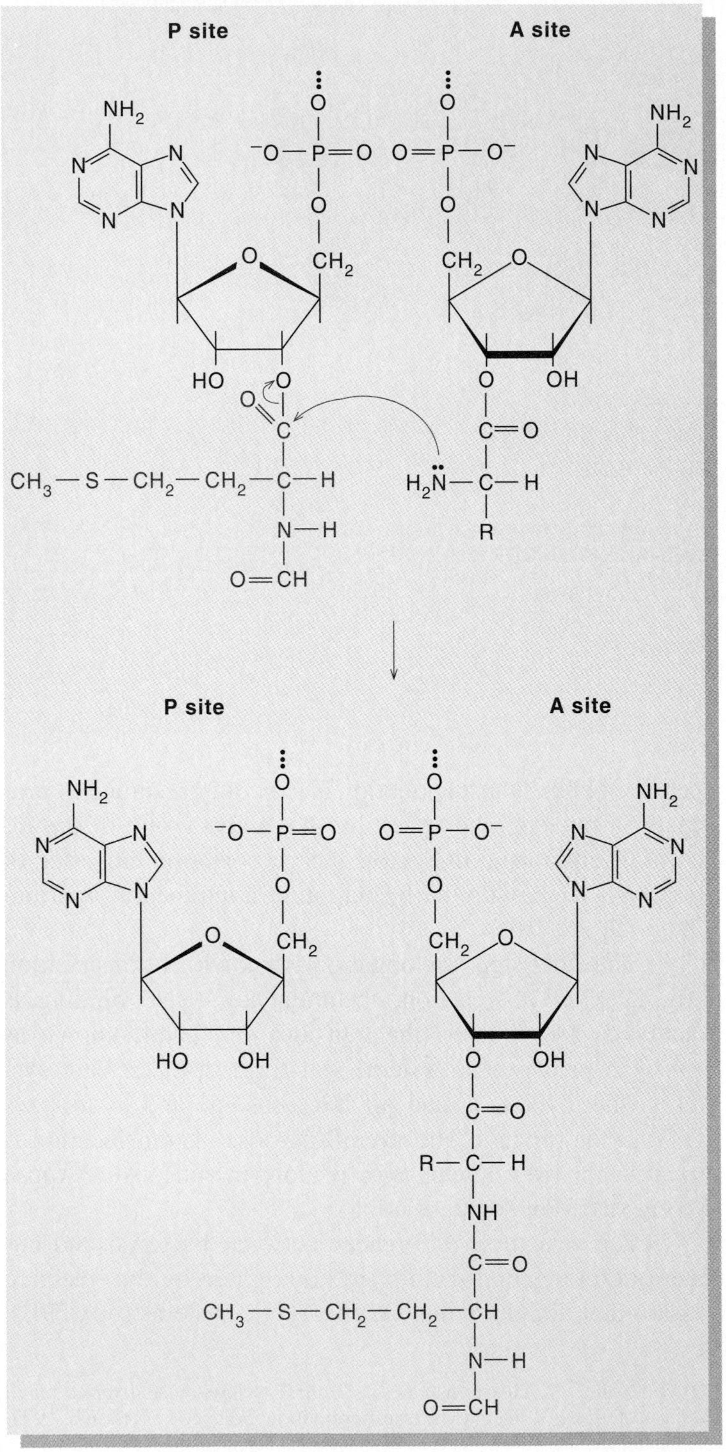

Two GTPs Are Required for Each Step in Elongation

The hydrolysis of GTP plays a conspicuous role in the translation process. Two equivalents of GTP are hydrolyzed during elongation with the incorporation of each amino acid. This hydrolysis accounts for about half of the total energy consumed during protein synthesis. The chemical and functional purposes of GTP hydrolysis are best understood in the case of *E. coli* EF-Tu and EF-G. The sites of GTP binding and hydrolysis are situated on these factors. The interaction of the factor–GTP complex with the ribosome is believed to activate the hydrolytic site. GTP hydrolysis leads to GDP. No covalent intermediates are formed. Rather, the change of bound ligand to GDP that results from hydrolysis and release of P_i is believed to change the conformation of the factor. The factor when bound to GTP is thought to be in the ''on'' configuration, able to bind to the ribosome and through binding to cause either aminoacyl-tRNA binding or translocation. The binding of GDP to the factor puts it in the ''off'' configuration and causes it to dissociate from the ribosome. Hydrolysis itself is not required for these changes to occur. This point is demonstrated by the fact that the nonhydrolyzable analog of GTP, GMPPCP, containing a methylene bridge instead of an oxygen between the β and γ phosphorus, can substitute for it in the reactions catalyzed by the elongation factors (see fig. 29.18). The use of this analog slows these reactions by delaying the dissociation of the factors from the ribosome.

The translation factors that interact with GTP are members of the so called G protein superfamily, which includes the signal-transducing G proteins that link membrane receptors with their intracellular targets (see chapter 24) and the Ras proteins that function as growth regulators (see supplement 4: Carcinogenesis and Oncogenes). All G proteins bind and hydrolyze GTP, and it is believed that they all act by the same general mechanism we have just described. Thus, when bound to GTP these proteins are in their ''active'' configuration, and when bound to GDP as a result of hydrolysis they are converted to their ''inactive'' configuration.

Figure 29.17

The translocation reaction in *E. coli.* The translocation reaction occurs immediately after peptide synthesis. It involves displacement of the discharged tRNA from the P site and concerted movement of the peptidyl-tRNA and mRNA so that the peptidyl-tRNA is bound to the P site and the same three nucleotides in the mRNA. The A site is vacated and ready for the addition of another aminoacyl-tRNA. Translocation in eukaryotes is similar except that the EF-2 factor is involved instead of the EF-G factor.

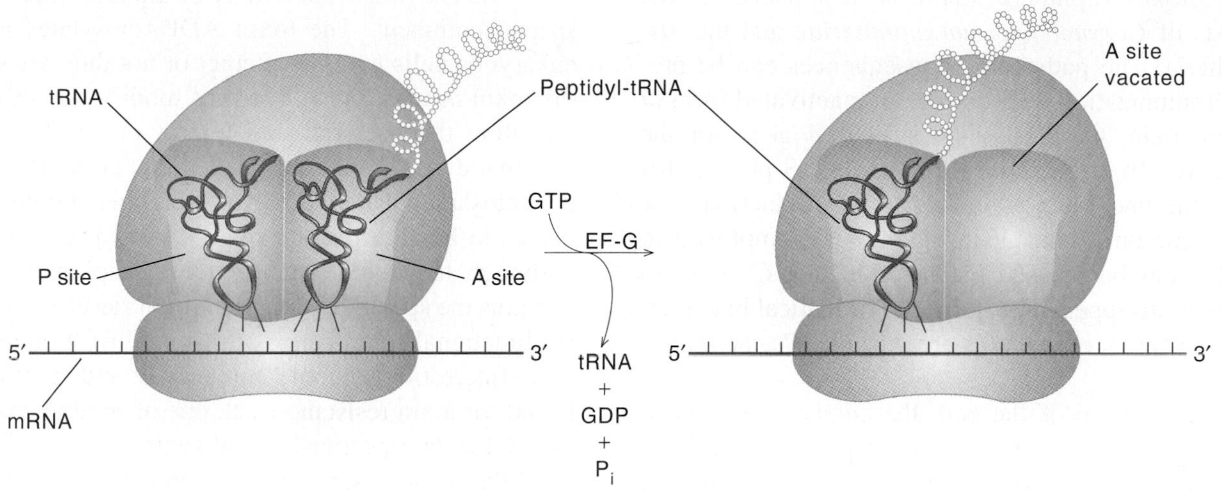

Figure 29.18

The structure of guanylyl methylene diphosphonate (GMPPCP).

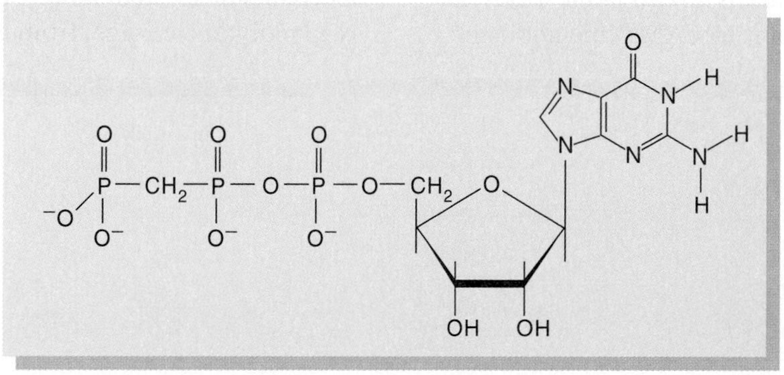

BOX

29B Mechanisms of Damage Produced by Certain Toxins

A single exotoxin, diphtheria toxin, is responsible for the pathogenesis of *Corynebacterium diphtheriae* and the disease diphtheria. The pathogenic consequences can be prevented by immunization with toxoid, an inactivated form of the purified toxin. Curiously, the structural gene for the toxin is carried by a bacterial virus, called β phage, that must infect the bacterium to induce toxin production. The widespread immunization against diphtheria employed in the United States has caused β phage, but not *C. diphtheriae,* largely to disappear. A catalytically identical but structurally very different exotoxin is produced by *Pseudomonas aeruginosa.*

In the cytoplasm of the cell, the catalytic portion of diphtheria toxin acts as a very specific protein-modifying enzyme. It catalyzes the ADP-ribosylation and consequent inactivation of EF-2 by the following reaction:

$$\text{EF-2} + \text{NAD}^+ \longrightarrow \text{ADP-ribosyl-EF-2} + \text{nicotinamide} + \text{H}^+$$

This reaction is reversible when conducted *in vitro,* but under the conditions of pH and nicotinamide concentration that exist in the cell, it is irreversible. Thus, diphtheria toxin kills cells by irreversibly destroying the ability of EF-2 to participate in the translocation step of protein synthesis elongation. A number of other protein toxins have subsequently been found to ADP-ribosylate and inactivate cellular proteins involved in other essential cellular pathways. For example, cholera and pertussis toxins ADP-ribosylate and inactivate proteins important to cAMP metabolism.

The enzymatic specificity of diphtheria toxin deserves special comment. The toxin ADP-ribosylates EF-2 in all eukaryotic cells *in vitro* whether or not they are sensitive to the toxin *in vivo,* but it does not modify any other protein, including the bacterial counterpart of EF-2. This narrow enzymatic specificity has called attention to an unusual posttranslational derivative of histidine, diphthamide, that occurs in EF-2 at the site of ADP-ribosylation (see fig. 1). Although the unique occurrence of diphthamide in EF-2 explains the specificity of the toxin, it raises questions about the functional significance of this modification in translocation. Interestingly, some mutants of eukaryotic cells selected for toxin resistance lack one of several enzymes necessary for the posttranslational synthesis of diphthamide in EF-2 that is necessary for toxin recognition, but these cells seem perfectly competent in protein synthesis. Thus, the *raison d'être* of diphthamide, as well as the biological origin of the toxin that modifies it, remains a mystery.

Some other toxins inhibit protein synthesis by inactivating the ribosome through alterations of rRNA. A fungal toxin, α-sarcin, inactivates the ribosome by specific nuclease cleavage of a single phosphodiester bond in a purine-rich region found in the 23–28S RNA of all ribosomes. A second group of toxins, known as the ribosome-inactivating proteins, is abundant in plants of many species. The best known example of this type of toxin is ricin, the toxic agent in the castor bean. These proteins all act by the curious mechanism of removing a single adenine residue, by N-glycolytic cleavage, from the site in rRNA adjacent to the

Figure 1

Inactivation of the EF-2 factor by diphtheria toxin through the ADP-ribosylation of a modified histidine side chain.

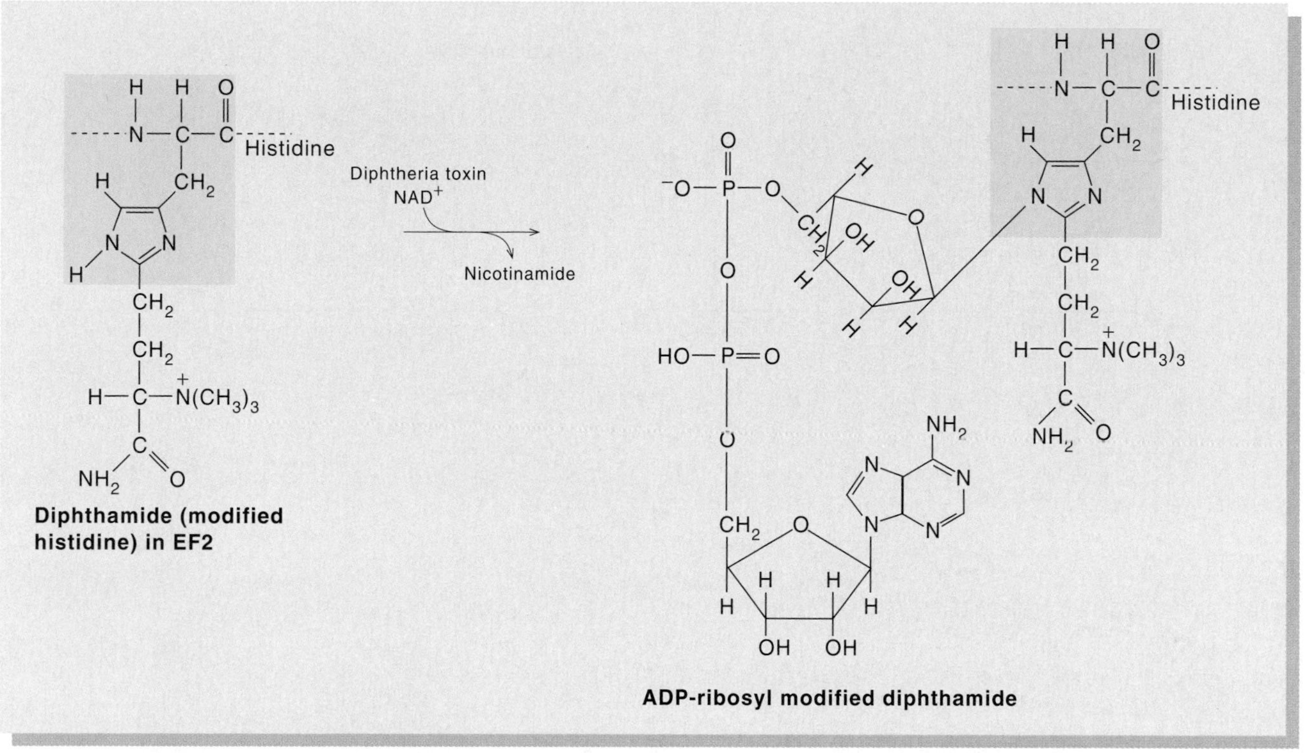

one that is cleaved by α-sarcin. Both of these modifications of rRNA alter the ribosome's ability to interact with factors, a fact suggesting that this region of rRNA is part of the factor interaction site on the large subunit. One of the ribosome-inactivating proteins, trichosanthin, was originally isolated from a plant tuber as the active principle of an Oriental folk medicine used to induce abortions. More recently, trichosanthin has been found to selectively inhibit viral protein synthesis in HIV-infected T cells, and it is now undergoing clinical trials as a therapeutic agent for the treatment of AIDS. The biological basis of this selective inhibition is unknown.

Figure 29.19

The release reaction in *E. coli.* The release reaction occurs when the codon adjacent to the anticodon–codon complex is one of the stop codons, for example, UAA. The stop is recognized by release factor proteins that cause the peptidyl transferase to transfer the nascent polypeptide to water, forming a free polypeptide. Following the release of the polypeptide, the final tRNA and the mRNA dissociate from the ribosome, and the ribosome dissociates into its constituent subunits. RF is an abbreviation for protein release factor.

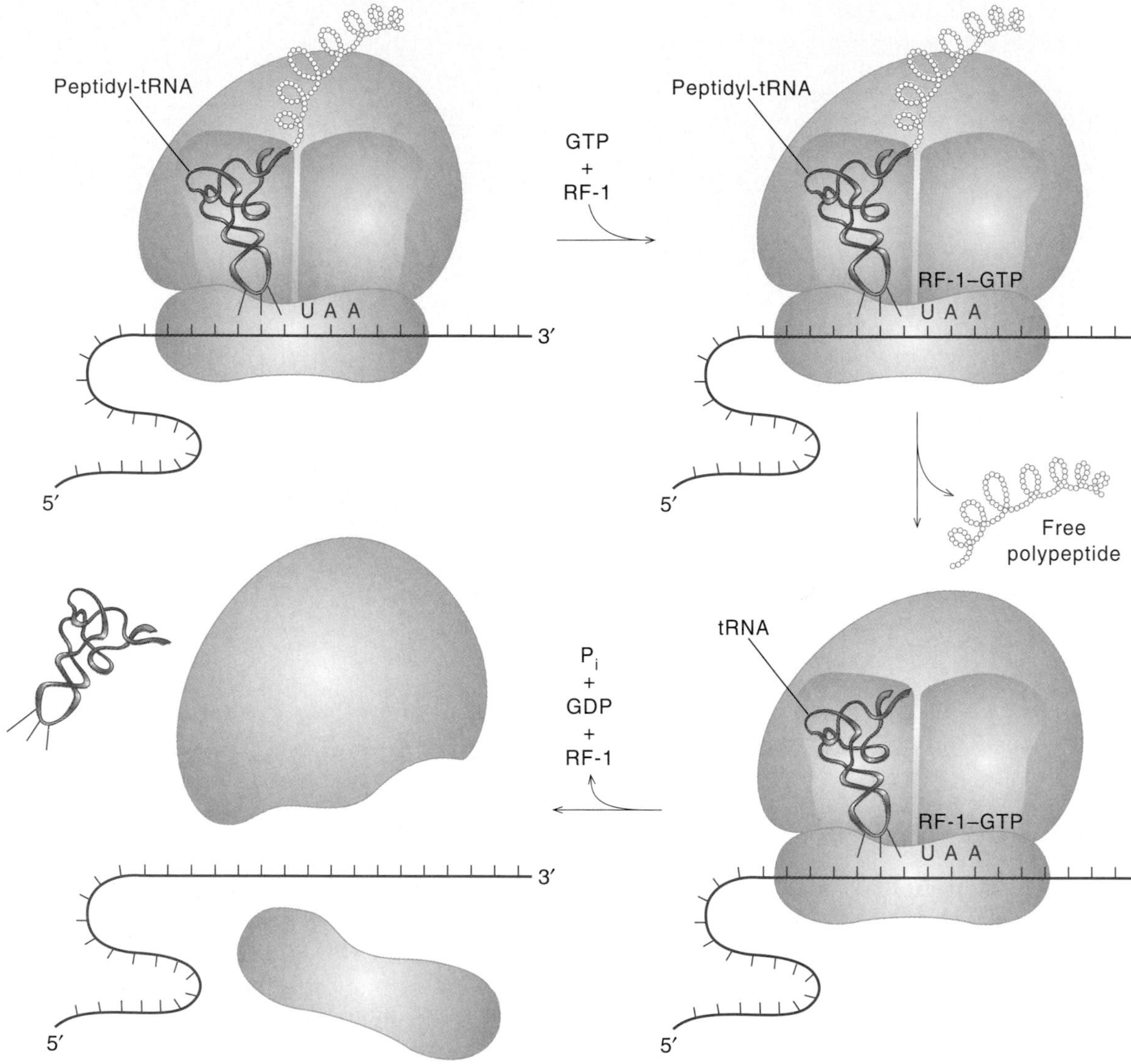

Termination of Translation Requires Release Factors and Termination Codons

The last step in translation involves the cleavage of the ester bond that joins the now complete peptide chain to the tRNA corresponding to its C-terminal amino acid (fig. 29.19). Termination requires a termination codon, mRNA and at least one protein release factor (RF). The freeing of the ribosome from mRNA during this step requires the participation of a protein called ribosome releasing factor (RRF).

In *E. coli,* termination codons that arrive at the A site on the ribosome are recognized by one of three protein release factors, RF-1 recognizes UAA and UAG, and RF-2 recognizes UAA and UGA. The third release factor, RF-3, does not itself recognize termination codons but stimulates the activity of the other two factors.

The consequence of release factor recognition of a termination codon in the A site is to alter the peptidyl transferase center on the large ribosomal subunit so that it can accept water as the attacking nucleophile rather than requiring the normal substrate, aminoacyl-tRNA (fig. 29.20). In other

Figure 29.20

An *in vitro* assay for release factors.

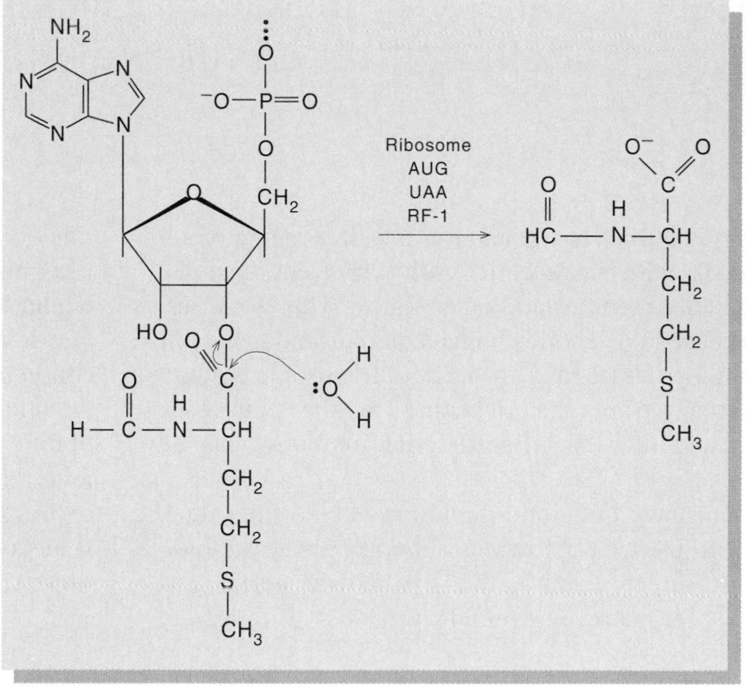

words, the termination reaction serves to convert the peptidyl transferase into an esterase. This feature of the termination reaction is clearly seen in the simple *in vitro* reaction that occurs when *E. coli* ribosomes are combined with fMet-$tRNA_i^{Met}$, RF-1, and two separate nucleotide triplets: AUG and UAA (see fig. 29.20). Formylmethionine is produced by hydrolysis, and this reaction is specifically inhibited by antibiotics such as sparsomycin that inhibit peptidyl transferase.

Ribosomes Can Change Reading Frame during Translation

Normally, ribosomes march in lock step down the mRNA, one codon at a time, beginning at the initiating AUG and ending at a termination codon. In this way translation is rigidly maintained in a single reading frame. In the last few years, a number of specific exceptions to this seemingly rigid rule have been discovered in both prokaryotic and eukaryotic systems. In these cases, the correct expression of specific messages requires the ribosome to violate the rule of lock-step translation. One way that this occurs is known as translational frameshifting. Frameshifting generally occurs one base at a time; this change, or slipping, can be in either the +1 or −1 direction. Another way involves skipping as many as 50 bases downstream in the mRNA. This process is known as translational jumping. From an informational point of view, translational jumping is analogous to mRNA splicing except that the information is not removed but is simply ignored by the translation system. Both frameshifting and jumping are favored at specific sites on the mRNA. In some cases these sites are known to require the presence of specific sequences or specific secondary structures in the mRNA. Although the nature of these sequences and structures have been identified in several cases, the mechanisms by which they are read have not been determined.

A notable instance of frameshifting occurs in the gene for the release factor RF-2, which recognizes the UGA stop codon in *E. coli.* This gene does not make RF-2 unless a frameshift takes place at an in-frame UGA sequence. When there is an abundance of the release factor, the frameshift is unlikely to occur because release occurs when the ribosome reaches this point on the messenger. However, when the release factor is in short supply, the ribosome is likely to pause at this site and then undergo a frameshift to make functional release factor. In this example, frameshifting results in a negative feedback mechanism for regulating the amount of release factor that is synthesized.

Translational fidelity of bacterial ribosomes is influenced in a more general way by the binding of antibiotics to the ribosome (box 29C).

BOX

29C Antibiotics Inhibit by Binding to Specific Sites on the Ribosome

Many antibiotics (generally, small organic compounds with therapeutic utility) prevent bacterial growth by inhibiting translation (see fig. 1). This is not surprising, because translation is both a complex and metabolically essential process. Also, the bacterial ribosome is structurally distinct from the ribosome in the eukaryotic cytoplasm, and thus specific bacterial inhibitors can be found.

A great many antibiotic inhibitors of ribosome function belong to the class known as aminoglycoside antibiotics. Of these, streptomycin is the best known and the best investigated. Streptomycin binding produces a variety of functional alterations in the ribosome. One of the first to be recognized was the loss of translational fidelity. When bound to the small subunit, streptomycin distorts its structure so as to allow altered codon–anticodon pairing and the consequent incorporation of incorrect amino acids. Indeed, mutations to streptomycin resistance frequently prevent antibiotic binding and involve alterations in ribosomal proteins that are known to play a role in maintaining the fidelity of translation. Streptomycin binding also alters the ribosome's ability to participate properly in the initiation reactions.

Figure 1

The structures of some antibiotic inhibitors of protein synthesis. All of the inhibitors shown function by binding to specific sites on the ribosome.

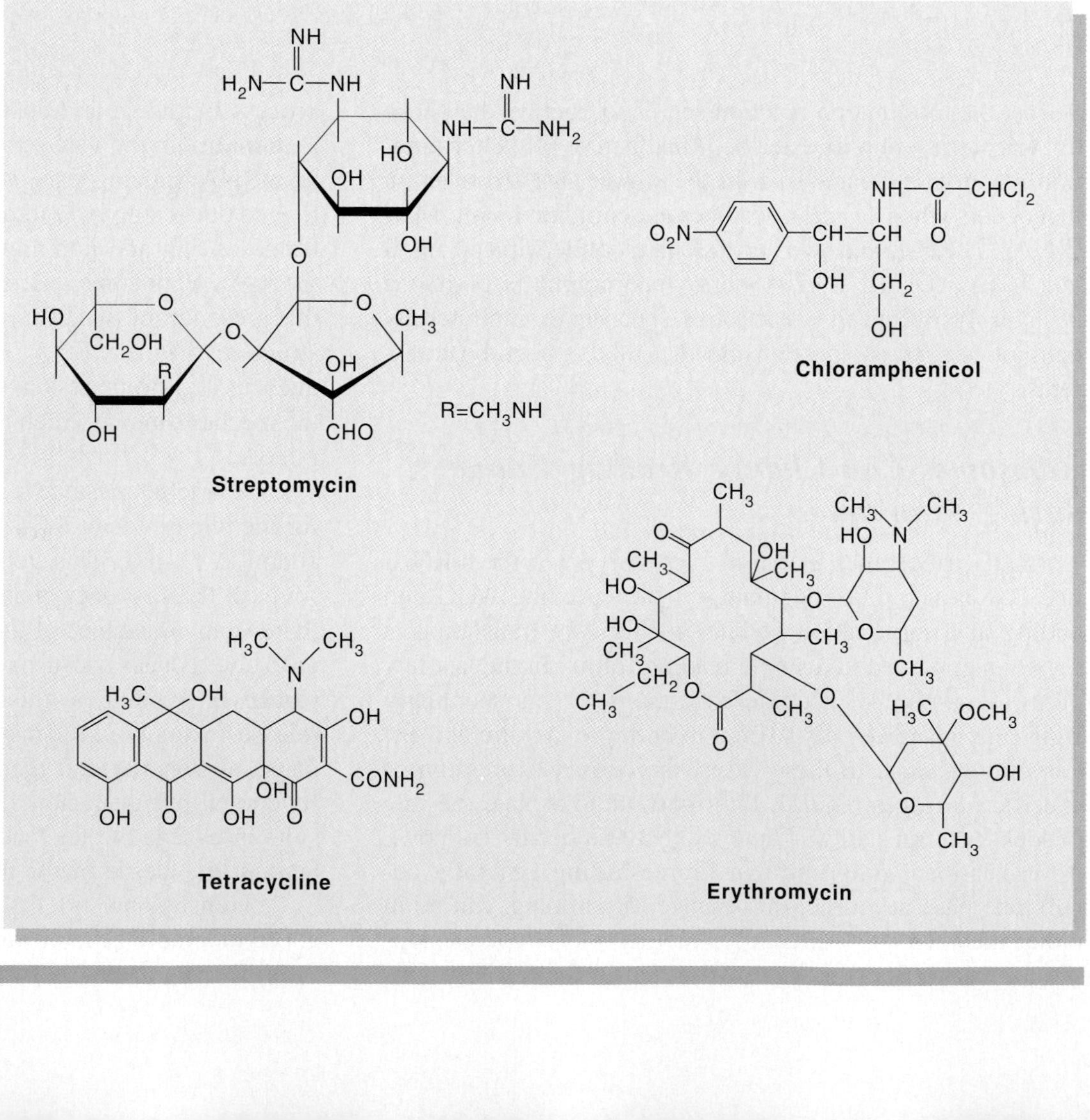

Targeting and Posttranslational Modification of Proteins

Despite the fact that only 20 amino acids (plus selenocysteine and formylmethionine in prokaryotic systems) are known to be directly specified by the genetic code, chemical analysis of mature proteins has revealed hundreds of different amino acids, all of them structural variants on the original 20. This structural diversity, which greatly expands the chemical lexicon of proteins, results from posttranslational modification of the primary products of translation. Our knowledge of the nature and significance of enzymatic reactions that bring about these important alterations is still very incomplete.

In addition to the modification of amino acid side chains many cases occur in which parts of the originally synthesized polypeptide chain are removed during the process of maturation. The types of modification and processing that polypeptide chains undergo is strongly related to the site of protein synthesis and to the mechanisms that are involved in targeting the polypeptide chain to its final destination. In fact, processing frequently begins during polypeptide synthesis and continues for some time thereafter. A major division involving the types of processing reactions that polypeptide chains undergo is related to the site of protein synthesis and the final destination of the protein. Proteins synthesized on free polysomes either remain in the cytosol or are targeted to the mitochondria, the chloroplasts, or the nucleus. Those that remain in the cytosol are often ready to function as soon as they are released from the ribosome. Those that must be transported to another site usually undergo substantial modification during the transport process. Many proteins, however, are synthesized on membrane-bound polysomes rather than on free polysomes. In bacteria such proteins are synthesized on polysomes associated with the inner plasma membrane; in eukaryotes, they are synthesized on polysomes associated with the endoplasmic reticulum. Proteins that are synthesized on ribosomes bound to the endoplasmic reticulum (ER) either remain in the ER or are targeted to the Golgi apparatus, secretory granules, the plasma membrane, or lysosomes. As a rule proteins that are synthesized on membrane-bound polysomes undergo more extensive modification before they reach their final destination and become fully functional (see chapter 16).

Proteins Are Targeted to Their Destination by Signal Sequences

Despite the apparent complexity of the problem, protein transport in all systems is accomplished by a single, rather simple underlying mechanism; each polypeptide destined for transport contains an amino acid sequence known as a signal, or leader sequence that identifies the polypeptide to the appropriate transporting system. This generality was first recognized in the middle 1970s by Gunter Blobel, who articulated the underlying signal hypothesis (see fig. 16.8). Frequently, the signal sequence is cleaved from the parent polypeptide during the transport process. A protein containing a signal sequence that is cleaved on transport is known as a preprotein. A protein containing a peptide sequence that must be removed for the protein to be active is known as a proprotein, and a protein that contains both of these sequences is known as a preproprotein.

Many protein hormones are synthesized in a preproprotein form. For example, insulin mRNA is translated into a single polypeptide chain, preproinsulin, that contains 84 residues of proinsulin plus a 23-residue signal peptide (fig. 29.21). Within the islets of Langerhans of the pancreas, the N-terminal sequence is removed cotranslationally in targeting proinsulin to the Golgi apparatus, and the two disulfides joining the ends of the molecule are formed. Following this, the C-peptide region of proinsulin is removed to yield the mature circulating form of insulin with 51 residues in two disulfide-linked peptides.

Signal sequences usually occur at the amino terminus of polypeptides to be transported. These sequences vary in length from 10 to 40 amino acid residues and are generally characterized by the occurrence of a block of hydrophobic residues. The presence of hydrophobic residues reflects the fact that protein transport invariably involves the movement of proteins across lipid membranes. The hydrophobic residues both target the protein to the appropriate compartment and initiate penetration into the membrane.

Some Mitochondrial Proteins Are Transported after Translation

Some proteins, especially those destined for the eukaryotic mitochondria and chloroplasts, are transported after their synthesis on free polysomes is complete. Such transport is known as posttranslational transport. In the case of posttranslational transport it is believed that the polypeptide to be transported must be unfolded from its native folded configuration by a system of polypeptide-chain-binding proteins (PCBs) before it can pass through the membrane. Posttranslational transport into the mitochondrion requires both ATP and a proton gradient. Presumably the energy from one or both of these sources is used to unfold the protein or separate it from the PCB system so that it can pass through the membrane.

The transport of proteins from the cytoplasm to the mitochondrion is further complicated by the fact that mito-

Figure 29.21

Processing of insulin. Insulin is synthesized by membrane-bound polysomes in the β cells of the pancreas. The primary translation product is preproinsulin, which contains a 24-residue signal peptide preceding the 81-residue proinsulin molecule. The signal peptide is removed by signal peptidase, cutting between Ala (−1) and Phe (+1), as the nascent chain is transported into the lumen of the endoplasmic reticulum. Proinsulin folds and two disulfide bonds cross-link the ends of the molecule as shown. Before secretion, a trypsinlike enzyme cleaves after a pair of basic residues 31, 32 and 59, 60; then a carboxypeptidase B-like enzyme removes these basic residues to generate the mature form of insulin.

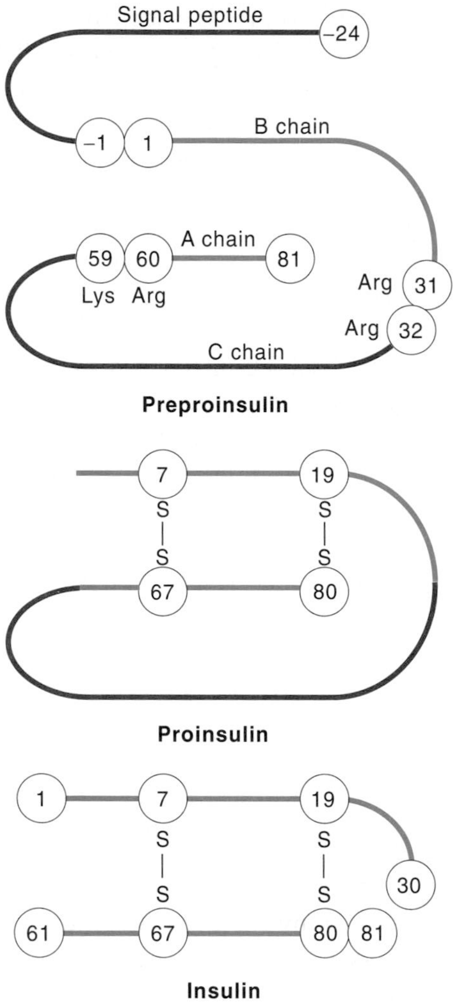

chondria themselves are compartmentalized. They have both an outer membrane and an inner membrane (see chapter 14). Some proteins are targeted to either of these membranes, some are targeted to the fluid layer enclosed by the inner membrane, called the matrix, and others such as cytochrome c_1 are targeted to the fluid layer bounded by the inner membrane and the outer membrane, known as the intermembrane space (fig. 29.22). Cytochrome c_1 has two signal sequences located in series at the amino-terminal end of the preprotein. The first is recognized by a specific receptor protein attached to the outer membrane of the mitochondria. This receptor protein guides the polypeptide chain to a transport channel, across which it is transported into the matrix. Once in the matrix, the first signal sequence is removed by a specific protease. A second signal sequence, which is exposed by this cleavage reaction, results in the transport of the polypeptide chain by a similar mechanism across a transport channel into the intermembrane space. In intermembrane space the second signal sequence is removed, and the protein folds into its mature configuration, which includes associating with a heme group.

As in the case of mitochondrial proteins, some of the proteins of chloroplasts are synthesized directly in the organelle, whereas others are synthesized in the cytosol and must be transported. The mechanisms for transport are very similar to those observed for transported mitochondrial proteins.

All nuclear proteins are synthesized on free polysomes in the cytosol. In contrast to the transport processes that involve specific amino-terminal sequences that are cleaved, more complex structural features are recognized in nuclear proteins, and they are somehow selectively transported in an intact state into the nucleus.

Eukaryotic Proteins Targeted for Secretion Are Synthesized in the Endoplasmic Reticulum

The endoplasmic reticulum (ER) is the largest membrane-bounded organelle in a typical eukaryotic cell (see fig. 16.9). The ER consists of a continuous network of tubules and cisternae extending throughout the cytoplasm, with a total surface area many times that of the plasma membrane. Most of the ER is studded with ribosomes to form the rough endoplasmic reticulum (RER). The ribosomes of the RER are the site of synthesis of membrane and secretory proteins and are the starting point for the protein secretory pathway. The membrane and lumen of the ER contain a characteristic set of proteins that function to process secretory and membrane proteins. After only a short time in the RER, secretory proteins are transported, by a process of vesicle budding and fusion, to the Golgi apparatus and from there to the cell surface, secretory granules, or lysosomes. As you will recall, the Golgi apparatus is a flattened stack of membranes that is the primary site of protein targeting as well as a principal site of carbohydrate addition to form glycoproteins (see chapter 16).

Let us review the way in which proteins that are destined for assembly into membranes, for secretion, or for targeting to other areas of the cell enter the lumen of the ER (see fig. 16.8). The initial targeting of nascent polypeptides

Figure 29.22

Two successive translocations are required to target proteins such as cytochrome c_1 to the intermembrane space. The precursors of cytochrome c_1 have two uptake-targeting sequences at its N terminus. The first targets the polypeptide to the matrix. In the matrix the first target sequence is cleaved by a specific protease. The second target sequence is thereby exposed and targets the polypeptide to intermembrane space, where the second target sequence is removed by another protease. The molecule folds and adds heme to become fully functional. (Source: After F. U. Hartl, J. Ostenmann, B. Guiard, and W. Neupert, *Cell* 51:1021–1027, 1987; and E. C. Hurt and A. P. G. M. van Loon, How proteins find mitochondria and intramitochondrial compartments, *Trends Biochem. Sci.* 11:204–207, 1986.)

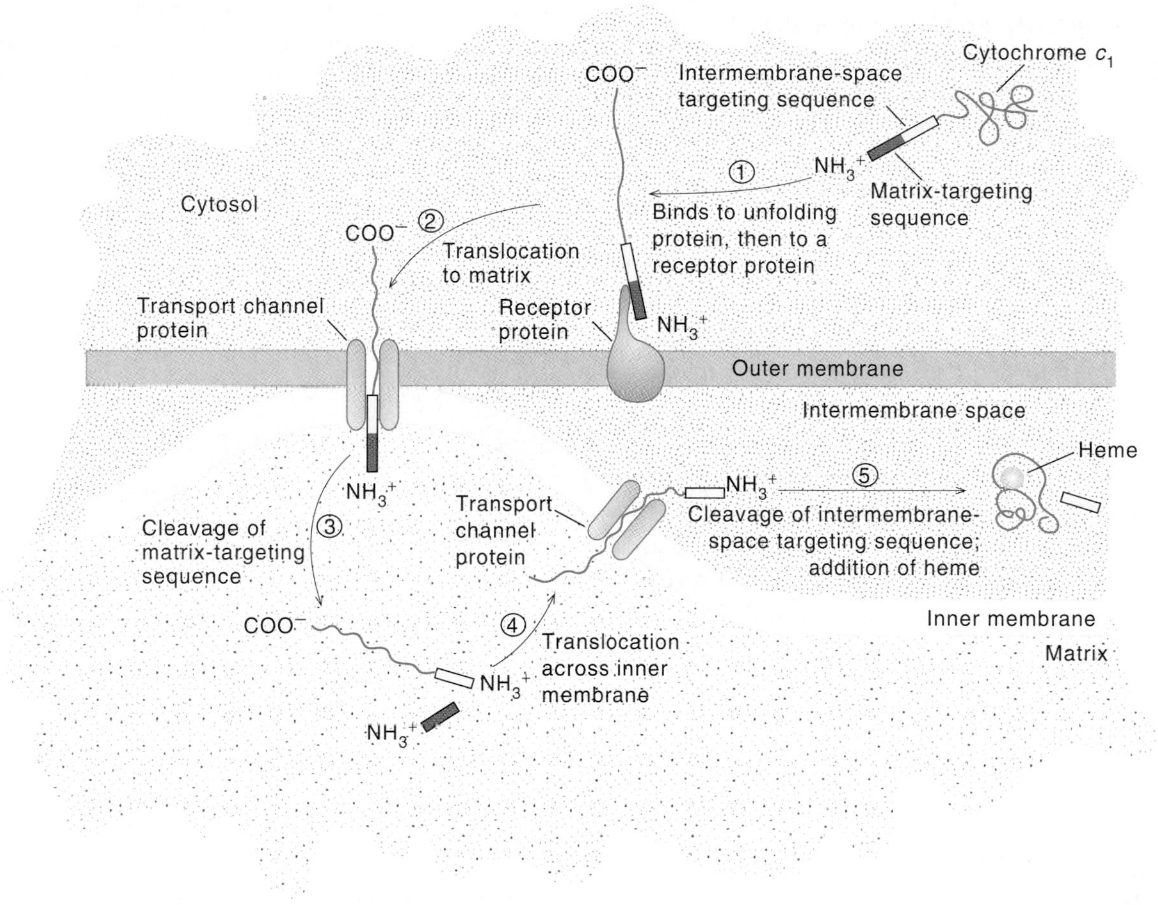

to the ER membrane results from the cotranslational recognition of a signal sequence by a ribonucleoprotein complex called the signal recognition particle, or SRP. The SRP is a complex that consists of six different proteins and a single 300-residue RNA molecule designated 7S RNA. This complex is especially adapted to recognize and bind to the signal sequence when the total nascent chain has reached a length of about 90 amino acid residues. On binding, the SRP arrests translation of the nascent chain while it searches for a specific receptor known as the SRP receptor, or "docking protein." The SRP receptor is an integral membrane protein that protrudes on the inner face of the ER membrane. When docking is complete, the SRP dissociates from the ribosome, which is now bound by the signal sequence of its nascent polypeptide to the SRP receptor. After dissociation of the SRP, the translation resumes, and translocation of the nascent polypeptide across the membrane begins. It has recently been discovered that GTP is essential for the docking maneuver and that its hydrolysis is probably essential for the subsequent release of SRP. Furthermore, sequencing of one of the protein subunits of SRP, SRP54, has revealed that it is homologous to the GTP-binding domain of the G proteins. Thus, signal sequence recognition by SRP may operate in a manner analogous to the GTP-dependent functioning of members of the G protein family.

Numerous nonmembranous proteins are retained by and function in the ER. Retention of these proteins is accomplished by a surprisingly simple mechanism. Soluble ER proteins all share the common C-terminal sequence. Lys-Asp-Glu-Leu (or KDEL in single-letter language). A protein receptor appears to be bound to the ER membrane that recognizes and binds this sequence. The passive nature of this retention mechanism is illustrated by the fact that if the four C-terminal residues are removed from an ER protein, it is no longer retained by the organelle.

Proteins That Pass through the Golgi Apparatus Become Glycosylated

Glycosylation presents a conspicuous modification of proteins as they pass through the Golgi apparatus (see chapter 16). Many of the cell surface and secretory proteins produced here are glycoproteins. Because of the diversity of the glycosylation reactions that occur in the Golgi and the fact that this organelle is the principal site of protein transport, it seems likely that the carbohydrate moieties attached to proteins are responsible for targeting them to their destinations.

One clear-cut example supporting this conclusion is the role of mannose-6-phosphate residues in targeting proteins to the lysosome. Digestive enzymes destined for the lysosome are identified by the presence of mannose-6-phosphate by binding to a specific mannose-6-phosphate receptor protein. The critical step in targeting glycoproteins from the Golgi to the lysosome is their recognition by the enzymes that catalyze the two-step phosphorylation of terminal mannose residues in their N-linked oligosaccharide side chains. Failure to phosphorylate the mannose residues results in the secretion of lysosomal enzymes. This malfunction, you may remember, is associated with the syndrome known as I-cell disease, a condition that leads to the crowding of lysosomes with damaged proteins that it cannot degrade (see chapter 16).

Processing of Collagen Does Not End with Secretion

Recall that collagen is an extracellular matrix protein that serves as a major constituent of many connective tissues (see figs. 4.10 to 4.13). Collagen fibrils have a distinctive banded pattern with a periodicity of 680 Å. Individual fibrils are composed of three polypeptide chains wound around one another in a right-handed helix with a total length of 3,000 Å. Each of the polypeptide chains in the triple helix has a repetitious tripeptide sequence, Gly-X-Y, where X is frequently a proline and Y is frequently a hydroxyproline. The latter amino acid is not one of the 20 that are specified genetically, so it must be formed posttranslationally by a modification of some of the prolines.

Since collagen is a secreted protein, we know that it must follow the route of synthesis that starts on a ribosome bound to the endoplasmic reticulum (fig. 29.23). This is where the fun begins, as we find that the nascent collagen polypeptide has extensive N and C termini (150 and 250 amino acids, respectively, for type I collagen) that are not found in mature collagen. The function of these extensions appears to be to facilitate the initial interaction of chains in triplets and to stabilize the triple helices once they have been formed. The first modification reaction to take place in the endoplasmic reticulum is hydroxylation of specific proline residues and some lysine residues by two different enzyme systems. Glycosylation begins soon thereafter with O-glycosylation of certain hydroxylysine residues and N-glycosylation of certain asparagine residues. Next, the modified polypeptide chains, in clusters of three, form interchain disulfides near the C termini. This brings the chains in close proximity, thereby facilitating the winding reaction that leads to triple-helix formation. The winding reaction proceeds in the C to N direction. The triple-helix procollagen molecule is packaged into a secretory vesicle. Following exocytosis, the ends of the procollagen molecule are removed by extracellular proteases. Collagen fibrils form by the spontaneous association of the mature collagen molecules.

Bacterial Protein Transport Frequently Occurs during Translation

The problem of protein transport in bacterial systems is relatively simple. Polypeptides synthesized in the cytoplasm may function there, may be inserted into the plasma membrane, or may be secreted by being passed through this membrane.

Most noncytoplasmic bacterial proteins are targeted into or across the plasma membrane while they are being synthesized on the ribosome. This process is known as cotranslational transport. Cotranslational secretion of proteins in *E. coli* involves a group of proteins encoded by the *sec* (for secretion) genes. Some of these proteins are membrane proteins that function by recognizing and binding the signal sequence of nascent polypeptide chains as they emerge from the ribosome. As a consequence of cotranslational transport and the action of the sec proteins, most bacterial ribosomes engaged in the synthesis of secreted proteins are tethered to the cytoplasmic face of the plasma membrane by their nascent peptide chains. A leader peptidase that is an integral membrane protein cleaves the leader sequence from the secreted polypeptide.

Protein Turnover

Cellular proteins are continuously being formed and degraded. At first glance, continuous degradation appears to be wasteful. However, protein degradation is of major biological importance in regulating protein levels, in protecting

Figure 29.23

Major events in the posttranslational processing of collagen.

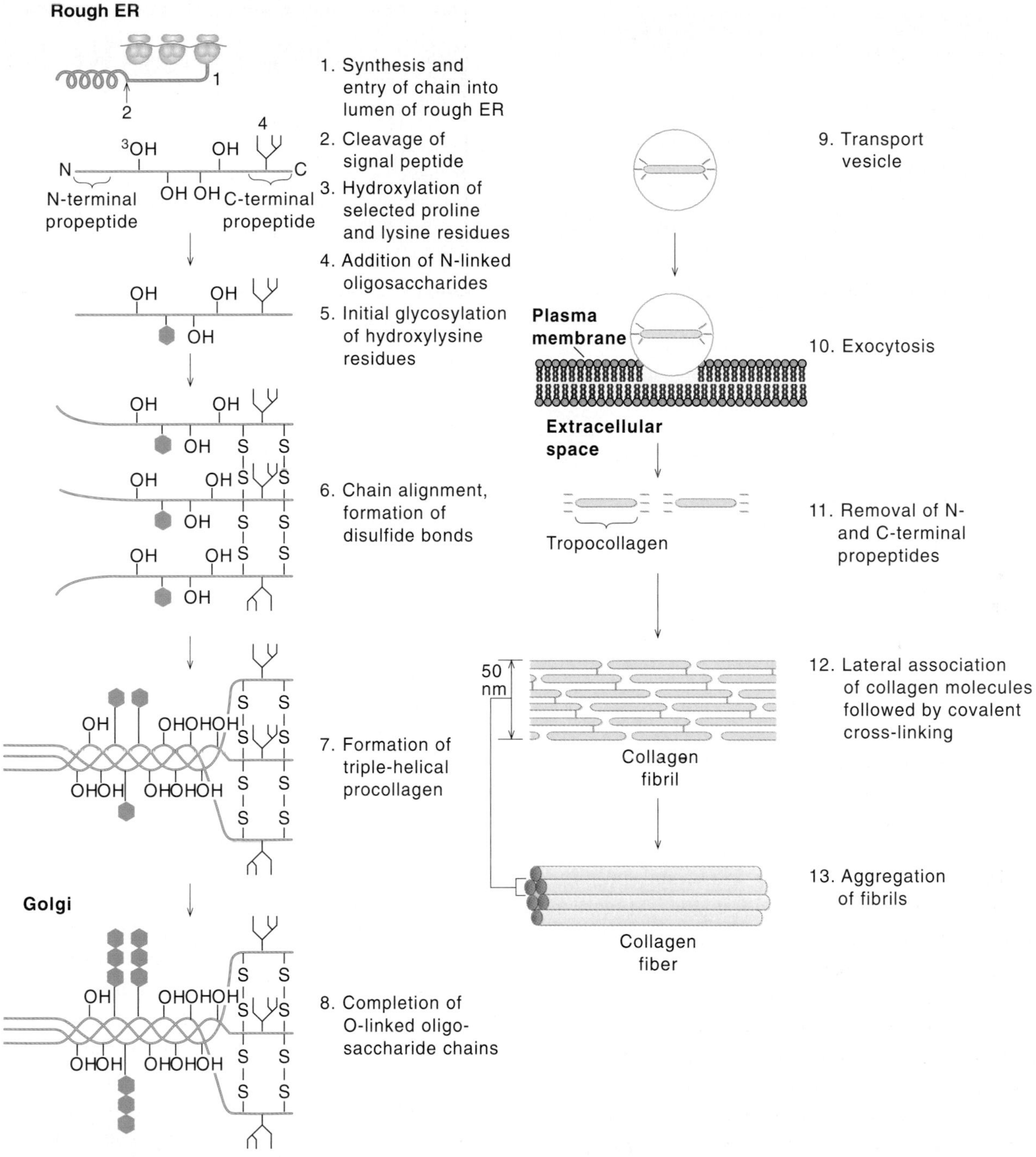

against the accumulation of abnormal proteins, in controlling growth and development, and in allowing adaptation to changing environmental conditions.

Although many proteolytic enzymes are known to exist, we are just beginning to understand how they operate to govern the levels and types of proteins within cells—to distinguish those proteins that must be degraded from those that must be preserved.

The Lifetimes of Proteins Differ

The level of a protein within a cell is determined by the balance between its rates of synthesis and degradation. As a consequence, changes in protein levels can be brought about by changes either in synthetic or degradative rates. Moreover, a rapid rate of degradation ensures that the concentration of a protein rises or falls rapidly when its synthetic rate changes.

Table 29.6

Half-Lives of Some Proteins in Mammalian Cells

Enzyme	Half-Life (h)
Rapidly degraded	
1. c-myc, c-fos, p53 oncogenes	0.5
2. Ornithine decarboxylase	0.5
3. δ-Aminolevulinate synthase	1.1
4. RNA polymerase I	1.3
5. Tyrosine aminotransferase	2.0
6. Tryptophan oxygenase	2.0
7. β-Hydroxyl-β-methylglutaryl coenzyme A reductase	2.0
8. Deoxythymidine kinase	2.6
9. Phosphoenolpyruvate carboxykinase	5.0
Slowly degraded	
1. Arginase	96
2. Aldolase	118
3. Cytochrome b_5	122
4. Glyceraldehyde-3-phosphate dehydrogenase	130
5. Cytochrome *b*	130
6. Lactic dehydrogenase (isoenzyme 5)	144
7. Cytochrome *c*	150

Table 29.7

Correlation between Half-Lives of Cytosolic Proteins and Amino Acid Residue at the N Terminal

Amino-Terminal Residue	Half-Life
Stabilizing	
Methionine	
Glycine	
Alanine	> 20 h
Serine	
Threonine	
Valine	
Destabilizing	
Isoleucine	$\approx$30 min
Glutamate	
Tyrosine	
Glutamine	$\approx$10 min
Proline	$\approx$7 min
Highly Destabilizing	
Leucine	
Phenylalanine	
Aspartate	$\approx$3 min
Lysine	
Arginine	$\approx$2 min

(*Source:* From A. Bachmjuir et al., *In vivo* half-life of a protein is a function of its amino-terminal residue, *Science* 234:179, 1986. Copyright © 1986.)

The half-lives of eukaryotic proteins are different and distinct. Representative examples are listed in table 29.6. It is apparent that degradative rates of individual proteins within a single cell can vary over a wide range. At least in the mammalian liver, enzymes that occupy important metabolic control points are degraded most rapidly, whereas especially long-lived proteins are rarely the sites of metabolic control. In addition, it has been established that degradative rates of specific proteins can vary with changes in physiological conditions. Thus, the protein degradative mechanism is in some way tuned to metabolic control.

How are individual proteins selected for hydrolysis by the proteolytic machinery? Two sequence characteristics have been identified as correlating with rates of protein degradation. Several years ago it was observed through sequence analysis that rapidly degraded liver proteins, those with half-lives of less than 2 h, nearly all contain regions of their sequence that are rich in the amino acids proline, glutamate, serine, and threonine. These regions, involving from 10 to 60 amino acid residues, have been designated PEST sequences on the basis of the single-letter designations of their constituent amino acids. Presumably, the PEST sequences create structural domains that are recognized by proteolytic enzymes.

A second structural feature that has been correlated with protein degradative rates is the N-terminal residue of the mature form of cytoplasmic proteins. This relationship, summarized in table 29.7, has become known as the N-terminal rule. Here the presumption is that the amino-terminal residue is at least partly responsible for recognition by the degradative machinery. Proteins that are degraded according to the N-terminal rule are believed to be recognized by the ubiquitin-ATP-dependent pathway, which we describe shortly.

Abnormal Proteins Are Selectively Degraded

A very important function of protein degradation is to protect the organism against the consequences of intracellular accumulation of abnormal proteins. All cells possess a degradative system that is capable of recognizing "abnormal" proteins.

The features of this degradative system are best seen in *E. coli,* in which altered intracellular proteins can be readily manipulated. For example, it has been known for many years that incomplete chains of β-galactosidase are rapidly degraded even though the completed protein is very stable. In a similar way, many protein alterations that result from miscoding induced by streptomycin, or by the incorporation of amino acid analogs, also fail to accumulate because the protein products that contain them are rapidly degraded. This system presumably limits expression of heterologous proteins in bacteria that are degraded because they cannot form their native structure.

Another manifestation of the defense against abnormal proteins is seen in the heat shock response. This defense reaction, involving programmed changes in gene transcription and translation, is exhibited by essentially all cells under stressful conditions. The cells reduce their overall rates of gene transcription and translation and for a brief time produce a small repertoire of proteins called heat shock proteins (hsp). Some hsp play a role in the folding and transport of proteins during normal cellular function. In the heat shock response it is believed that these proteins protect against the presence of damaged proteins by binding to them and promoting either their refolding or their proteolytic degradation.

Proteolytic Hydrolysis Occurs in Mammalian Lysosomes

The classic studies of Christian DeDuve in the 1960s established that mammalian cells contain a degradative organelle, the lysosome, that is produced in the Golgi apparatus and contains a large number of proteases and other hydrolytic enzymes. A similar vacuole containing hydrolytic enzymes is also present in yeast and higher plants. Together, the hydrolytic enzymes in these organelles are capable of completely degrading many macromolecules and delivering their monomeric units to the cytoplasm for further metabolism. The metabolic importance of the lysosome is demonstrated by the existence of various lysosomal storage diseases, in which one or more hydrolytic enzymes is missing. An accumulation of undegraded molecules results from these genetic diseases, frequently with devastating consequences.

A primary function of the lysosome is to digest protein-containing particles derived from the extracellular space. One mechanism of delivery is the process of endocytosis. Endocytosis is the invagination of a group of occupied receptors on the plasma membrane. Most mammalian cells can also engulf large extracellular particles by the less specific processes of pinocytosis and phagocytosis. The endocytic vesicles formed by these processes fuse with lysosomes to form secondary lysosomes where hydrolysis occurs. Lysosomes also function to degrade intracellular proteins, especially under conditions of nutritional deprivation. Under these circumstances, lysosomes can engulf cytoplasmic contents to form autophagic vacuoles and thus recycle cellular proteins.

Lysosomal proteases are called cathepsins, a name derived from the Greek term meaning "to digest." The interior of the lysosome is acidic and the cathepsins, like all lysosomal hydrolases, possess acidic pH optima and exhibit little enzymatic activity at neutral pH. This characteristic protects the cell from autolytic breakdown that might result from leakage of lysosomal contents into the neutral cytoplasm.

Ubiquitin Tags Proteins for Proteolysis

A number of different proteolytic systems are thought to be responsible for the degradation of soluble proteins in the cytoplasm of eukaryotic cells. One of the best understood is that which involves ATP and the protein ubiquitin. Ubiquitin is a small protein of only 76 residues. It occurs universally in eukaryotic cells and is highly conserved in sequence; only three residues distinguish the ubiquitin in yeast and humans. The covalent attachment of ubiquitin to proteins is thought to "tag" them for subsequent hydrolysis by cellular proteases.

Three enzymes participate in the ATP–ubiquitin system that prepares proteins for proteolysis (fig. 29.24). In the first step of this series of reactions, the C-terminal glycine residue of ubiquitin is activated by forming a thiol ester with a specific activating enzyme designated E1. This reaction is driven by the hydrolysis of ATP to AMP and PP_i. Ubiquitin is then transferred to a sulfhydryl group of a second protein (E2) that is a substrate for a group of ubiquitin-targeting proteins designated E3. The E3 proteins are responsible for identifying proteins for degradation (some on the basis of the identity of their N-terminal residue and some because they are "abnormal") and catalyzing the covalent transfer of ubiquitin from E2 to these proteins. This attachment is via isopeptide bonds that join the carboxyl group of the previously activated C-terminal glycine residue of ubiquitin with ϵ-amino groups of lysine side chains of the targeted proteins. Proteins tagged in this way are then recognized and degraded by specific proteases that release ubiquitin so that it can recycle.

ATP Plays Multiple Roles in Protein Degradation

ATP plays a conspicuous and rather surprising role in the degradation of proteins. Clearly, the hydrolysis of peptide bonds is a reaction that in and of itself does not require the input of energy. Nonetheless, ATP is required for the action of many proteolytic enzymes, independent of its role in ubiquitination that we have just seen.

One particularly well-understood ATP-dependent protease is the La protease that is the product of the *lon* gene of *E. coli.* This protease, like many ATP-dependent proteases, is a large protein, and its ability to hydrolyze proteins is tightly coupled to its ability to hydrolyze ATP. Approximately two ATPs are hydrolyzed to ADP and P_i for each peptide bond that is hydrolyzed. It appears that the hydrolysis of ATP is required to activate the proteolytic active site of La. Other proteases seem to require ATP as an allosteric effecter to activate their hydrolytic sites, but this ATP is not hydrolyzed.

Thus, ATP serves at least three roles in intracellular proteolysis: (1) It functions in the tagging of proteins through covalent attachment of ubiquitin, (2) it activates proteases such as La through hydrolysis, and (3) it serves as a positive allosteric effecter of other proteases without being hydrolyzed. The presumed function of this energetic requirement is to ensure the fidelity of protein degradation. It is, after all, nearly as important to cellular function to degrade proteins correctly as it is to synthesize them correctly.

Figure 29.24

The ubiquitin marking system targets certain proteins for degradation. At least three enzymes, E1, E2, and E3, are involved in addition to ubiquitin-specific proteases.

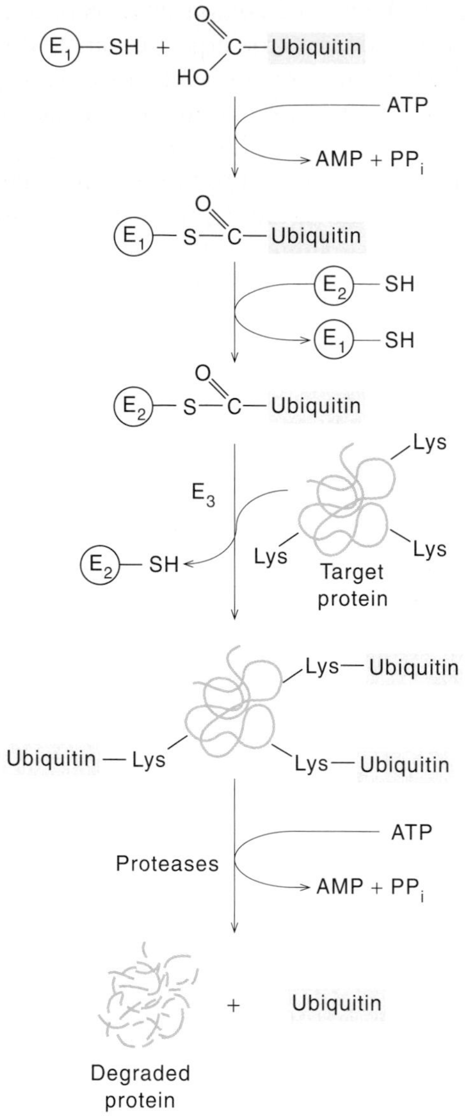

Summary

We focused in this chapter on the complex mechanisms of protein synthesis. The following points are central to this subject.

1. Three types of RNA carry out protein synthesis: Ribosomal RNA, transfer RNA, and messenger RNA. Ribosomal RNA is invariably complexed with many proteins to form ribosomes, on which amino acids are assembled into polypeptides. The amino acids are brought to the ribosomes attached to transfer RNAs. The messenger RNA contains the instructions for translation in the form of the genetic code. Messenger RNAs form transient complexes with ribosomes. Individual aminoacyl-tRNAs bind to specific sites on the messenger RNAs. The interacting site on the messenger is the codon; the interacting site on the tRNA is the anticodon.

2. The part of the messenger that is translated is the reading frame. Eukaryotic messages carry only one reading frame, whereas prokaryotic messengers may carry more than one. In prokaryotes the initiation codons are recognized by a ribosome-binding site upstream of the start codon.
3. Most transfer RNAs have common parts and uncommon parts. The common parts facilitate binding of the aminoacyl-tRNAs to common sites on the ribosome. The uncommon sites permit specific reactions with charging enzymes that covalently attach the correct amino acids to the correct tRNA. Another uncommon site on the tRNAs is the anticodon, which leads to specific complex formation with the complementary codon site on the messenger.
4. Attachment of the amino acid to the tRNA is catalyzed by a specific aminoacyl synthase, which recognizes all the cognate tRNAs for a specific amino acid.
5. A unique methionyl-tRNA binds to the initiation codon on all messages.
6. The genetic code is the sequence relationship between nucleotides in the messenger RNA and amino acids in the proteins they encode. Triplet codons are arranged on the messenger in a nonoverlapping manner without spacers.
7. The code was deciphered with the help of synthetic messengers with a defined sequence, by analyzing the types of polypeptide chains that were made when these messengers were used in an *in vitro* protein-synthesizing system.
8. The genetic code is highly degenerate, with most amino acids represented by more than one codon. In many cases the 3′ base in the codon may be altered without changing the amino acid that is encoded.
9. The codon–anticodon interaction is limited to Watson–Crick pairing for the first two bases in the codon but is considerably more flexible in the third position.
10. Translation begins with the binding of the ribosome to mRNA. A number of protein factors transiently associate with the ribosome during different phases of translation: Initiation factors, elongation factors, and termination factors.
11. Initiation factors contribute to the ribosome complex with the messenger RNA and the initiator methionyl-tRNA. Elongation factors assist the binding of all the other tRNAs and the translocation reaction that must occur after each peptide bond is made. Termination factors recognize a stop signal and lead to the termination of polypeptide synthesis and the release of the polypeptide chain and the messenger from the ribosome.
12. A large number of antibiotics have been characterized that inhibit protein synthesis. These antibiotics are usually made by a particular microorganism, and they inhibit protein synthesis in a broad family of other organisms, mostly bacterial.
13. Specific enzymes catalyze folding after polypeptide synthesis.
14. Proteins are targeted to their destination by signal sequences built into the polypeptide chain. These signals are usually located at the N-terminal end of the protein and are generally cleaved during protein maturation.
15. Posttranslational modifications include many covalent alterations: Polypeptide processing, attachment of carbohydrate or lipid groups to specific side chains, and addition of many other low-molecular-weight ligands to side chains.
16. Intracellular protein degradation is not random. Different proteins have quite different half-lives, which are related to specific structural features. Imperfectly folded proteins and polypeptide fragments are frequently degraded most rapidly. In eukaryotes, lysosomes play a major role in protein degradation.

Selected Readings

Bachmjuir, A., and A. Varshavsky, The degradation signal is a short-lived protein. *Cell* 56:1019–1032, 1989.

Beasley, E. M., and G. Schatz, Import of proteins into mitochondria. *Chemtracts* 2:305–317, 1991.

Bjork, G. R., J. U. Ericson, C. E. D. Gustafsson, T. G. Hagervall, Y. H. Jonsson, and P. M. Wilkstrom, Transfer RNA modification. *Ann. Rev. Biochem.* 56:263–288, 1987.

Böck, A., K. Forchhammer, J. Heider, and C. Baron, Selenoprotein synthesis: An expansion of the genetic code. *Trends Biochem. Sci.* 16:463–467, 1991.

Bond, J. S., and P. E. Butler, Intracellular proteases. *Ann. Rev. Biochem.* 56:333, 1987. An overview of the types of proteolytic enzymes that are found in cells and how they may function in biologically important cleavages of proteins.

Brunori, M., M. C. Silvestrini, and M. Pocchiari, The scrapie agent and the prion hypothesis. *Trends Biochem. Sci.* 13:309–313, 1988.

Burgess, T. L., and R. B. Kelly, Constitutive and regulated secretion of proteins. *Ann. Rev. Cell. Biol.* 3:243–294, 1987.

Craig, E. A., Chaperones: Helpers along the pathways to protein folding. *Science* 260:1902–1903, 1993.

Crick, F. H. C., Codon–anticodon pairing: The wobble hypothesis. *J. Mol. Biol.* 19:548–555, 1966. A classic paper.

Ellis, R. J., and S. M. van der Vies, Molecular chaperones. *Ann. Rev. Biochem.* 60:321–348, 1991.

Englander, S. W., In pursuit of protein folding. *Science* 262:848–850, 1993.

Englesberg-Kulka, H., and R. Schoulakerk-Schwarz, A flexible genetic code, or why does selenocysteine have no unique codon? *Trends Biochem. Sci.* 13(11):419–421, 1988.

Ferguson, M. A. J., and A. F. Williams, Lipids as membrane tethers for proteins. *Ann. Rev. Biochem.* 57:285, 1988.

Fessler, J. H., and L. I. Fessler, Biosynthesis of procollagen. *Ann. Rev. Biochem.* 47:129–162, 1978.

Finley, D., and A. Varshavsky, The ubiquitin system functions and mechanisms. *Trends Biochem. Sci.* 10:343–347, 1985.

Fox, T. D., Natural variation in the genetic code. *Ann. Rev. Gen.* 21:67, 1987. A review of the exceptions to the ''universal'' genetic code.

Gold, L., Posttranslational regulatory mechanisms in *E. coli. Ann. Rev. Biochem.* 57:199, 1988. A summary of the mechanisms of initiation and regulation of translation in prokaryotic systems.

Hershey, J. W. B., Protein phosphorylation controls translation rates. *J. Biol. Chem.* 264:20823, 1989. Describes how protein kinases are thought to regulate translation in eukaryotic systems.

Hershko, A., Ubiquitin-mediated protein degradation. *J. Biol. Chem.* 263:15237–15240, 1988.

Hoagland, M. B., M. L. Stephenson, J. F. Scott, L. I. Hecht, and P. Zamecnik, A soluble ribonucleic acid intermediate in protein synthesis. *J. Biol. Chem.* 231:241–257, 1958. Describes the pioneering tracer studies that chart the course of amino acid into polypeptide chain.

Kleinkauf, H., and H. Dohren, Nonribosomal polypeptide formation on multifunctional proteins. *Trends Biochem. Sci.* 8:281–283, 1983.

Kozak, M., The scanning model for translation: An update. *Mol. Cell. Biol.* 8:2737, 1989. The current view of the way in which the eukaryotic ribosome selects the initiating codon, by the originator of the scanning model.

Moore, P. B., The ribosome returns. *Nature* 33:223–227, 1988.

Nirenberg, M. W., and J. H. Mattaei, The dependence of cell-free protein synthesis in *E. coli* upon naturally occurring or synthetic polyribonucleotides. *Proc. Natl. Acad. Sci. USA* 47:1588–1602, 1961. The landmark paper reporting the finding that poly(U) stimulates the synthesis of polyphenylalanine.

Noller, H. F., V. Hoffarth, and L. Zimniak, Resistance of peptidyl transferase to protein extraction procedures. *Science* 256:1416–1418, 1992. Demonstration that peptidyl transferase is an RNA.

Normanly, J., and J. Abelson, tRNA identity. *Ann. Rev. Biochem.* 58:1029, 1989. A summary of the chemical features of tRNA.

Pain, V. M., Initiation of protein synthesis in mammalian cells. *Biochem. J.* 235:625, 1986. A comprehensive review of the complex process of translational initiation in mammalian systems, emphasizing the mechanism of the process and its regulation.

Parker, J., Errors and alternatives in reading the universal genetic code. *Microbiol. Rev.* 53:273, 1989. A summary of the current knowledge of alternative mechanisms of translation.

Pelham, H. R. B., Control of protein exit from the endoplasmic reticulum. *Ann. Rev. Cell. Biol.* 5:1, 1989. Describes the current picture of how proteins are sorted and transported through the endoplasmic reticulum.

Pfeffer, S. R., and J. E. Rothman, Biosynthetic protein transport and sorting by the endoplasmic reticulum and Golgi. *Ann. Rev. Biochem.* 56:829, 1987. An excellent overview of the major features of protein targeting in eukaryotic cells.

Proud, C. G., Guanine nucleotides, protein phosphorylation and control of translation. *Trends Biochem. Sci.* 11:73–77, 1986.

Rechsteiner, M., S. Rogers, and K. Rote, Protein structure and intracellular stability. *Trends Biochem. Sci.* 12:390–394, 1987.

Rothman, J. E., Polypeptide chain binding proteins: Catalysts of protein folding and related processes in cells. *Cell* 59:591, 1989. A description of the proteins that are thought to be involved promoting the formation of three-dimensional structure in proteins.

Rould, M. A., J. J. Perona, and T. A. Steitz, Structural basis of anticodon loop recognition by glutaminyl-tRNA synthetase. *Nature* 352:213–218, 1991.

Saks, M. E., J. R. Sampson, and J. N. Abelson, The transfer RNA identity problem: A search for rules. *Science* 263:191–197, 1994.

Siegel, V., and P. Walter, Each of the activities of signal recognition particle (SRP) is contained within a distinct domain. *Cell* 52:39–49, 1988.

Thompson, R. C., EFTu provides an internal kinetic standard for translational accuracy. *Trends Biochem. Sci.* 13:91–93, 1988.

Tobias, J. W., T. E. Shrader, G. Rocap, and A. Varshavsky, The N-end rule in bacteria. *Science* 154:1374–1377, 1991.

Von Figura, K., and A. Hasilik, Lysosomal enzymes and their receptors. *Ann. Rev. Biochem.* 55:167–193, 1986.

Webb, R., and L. A. Sherman, Chaperones classified. *Nature* 359:458–486, 1992.

Problems

1. Compare the translation initiation signals in prokaryotic and eukaryotic systems, and describe those features of each type of mRNA that determine the frequency with which a particular message is translated. What consequences do these differences have for gene organization in the two systems?
2. The relationship between tRNAs and their synthases is sometimes called the "second genetic code." Explain.
3. A single tRNA can insert serine in response to three different codons: UCC, UCU, or UCA. What is the anticodon sequence of this tRNA?
4. How much energy is required to synthesize a single peptide bond in protein synthesis? How does this compare with the free energy of formation of the peptide linkage, which is about 5 kcal/mole?
5. Explain this statement: "The universal genetic code is not quite universal."
6. Explain why the use of GUG and UUG as initiation codons in place of AUG was not expected, even based on Crick's wobble hypothesis.
7. Assuming that translation begins at the first codon, deduce the amino acid sequence of the polypeptide encoded by the following mRNA template:

   AUGGUCGAAAUUCGGGACACCCAUUUGAA–
   –GAAACAGAUAGCUUUCUAGUAA

8. Assume that you have a copolymer with a random sequence containing equimolar amounts of A and U. What amino acids would be incorporated and in what ratio, when this copolymer is used as an mRNA?
9. Researchers often design degenerate oligonucleotides based on a protein sequence for use as hybridization probes to isolate the corresponding gene. (A degenerate oligonucleotide is actually a mixture of oligonucleotides, the sequences of which differ at positions corresponding to degeneracies in the genetic code.) The N-terminal amino acid sequence of a protein is:

   Met-Val-Asp-Ser-Asn-Trp-Ala-Gln-Cys-Asp-Pro-Ala-Thr

   Give the sequence of the least degenerate 20-residue-long oligonucleotide that hybridizes to the gene encoding this protein.
10. The effect of single-point mutations on the amino acid sequence of a protein can provide precise identification of the codon used to specify a particular residue. Assuming a single base change for each step, deduce the wild-type codon in each of the following cases.

    (a) Gln → Arg → Trp

    (b) Glu → Lys → Ile

    (c) Leu → Ser; Leu → Val; Leu → Met

    (d) Thr → Ile; Thr → Pro; Thr → Lys

11. Even though the roles of IF-2, EF-Tu, EF-G, and RF-3 in protein synthesis are quite different, all four of these proteins share a domain with significant amino acid sequence similarity. Suggest a role for this conserved domain.
12. The antibiotic fusidic acid inhibits protein synthesis by preventing EF-G from cycling off of the ribosome. Fusidic-acid-resistant mutants of EF-G have been isolated. Fusidic acid resistance is recessive to sensitivity. In other words, an *E. coli* cell containing two EF-G genes, one resistant and one sensitive, is still sensitive to the antibiotic. Why? (*Hint:* Look at fig. 29.2.)
13. What are the major differences between the mechanisms of protein import into endoplasmic reticulum compared to protein import into mitochondria?
14. Scientists have tried to isolate the peptidyl transferase from ribosomes for many years without success. It is now thought that this activity is part of the ribosome (large subunit). Discuss this point, in view of what you know about other catalytic RNP complexes (RNA-protein complexes).
15. What are the possible amino acid changes that can result from a single nucleotide change in a GAA codon? Knowing the structures of the amino acids, what do you predict are the effects of the altered amino acids?

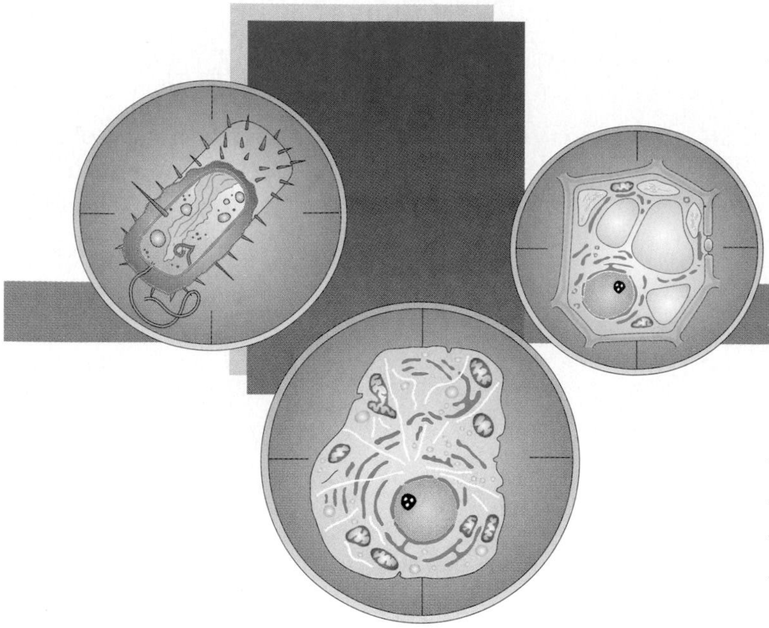

CHAPTER

30

Regulation of Gene Expression in Prokaryotes

Chapter Outline

In bacteria the level of expression of a particular messenger RNA is a function of the affinity of the RNA polymerase for a DNA promoter; this affinity is modulated by regulatory proteins that bind to the DNA or the RNA polymerase.

In all biological systems, gene expression is regulated so that gene products are produced either before or as they are needed. In this chapter we examine the mechanisms that ensure efficient regulation in the bacterium *Escherichia coli* and the bacteriophage λ.

E. coli maintains all of its genes in a state where they can be turned on or turned off on short notice. The short messenger lifetime makes it possible to control gene expression from the transcription level. The lack of separate compartments for RNA and protein synthesis has fostered mechanisms where translation actually exerts a direct role on transcription. These are some of the special features that have influenced the evolution of regulatory systems in *E. coli*.

Control of Transcription Is the Dominant Mode of Regulation in *Escherichia coli*

The *E. coli* chromosome contains about 3,000 genes. This system is regulated so that under conditions of active growth, only about 5% of the genome is actively transcribed at any given time. The remainder of the genome is either silent or transcribed at a very low rate. When growth conditions change, some active genes are turned off, and other, inactive genes are turned on. The cell always retains its totipotency, so that within a short time (seconds to minutes in most cases) and given appropriate circumstances, any gene can be fully turned on. The fully expressing rRNA gene makes one copy per second, a fully turned-on β-galactosidase gene makes about one copy per minute, and a fully turned-on biotin synthase gene makes about one copy every 10 min. In the maximally repressed state, all these genes express less than one transcript every 10 min.

The level of transcription for any particular gene usually results from a collection of control elements organized into a hierarchy that coordinates all the metabolic activities of the cell. For example, when the rRNA genes are highly active, so are the genes for ribosomal proteins, and the latter are regulated in such a way that stoichiometric amounts of most of the ribosomal proteins are produced. When glucose is abundant, most genes involved in processing more complex carbon sources are turned off by a process called catabolite repression. If the glucose supply is depleted and lactose is present, then the genes involved in lactose catabolism are expressed. In *E. coli,* the production of most RNAs and proteins is regulated exclusively at the transcriptional level, although notable exceptions occur. Rapid response to changing conditions is ensured partly by a short mRNA lifetime—on the order of 1–3 min for most mRNAs. Some mRNAs have appreciably longer lifetimes (10 min or longer) and the consequent potential for much higher levels of protein synthesis per mRNA subject to translational control. Examples of all these situations are considered later. Finally, the fine-level control for any particular enzyme system is subject to regulation by activators or inhibitors.

The Initiation Point for Transcription Is a Major Site for Regulating Gene Expression

The rate of initiation of transcription can be regulated in several ways, most of which influence the rate of formation of the RNA polymerase–DNA promoter complex. The primary sequence of nucleotides in the promoter region is the first factor to be considered. The closer this sequence is to the consensus sequence, the greater the affinity of the polymerase for the promoter (fig. 30.1).

The rate of initiation of transcription also can be altered by changes in the RNA polymerase structure. This can occur by subunit replacement, subunit covalent modification, or small-molecule-induced allosteric transition. During a temperature upshift (30 → 42°C), the usual σ subunit (σ^{70}) is partially replaced by an alternative σ factor (σ^{32}), changing the types of promoters recognized by the polymerase. In bacteriophage T4 infection, the subunits of the polymerase become ribose-adenylated, lowering the affinity of polymerase for bacterial promoters and raising the affinity for phage promoters. Binding of guanosine tetraphosphate (ppGpp) to RNA polymerase changes the structure of the polymerase so that it has a greatly lowered affinity for rRNA, tRNA, and ribosomal protein promoters and at the same time a somewhat greater affinity for some other promoters.

Finally, the rate of initiation of RNA synthesis can be controlled by auxiliary regulatory proteins that affect the rate of formation of the polymerase–promoter complex. According to their positive or negative action on gene expression, regulatory proteins are known as activators or repressors, respectively. Activators augment polymerase binding to the promoter, whereas repressors have the opposite effect.

Figure 30.1

Schematic diagram of DNA conformation in the rapid-start complex. Two regions most important in polymerase binding are lettered with most favored sequences. Transcription starts at the +1 base pair. Upstream of the start bases are numbers starting with −1.

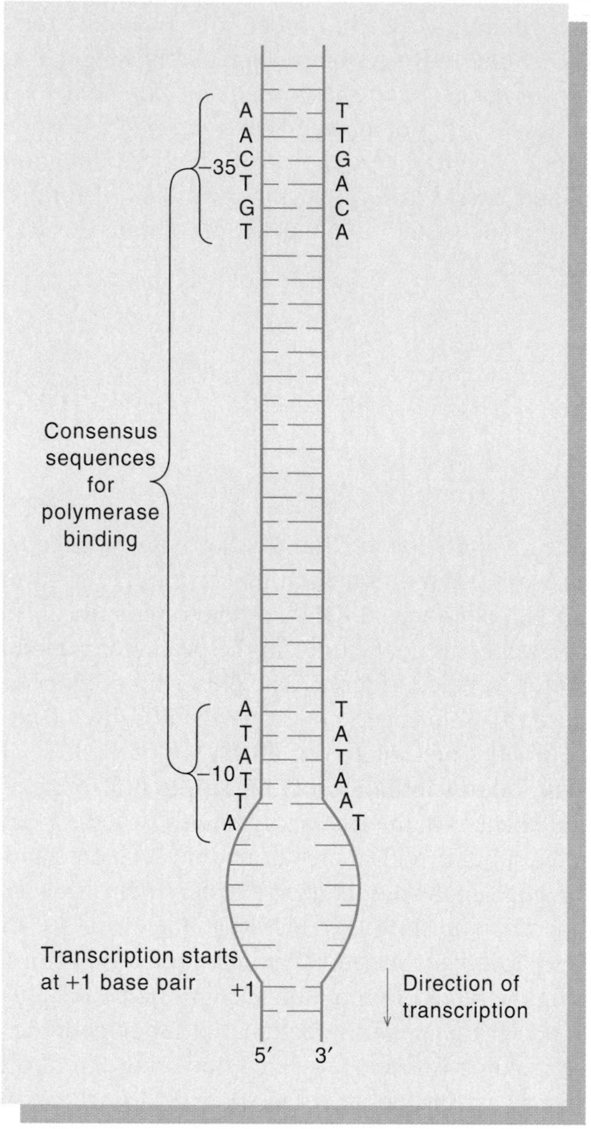

Figure 30.2

Different genetic elements of the *lac* operon. The operon contains a control region, the promoter–operator region, and three structural genes, *z, y,* and *a.* The *i* gene, a repressor, is also shown. It is not part of the operon, but it is located at an adjacent site on the genome with its own promoter.

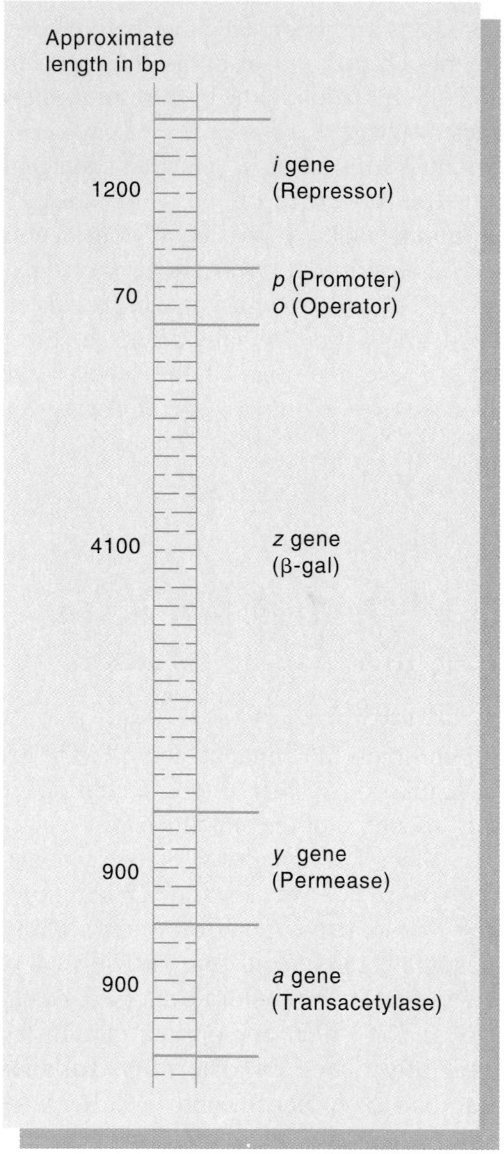

Regulation of the Three-Gene Cluster Known as the *Lac* Operon Occurs at the Transcription Level

In bacteria it is common to find units of expression that contain clusters of two or more functionally related genes. Such is the case with the lac operon, a three-gene cluster associated with the metabolism of the dissacharide lactose (fig. 30.2). The first of these genes, the *z* gene, encodes β-galactosidase, which hydrolyzes β-galactosides, in particular lactose, to produce the monosaccharides, glucose and galactose. The middle gene, *y*, encodes lactose permease, which is associated with the active transport of lactose into the cell. The third gene, *a*, encodes thiogalactoside transacetylase. A useful function has yet to be found for this gene.

Figure 30.3

Effect of inducer on β-galactosidase synthesis. Differential plot expressing accumulation of β-galactosidase as a function of increase in mass of cells in a growing culture of *E. coli.* Because the abscissa and ordinate are expressed in the same units (micrograms of protein), the slope of the straight line gives galactosidase as the fraction (*P*) of total protein synthesized in the presence of inducer. (Source: After Melvin Cohn, *Bact. Rev.* 21:140, 1957.)

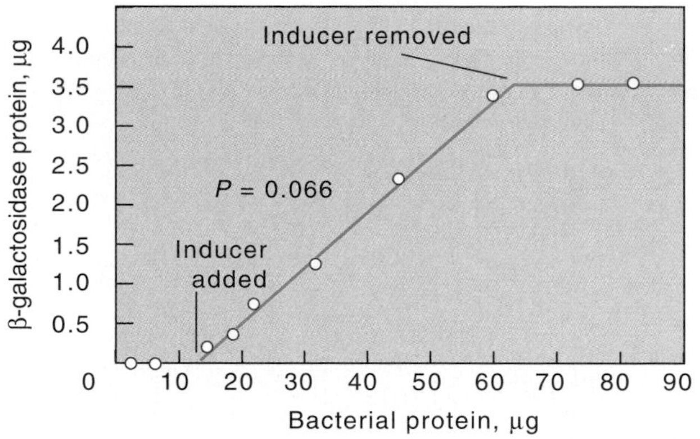

Figure 30.4

Inducers of the *lac* operon. (*a*) All inducers have the β-galactoside structure shown, in which R can be a variety of substituents. (*b*) Allolactose is the natural inducer when cells are grown on lactose. (*c*) Isopropyl-β-D-thiogalactoside (IPTG) is a synthetic inducer useful in the laboratory; the β-oxygen is replaced by a β-sulfur atom. This change prevents hydrolysis by β-galactosidase.

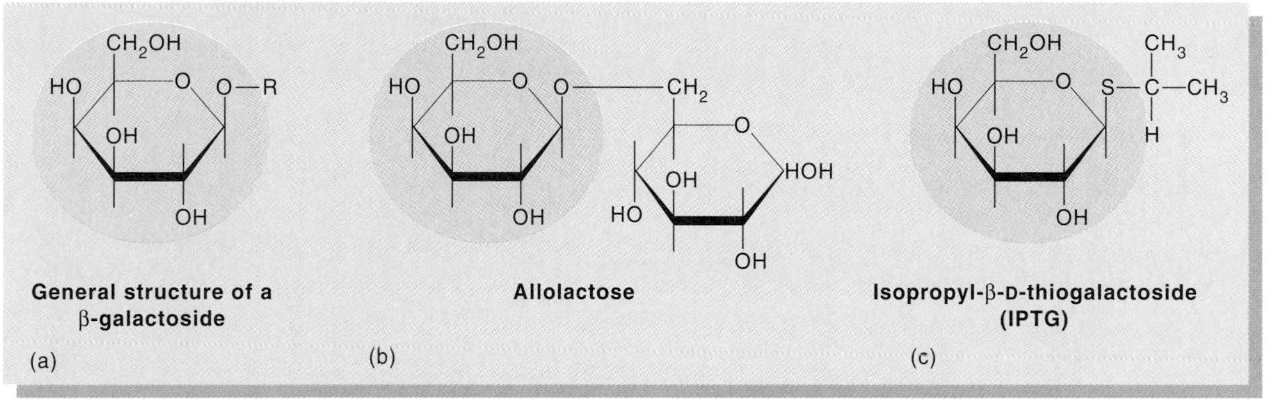

Expression of the *lac* operon is regulated by controlling elements, which are separate from the structural genes. The controlling elements consist of a promoter locus, which is the site where RNA polymerase binds and initiates transcription; the promoter locus also contains sites for the binding of a repressor and an activator. The *i* gene encodes the repressor, and the *crp* gene encodes the activator.

β-Galactosidase Synthesis Is Augmented by a Small-Molecule Inducer

Wild-type *E. coli* cells grown in the absence of lactose contain an average of 0.5–5.0 molecules of β-galactosidase per cell, whereas bacteria grown in the presence of an excess of lactose or certain lactose analogs contain 1,000–10,000 molecules per cell. Radioactive amino acid has been used as a tracer to show that the increase in enzyme activity observed on induction results from *de novo* protein synthesis. When excess β-galactoside inducer is added, enzyme activity increases at a rate proportional to the increase in total protein within the culture (fig. 30.3). Enzyme formation reaches its maximum rate within 3 min after inducer is added. Removal of inducer leads to cessation of enzyme synthesis in about the same amount of time.

A large number of compounds have been tested for their capacity to induce β-galactosidase. All inducers contain an intact, unsubstituted galactosidic residue (fig. 30.4). Many compounds that are not themselves substrates for β-galactosidase such as thiogalactosides are good inducers

Figure 30.5

Conversion of lactose to allolactose, the natural inducer of the *lac* operon. Ultimately, lactose is broken down to its constituent monosaccharides, galactose and glucose.

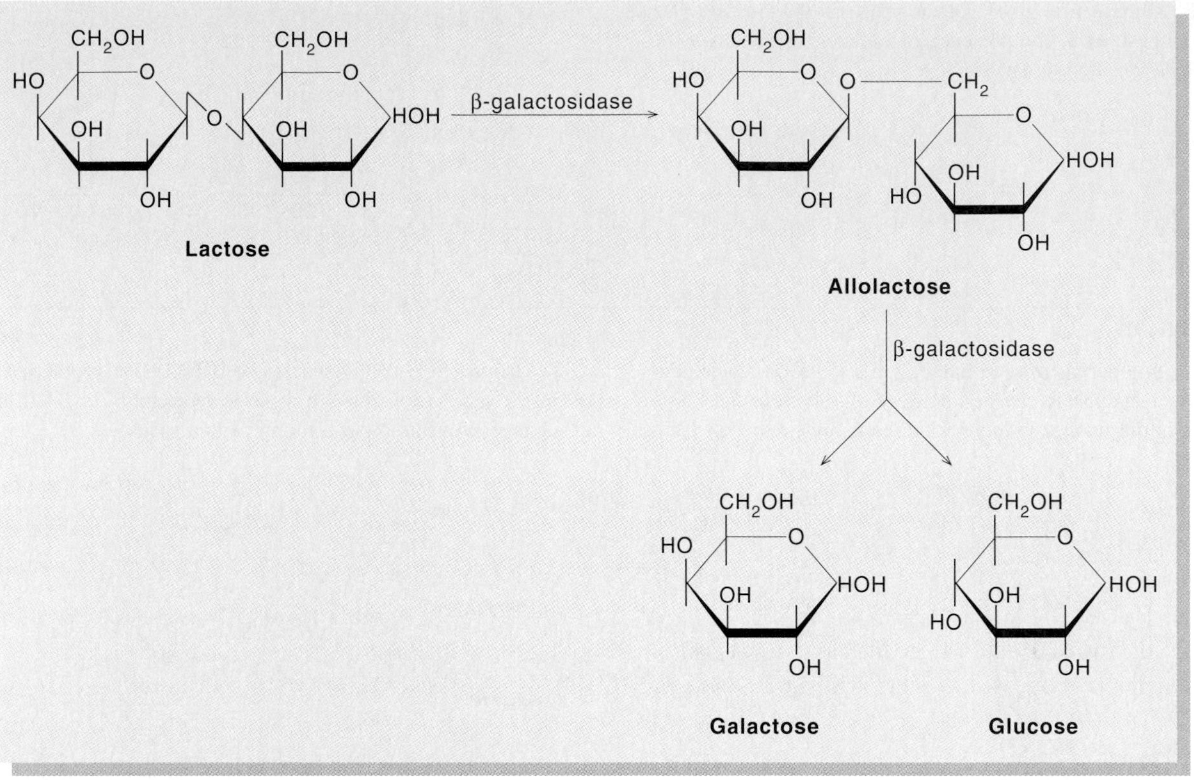

(such compounds are called gratuitous inducers because they are not substrates for the enzyme). No correlation exists between affinity for β-galactosidase and the capacity to induce. Lactose, the natural substrate of the operon, is not an inducer *in vivo*. Rather, allolactose, which is formed as an intermediate in lactose metabolism in the presence of the very limited amount of β-galactosidase that exists in uninduced cells, is believed to be the natural inducer (fig. 30.5). The three proteins of the *lac* operon are coordinately induced, that is, they are induced to the same extent by the same inducer. These results suggest that the receptor for the inducer is distinct from the proteins encoded by the operon, and the inducer acts at one site.

A Gene Was Discovered That Leads to Repression of Synthesis in the Absence of Inducer

Two distinct types of mutations have been observed in the genes associated with the *lac* operon. One class of mutations includes structural gene mutations: (1) β-galactosidase mutations ($z^+ \rightarrow z^-$), expressed as the loss of the capacity to synthesize active β-galactosidase; (2) permease mutations ($y^+ \rightarrow y^-$), expressed as the loss of the capacity to concentrate lactose; and (3) transacetylase mutations ($a^+ \rightarrow a^-$), expressed as the loss of the capacity to form thiogalactoside transacetylase. The other class of mutations involves controlling elements of the operon such as *i* gene mutations ($i^+ \rightarrow i^-$), expressed as the capacity to synthesize large amounts of β-galactosidase even in the absence of inducer. This type of *i* gene mutation is called a constitutive mutation. Structural mutations usually affect only the enzyme in whose gene the mutation occurs. In contrast, constitutive mutations invariably affect the amounts of all three structural gene products but not their structures (at this point you may wish to refer to box 30A for more information on genetic concepts and notation).

The most informative genetic studies were performed on cells containing two copies of the *lac* operon with point mutations in different genes. Partial diploids (merodiploids) of this sort are constructed by mating experiments in which a second copy of the *lac* region is incorporated into the test cell by mating. For this purpose the F factor plasmid is a

BOX

30A Genetic Concepts and Genetic Notation

Much of the early work on the *lac* operon was purely genetic. It is essential that certain aspects of genetics be understood. The information in this box should be adequate for those with no prior exposure to genetics except for what they have already encountered in this text.

Genes are specified by one or more small letters in italic. Thus, *z* indicates the gene for β-galactosidase, and *lac* indicates the operon. Frequently a superscript is appended to the genetic symbol. The two most common superscripts are +, indicating a normal (wild-type) gene, and −, indicating a nonfunctioning (mutant) gene. Different representations of the same gene are referred to as alleles. Thus, z^+ and z^- are both alleles of the *z* gene.

Cells that carry a single copy of each gene are referred to as haploids. Cells that carry two copies of each gene are referred to as diploids. Bacteria are haploid cells because they carry a single chromosome with a unique representation for each gene. Bacterial cells that are partial diploids (merodiploids) may occur naturally, or they may be selected for by genetic techniques.

A favored method for constructing merodiploids is to infect the bacterial cell with a virus or a plasmid DNA that carries the extra genes of interest. The F plasmid is commonly used for this purpose. In strict usage, the genetic representation for a cell carrying the *lac* operon on the chromosome and the F plasmid would be $z^+y^+a^+$//F$z^+y^+a^+$, where the diagonal lines separate the host chromosome to the left and the plasmid chromosome to the right. As a rule, however, only one diagonal line is used to separate the genetic symbols. Also, for convenience, if all the alleles for a given gene are wild type, they may not be shown. Thus, $z^+y^+a^+$/F$z^-y^+a^+$ and z^+/Fz^- may be taken as representations of the same genetic state in cases in which it is understood that the *lac* operon is present on both the host and the plasmid chromosome.

Two genetic elements located on the same chromosome are said to be in the *cis* orientation. Two genetic elements located on different chromosomes in the same cell are said to be in the *trans* orientation. In the merodiploid $z^-y^-a^+$/F$z^+y^+a^+$, the two mutant genes are in the *cis* orientation. In the merodiploid $z^-y^+a^+$/F$z^+y^-a^+$, they are in the *trans* orientation.

A major reason for using merodiploids is to study the interaction between different alleles of the same gene. This often tells us a great deal about how a gene or the gene product functions. The two simplest types of interactions are dominant and recessive. A cell that is z^+/Fz^- behaves like a z^+ cell as far as the metabolism of β-galactosidase is concerned. Therefore, the z^+ allele is dominant to the z^- allele, or conversely, the z^- allele is recessive to the z^+ allele.

---

favorite. Merodiploids of the type $z^+y^-a^-$/F$z^-y^+a^+$ or $z^-y^+a^+$/F$z^+y^-a^-$ behave like normal wild-type cells with respect to the expression and metabolism of the *lac* operon. This demonstrates that the distribution of normal genes and mutant genes on the two chromosomes does not influence the phenotype as long as there is at least one functional gene of each type. This tells us that each of the structural genes behave as an independent entity not affected by other genes on the same operon.

The study of merodiploids of the types i^+z^-/Fi^-z^+ give the same inducible phenotype. This demonstrates that the i^+ inducible allele is dominant to the i^- constitutive allele, and that it is active on the same chromosome *(cis)*, or on a different chromosome *(trans)* with respect to the structural gene it influences (table 30.1). In fact the i^+ gene could be moved to any location on the chromosome and it would still show dominant behavior. The fact that the influence of the *i* gene is not sensitive to location suggests that *i* gene action results from a diffusible gene product. The dominance of the inducible i^+ allele to the constitutive i^- allele suggests that the former corresponds to the active form of the *i* gene.

Further understanding of *i* gene function has come from study of rare mutations designated i^s. Mutants bearing

Table 30.1

Expression of β-Galactosidase as a Function of Genotype

Genotype[a]	Phenotype	
	−Inducer	*+Inducer*
$i^+o^+z^+$	−	+
$i^-o^+z^+$	+	+
$i^+o^+z^-$	−	−
$i^+o^+z^+/i^-o^+z^+$	−	+
$i^so^+z^+/i^+o^+z^+$	−	−
$i^+o^cz^+$	+	+
$i^+o^cz^+/i^+o^+z^+$	+	+
$i^+o^cz^+/i^+o^+z^-$	+	+
$i^+o^cz^-/i^+o^+z^+$	−	+

[a] The diagonal indicates that two *lac* operons, including the i^+, are present in the same cell. Under phenotype a + indicates a high level expression of β-galactosidase, a − indicates low level or the absence of expression.

this allele are noninducible, meaning that they have lost their capacity to express the structural gene products of the operon. In merodiploids of the constitution i^+/i^s, the i^s allele is dominant, that is, the merodiploids cannot synthesize structural gene products even in the presence of inducer (see table 30.1). The most likely explanation for the i^s mutant is that it is an allele of *i* in which the repressor is not influenced by the inducer so that it always represses.

A Locus Adjacent to the Operon Is Found to Be Required for Repressor Action

The vast majority of constitutive mutants result from i^- mutations. However, occasional constitutive mutants have been mapped outside of the *i* gene in the region of the *lac* operon promoter. Rare mutants of this type designated O^c are much easier to isolate by selection for constitutivity in cells diploid for the i^+ gene. This selection procedure minimizes the chance of finding constitutive mutants that result from *i* gene mutations because both copies of the i^+ gene would have to mutate to i^- simultaneously to give a constitutive phenotype. If the probability of an $i^+ \rightarrow i^-$ mutation is 10^{-6}, the probability of two such events occuring simultaneously in the same cell is 10^{-12}, which is extremely unlikely. As a result, constitutive mutations obtained under these circumstances are almost always of the o^c type. Like i^- mutations, o^c mutations affect the quantity of β-galactosidase synthesized but not its structure.

In merodiploids of the type o^+z^+/o^cz^+. β-galactosidase is constitutively expressed, showing that the o^c allele is dominant to o^+ in this situation. In merodiploids of the type o^cz^+/o^+z^- the o^c-allele is dominant also; but in o^cz^-/o^+z^+, it is recessive. Thus, the o^c-allele is dominant only when it is located *cis* to the structural genes it influences. From this result, François Jacob and Jacques Monod inferred that $o^+ \rightarrow o^c$ mutations correspond to a modification of the DNA structure that affects the ability of the repressor to bind.

Genetic Studies on the Repressor Gene and the Operator Locus Lead to a Model for Repressor Action

The behavior of the various mutations we have just discussed led Jacob and Monod to propose a model for the regulation of protein synthesis. The genetic elements of this model consist of a structural gene or genes, a regulator gene, and an operator locus (fig. 30.6).

1. The structural gene produces an mRNA that serves as a template for protein synthesis.
2. The regulator gene (not itself part of the operon) produces a repressor that can interact with the operator locus.
3. The operator is always adjacent to the structural genes it controls.
4. The operator and its associated structural genes are referred to as the operon.
5. The repressor molecule combines with the operator locus to prevent the structural gene(s) from synthesizing mRNA. In induction, the inducer combines with the repressor, changing its structure so that it no longer binds to the operator; this region of the genome is then free to combine with RNA polymerase.

The Jacob-Monod operon hypothesis has provided a tremendous stimulus for investigations directed toward understanding not only the *lac* system but other genetic regulatory systems as well.

Biochemical Investigations Verify the Operon Hypothesis

Genetic studies led to the operon hypothesis. Biochemical investigations were essential to provide direct evidence for the hypothesized properties of the repressor. The first task was to isolate the repressor.

Figure 30.6

Schematic model illustrating the operon hypothesis. This diagram is modified from the original proposed by Jacob and Monod, who thought *i* gene repressor was an RNA rather than a protein. (*a*) The *i* gene encodes a repressor that binds tightly to the operator *o* locus, thereby preventing transcription of the mRNA from the *z*, *y*, and *a* structural genes. (*b*) When inducer is present, it combines with repressor, changing its structure so it can no longer bind to the operator locus. Inducer also can remove repressor already complexed with the *o* locus.

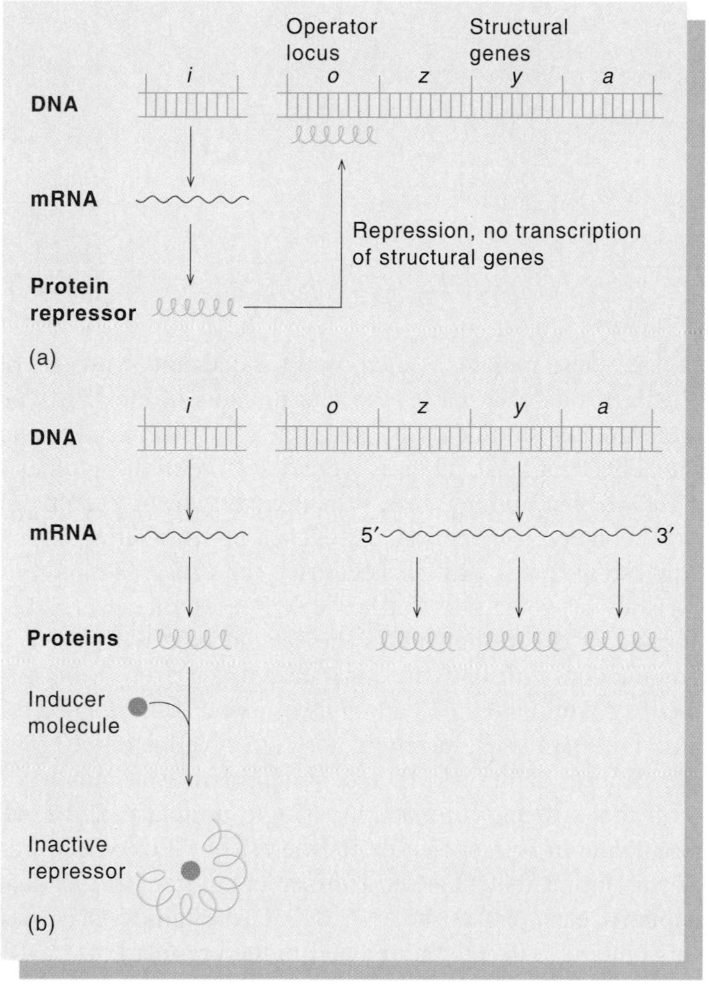

According to the operon hypothesis, inducer is supposed to bind to repressor. Walter Gilbert and Benno Muller-Hill used ^{14}C-labeled isopropyl-β-D-thiogalactoside isopropylthiogalactopyranoside (IPTG), the strongest known inducer (see fig. 30.4), to monitor repressor purification from a crude cell extract. IPTG has a further advantage for such studies in that it is completely stable. A crude extract of disrupted cells was fractionated by standard protein purification procedures, and the fraction containing the ^{14}C-IPTG-labeled product was isolated.

Several properties of normal and abnormal repressor were studied. The repressor was found to bind strongly to the *lac* promoter. This binding was disrupted by adding inducer. Promoter DNA containing an o^c mutation was not effective in binding repressor. Only repressor prepared from i^+ cells was effective in binding to DNA.

A more detailed characterization of the operator binding site was made. Eight o^c mutations were found to involve base replacements near the center of this region. The reactivity of wild-type operator to various chemical agents was determined in the presence and in the absence of repressor. Repressor binding substantially decreases the reactivity toward dimethylsulfate of several purine bases in the operator region (fig. 30.7). Dimethylsulfate reacts with N-7 of guanine and the N-3 of adenine in the double helix. Finally, if all the thymines in the DNA are replaced by bromouracil, a number of the bromouracil bases become cross-linked to the bound repressor in the presence of ultraviolet light. Taking into account the normal twist of the double helix, all the groups shown by chemical methods to be in the vicinity of the repressor are situated on one side of the double helix.

The sequence of bases in the operator region shows a remarkable symmetry property. Twenty-eight out of 36 of the base pairs in this region of the promoter are located on a twofold (dyad) axis of symmetry (see fig. 30.7). We shall see that dyad symmetry is a common property for repressor or activator binding sites and that it relates to the way in which DNA and regulatory proteins interact.

Biochemical proof that repressor inhibits operon expression was shown in a cell-free system in which most of the components were prepared from an i^- extract. Addition of repressor to such an extract inhibited the synthesis of β-galactosidase. This inhibition was reversed by addition of IPTG inducer.

The detailed biochemical studies not only confirmed the Jacob-Monod hypothesis, they gave further details about the nature of repressor interaction and a detailed characterization of the repressor binding site. The repressor binds mainly to one side of the double helix, over a 36-base region covered by the symmetry axis. It inhibits expression because it binds to a site that overlaps the polymerase binding site.

An Activator Protein Is Discovered That Augments Operon Expression

Although the genetic and biochemical studies on the action of repressor on the *lac* operon answered many questions about gene expression of the *lac* operon, they left equally important questions unanswered. It had been known since the turn of the century that the *lac* operon expresses at a

Figure 30.7

The operator locus (the presumptive repressor binding site). Bases are numbered +1 for the first base transcribed and −1 for the base before that. Regions showing dyad symmetry are underlined and overlined. Arrows indicate point mutations leading to the constitutive phenotype (o^c). Circled bases are those groups that are strongly protected against reaction with dimethylsulfoxide when *lac* repressor is bound. Shaded circles indicate those groups that become cross-linked to repressor in the presence of ultraviolet light when thymine in the DNA is replaced by 5-bromouracil.

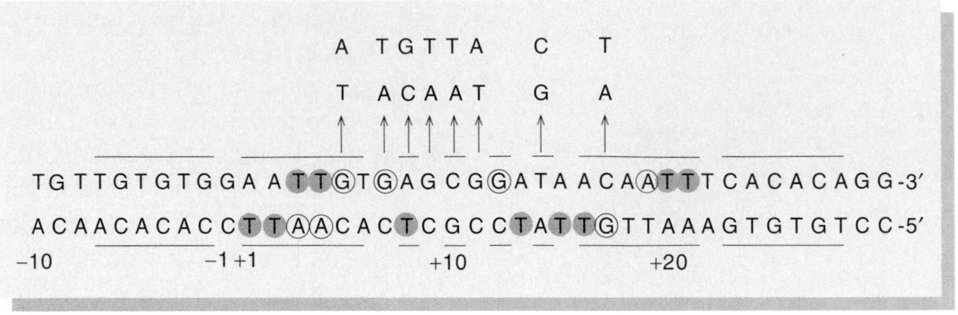

greatly reduced level if lactose and glucose are present simultaneously. Either of these sugars can be used by the bacterium as a source of carbon compounds and energy, but the lactose is not utilized to any appreciable extent until the glucose supply has been exhausted. This effect is called catabolite repression. As long as glucose is available, lactose is underutilized.

A turning point in our understanding of catabolite repression was provided by Earl Sutherland, who found that when glucose was added to growing *E. coli* cells, the level of 3′,5′-cAMP (cAMP) was drastically reduced. Could the lack of cAMP be responsible for the poor expression of the *lac* operon in the presence of glucose? In support of this I. Pastan and R. Perlman found that large quantities of cAMP added to the growth medium could partially reverse the glucose catabolite repression effect. In a cell-free system containing crude extracts from *E. coli* and DNA containing the lac operon, G. Zubay found that the low-level expression of the *lac* operon could be greatly increased by addition of cAMP. This provided support for the notion that cAMP was playing a direct role in activating the *lac* operon. Further investigations were facilitated by Jonathan Beckwith's genetic studies and the isolation of key mutants relating to the action of cAMP.

Beckwith and his colleagues isolated a large family of mutants that were permanently catabolite-repressed. These mutants fell into two categories: Those that could be phenotypically corrected by growing in the presence of cAMP and those that could not. The first class of mutants were believed to be defective in the synthesis of cAMP, and the latter class were presumed to be defective in the protein receptor for cAMP. Cell-free extracts were prepared from both of these mutants. When used for cell-free synthesis of β-galactosidase, it was found that mutants of the first type were greatly stimulated by addition of cAMP, confirming the belief that these mutants were defective in the synthesis of cAMP but nothing else. When extracts from mutants of the second type were used, cAMP had no stimulating effect, suggesting that a protein necessary for cAMP action was missing or defective. Further cell-free studies were performed in which mutants of the second type were used in conjunction with partially purified extracts from a normal strain. Addition of small amounts of extracts from a normal strain reestablished the stimulatory effect of the cAMP. The purification of the cAMP receptor protein was monitored with this system. Ultimately, a single protein called CAP was found to be responsible for the effect. Soon afterwards it was found that CAP is a dimer composed of identical subunits, each with an M_r of 22,000. CAP binds to DNA and this binding is greatly stimulated in the presence of cAMP. The cAMP apparently alters the conformation of CAP so that it can form a strong complex with DNA at the *lac* promoter region.

A series of genetic deletions was used to demonstrate that the site necessary for CAP stimulation of the *lac* operon is in the −50–−80 region of the *lac* operon. A 14-bp segment between −53 and −68 shows dyad symmetry for 12 of the 14 bp (fig. 30.8).

In the absence of bound repressor, the full sequence of reactions leading to initiation of transcription is summarized by the following set of equations. First, the coactivator cAMP combines with the activator CAP, which then binds in the −60 region of the promoter. This complex stimulates the binding of RNA polymerase to an adjacent site on the promoter. CAP stimulates binding of RNA polymerase by an affinity between CAP and polymerase.

$$\text{cAMP} + \text{CAP} \rightleftharpoons \text{cAMP-CAP}$$

$$\text{cAMP-CAP} + \text{DNA} \rightleftharpoons \text{cAMP-CAP-DNA}$$

$$\text{cAMP-CAP-DNA} + \text{polymerase} \rightleftharpoons \text{cAMP-CAP-DNA-polymerase}$$

The overall strategy in creating a promoter sequence responsive to cAMP-CAP activation can be summarized as follows. The nucleotide sequence in the RNA polymerase-binding site is adjusted so that polymerase by itself produces a low level of transcription. An adjacent site for binding CAP is created so that when CAP is binding, the additional affinity contributed by favorable contacts between the CAP and the polymerase convert this into a high-level promoter. The crucial importance of the promoter sequence in eliciting the polymerase response to CAP is shown by the fact that certain mutations that result in producing the consensus sequence for the promoter (see fig. 30.1) eliminate the catabolite repression effect for the *lac* operon *in vivo* as well as the need for cAMP in the cell-free system.

Figure 30.8

CAP-binding region in the *lac* promoter. Regions showing dyad symmetry that are believed to interact strongly with CAP are overlined and underlined. Circled bases are those that are strongly protected from reaction with dimethylsulfate when CAP is bound. Point mutations L8 or L29 result in a promoter that is not stimulated in transcription by cAMP. Red dots indicate those phosphate positions that, if ethylated by ethylnitrosourea, block CAP binding. If the area of interest exists as a normal DNA double helix when binding CAP, then most of the groups implicated in CAP binding appear on one side of the DNA.

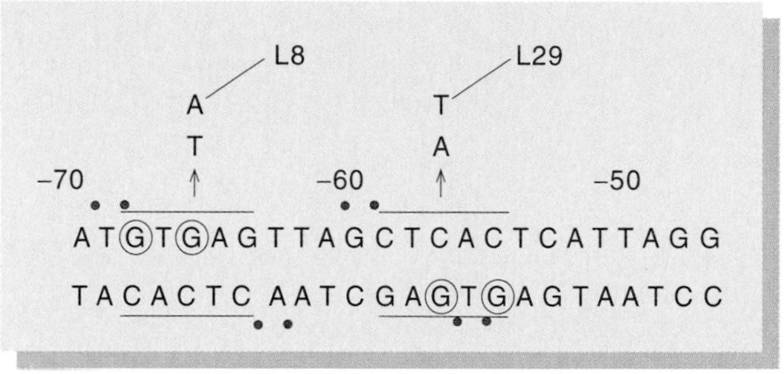

Enzymes That Catalyze Amino Acid Biosynthesis Are Regulated at the Level of Transcription Initiation

Among the genes concerned with biosynthetic processes, those uniquely involved in synthesis of the amino acid tryptophan are possibly the best understood. This is mostly a result of the efforts of Charles Yanofsky and his colleagues, who have used a wide variety of genetic and biochemical techniques to probe the complexities of this system.

Wild-type *E. coli* can synthesize all 20 types of amino acids from simpler substrates, but for most of them, it does so only when they are not available in adequate amounts in the growth medium. For example, synthesis of enzymes for tryptophan synthesis is sharply reduced when the external tryptophan supply is high. Lowering available L-tryptophan selectively stimulates synthesis of the mRNA and associated enzymes (fig. 30.9). The five contiguous structural genes are transcribed as a single polycistronic mRNA. Initiation of transcription is regulated in part by the interaction of the tryptophan repressor, the protein product of the *trpR* gene, with its target site on the DNA, the *trp* operator, *trpO* (fig. 30.10). Binding of L-tryptophan to the repressor causes a structural alteration essential for strong specific binding to the *trpO* locus. Like *lac* repressor, the *trp* repressor binds at a site that overlaps the RNA polymerase-binding site, and the region where the repressor binds contains a number of bases arranged with dyad symmetry (fig. 30.11).

The most significant difference between the action of the *trp* and *lac* repressors relates to the function of the small-molecule effector. In the case of *lac*, the effector molecule allolactose acts as an antirepressor (inducer), causing release of repressor from the operator; in the case of *trp*, the effector molecule L-tryptophan acts as a corepressor, stimulating the binding of repressor to the operator. It should be obvious that the difference in action of these small-molecule effectors, the concentrations of which dictate the level of operon activity, is well suited to the different metabolic needs of the cell satisfied by the two operons.

The trp *Operon Is Also Regulated after the Initiation Point for Transcription*

From this point on, the close parallel between the regulation of the *trp* operon and the *lac* operon ends. The *trp* operon has no positive control system like cAMP-CAP, but it does have another means for regulating transcription at a site downstream from the initiation point. The existence of a second signal for regulation was first suspected when it was discovered that a genetic deletion of some of the bases between the initiation site for transcription and the first structural gene (*trpE*) raised the level of expression of the operon 8- to 10-fold. This was true even in strains with a defective repressor gene (*trpR*$^-$ strains). How could this be, and what could be the mechanism of action? An important clue was

Figure 30.9

The tryptophan operon, indicating the location of the different genes, the polypeptide chains, the resulting enzyme complexes, and the reactions catalyzed by the enzyme complexes.

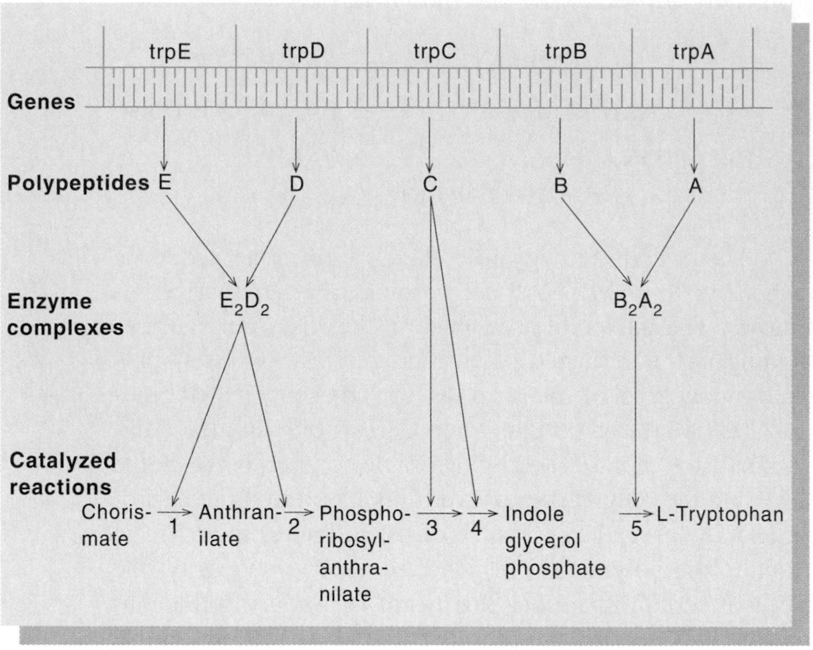

Figure 30.10

Schematic diagram of the repressor control of *trp* operon expression. The *trp* promoter (*P*) and *trp* operator (*O*) regions overlap. The *trp* aporepressor is encoded by a distantly located *trpR* gene. L-Tryptophan binding converts the aporepressor to the repressor that binds at the operator locus. This complex prevents the formation of the polymerase–promoter complex and transcription of the operon that begins in the leader region (*trpL*). Only a fraction of the transcripts extends beyond the attenuator locus in the leader region. The regulation of this fraction is discussed in the text.

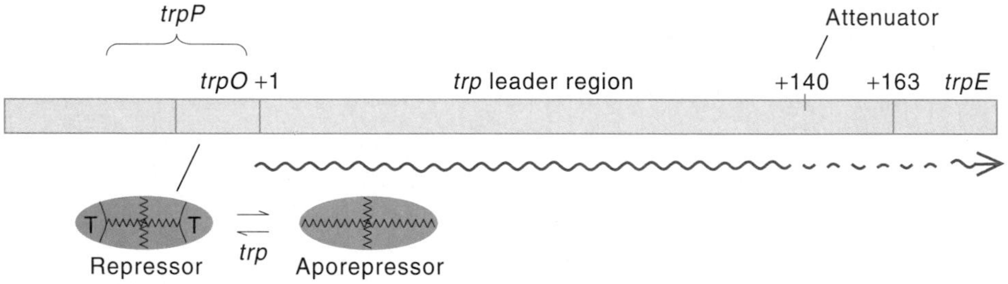

Figure 30.11

The promoter–operator region of the tryptophan operon. Two regions, PBS1 and PBS2, where polymerase binds are bracketed. Regions within the repressor-binding site showing dyadic symmetry are underlined and overlined. Single-base changes that lead to operator constitutive (o^c) mutants are indicated below the duplex.

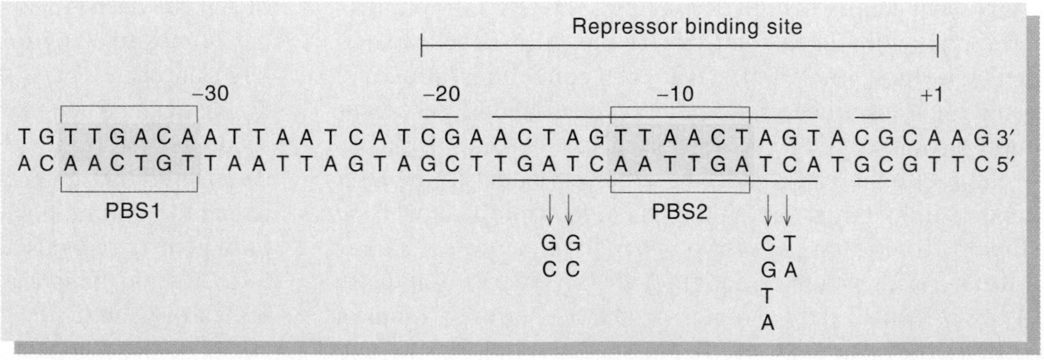

Figure 30.12

The leader region for the tryptophan operon. The region of the leader RNA containing the hypothesized leader polypeptide is shown. The translation start of the trpE protein is also shown.

Leader polypeptide

Leader start site +27 +54 +69

MET-LYS-ALA-ILE-PHE-VAL-LEU-LYS-GLY-TRP-TRP-ARG-THR-SER-

pppAAGUU ~ AUG AAA GCA AUU UUC GUA CUG AAA GGU UGG UGG CGC ACU UCC UGAAACGGGC ~

trpE start site +163 trpE protein

MET-GLN-THR-GLN-LYS-PRO-

+141

GCGUAAAGCAAUCAG ~ CCUAAUGAGCGGGCUUUUUUUUGAACAAAAUUAGAGAAUAACA AUG CAA ACA CAA AAA CCG

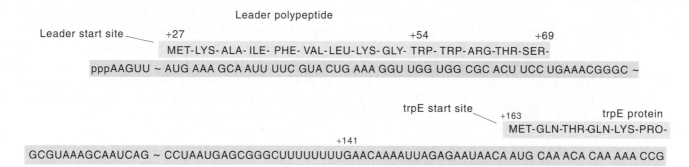

provided by sequence analysis of the 162 bases in the *trp* leader region, that is, the region between the initiation site for transcription and the initiation site for translation of the first structural gene (fig. 30.12). This leader region contains a potential initiation codon (bases 27–29), two tandem *trp* codons (bases 54–59), and a terminator codon (bases 69–71). A so-called leader peptide of 14 amino acids would result from translation of this region.

There are numerous reasons for believing that translation of the leader peptide up to or through the *trp* codons regulates attenuation of mRNA transcription. First of all, selective starvation of cells for tryptophan relieves attenuation and permits most RNA polymerase molecules to read through the leader region. The only other amino acid that relieves attenuation of the *trp* operon when it is lacking is arginine; arginine starvation is about 80% as effective as tryptophan starvation. It should be noticed that an *arg* codon is located adjacent to the two *trp* codons in the leader region. Most telling of all was the finding that a mutation resulting in the replacement of the AUG start codon by AUA, which should eliminate translation of the leader peptide, also prevents transcription beyond the attenuator.

Other experiments indicated that the fraction of tRNA that is charged with an amino acid is a crucial factor in the attenuation response. This has been examined *in vivo* by comparing the *trp* operon enzyme levels in *trpR*$^-$ strains that are otherwise normal with strains that are defective in some respect in charged tRNA^{Trp}. Such structural defects in tRNA^{Trp} or in the charging enzyme elevates expression, probably by permitting polymerase to transcribe through the attenuator.

These results support the hypothesis that transcription read-through requires partial translation of the leader sequence. However, only if the translation pauses or stops in the region where the *trp* or *arg* codons occur is read-through favored. A careful examination of the secondary-structure possibilities in the attenuator region suggests why this is so. The leader region RNA between bases 50 and 141 has the potential to form a variety of base-paired conformations. Figure 30.13 illustrates the most likely secondary structures that form in terminated *trp* leader RNA. These are based on analysis of regions of the transcript that show resistance to RNase T1 digestion under mild conditions and the base pairing established by studies of defined oligonucleotides. Four regions of base pairing that can form three stem-and-loop structures have been proposed. Region 1, which includes the tandem *trp* codons and the leader peptide translation stop codon (bases 54–68), can base-pair with region 2 (bases 76–91). Although region 2 (bases 74–85) also should be able to base-pair with region 3 (bases 108–119), stem-and-loop 2 · 3 has not been observed *in vitro,* presumably because stem-and-loop 3 · 4 and stem-and-loop 1 · 2 form preferentially. Region 3 (bases 114–121) can base-pair with region 4 (bases 126–134). The existence of this stem and loop is inferred from the T_1 RNase-resistance of the GC-rich region from residue 107 to the 3′ end of the transcript. The stem-and-loop structure formed between regions 3 and 4, followed by a sequence of U residues, is a common structure for a transcription terminator (see chapter 28). Hence, it is expected that conditions under which this structure is preserved would favor transcription termination. In support of this, a number of single-base replacement mutations have been isolated that lead to mispairing in the 3 · 4 stem; all of these lower the level of transcription termination in the leader to some extent.

The absence of translation of the leader would not perturb this 1 · 2 and 3 · 4 structure (fig. 30.14*a*). Consistent with this, changing the initiator codon of the leader peptide by a single base prevents read-through, as discussed earlier. On the other hand, selective starvation resulting from either

Figure 30.13

Proposed secondary structures in the leader RNA. Four regions, labeled 1, 2, 3, and 4 at the left, can base-pair to form three stem-and-loop structures (1.2, 2.3, and 3.4). The arrows in the main figure mark the RNAse T1 cleavage sites.

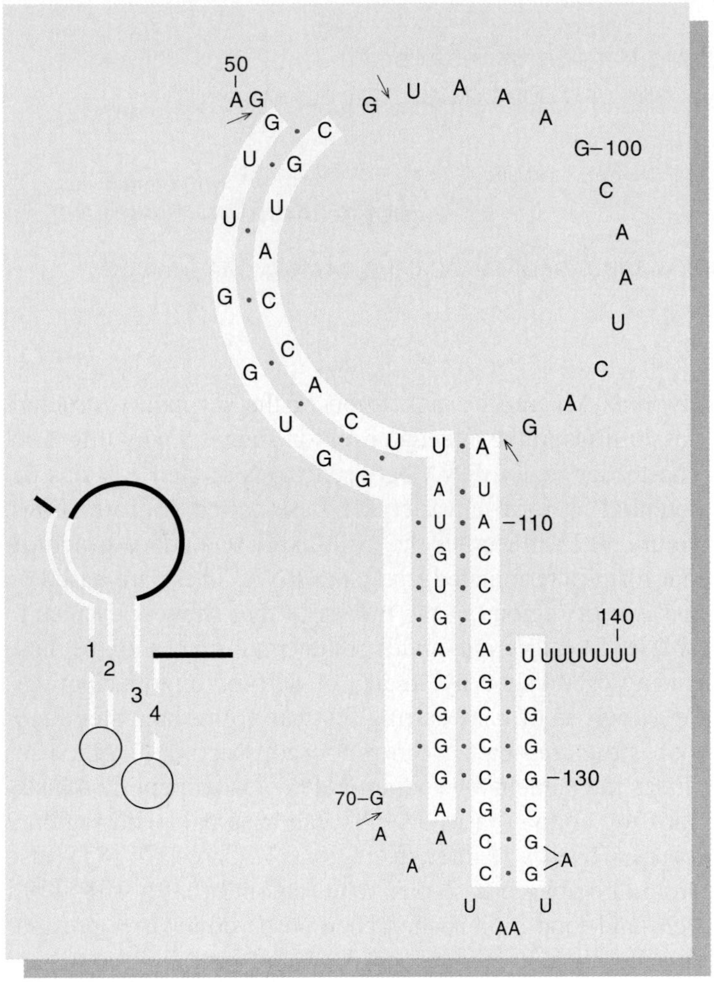

tryptophan or arginine deprivation stimulates transcription read-through. Most likely, this is because the ribosome stalls in the region of the *trp* or *arg* codons (bases 54–62). The resulting rupture of the base-paired 1 · 2 structure would make region 2 available for pairing with region 3. This would encourage disruption of the stem-and-loop 3 · 4 structure, resulting in transcription read-through (see fig. 30.14*b*). In the presence of an adequate supply of all amino acids, translation would proceed beyond this critical region, so that region 2 would not be available for base pairing, and the 3 · 4 loop would be maintained, favoring transcription termination at the attenuator (see fig. 30.14*c*).

This attenuator mechanism of control is amazingly simple because it requires no proteins other than those normally used for transcription and translation. One might expect such a simple and effective mechanism to be used repeatedly for other operons involved in amino acid biosynthesis. Indeed, for several other amino acid biosynthetic pathways in *E. coli* for which tRNA charging is involved in regulation, attenuator mechanisms have been found.

Genes for Ribosomes Are Coordinately Regulated

Over 100 genes in *E. coli* participate in the synthesis of the RNAs and proteins that constitute the enzymatic machinery for translation. The relevant gene products make up between 20% and 40% of the dry cell mass. In rapidly growing cells about 85% of the RNA is ribosomal, 10% is tRNA, and most of the remainder is mRNA. The various RNAs and proteins are produced according to need. For ribosomes and tRNAs, this results in a synthesis rate that is roughly proportional to the cell growth

Figure 30.14

Model for attenuation in the *trp* operon, showing ribosome and leader RNA. (*a*) Where no translation occurs, as when the leader AUG codon is replaced by an AUA codon, stem-and-loop 3.4 is intact, and termination in the leader is favored. (*b*) Cells are selectively starved for tryptophan so that the ribosome stops prematurely at the tandem *trp* codons. Under these conditions, stem-and-loop 2.3 can form, and this is believed to lead to the disruption of stem-and-loop 3.4. (*c*) All amino acids, including excess tryptophan, are present so that stem-and-loop 3.4 is present.

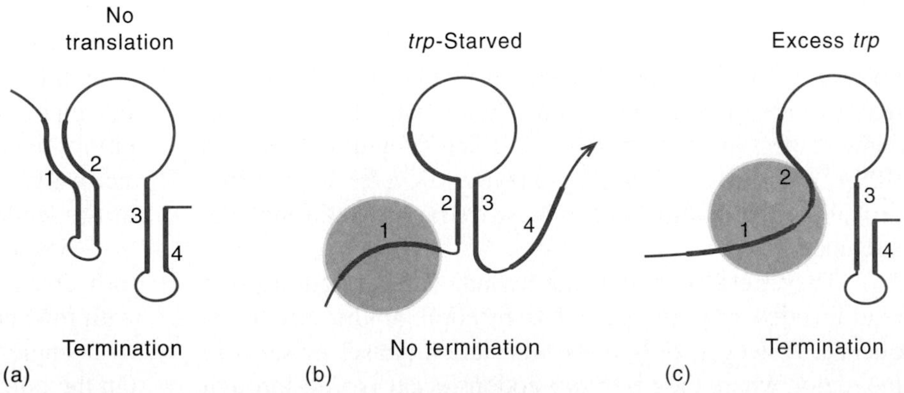

Figure 30.15

A typical rRNA (*rrn*) operon contains two promoters and genes for 16S, 23S, and 5S rRNA and a single 4S tRNA gene. The four fully processed RNAs are derived from a single intact 30S primary transcript.

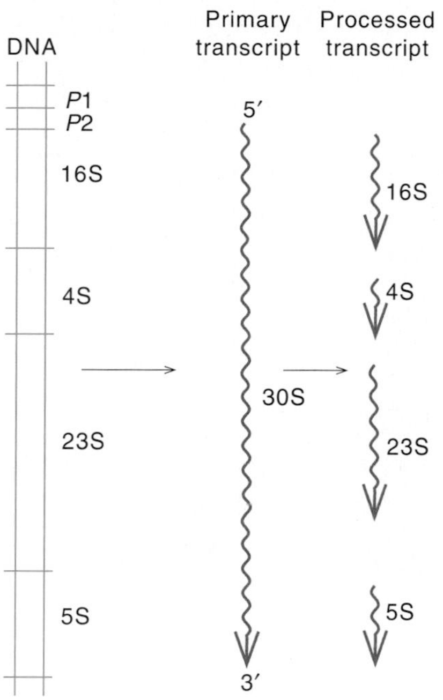

Figure 30.16

Guanosine tetraphosphate (ppGpp) concentration under normal conditions, after amino acid starvation, and after readdition of amino acids (● = wild-type cells; ▲ = *relA* cells; and ■ = *spoT* cells).

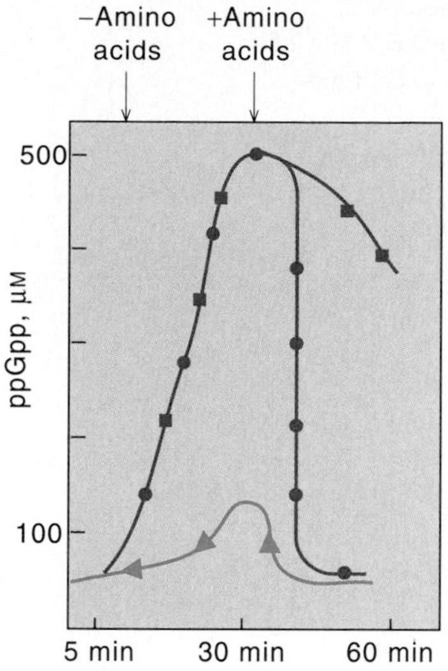

rate; the relative amounts of the three rRNAs (16S, 23S, and 5S), the 60 or so tRNAs, and the 50 ribosomal proteins, are consistent with the stoichiometric needs for making ribosomes.

Control of rRNA and tRNA Synthesis by the rel *Gene*

Under conditions of rapid growth, *E. coli* cells contain about 10^4 ribosomes. The maximum rate of reinitiation at the ribosomal gene promoter is about one per second. In rapid growth, *E. coli* can duplicate once every 20 min, which would allow for the synthesis of only about 1,200 molecules of rRNA if there were only one gene for ribosomal RNA. In fact, there are seven copies for ribosomal RNA operons in the bacterial chromosome, which makes it possible for rRNA synthesis to maintain the necessary pace under conditions of rapid growth. These operons are dispersed at seven locations around the circular *E. coli* chromosome. Each operon is transcribed into a single transcript, which is processed into four or five RNAs (fig. 30.15). The order of RNAs in the original transcript starting from the 5′ end is 16S, 4S, 23S, and 5S. In some operons additional 4S genes for tRNA are located downstream from the 5S gene. All the known ribosomal RNA operons appear to have two strong promoters located in tandem; this probably permits a more rapid rate of reinitiation for the genes than would be possible with only one promoter.

As stated earlier, rRNA synthesis is usually maintained at a rate proportional to the gross rate of protein synthesis. In a normal wild-type cell, when protein synthesis is limited (e.g., by amino acid availability), M. Cashel and J. Gallant have shown that the ppGpp concentration rises rapidly from about 50 μM to 500 μM (fig. 30.16). Concomitantly, rRNA synthesis ceases abruptly. This is part of the phenotype known as the stringent response: First, amino acid deprivation or other factors that slow down protein synthesis provoke an increased rate of accumulation of ppGpp; this in turn leads to an inhibition of rRNA synthesis.

Observations on different mutants indicate that the concentration of ppGpp is regulated by a careful balance between its rate of synthesis (controlled by the *rel* gene product) and rate of breakdown (controlled by the *spoT* gene product). Correlated observations indicate that the synthesis of rRNA is inversely proportional to the ppGpp concentration. In a *relA* mutant cell, neither the rapid rise in ppGpp concentration nor the cessation of rRNA synthesis is seen when amino acids are removed. The *relA* gene encodes a protein that is required for ppGpp synthesis. If

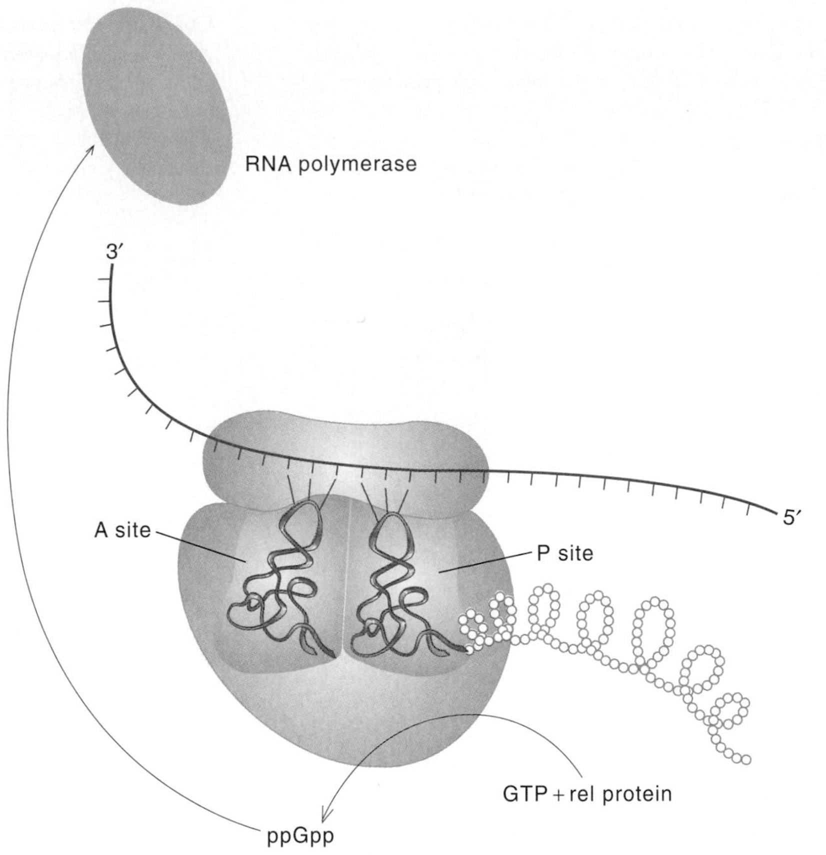

Figure 30.17

Schematic diagram of ppGpp synthesis and the hypothesized mechanism for its action. ppGpp is synthesized on the ribosome when there is a peptidyl-tRNA on the P site of the ribosome and uncharged tRNA on the A site. The ppGpp probably inhibits rRNA synthesis by complexing with the RNA polymerase.

amino acids are reintroduced into the growth medium of a wild-type culture, the ppGpp concentration falls rapidly (half-life about 20 s), and the rate of rRNA synthesis rises rapidly. In a mutant called *spoT*$^-$, the normal rise in ppGpp level is observed on amino acid starvation, but the level of ppGpp falls much more slowly on readdition of amino acids to the growth medium. Correlated with this, the rate of rRNA synthesis in a *spoT*$^-$ mutant also increases very slowly on readdition of amino acids.

The synthesis of ppGpp has been studied both in crude cell-free extracts of *E. coli* and in a partially purified system to determine what factors influence its rate of synthesis. It was found that ppGpp is synthesized on the ribosome from GTP in the presence of the protein encoded by the wild-type *relA* gene. Maximum ppGpp synthesis occurs in the presence of ribosomes associated with mRNA and uncharged tRNA with anticodons specified by the mRNA. If the uncharged tRNA bound to the ribosome acceptor site (A site) is replaced by charged tRNA, the rate of ppGpp synthesis is greatly lowered. If uncharged tRNA anticodons are not complementary to the mRNA codons exposed on the ribosome for protein synthesis, ppGpp synthesis does not occur.

Cell-free synthesis studies also strongly support the notion that ppGpp directly inhibits RNA synthesis. Thus, in a cell-free system the DNA-directed synthesis of rRNA with *E. coli* RNA polymerase is strongly inhibited by 100–200 μM ppGpp. Such experiments have led to the hypothesis that ppGpp, by binding to RNA polymerase, alters its structure so that it has a lowered affinity for rRNA promoters. The *in vivo* and *in vitro* studies on ppGpp and rRNA have resulted in a model for how the level of amino acid charging of tRNA controls the rate of rRNA synthesis (fig. 30.17). First, uncharged tRNA that is codon-specific for the exposed codons on the mRNA becomes bound to the ribosome acceptor site, creating a situation unfavorable for protein synthesis but favorable for ppGpp formation. Second, the ppGpp diffuses and binds to RNA polymerase, thereby lowering its affinity for rRNA promoters.

This alteration of the RNA polymerase by ppGpp affects the ability of RNA polymerase to interact with promoters in a differential way. For the promoters of the rRNA operons, the polymerase–promoter interaction is strongly inhibited by ppGpp. For some other promoters the effect of ppGpp is actually stimulating. As first observed in the cell-free system, ppGpp stimulates expression from the *lac* and *trp* operons. In more recent *in vivo* studies Cashel has found that the biosynthetic operons associated with arg, gly, his, leu, lys, phe, ser, thr, and val biosynthetic enzymes have an absolute requirement for ppGpp. This is a most appropriate response because an absence of an amino acid leads to ppGpp buildup, which then stimulates the genes required for biosynthesis of the amino acid.

The observations on ppGpp's role in rRNA synthesis show that this nucleotide is an important control factor regulating rRNA synthesis, but it does not eliminate the possibility that other factors also affect the level of rRNA. *In vivo* and *in vitro* evidence indicates that the inhibitory effect of ppGpp on transcription extends to most tRNA and ribosomal protein genes. Ribosomal protein gene expression also appears to be regulated at the translational level.

Translational Control of Ribosomal Protein Synthesis

In exponentially growing *E. coli* cells, synthesis rates of most ribosomal proteins are nearly identical and coordinately regulated. M. Nomura and his coworkers have suggested that free ribosomal proteins inhibit the translation of their own mRNA and that as long as the assembly of ribosomes removes ribosomal proteins, the corresponding mRNA escapes this feedback inhibition. This hypothesis has been tested *in vitro* using a protein-synthesizing system with various template DNA molecules carrying ribosomal protein genes; it has been tested *in vivo* by examining the effect of overproduction of certain ribosomal proteins on the synthesis of other ribosomal proteins using various recombinant plasmids. By these means it was found that certain ribosomal proteins selectively inhibit the synthesis of other ribosomal proteins whose genes are part of the same operon; this autogenous, or self-imposed, inhibition occurs at the level of the translation of the mRNA rather than at the level of the transcription of mRNA. Figure 30.18 illustrates how this scheme works for the regulatory protein L1. L1 and L11 are encoded by the P_{L11} operon. L1 can form a complex with either the 5′ end of its own mRNA or with 23S rRNA. It binds more strongly to the 23S rRNA. However, if an excess of L1 occurs over the available 23S rRNA, then L1 binds to the 5′ end of the mRNA, thereby inhibiting the synthesis of both L1 and L11. In this way the amounts of L1 and L11 proteins synthesized are kept in register with the amounts of rRNA synthesized.

Figure 30.18

Schematic diagram explaining autogenous inhibition by L1. L1 can form a complex either with the 5′ end of its own mRNA or with 23S rRNA. If there is an excess of L1 over the available 23S rRNA, then L1 binds to the 5′ end of the mRNA and inhibits both L1 and L11 synthesis. The inhibition of L1 translation by this binding depends on the fact that the translation of L1 and L11 is somehow coupled.

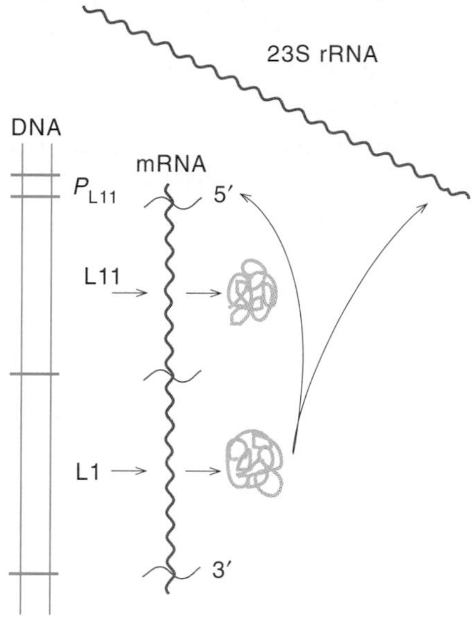

Other operons carrying ribosomal protein genes are regulated in a similar manner. Most ribosomal protein genes exist in operons with promoters at one end. In some cases, individual operons are regulated by one of the encoded ribosomal proteins, in other cases, operons appear to be subdivided into units of regulation, and individual units of regulation are regulated at the translation level by one of the translation products.

Regulation of Gene Expression in Bacterial Viruses

Bacterial viruses rarely coexist indefinitely with the host cell. Temperate viruses can adopt either an active, replicating state or a dormant, prophage state. In the prophage state, the viral genome exists at a low copy number, sometimes in a host-integrated form as with λ bacteriophage, and other times in an independent plasmidlike chromosome, as with P1 bacteriophage. In the active, replicating state, temperate viruses duplicate rapidly so that within about an hour they kill the host cell and release infectious particles. Most temperate viruses in the prophage state utilize the unmodified host RNA polymerase. But once they enter the lytic cycle, whether they are temperate or virulent, they either modify or replace the host polymerase to enhance their own transcription. The lytic life cycle of viruses employs multiple regulatory proteins that function in a cascade. The pattern of regulation is unidirectional and irre-

Table 30.2

Regulatory Proteins of λ

Protein	Function
cI	At low concentrations represses P_R and P_L and activates P_{RM}; at high concentrations represses P_R, P_L, and P_{RM}
cII	Activator for P_{int} and P_{RE}
cIII	Stabilizes cII
cro	At low concentrations represses P_{RM}; at high concentrations represses P_{RM}, P_R, and P_L
N	An antiterminator at t_{L1}, t_{R1}, and t_{R2}
Q	An antiterminator at t_{6S}

versible. A small percentage of virus genes designated as early genes are expressed first. Subsequently, late genes are turned on, and early genes are sometimes turned off, until mature virus particles are assembled and exit the cell. The pattern of gene expression observed for a bacteriophage has been likened to the pattern of expression observed in the development of differentiated cells in multicellular organisms (see chapter 31).

In this chapter we consider the patterns of gene expression found for λ, the best understood of the temperate phages.

λ Metabolism Is Directed by Six Regulatory Proteins

λ encodes six regulatory protein: cI, cII, cIII, cro, N, and Q (table 30.2). The developmental processes accompanying infection center on the roles of these six regulatory proteins and the factors that govern the level of their expression.

The Dormant Prophage State of λ Is Maintained by a Phage-Encoded Repressor

About half of the time when λ infects a cell it adopts a dormant lysogenic state in which the virus is linearly integrated into the host genome. This state is maintained by moderate amounts of the λ-encoded cI repressor. The cI repressor prevents the lytic cycle from developing by inhibiting two promoters: the P_L promoter for early leftward transcription and P_R the promoter for early rightward transcription (fig. 30.19). It does this by binding to sites that prevent the polymerase from binding, just like the bacterial repressor proteins we have already discussed. The repressor binding site closest to the P_R promoter is actually a composite of three adjacent repressor binding sites, O_{R1}, O_{R2}, and O_{R3}, which have different affinities for repressor. At moderate concentrations (about 100 copies of cI repressor per cell) O_{R1} and O_{R2} are occupied. At higher concentrations of cI the O_{R3} site is also occupied.

Binding of repressor at O_{R1} and O_{R2} has two effects: It inhibits transcription from the P_R promoter while it stimulates the promoter for repressor maintenance P_{RM}. Thus, at moderate concentrations the cI repressor functions simultaneously as a repressor and as an activator. The stimulation of leftward transcription from the P_{RM} promoter is due to favorable contacts made between the repressor bound at O_{R2} and the polymerase bound to the P_{RM} promoter. At more elevated concentrations of cI when the O_{R3} binding site is also occupied by repressor, the P_{RM} promoter is inhibited. This prevents overproduction of cI.

As long as a moderate concentration of the cI repressor is present, λ remains dormant. Exposure of lysogenic cells to ultraviolet light or DNA-damaging drugs such as mitomycin C leads to wholesale destruction of the cI protein. This is because DNA damage activates the host recA protease, which destroys the cI repressor (the activated *recA* protease cleaves the cI repressor in much the same way that it cleaves the *lexA* repressor; see chapter 26). With *cI* gone the prophage spontaneously enters the lytic cycle. This mechanism of prophage activation seems to be an instruction to the prophage that when conditions are unfavorable in the cell, it is time to replicate and find a new host cell to infect or lysogenize.

Events That Follow Infection of Escherichia coli by Bacteriophage λ Can Lead to Lysis or Lysogeny

Infection of *E. coli* by λ starts by attachment of the virus to the bacterial membrane and injection of the viral genome (fig. 30.20). This is followed by circularization catalyzed by the cellular DNA ligase enzyme. Only three genes are expressed at this time: The regulatory gene *N* from the P_L promoter and the regulatory genes *cro* and *cII* from the P_R promoter. The N protein combines with the host RNA polymerase, permitting extension of the early right and early left transcripts. This leads to expression of the other regulatory genes: *cII*, *cIII*, and *Q*, and replication and recombination genes. At this point the infection process can proceed in either of two, mutually exclusive directions: The lytic direc-

Figure 30.19

A segment of the λ genome containing the P_{RM} and nearby P_L and P_R promoters. Transciption from P_{RM} results in cI synthesis. An expanded view is shown of the three cI binding sites flanking the P_{RM} and P_R promoters. Transcripts that originate from the three promoters are shown by the wavy red lines. The O_{R1}, O_{R2}, and O_{R3} sites are binding sites for the cI repressor.

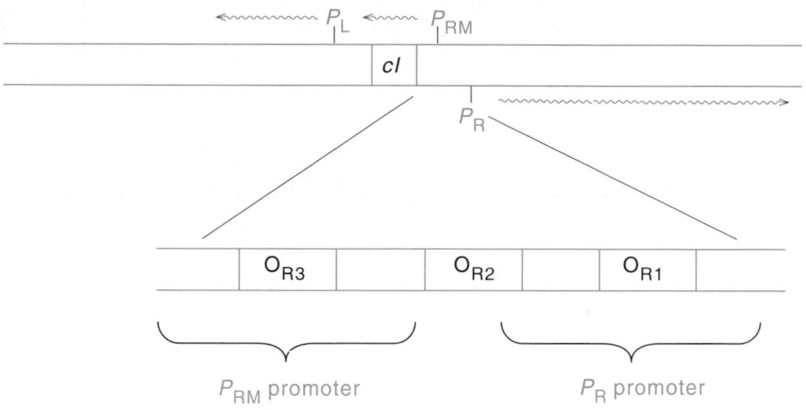

tion or the lysogenic direction. First we consider the events that take place in the lytic direction.

The N Protein Is an Antiterminator that Results in Extension of Early Transcripts

The N protein produced by early leftward transcription influences the RNA polymerase by causing it to ignore certain termination signals. For N to act it must first bind to the polymerase. Remarkably N can only attach to a transcribing polymerase at two polymerase pause sites, which are designated nut_L and nut_R (fig. 30.21). At these points during transcription the polymerase pauses and can pick up the N protein. The polymerase structure is altered after binding N so that it no longer recognizes the early termination sites, t_{R1}, t_{R2}, and t_{L1}, as stop signals. As a result, polymerase transcribes through these termination sites, producing longer transcripts from which additional gene products are expressed (see figs. 30.21 and 30.22).

Another Antiterminator, the Q Protein, Is the Key to Late Transcription

Extension of the rightward transcript stimulated by the N protein leads to expression of the *Q* gene protein, a key regulatory protein for late transcription. The Q protein functions like N as an antiterminator; in this case antitermination occurs at the t_{6S} terminator (see fig. 30.22). This leads to transcription of all of the late genes, including the lysis gene and the genes for the head and tail proteins of the mature phage. At late times other factors encourage rightward transcription from the $P_{R'}$ promoter.

Cro Protein Prevents Buildup of cI Protein during the Lytic Cycle

As the lytic cycle progresses, the phage DNA replicates. The increase in the number of gene copies can result in overexpression of the cI, which could shut down the lytic cycle. This effect is overcome by the regulatory protein cro, which is specifically designed to inhibit cI synthesis during late infection.

The cro protein and the cI protein bind to exactly the same sites on the DNA. Despite this fact, the cI protein is required for lysogeny, whereas the cro protein is required for lysis. These requirements can be shown with mutants. A cI^- mutant invariably undergoes lysis, whereas a cro^- mutant can lysogenize but cannot complete the lytic cycle. This remarkable difference in the behavior of *cro* and *cI* results from the fact that although they bind to the same sites, they do so with totally different relative affinities (fig. 30.23). *Cro* binds preferentially to O_{R3} and less strongly to O_{R1} and O_{R2}.

The binding of *cro* to O_{R3} turns off *cI* expression originating from the P_{RM} promoter. At high concentrations *cro* binds to one or more of the other sites, turning off the P_R promoter. The P_L promoter is also turned off at high concentrations of *cro*. The most important physiologic effect of *cro* is the turning off of the P_{RM} promoter. In the absence of *cro*, *cI* expression increases because of the increase of the number of *cI* gene copies resulting from early replication of viral DNA during the lytic cycle. This buildup shuts down transcription from the P_R and P_L promoters before sufficient transcripts have been made to ensure the lytic pathway. The binding properties of *cI* and *cro* to the tripartite operator shared by P_{RM} and P_R are shown in fig. 30.23.

Figure 30.20

Overview of λ development.

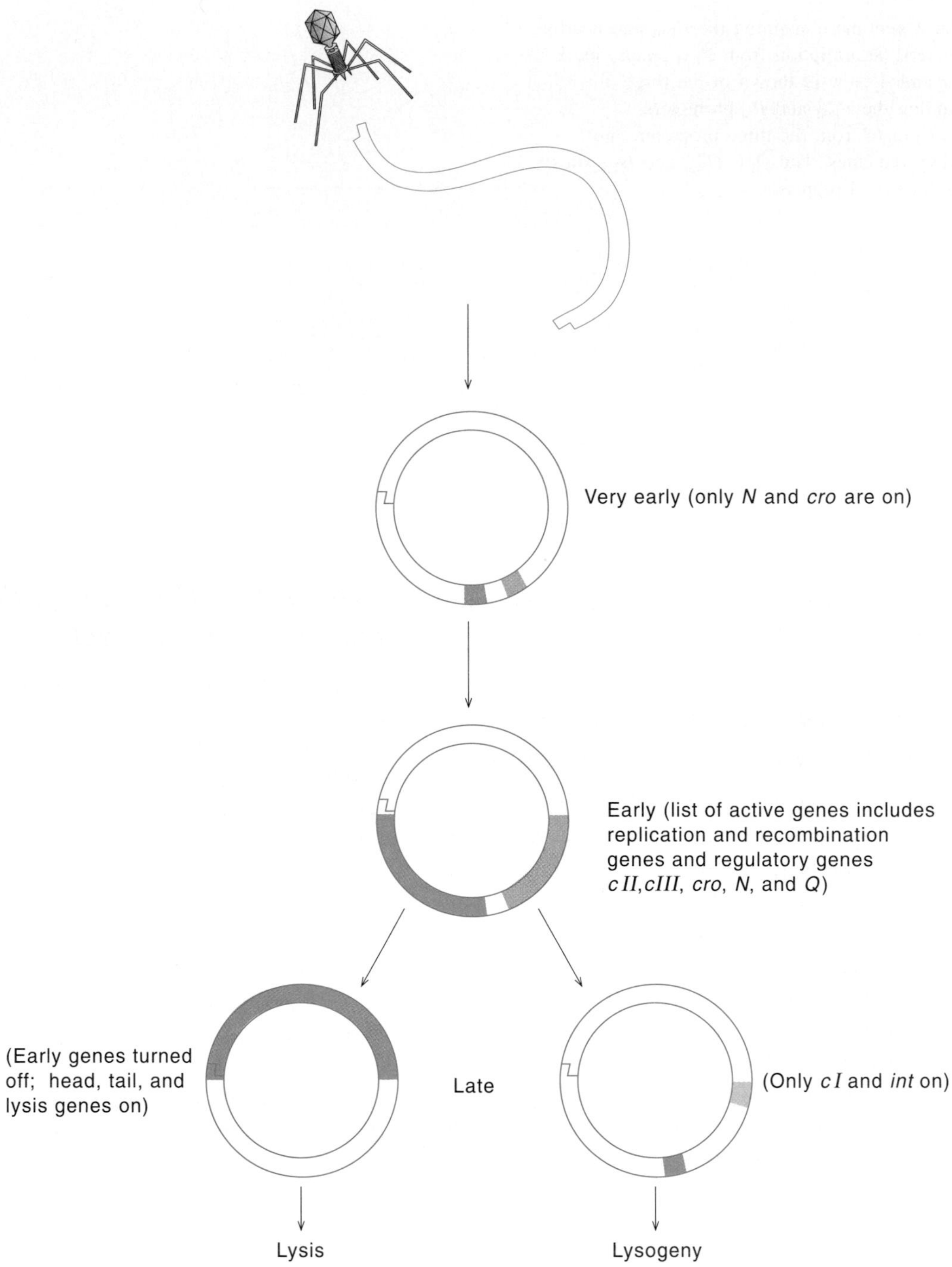

Late Expression along the Lysogenic Pathway Requires a Rapid Buildup of the cII Regulatory Protein

Thus far we have considered the events that take place after infection or after activation of the prophage that lead to phage replication and ultimately lysis. As already indicated, when λ infects a cell, the cells can follow this lytic pathway, or alternatively they can follow the lysogenic pathway, in which case the λ genome becomes dormant and integrates into the host genome. The choice appears to depend primarily on the level of cII protein formed during the early stages of infection. If this level is sufficiently high, the lysogenic pathway is followed; otherwise the lytic pathway is followed. Although we do not understand precisely what controls the level of cII expression, it is easy to appreciate the importance of cII to elaboration of the lysogenic pathway. Thus, cII protein is an activator for two promoters: P_{RE} and

Figure 30.21

Early transcripts. The early left transcript is initiated from the P_L promoter. In the absence of N protein this transcript terminates at t_{L1}. In the presence of N protein the polymerase picks up an N protein at the N utilization site, *nut*$_L$. This makes it possible for the polymerase to transcribe through t_{L1}. The early right transcript tends to terminate at t_{R1} unless N protein is present. In this case the N protein becomes bound to the polymerase at the *nut*$_R$ site.

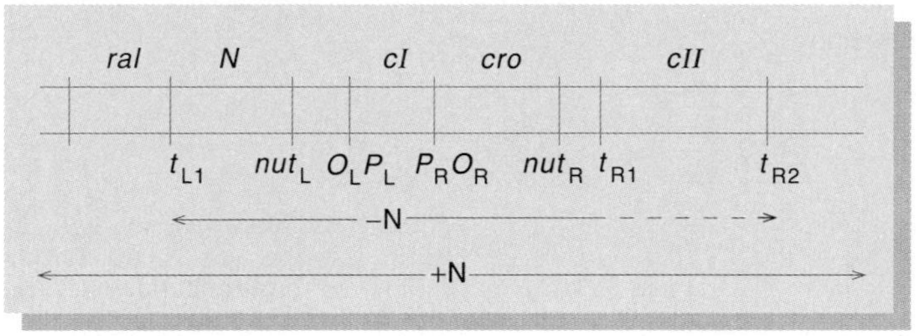

Figure 30.22

Bacteriophage λ DNA. *Top:* Circular map indicating locations of main control genes and early and late functions. *Bottom:* Expanded region containing control genes and main promoters and terminators. Arrows indicate main transcripts and conditions under which they are active. In the absence of N protein, about half the transcription beginning at P_R reads through t_{R1} to t_{R2} (indicated by dashed portion of arrow). More information on regulatory proteins and promoters is given in table 30.2.

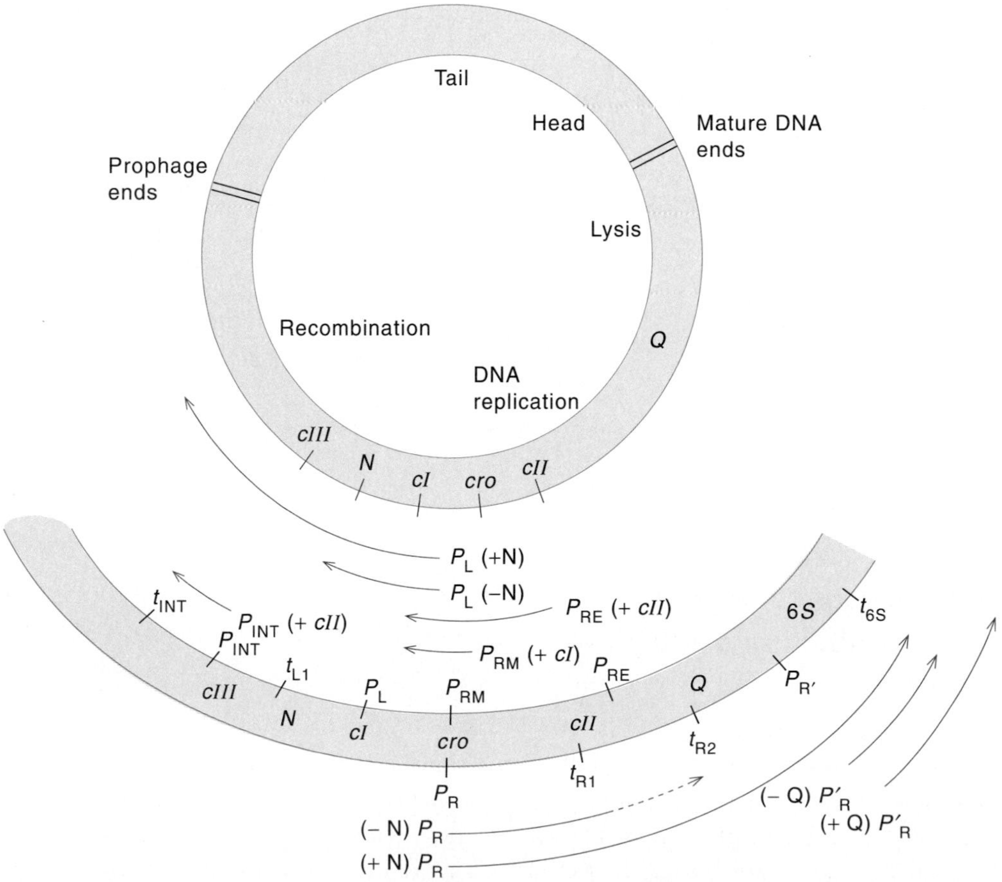

Figure 30.23

Segment of the λ genome showing the three operators O_{R1}, O_{R2}, and O_{R3} around the P_{RM} and the P_R promoters. The cI and cro regulatory proteins bind to these operators with different relative affinities. The net result of these differing affinities is that cI is required for lysogeny and cro is required for the lytic cycle.

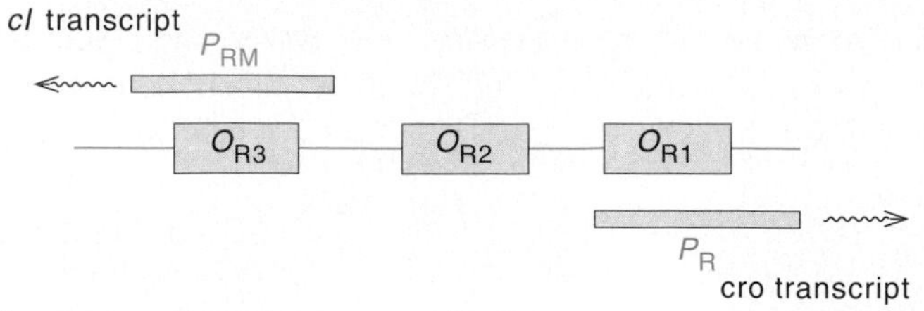

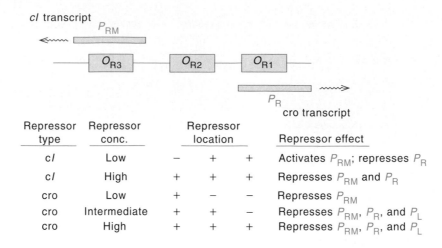

Repressor type	Repressor conc.	Repressor location			Repressor effect
cI	Low	−	+	+	Activates P_{RM}; represses P_R
cI	High	+	+	+	Represses P_{RM} and P_R
cro	Low	+	−	−	Represses P_{RM}
cro	Intermediate	+	+	−	Represses P_{RM}, P_R, and P_L
cro	High	+	+	+	Represses P_{RM}, P_R, and P_L

Figure 30.24

A segment of the λ genome containing the P_{RM} and nearby P_L, P_R, and P_{RE} promoters. Transcripts that can originate from the three promoters are shown by the wavy lines. The P_{RE} and P_{RM} require different activators for expression. The transcript from P_{RE} is 10 times more effective in cI expression because it has a Shine-Dalgarno sequence for ribosome attachment.

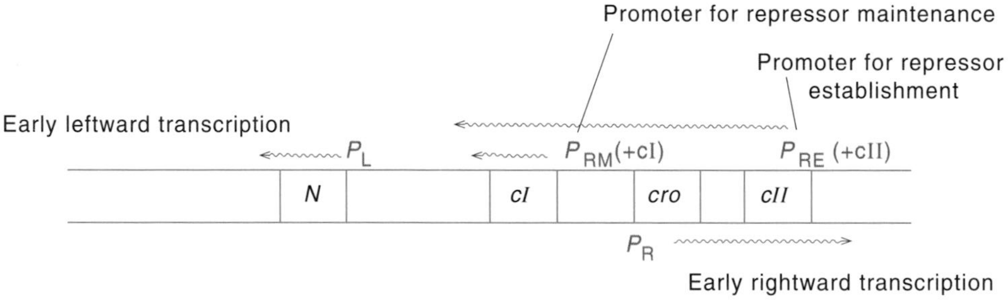

P_{int} (see fig. 30.22). The P_{RE} promoter leads to transcripts that are very efficient in producing high levels of the cI repressor (fig. 30.24). High levels of cI can shut down the P_L and P_R promoters before the lytic cycle gets underway. The other promoter activated by cII is P_{int}, which leads to int protein synthesis. This protein is required for the integration of the phage genome into the host genome, a process that completes the lysogenic process. After integration, moderate levels of cI resulting from transcription initiated at the P_{RM} promoter (as described above) keep both the P_L and the P_R promoters in the repressed state.

Interaction between DNA and DNA-Binding Proteins

So far we have focused on where regulatory proteins bind to the DNA and under what conditions they bind. The ability to isolate and sequence DNA promoters and to isolate and characterize regulatory proteins has greatly increased our appreciation of how specific complexes are formed between DNA and regulatory proteins.

The three-dimensional structures of about 20 regulatory proteins have been determined; in four cases the structures of the specific DNA–regulatory protein complexes have also been determined.

Recognizing Specific Regions in the DNA Duplex

The problem of regulation in prokaryotes such as *E. coli* is far simpler than in complex eukaryotes because of the smaller number of genes. It seems likely that a system of this complexity (about 3,000 genes) should contain no more than 100–300 regulatory proteins because genes with related functions are often under the control of the same regulatory proteins.

Each member of this diverse family of regulatory proteins must be able to scan the DNA structure and recognize control regions with a high degree of specificity. How is this possible, particularly as the specific regions of DNA are embedded in a core of base pairs that are already hydrogen-bonded to each other? Present indications are that the secondary structure of DNA is left intact when these regions interact with their regulatory proteins. Thus, the base pairs must be identifiable by the groups that are exposed in the core of an intact duplex structure. As in most specific interactions between nucleic acids and proteins, it seems likely that the specificity of interactions between DNA and regulatory proteins must result primarily from hydrogen bond interactions. Inspection of the DNA duplex in the major and minor grooves reveals that many of the hydrogen-bonding groups of the base pairs are available for such interactions (see fig. 25.6). In the major groove the GC and AT base pairs both have three hydrogen-bonding groups accessible for interaction with other molecules. In the minor groove the GC base pair also has three, but the AT base pair has only two. There are other considerations. The major groove can accommodate a larger protein element than the minor groove. An α helix or a two-chain β structure can fit comfortably into the major groove so that their polar side chains have no difficulty making hydrogen bonds with a core of base pairs. Such structures cannot be fitted into the minor groove, the contacts of which are probably limited to those that can be made with an extended polypeptide chain. Based on size considerations alone it seems likely that regulatory proteins make most of their specific contacts with DNA in the major groove. Quite remarkably almost all regulatory proteins involve α-helical elements interacting in the major groove.

The Helix-Turn-Helix Is the Most Common Motif Found in Prokaryotic Regulatory Proteins

In prokaryotes such as *E. coli,* the helix segment recognized by the DNA is part of a larger domain known as the helix-turn-helix motif. A protruding recognition helix is supported by a second segment of helix, which stabilizes the recognition helix and fixes its orientation with respect to the remainder of the regulatory protein.

The primary sequence for the helix-turn-helix motif consists of a 20-amino-acid sequence in which 6 of the amino acids tend to be conserved (residues 4, 5, 8, 9, 10, and 15 in fig. 30.25). Four of these residues, 4, 5, 10, and 15, usually have hydrophobic side chains; they make stabilizing contacts between the two helices, ensuring the preservation of their mutual orientation. Two other residues, 8 and 9, are important for making the bend between the two segments of α helix. Residue 9 is frequently a glycine because its small side chain (—H) is often necessary to make a bend. The remaining amino acid residues vary considerably in different proteins. These are the residues that either face the DNA or the remaining portion of the regulatory protein.

Figure 30.25

Helix-turn-helix motif. Helix 2 is the recognition helix. Individual amino acids are numbered. Residues 4, 5, 8, 9, 10, and 15 tend to be conserved in different regulatory proteins. (From Carl Branden and John Tooze, *Introduction to Protein Structure,* Garland Publishing, Inc., New York, 1991, p. 102.)

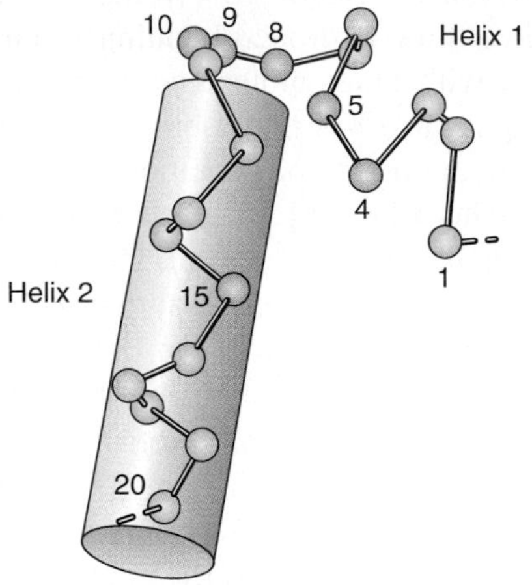

The simplest way for a segment of protein helix to make contact in the major groove is for it to run parallel to the groove. Although it adopts this orientation in some cases, it frequently orients itself in other ways in different regulatory proteins. In all known orientations ample opportunity arises for many of the amino acid side chains to make contact with the hydrogen-bonding groups in the DNA base pairs.

Helix-Turn-Helix Regulatory Proteins Are Symmetrical

Three helix-turn-helix regulatory proteins are depicted in figure 30.26: cro repressor, λ repressor, and CAP activator. Of these, only in cro is the recognition helix oriented so that it runs parallel to the major groove when bound to the DNA. In all three cases the regulatory proteins are homodimers containing monomers with recognition helices spaced precisely 34 Å apart along the direction of the DNA helix's axis so that they can make identical contacts with adjacent major grooves of the DNA duplex. The strategy seems clear; the regulatory protein contains two identical half-sites for interaction with two virtually identical half-sites in the DNA.

Beyond helix 2 in the helix-turn-helix protein we see a great deal of variability in overall structure in the different regulatory proteins. This variability is not surprising because the remainder of the structure serves quite different functions in different regulatory proteins.

The symmetrical aspect of the regulatory protein-binding sites is mirrored in the DNA sequence of the binding site. Recall that the sequence in this region for the *lac* repressor binding site (fig. 30.7), the CAP binding site (fig. 30.8), and the *trp* repressor binding site (fig. 30.14) all show a dyad axis of symmetry with respect to the arrangement of base pairs.

DNA–Protein Cocrystals Reveal the Specific Contacts between Base Pairs and Amino Acid Side Chains

Examination of DNA–regulatory protein complexes have permitted reasonable guesses to be made about the precise nature of the contacts between amino acid side chains and DNA in many cases. Figure 30.27 illustrates three examples: One for the 434 phage repressor (fig. 30.27*a*), one for the λ cI repressor (fig. 30.27*b*), and one for the trp repressor (fig. 30.27*c*). In all cases only half-sites are depicted because symmetry considerations dictate that the two half-sites should have virtually identical structures.

In the case of the two phage repressors, several contacts are observed between individual amino acid side chains and individual base pairs. In addition, one or two hydrogen bonds are observed to nearby phosphate groups. In both of these cases the side chain that is contacting a base participates in the hydrogen bond to a phosphate group, either directly or indirectly.

In these two examples we can see that the amino acids participating in direct interaction are frequently glutamine or asparagine, the two amino acids with amide side chains. The presence of both a hydrogen bond donor and a hydrogen bond acceptor gives the amide side chain the advantage of greater versatility than most amino acid side chains for forming hydrogen bonds. Indeed, in one case we can see that glutamine makes two hydrogen bonds with an adenine. In four of these five examples glutamine is used. The greater length of the glutamine side chain gives it more ''reaching power'' for making contact with the base pairs.

Despite the apparent advantages in the amide-containing amino acids, we do not always find amides used as part of the recognition sequence. For example, the trp repressor does not use any amides. The basic amino acids, lysine and arginine, and the hydroxylic amino acid, threonine, are the main interacting amino acids in the trp repressor. The side

Figure 30.26

The structures of three regulatory proteins. They all possess twofold axes of symmetry, and the protruding helical cylinders that interact with adjacent major grooves on the DNA (red) are spaced about 34 Å apart. N and C labels indicate the N and C termini of the polypeptide chains.

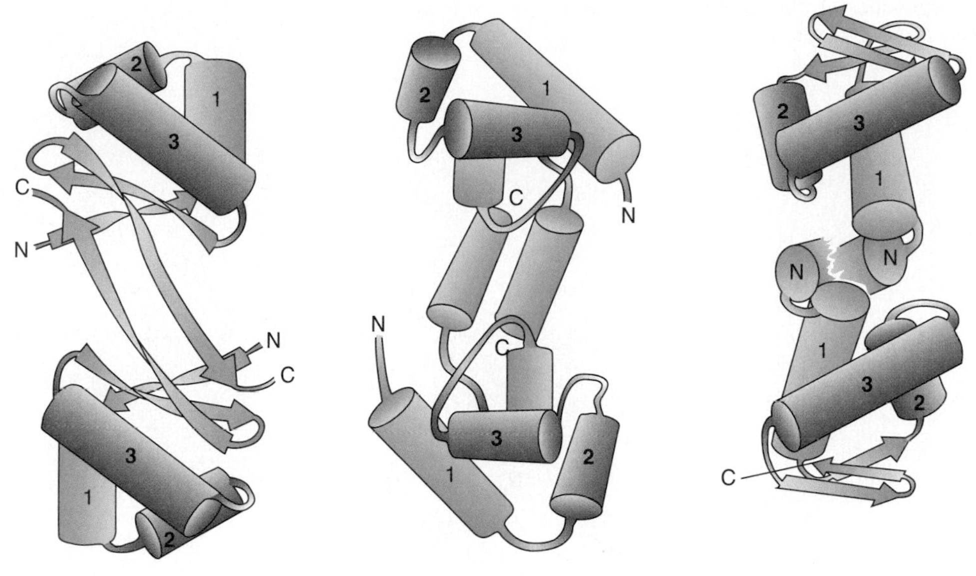

chain in arginine can make a pair of hydrogen bonds with a guanine, which should result in a particularly strong interaction.

The trp repressor presents an unusual variation for interaction with the base pairs in its operator. Except for the arginine–guanine interaction, the interactions are mediated by a water molecule (see fig. 30.27*c*). For example, a threonine hydroxyl group makes a hydrogen bond with a water molecule, which in turn makes two hydrogen bonds with an adenine.

Some Regulatory Proteins Use the β-Sheet Motif

Despite the dominance of the α-helical motif as a recognition unit, the β-sheet motif has been used in a limited number of cases. In these cases (three are known) two chains in the anti-parallel orientation interact with half-sites in adjacent large grooves of the DNA (see fig. 30.28 for an example). (The β-sheet motif obviously works in these cases despite the fact that it is more difficult to stabilize than a segment of α helix.)

Involvement of Small Molecules in Regulatory Protein Interaction

A common but not universal feature of regulatory proteins is their sensitivity to small-molecule effectors. The lambdoid phage repressors are examples of molecules that almost always function as repressors; their action is controlled merely by their concentration. When they are needed, they are synthesized; when they are no longer needed or their presence is undesirable, they are selectively removed by degradation.

For many other bacterial regulatory proteins the action of the regulatory protein depends on specific small-molecule effectors. As we have seen in the case of the lac repressor, the binding of allolactose results in release of the repressor from the operator. By contrast the binding of cAMP

Figure 30.27

Specific interactions between three different repressors and their operator binding sites. Only half the operator binding site is shown because identical contacts are made with the other half. The numbers associated with the amino acid side chains refer to the distance of amino acids from the amino-terminal end of the protein. Nucleotides are numbered from the central dyad at the operator. (*a*) The 434 phage repressor; (*b*) The λ repressor. (*c*) The trp repressor. (Source: Adapted from T. Steitz, *Q. Rev. Biophys.* 23:236, 1990.)

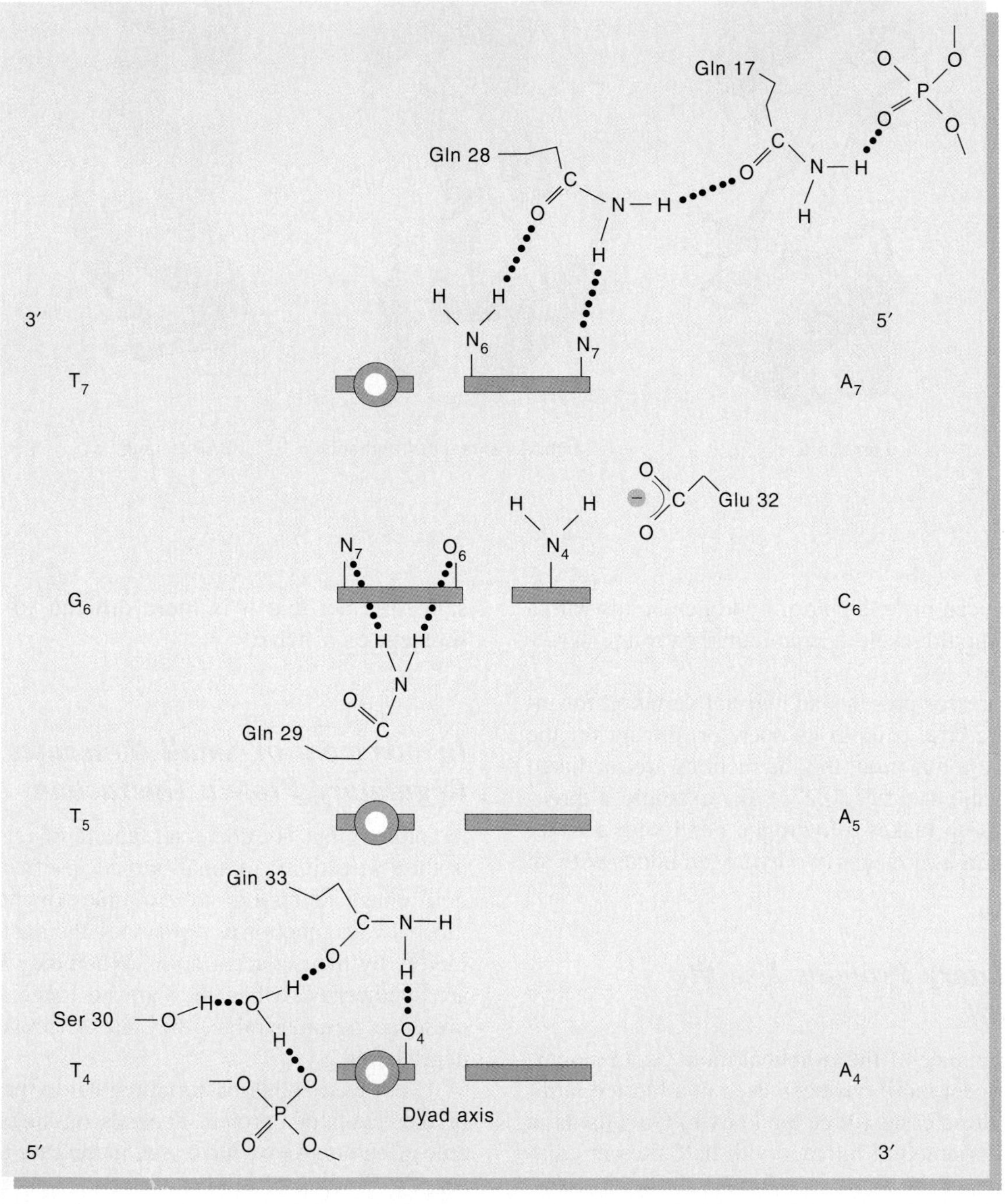

(a)

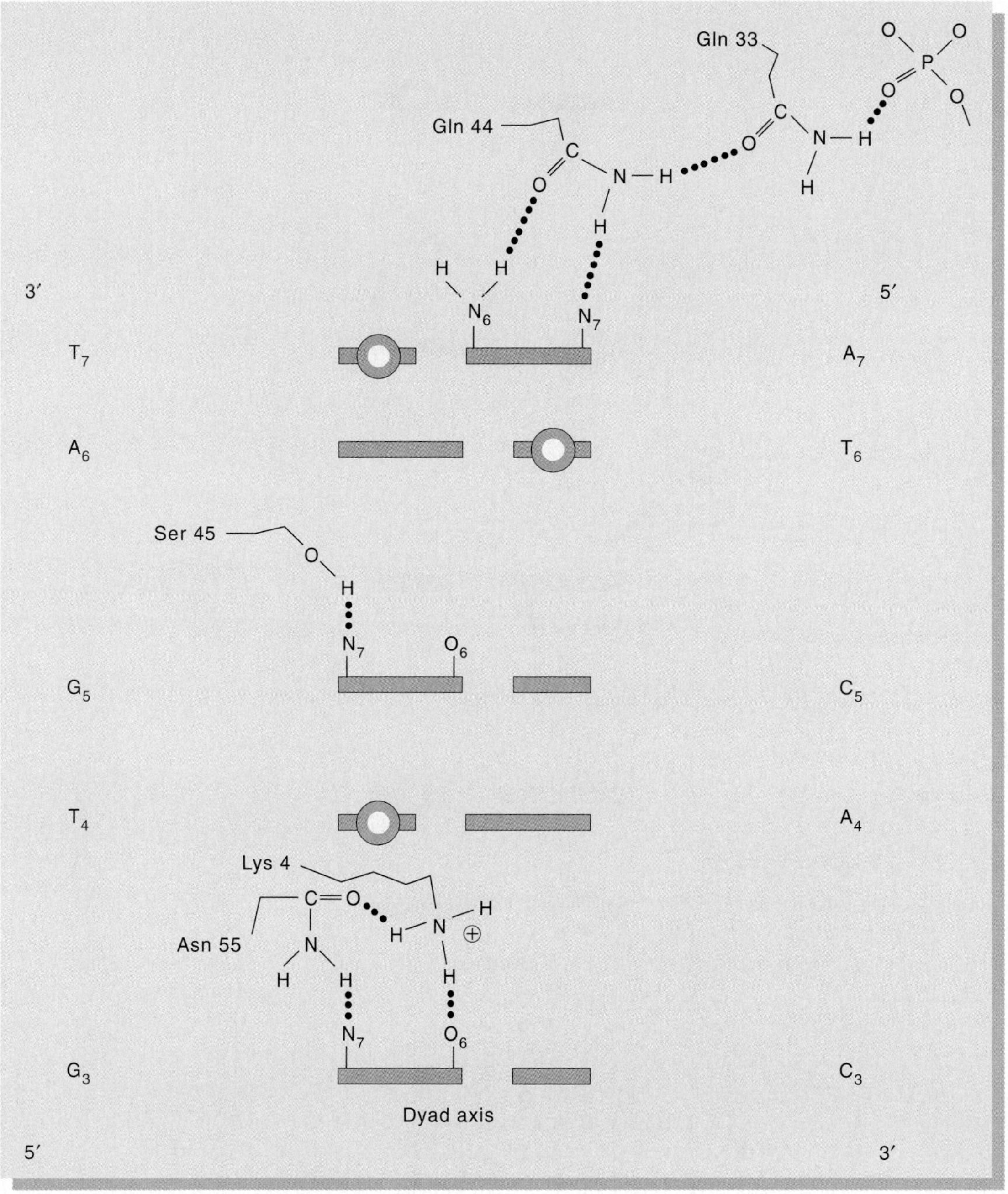

Gln 33
Gln 44
O
P
C
N—H
H
3′
5′
N_6
N_7
T_7
A_7
A_6
T_6
Ser 45
O_6
G_5
C_5
T_4
A_4
Lys 4
Asn 55
C=O
⊕
G_3
C_3
Dyad axis
5′
3′

(b)

(cont.)

Figure 30.27 *(cont.)*

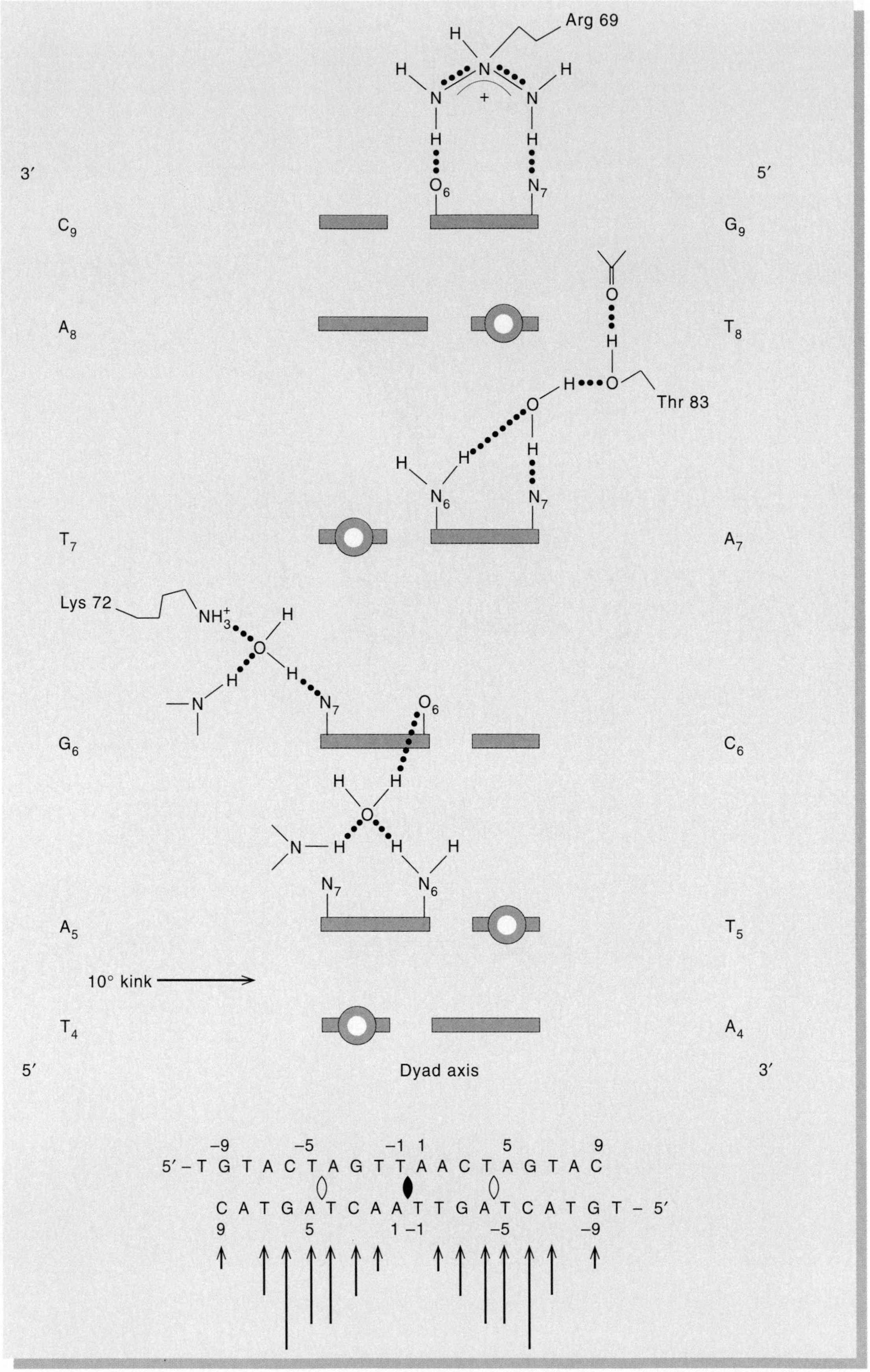

(c)

Figure 30.28

Diagram of the met repressor–DNA structure illustrating the regions of the dimeric met repressor that contact DNA. The two-stranded β sheet of the repressor is bound in the major groove of DNA, where it forms the sequence-specific interactions.

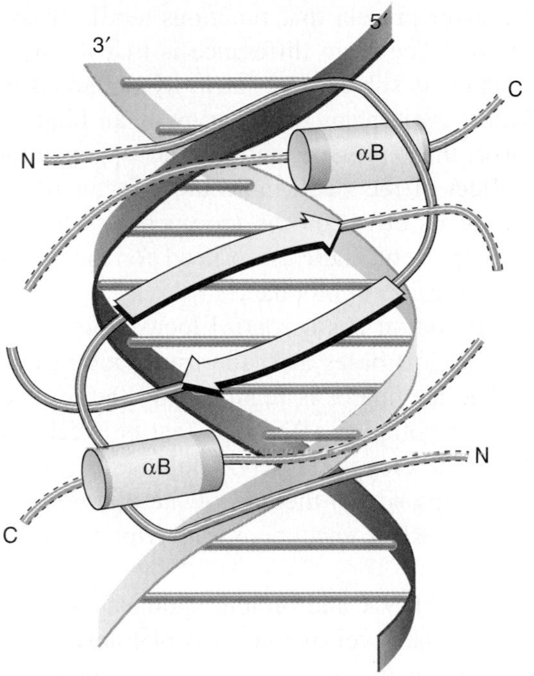

is necessary for the binding of the CAP protein to its DNA-binding site. The small molecule signals a particular metabolic state, and the response of the regulatory protein is appropriate to the metabolic needs of the cell.

It is often difficult to see the change in regulatory protein structure as a result of the binding of its small-molecule effector. In spite of this the theory is that the small-molecule effector changes the structure of the regulatory protein sufficiently to cause release or binding to the appropriate DNA-binding site as the case may be.

In the case of the trp repressor the binding of tryptophan to the repressor converts the inactive repressor (aporepressor) into an active repressor by a slight shift in the orientation of the recognition helices (fig. 30.29).

Figure 30.29

Docking of the trp repressor to the trp operator. The repressor does not bind unless it first complexes with tryptophan. Note how the angle of the recognition helices changes when tryptophan is bound. With regard to the trpR/O interaction, the bound tryptophan not only causes a readjustment of the so-called recognition helix, but it also shapes the side chains in its immediate environment in a very special way.

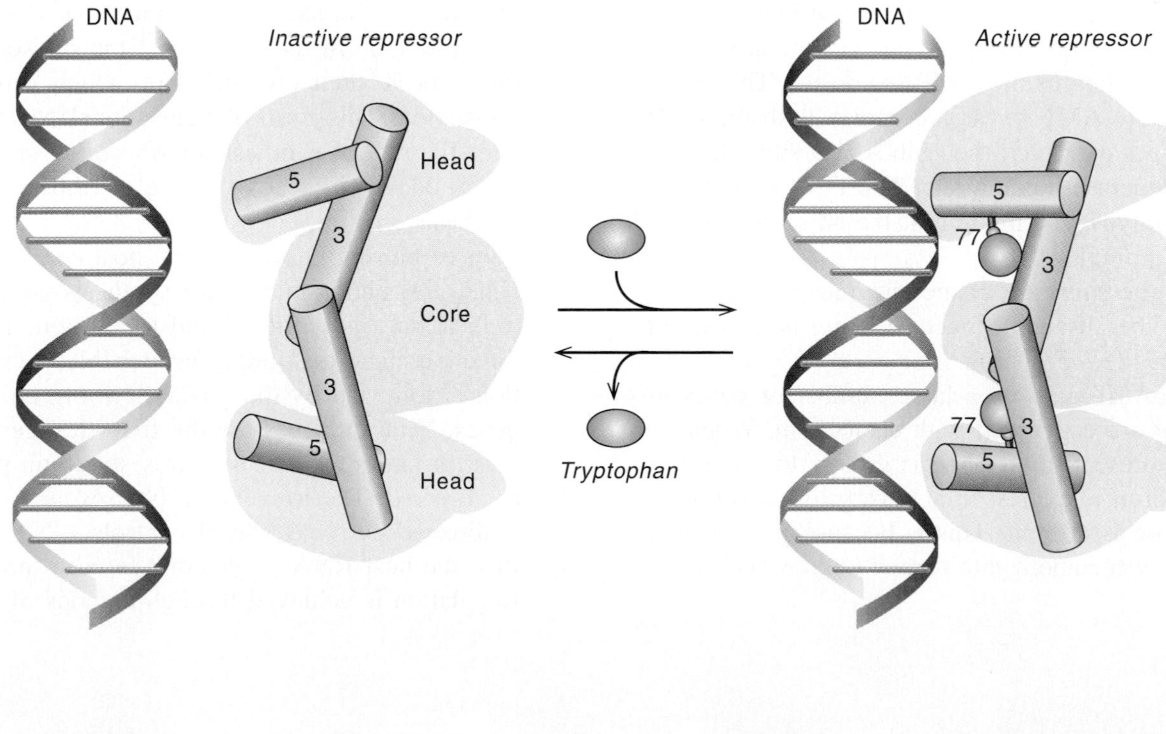

Summary

In this chapter we discussed the regulatory systems of the *E. coli* bacterium and the λ bacteriophage. The main points in our presentation are as follows.

1. *Escherichia coli* carries about 3,000 genes. Only a small fraction of the genome is actively transcribed at any given time. But all of the genes are in a state where they can be readily turned on or turned off in a reversible fashion. The level of transcription is regulated by a complex hierarchy of control elements.
2. In the most common form of control, expression is regulated at the initiation site of transcription. There are several ways of doing this, all of them revolving around protein or small-molecule factors that influence the binding of RNA polymerase at the transcription start site.
3. The *lac* operon, a cluster of three genes involved in the catabolism of lactose, exemplifies both positive and negative forms of control that influence the rate of initiation of transcription. Jacob and Monod identified the repressor as a negative control element that is *trans*-dominant and the operator as a *cis*-dominant site for binding the repressor. Transcription is initiated by RNA polymerase at the promoter, which overlaps the operator site on one side of the three structural genes of the *lac* operon. The tight complex between repressor and operator prevents initiation, and it is broken when lactose is present. The lactose is readily converted to allolactose, which binds to the lac repressor. This changes the structure of the repressor so that it dissociates from the DNA.
4. Initiation of transcription proceeds at a greatly increased rate when cyclic AMP is present. This is because cyclic AMP forms a complex with the CAP activator protein, which then binds at a site adjacent to the polymerase-binding site. The CAP protein enhances polymerase binding at the adjacent site by cooperative binding.
5. Lactose is the substrate of the enzymes of the *lac* operon. In the absence of lactose, there is no use for enzymes of the *lac* operon.
6. CAP and cAMP activate a large number of genes in *E. coli* that are concerned with catabolism. When glucose is present, the cAMP is greatly lowered and the *lac* operon is expressed at a very low level, even when lactose is present. This is because glucose is a more readily metabolizable carbon source than lactose.
7. The *trp* operon contains a cluster of five structural genes associated with tryptophan biosynthesis. Initiation of transcription of the *trp* operon is regulated by a repressor protein that functions similarly to the lac repressor. The main difference is that the trp repressor action is subject to control by the small-molecule effector, tryptophan. When tryptophan binds the repressor, the repressor binds to the *trp* operator. Thus, the effect of the small-molecule effector here is opposite to its effect on the *lac* operon. When tryptophan is present, there is no need for the enzymes that synthesize tryptophan.
8. The *trp* operon has a control locus called an attenuator about 150 bases after the transcription initiation site. The attenuator is regulated by the level of charged tryptophan tRNA, so that between 10% and 90% of the elongating RNA polymerases transcribe through this site to the end of the operon. Low levels of *trp* tRNA encourage transcription through the attenuator.
9. Ribosomal RNA and protein synthesis are both controlled at the level of initiation of transcription. This is a result of the direct binding of guanosine tetraphosphate, ppGpp, to the RNA polymerase. This binding decreases the affinity of RNA polymerase for the initiation sites of transcription. Guanosine tetraphosphate is synthesized when the general level of amino-acid-charged tRNA is low.
10. The synthesis of ribosomal proteins is regulated at the level of translation. Certain ribosomal proteins bind to specific sites on the ribosomal RNAs or their own mRNAs. In the absence of the ribosomal RNAs, they bind to their own mRNAs, which inhibits their translation. This form of translational control regulates the rate of synthesis of ribosomal proteins so that it does not exceed the rate of ribosomal RNA synthesis.
11. Viruses borrow heavily on the host enzymatic machinery to obtain energy for synthesis, as well as for replication, transcription, and translation. The virus infective cycle is strongly irreversible. Virus infection is followed by the gradual turning on of viral genes. Viral enzymes are the first viral gene products; in late infection, the virus structural proteins are favored. The irreversible lytic cycle of the virus is directed by a cascade of controls.
12. In λ the host RNA polymerase is used throughout. Regulation is achieved through a series of repressors

and activators, as well as two viral proteins that bind directly to the RNA polymerase. The viral proteins that bind to the polymerase modify it so that it can transcribe through provisional stop signals.

13. When λ phage infects an *E. coli* cell, it does not always produce viral progeny. Sometimes it integrates its genome into the host genome and replicates only as the host genome replicates. This so-called lysogenic state can be disrupted by DNA-damaging conditions such as exposure to UV radiation. Under these conditions the dormant viral genome enters the active replication cycle.
14. Bacterial regulatory proteins are controlled by small-molecule effectors; viral regulatory proteins are not. Bacterial genes are regulated in a highly reversible manner; viral genes are usually turned on only once.
15. Proteins that regulate transcription usually bind to specific sites on the DNA. The recognition process involves specific hydrogen bonds formed between amino acid side chains of the protein and the base pairs of the DNA. Most of this interaction takes place in the major groove of the DNA, which is more accessible to the protein. The vast majority of regulatory proteins interact with the DNA from the side chains of a segment of α helix that fits snugly into the major groove. The binding site on the DNA usually consists of two half-sites, which are arranged on a dyad axis of symmetry that matches two half-sites on the regulatory protein. The two half-sites are situated in adjacent major grooves on one side of the DNA.

Selected Readings

Anderson, J. E., M. Ptashne, and S. C. Harrison, Structure of the repressor-operator complex of bacteriophage 434. *Nature* 306:846–852, 1987.

Brennan, R. G., and B. W. Matthews, The helix-turn-helix DNA binding motif. *J. Biol. Chem.* 264:1903–1906, 1989. A mini review.

Gilbert, W., and B. Muller-Hill, Isolation of the lac repressor. *Proc. Natl. Acad. Sci. USA* 56:1891–1898, 1966. First isolation of a repressor protein.

Gold, L., Posttranscriptional regulatory mechanisms in *Escherichia coli. Ann. Rev. Biochem.* 56:199–234, 1988.

Goodrich, J. A., and W. R. McClure, Competing promoters in prokaryotic transcription. *Trends Biochem. Sci.* 16:394–396, 1991. Two or more bacterial promoters are often found in close proximity and may compete for the binding of RNA polymerase.

Gralla, J. D., Transcriptional control—Lessons from an *E. coli* promoter data base. *Cell* 66:415–418, 1991.

Green, P. J., O. Pines, and M. Inouye, The role of antisense RNA in gene regulation. *Ann. Rev. Biochem.* 55:569–597, 1986.

Helman, J. D., and M. J. Chamberlain, Structure and function of bacterial sigma factors. *Ann. Rev. Biochem.* 57:839–872, 1988.

Jacob, F., and J. Monod, Genetic regulatory mechanisms in the synthesis of proteins. *J. Mol. Biol.* 3:318–356, 1961. A classic paper.

Kaiser, D., and R. Losick, How and why bacteria talk to each other. *Cell* 73:873–886, 1993.

Kustu, S., A. K. North, and D. S. Weiss, Prokaryotic transcriptional enhancers and enhancer-binding proteins. *Trends Biochem. Sci.* 16:397–401, 1991. First discovered in eukaryotes, enhancers have now been found to exist for a number of prokaryotic genes.

Losick, R., and P. Stragier, Crisscross regulation of cell-type-specific gene expression during development in *B. subtilis*. *Nature* 355:601–604, 1992.

Nomura, M., R. Gourse, and G. Baughman, Regulation of the synthesis of ribosomes and ribosomal components. *Ann. Rev. Biochem.* 53:75–117, 1984.

Pabo, C. O., and R. T. Sauer, Transcription factors: Structural families and principles of DNA recognition. *Ann. Rev. Biochem.* 61:1053–1095, 1992.

Parkinson, J. S., Signal transduction schemes of bacteria. *Cell* 73:857–872, 1993.

Ptashne, M., *A Genetic Switch: Gene Control and Phage λ.* Cambridge, Mass.: Cell Press, and Palo Alto, Calif.: Blackwell Scientific, 1987.

Ptashne, M., A. D. Johnson, and C. O. Pabo, A genetic switch in a bacterial virus. *Sci. Am.* 247(5):128–140, 1982.

Roberts, J. W., RNA and protein elements of *E. coli* and λ transcription antitermination complexes. *Cell* 72:653–656, 1993.

Simons, R. W., and N. Kleckner, Biological regulation by antisense RNA in prokaryotes. *Ann. Rev. Genet.* 22:87–600, 1988.

Steitz, T. A., Structural studies of protein–nucleic acid interaction: The sources of sequence-specific binding. *Quar. Rev.*

Biophys. 23:205–280, 1990. A very readable and very thorough review of the subject with excellent illustrations, written by one of the pioneers in the field.

Storz, G., L. A. Tartaglia, and B. N. Ames, Transcriptional regulator of oxidative stress-inducible genes: Direct activation by oxidation. *Science* 248:189–194, 1990.

Stwinowski, Z., R. W. Schevitz, R.-G. Zhang, C. L. Lawson, A. Joachimiak, R. Q. Marmorstein, B. F. Luisi, and P. B. Sigler, Crystal structure of trp repressor/operator complex at atomic resolution. *Nature* 335:321–329, 1988.

Weintraub, H., Antisense RNA and DNA. *Sci. Am.* 262(1):40–46, 1990.

Wolberger, C., Y. Dong, M. Ptashne, and S. C. Harrison, Structure of phage 434 Cro/DNA complex. *Nature* 335:789–795, 1988.

Yanofsky, C., Operon-specific control by transcription attenuation. *Trends Genet.* 3:356–360, 1987.

Zubay, G., M. Lederman, and J. DeVries, DNA-directed peptide synthesis III. Repression of β-galactosidase synthesis and inhibition of repressor by inducer in a cell-free system. *Proc. Natl. Acad. Sci. USA* 58:1669–1675, 1967. Showing that repressor works in a cell-free system.

Zubay, G., D. Schwartz, and J. Beckwith, Mechanism of activation of catabolite-sensitive genes: A positive control system. *Proc. Natl. Acad. Sci. USA* 66:104–110, 1970. First isolation of an activator protein.

Problems

1. What set of data originally led Jacob and Monod to suggest the existence of a repressor in *lac* operon regulation?
2. The *lac* promoter–operator region is found in many vectors used by molecular biologists to clone genes. Not infrequently, high levels of transcription driven by the *lac* promoter generate toxic levels of the cloned gene product, causing *E. coli* cells to grow poorly. Faced with such a situation, how could you minimize expression of the *lac* promoter?
3. With respect to β-galactosidase production in the presence and absence of inducer, what would be the phenotype of the following *E. coli* mutants:
   (a) $i^s o^+ z^+$
   (b) $i^s o^c z^+$
   (c) $i^s o^c z^- / i^+ o^+ z^+$
   (d) $i^+ o^c z^- / i^- o^+ z^+$
4. In a cell that is *lac*Z$^-$, what would be the relative thiogalactoside transacetylase concentration, compared with wild type, under the following conditions?
   (a) After no treatment
   (b) After addition of lactose
   (c) After addition of IPTG
5. Consider a negatively controlled operon with two structural genes (*A* and *B*, for enzymes A and B), an operator gene (*O*), and a regulatory gene (*R*). The first line of data in the table gives the enzyme levels in the wild-type strain after growth in the absence or presence of the inducer. Complete the table for the other cultures.

	Uninduced		Induced	
Strains	***Enz A***	***Enz B***	***Enz A***	***Enz B***
Haploid strains				
(1) $R^+O^+A^+B^+$	1	1	100	100
(2) $R^+O^cA^+B^+$				
(3) $R^-O^+A^+B^+$				
Diploid strains				
(4) $R^+O^+A^+B^+/R^+O^+A^+B^+$				
(5) $R^+O^cA^+B^+/R^+O^+A^+B^+$				
(6) $R^+O^+A^-B^+/R^+O^+A^+B^+$				
(7) $R^-O^+A^+B^+/R^+O^+A^+B^+$				

6. Explain why, when *E. coli* is grown in the presence of *both* glucose and IPTG, β-galactosidase protein levels are lower than when grown only in the presence of IPTG.
7. Although *E. coli* promoters generally conform to a rather well-defined consensus sequence, no perfect match to this consensus has ever been observed in a naturally occurring promoter. Suggest an explanation.
8. Do you expect a regulatory mechanism like the prokaryotic attenuator to be found in eukaryotes? Why or why not?
9. The *lac* repressor has an ''on'' rate constant for the binding of the *lac* operator (when cloned into λ) of about $5 \times 10^{10} \text{M}^{-1}\text{s}^{-1}$. This value is much greater

than the calculated diffusion-controlled process, which is about $10^8 \text{M}^{-1}\text{s}^{-1}$ for a molecule the size of the lac repressor. Explain why this repressor binding works better than expected.

10. A mutation in the *trp* leader region is found to result in a reduction in the level of *trp* operon expression when the mutant is grown in rich medium. However, when the mutant is grown in a medium lacking glycine, a stimulation in the level of *trp* enzymes is observed. Explain these observations. What do you anticipate is the effect of growing the mutant in a medium lacking both glycine and tryptophan?
11. In *E. coli* no pools of free rRNAs or ribosomal proteins are floating around in the cell even when the bacteria are grown at different growth rates. Explain how *E. coli* coordinates the biosynthesis of the ribosome.
12. Draw a graph of rRNA gene transcription levels under the conditions shown in figure 30.16.
13. The rRNA genes of *E. coli* are present in multiple copies to facilitate production of many copies of rRNA during periods of rapid growth. If ribosomal proteins and RNAs need to be assembled in a 1:1 ratio, then why are single-copy genes adequate for ribosomal protein expression?
14. Describe the principal differences between patterns of control of gene expression used by bacterial host and bacteriophage systems.
15. How does the clustering of genes on the bacteriophage λ genome (fig. 30.20) facilitate its genetic regulation?
16. How is the synthesis of the CAP protein regulated? What is unusual about this regulation?
17. Gene regulatory proteins in bacteria were predicted (before their precise structure was known) to interact in the major groove of DNA by a two-site model of binding. Describe the data that showed this model to be correct.
18. Referring to figure 30.27c, draw a detailed structure of the interaction of residue Arg69 of the *trp* repressor with G9 of its recognition sequence. Explain why binding of repressor proteins to DNA does not disrupt the DNA double helix.

CHAPTER

31

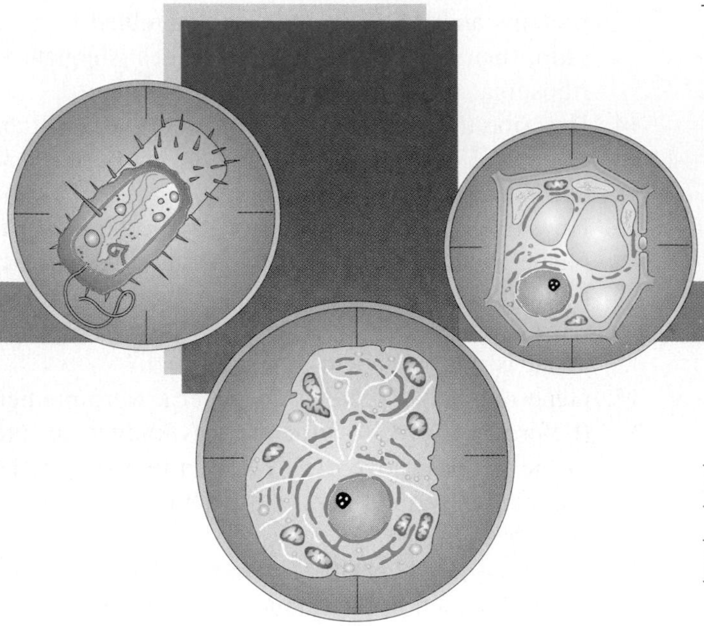

Regulation of Gene Expression in Eukaryotes

Chapter Outline

Gene expression in eukaryotes is usually regulated at the level of transcription initiation by a complex of RNA polymerase and an array of regulatory proteins that bind in and around the DNA promoter.

The structural and metabolic differences between prokaryotes and eukaryotes are reflected by many differences in the modes used to regulate gene expression (fig. 31.1). The physical separation of transcription and translation in eukaryotes permits more elaborate processing of messenger RNA while eliminating the possibility of regulatory processes that require strict coupling of the transcription and translation processes. The much greater size of the eukaryotic genome necessitates the existence of many more DNA-binding proteins for regulating transcription, and we will see that basically new types of regulatory systems have developed to fulfill this need. Finally, two kinds of regulatory phenomena occur, which are almost unique to multicellular eukaryotes. First, some regulatory processes facilitate communication (including metabolic regulation) between cells; second, other regulatory devices trigger changes during the course of development that lead to differentiated cells.

It was much easier to discuss regulation in prokaryotes, in which so much of what is known comes from work on *E. coli* and related bacteriophages. In eukaryotes the important information comes from many quarters, and the array of processes is wider by far. We, therefore, needed to be brief and had to leave out a great deal. In so doing we pondered what was most important for students to know. To compensate for the necessary omissions we included an extensive reference list. Consult this for topics we left out or discussed only briefly. The supplements on immunobiology and carcinogenesis are also of value on those subjects.

First we look at a unicellular organism, and then we examine mechanisms prevalent in multicellular eukaryotes.

Gene Regulation in Yeast: A Unicellular Eukaryote

The yeast, *Saccharomyces cerevisiae,* has a cell volume 10 times larger and a genome about 4 times larger than *E. coli.* Yeast cells are nearly spherical in shape, divide by budding, and are bounded by a cytoplasmic membrane that is surrounded by a thick polysaccharide cell wall (fig. 31.2). Structures visible in the electron microscope include a nucleus, mitochondria, and microsomes, as well as ribosomes, a Golgi apparatus with secretory vesicles, and several types of granular and vesicular inclusions. The number of mitochondria, microsomes, and ribosomes fluctuates widely with growth conditions, reflecting the presence of regulatory devices that control their numbers.

Galactose Metabolism Is Regulated by Specific Positive and Negative Control Factors in Yeast

Yeast mRNAs are usually translated so that only the 5′ proximal AUG is recognized as a translation start. This minimizes the value of polycistronic mRNAs. Even when functionally related genes are clustered, they usually give rise to separate transcripts. Three of the four genes (*GAL7, GAL10,* and *GAL1*) associated with galactose utilization are clustered on chromosome XI, whereas the fourth, for galactose transport is specified by a gene (*GAL2*) located on chromosome XII. (Note: In yeast the normal wild-type genotype is capitalized, and the mutant genotype is written in lower case.) Expression of the four structural genes is regulated by specific positive and negative controls. Transcription of the *GAL1, GAL7,* and *GAL10* genes is increased over 1,000-fold when galactose is present, suggesting that galactose is an inducer.

Each of the structural genes is associated with a distinct mRNA. The *GAL7* and *GAL10* genes are transcribed from the same DNA strand, whereas the *GAL1* gene, approximately 600 bp from *GAL10,* is transcribed from the complementary DNA strand (fig. 31.3).

Yeast is ideally suited for genetic analysis because it can be grown and examined in either the haploid or diploid states. Genetic investigations of the GAL system in diploids indicates two *trans*-acting gene products that regulate expression: *GAL4* and *GAL80.* Recall from observations on *E. coli* that *trans*-acting genetic loci usually signify that diffusible gene products are the active agent. Most *gal4* mutants are uninducible for the *GAL* structural genes in the haploid state but inducible in the diploid state when paired with a *GAL4* wild-type gene. This suggests that the active *GAL4* gene product is a positive control protein (an activator) like CAP in the *lac* operon. Rare $GAL4^{c}$ mutants result in constitutive expression of the *GAL* structural genes, implying that the structure of the $\mathrm{GAL4}^{c}$ protein in these unusual mutants is modified so that it is no longer repressible. Most *gal80* mutants also give rise to constitutive expression of the *GAL* genes, which suggests that GAL80 normally acts as a negative control protein (a repressor) of *GAL* gene expression. Rare $GAL80^{s}$ mutants are uninducible in the haploid or the diploid states where they are paired with wild-type *GAL80.* Based on these results it was proposed that the wild-type GAL80 protein binds the inducer galactose and that this binding converts the protein to an inactive form. The $\mathrm{GAL80}^{s}$ protein appears to have lost the site for galactose binding. As a result it functions as a repressor even in the presence of galactose. This is similar to the situation for i^{s} mutants in the *lac* operon of *E. coli* (see

Figure 31.1

Overview of some of the unique features associated with regulation of gene expression in higher organisms. Eukaryotes may be unicellular, as in the case of yeast, or multicellular, as in the case of vertebrates. Regulation of gene expression in all eukaryotes bears a close resemblance to regulation of gene expression in simple unicellular prokaryotes. Yet some striking differences exist. Thus, multicellular eukaryotes are composed of differentiated cells that can express only a limited number of the total genes in their chromatin. And most of the chromatin in any particular cell type is highly condensed (heterochromatic), but a small fraction of the chromatin is expanded, or swollen (euchromatic). It is the latter portion of the chromatin that has the potential to be expressed. The regions of the chromatin that are euchromatic are different in different cell types. To convert a region of the genome from the potentially active to the active state requires the binding of regulatory proteins to the promoter. The role of these proteins is to create an environment favorable for the binding of RNA polymerase. Here we come to one of the most striking differences between prokaryotes and eukaryotes. In prokaryotes the regulatory proteins usually bind in the immediate vicinity of the RNA polymerase. In eukaryotes the regulatory proteins may be bound to the DNA at several places, ranging from a point adjacent to the polymerase-binding site to points a thousand or more bases away. Although the explanation for how distantly bound regulatory proteins can influence gene expression is not totally clear, the consensus is that chromosome folding is the most likely way of bringing such proteins into the immediate proximity of the RNA polymerase-binding site. The function of all of the regulatory proteins binding to a single promoter of the DNA is to promote formation of the transcription initiation complex.

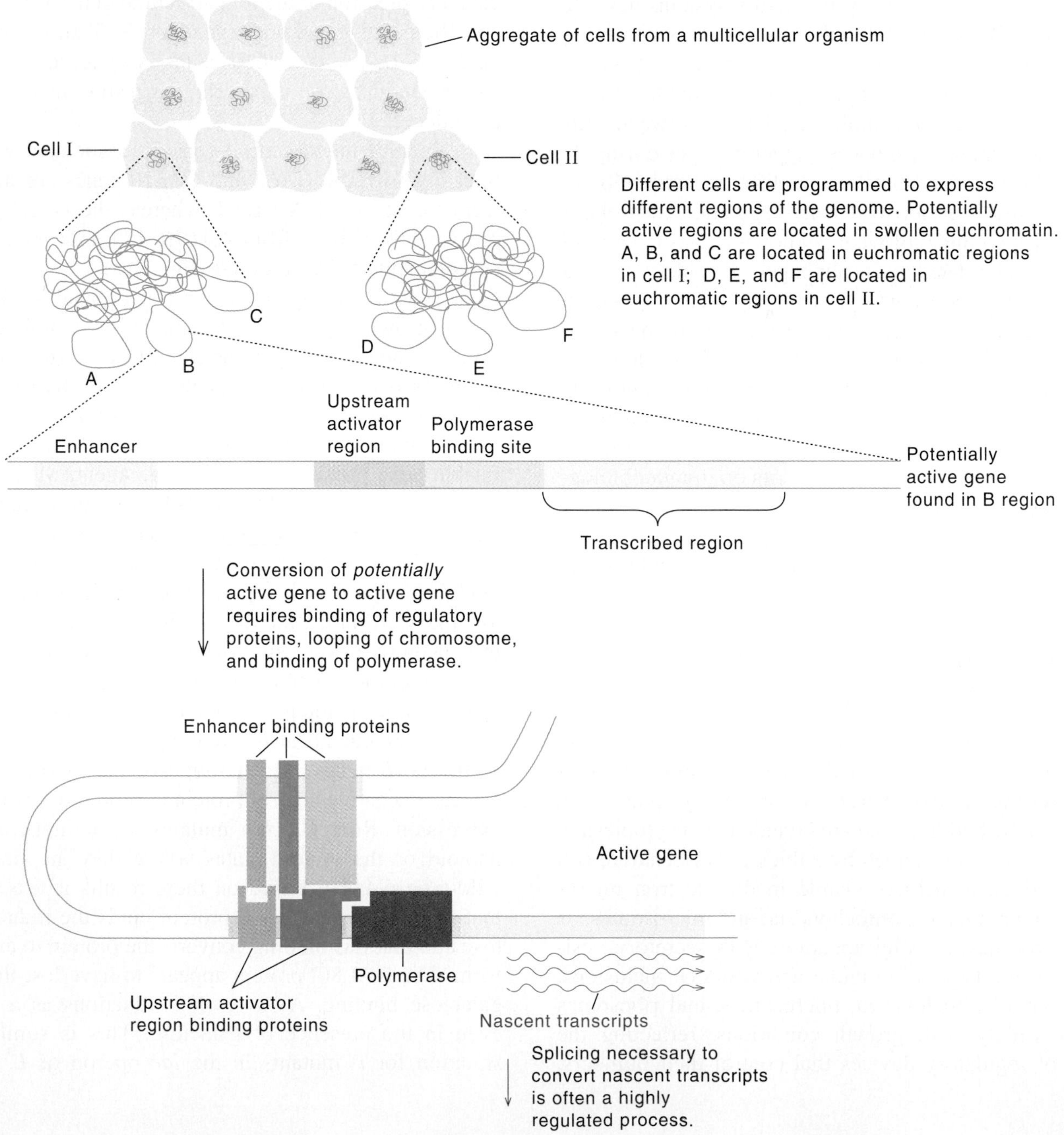

Figure 31.2

Diagram of a haploid yeast cell. A yeast cell contains many of the organelles characteristic of a typical eukaryotic cell. Duplication occurs by a budding process. The bud gradually grows until, just before pinching off, it contains a nuclear equivalent of chromosomes, as well as some mitochondria and other elements present in the cytoplasm. Mechanical strength of the yeast cell is guaranteed by a thick cell wall.

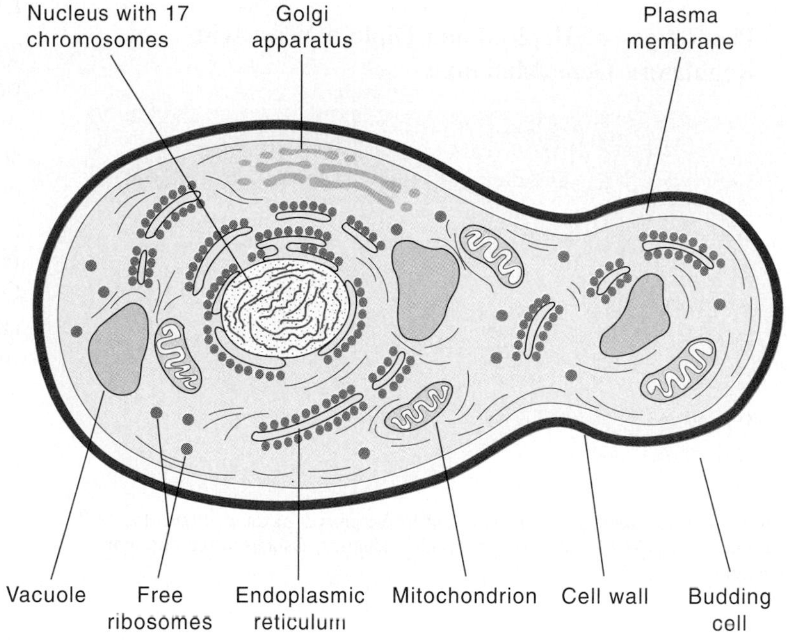

Figure 31.3

Model for the regulation of enzymes of the *GAL* system. (*a*) Three structural genes synthesize distinct mRNAs and enzymes (transferase, epimerase, and kinase). Arrows next to the genes indicate the direction of transcription. Synthesis requires the GAL4 protein. However, it is inactive in the absence of inducer because of complex formation with the GAL80 protein. The GAL80 protein does not prevent the GAL4 protein from binding to specific sites on the DNA, but it does prevent GAL4 protein from activating transcription once bound. (*b*) GAL4 protein becomes active when the GAL80 protein is removed by adding inducer, which binds to the GAL80 protein. Recent evidence suggests that GAL80 protein may not dissociate from GAL4 on induction.

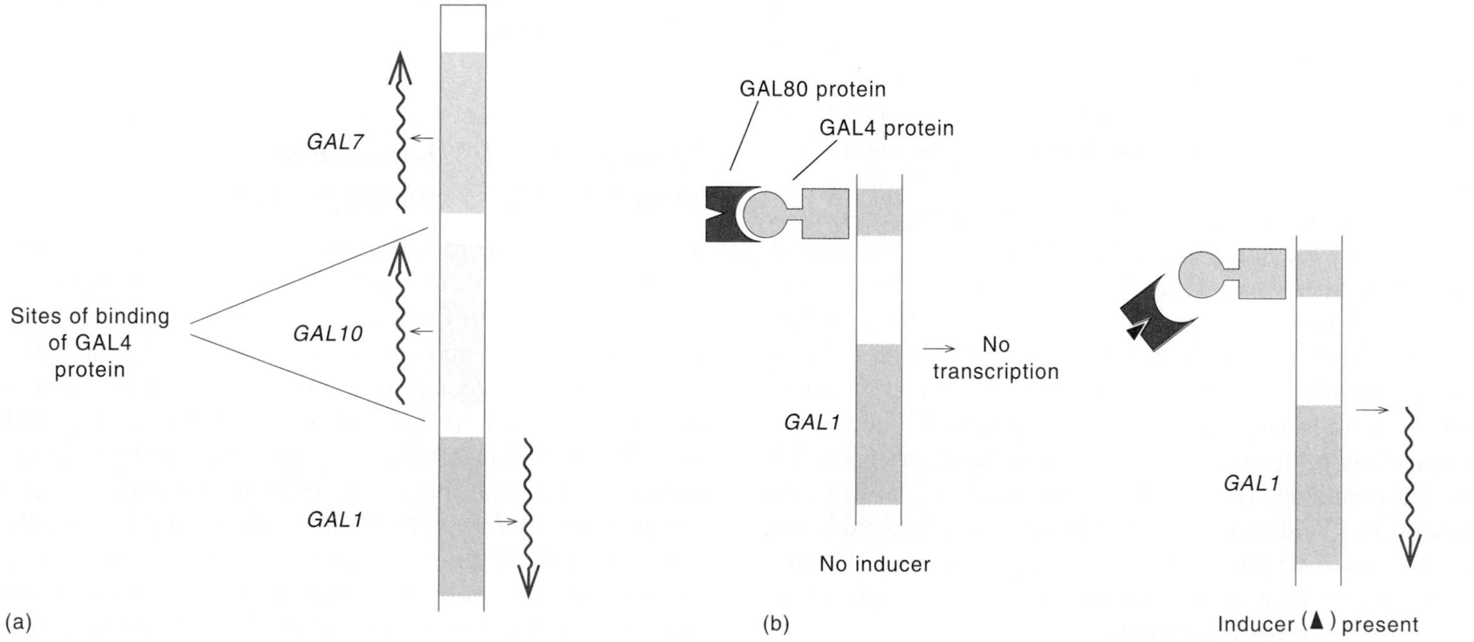

Table 31.1

Phenotypes of Haploid and Diploid Yeast with Regulatory Gene Mutations

Genotype	GAL10 (epimerase)[a] Noninduced	Induced	GAL1 (kinase) Noninduced	Induced
Wild type	−	+	−	+
gal4	−	−	−	−
gal4/GAL4	−	+	−	+
gal80	+	+	+	+
GAL4[c]	+	+	+	+
GAL80[s]	−	−	−	−
GAL80[s]*/GAL80*	−	−	−	−

[a] In the induced state, galactose is added to the growth medium. It should be noted that GAL10 and GAL1 respond in identical fashion to the different mutations.

chapter 30). The phenotypes for different regulatory gene mutants are summarized in table 31.1.

Based on these genetic studies a model was proposed for the mechanism of regulation of the *GAL* system (see fig. 31.3*b*). In this model the GAL4 protein binds to a site upstream of the gene(s) it regulates and promotes RNA polymerase II-dependent transcription of these genes. The GAL80 protein prevents GAL4 protein from functioning as an activator by binding to it. Induction by galactose results from the binding of the sugar to the GAL80 protein, which changes its structure so that the GAL4–GAL80 complex is altered. It is not clear if actual break-up of the complex is required to relieve the repression.

The proposed model postulates *cis*-acting sequences upstream of the target genes for GAL4 protein. The GAL4 protein binds to four related 17-bp sequences located between 150 and 200 bases upstream of the *GAL10* gene. The specific DNA sequences to which the GAL4 protein binds were determined by reacting the DNA with dimethylsulfate and then following the methylation pattern by nucleotide sequencing to determine which guanine bases are protected by the regulatory protein. The sequences that showed protection in a wild-type *GAL4* strain were not protected in the *gal4* mutant derivative. Further manipulations of the promoter region showed that only one of the 17-bp sequences is needed for normal regulation.

The GAL4 Protein Is Separated into Domains with Different Functions

The model for GAL4 protein specifies a multifunctional protein with binding sites for DNA and GAL80 protein as well as a site that somehow activates transcription. Deletion studies were performed to see if these different sites could be allocated to different parts of the 881-amino-acid GAL4 protein. A mutant protein containing only the first 98 N-terminal amino acids still binds DNA but cannot bind the GAL80 protein or activate transcription. Additional mutant studies show that amino acids 148–196 and 768–881 are required to activate transcription and that amino acids 851–881 are required for GAL80 protein building.

Further experiments indicate that different parts of the GAL4 protein and the lexA bacterial repressor protein (see chapter 26) are functionally interchangeable. In a ''domain swap'' experiment, the DNA-binding domain of the lexA protein was inserted in place of the GAL4 DNA-binding domain (fig. 31.4). As might be expected this hybrid protein could not activate the wild-type *GAL1* gene for transcription unless other changes were made in the DNA. If the DNA-binding site for the GAL4 protein was replaced by the DNA-binding site for the lexA protein, then the same hybrid protein was able to activate transcription of the *GAL1* gene.

The *GAL* system shows both similarities and differences to typical bacterial regulatory systems. In both systems DNA-binding regulatory proteins play a major role. However, in the yeast system the binding site for the regulatory proteins is often located some distance upstream from the RNA polymerase-binding site. In yeast such distant sites required for activation are referred to as upstream activator sequences (UASs).

Mating Type Is Determined by Transposable Elements in Yeast

Yeast has two haploid cell mating types: *MAT*α (or simply α) and *MATa* (or *a*); on contact, haploid cells of opposite mating types fuse to form a single diploid (a/α) cell. Diploid cells can grow and divide indefinitely as diploid cells, or they can sporulate, a process in which they undergo meiosis and give rise to two α and two *a* cells for each diploid cell. The haploid mating type is determined by specific sequences at the *MAT* locus (fig. 31.5). These sequences are found at the *HML*α or *HMRa* loci, where they are usually not expressed. When the sequences stored at the *HML*α locus are transposed to *MAT,* they express; similarly when sequences are transposed from *HMRa* to *MAT,* they express.

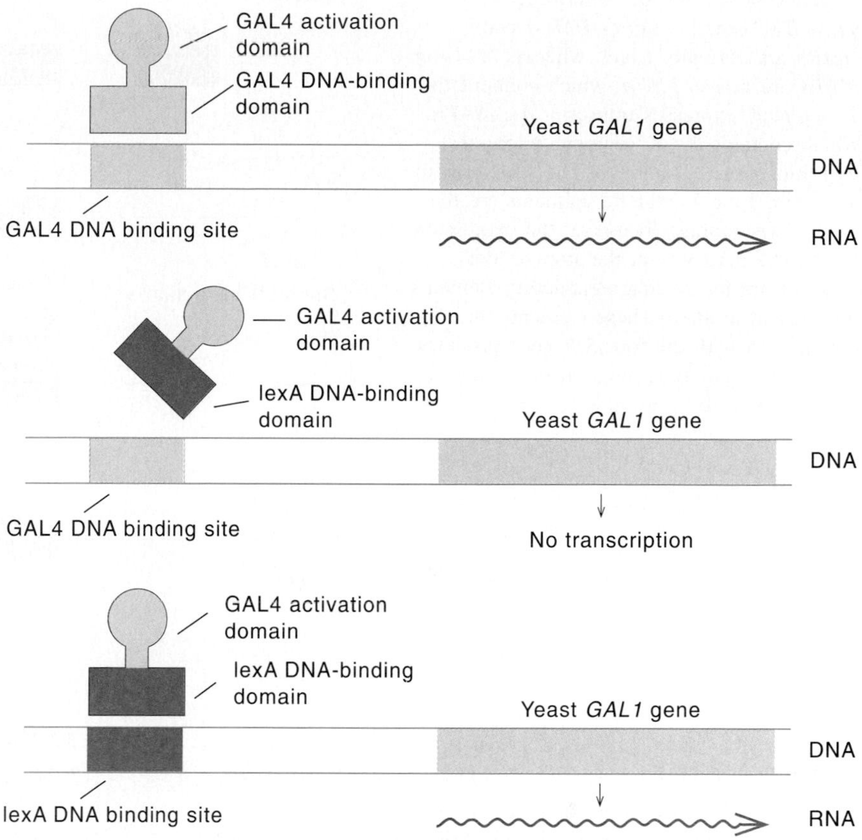

Figure 31.4

Ptashne's domain-swap experiment. The GAL4 protein normally binds at a site upstream from the *GAL1* gene. In the absence of GAL80 this activates transcription from the *GAL1* gene. If the DNA-binding domain of GAL4 is replaced by the DNA-binding domain of the lexA protein, transcription from the *GAL1* gene does not occur because the hybrid regulatory protein does not bind at the appropriate site. However, if the DNA-binding site is also changed to that of the DNA-binding site for the lexA protein, then the *GAL1* gene becomes active again.

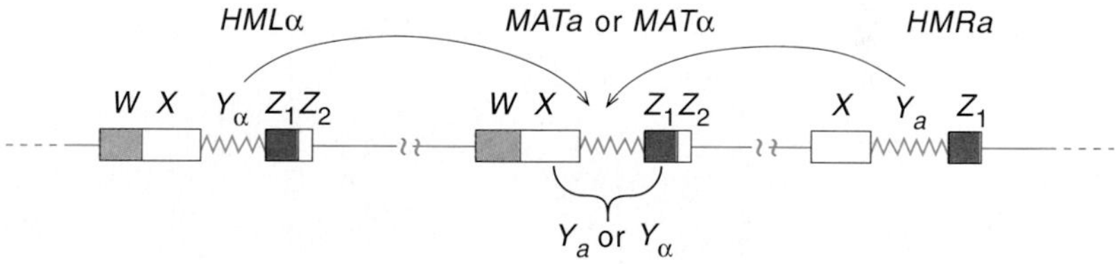

Figure 31.5

Yeast chromosome III, showing the *HML, MAT,* and *HMR* loci. *W* is a region common to *MAT* and *HML,* but not *HMR* (≈750 bp). *X* is a region found at *MAT, HMLα,* and *HMRa* (≈700 bp). Y_a is a specific substitution found at *MATa* and *HMRa* (≈600 bp). Y_α is a specific substitution found at *HMLα* and *MATα* (750 bp). Z_1 is a region found at *MAT, HMLα,* and *HMRa* (≈250 bp). Z_2 is a region found at *MAT* and *HMLα* (≈70 bp).

Why are the *α* and *a* sequences only expressed when they are present at the *MAT* locus? The explanation for this was suggested by the finding of four unlinked genes: *SIR1–SIR4.* If any of these genes are mutated, the *a* and *α* sequences at *HMLα* and *HMRa* express just as they do at *MAT*. Further experiments showed that *trans*-acting regulatory factors, encoded by the *SIR* genes act as repressors at "silencer" sites (*HMLE* and *HMRE*) adjacent to *HMRa* and *HMLα*. Remarkably, these silencer sites are located more than a kilobase away from the regions they control (fig. 31.6). The silencer sites are not found at or near the mating type locus (*MAT*), explaining why the *a* and *α* coding sequences are expressed when transposed to *MAT*.

An important question remaining is how does a single genetic locus determine the haploid yeast cell mating type? Haploid cells that carry *MATα* behave as *α* cells; similarly

Figure 31.6

Structure of mating loci determinants in yeast. The genetic regions *HMLα* and *HMRa* are normally silent, whereas *MATa* or *MATα* are active. *MATa,* which contains the Y_a segment, expresses transcript *a*1. *MATα*, which contains the Y_α segment, expresses two transcripts: α_1 and α_2. The structures in and around the Y_a and Y_α segments are the same at the storage locus and the expression locus. The inactivity at the storage loci results from far upstream *cis*-acting elements not present at *MAT*. These elements, in conjunction with the four *SIR* gene products, negatively regulate expression of mating type genes at the storage loci.

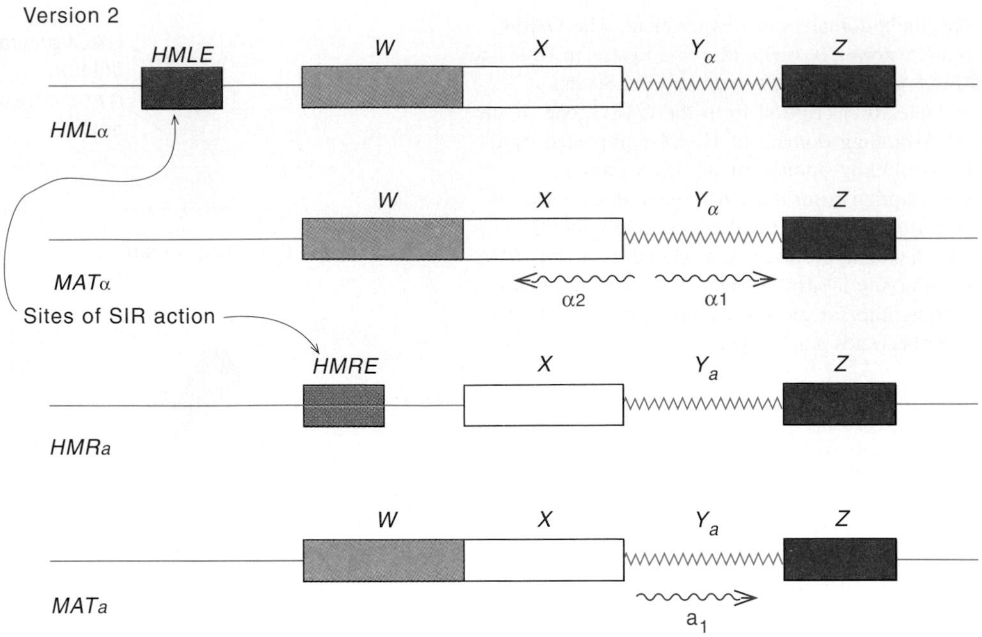

Figure 31.7

Regulation of *a*-specific genes (*a*sg) and α-specific genes (αsg) in α and *a* haploid cells. In an α cell, α2 inhibits *a*sg and α1 activates αsg. In an *a* haploid cell, the *a*1 transcript made at *MATa* does not appear to exert any regulatory function. (⊕ = positively regulated; ⊖ = negatively regulated.)

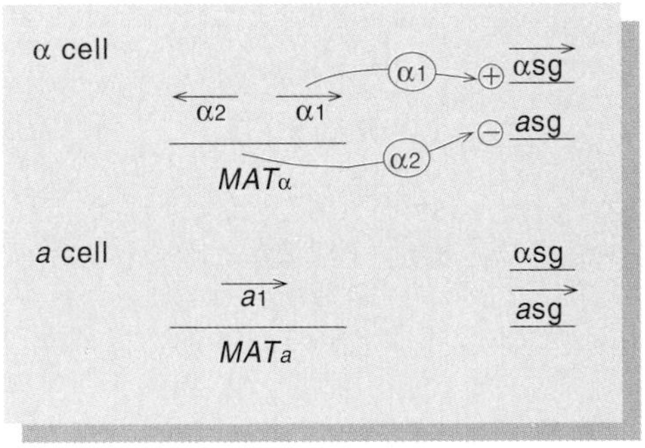

cells that carry *MATa* behave as *a* cells; and, finally, diploid cells that carry both *MATα* and *MATa* do not express any haploid-specific genes. To explain the key role of *MAT* it was proposed that the *MAT* locus encodes regulatory proteins that control unlinked genes that determine mating type. Subsequently it was shown that the *MATα* locus contains two genes: *MATα*1 and *MATα*2, which encode the α1 and α2 regulators, respectively. Ira Herskowitz and his coworkers found that α1 turns on expression of the α-specific genes, and α2 turns off expression of *a*-specific genes (fig. 31.7).

In addition to understanding the regulatory mechanisms that give the haploid cells their specific properties, considerable progress has been made in determining the mechanism that regulates meiosis in diploid cells. Diploid *a*/α cells undergo sporulation and meiosis when they are subjected to nutritional starvation. It is believed that the reactions leading to meiosis are triggered by a complex of the α2 and *a*1 proteins, which repress the haploid specific gene *RME*1 (regulator of meiosis), allowing the cells to initiate meiosis (fig. 31.8).

The control of mating type in yeast illustrates a specialized role of mobile or transposable genetic elements in which mobile elements hop from one site to another. Mobile genetic elements are commonly found in both prokaryotes and eukaryotes, but most of them show little site specificity. Indeed, their almost random insertion behavior has led to the proposal that their major function in most cases is to promote evolutionary change.

Mating type in yeast is much better understood than mating type of multicellular organisms such as vertebrates where sex type is determined by gene dosage. For example, in humans, males carry one X chromosome whereas females carry two X chromosomes. The mechanism used in yeast and other fungi is more responsive to factors in the external environment, which may be more suitable for a unicellular organism that is directly exposed to its surroundings.

Figure 31.8

Regulation of meiosis in yeast. Haploid cells are not able to initiate meiosis because they express *RME*1, which is a negative regulator of meiosis. Diploid cells are able to initiate meiosis because they make a complex of $\alpha 2$ and $a1$ that inhibits the synthesis of the *RME*1 product.

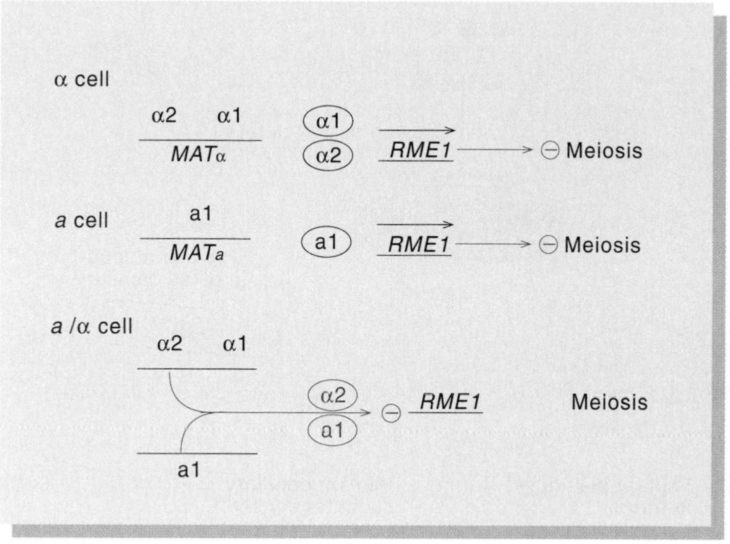

Figure 31.9

Nuclear transplantation technique. The nucleus of an unfertilized egg is mechanically removed with a micropipette (*left*), and a nucleus from a blastula cell is removed and microinjected into the enucleated egg.

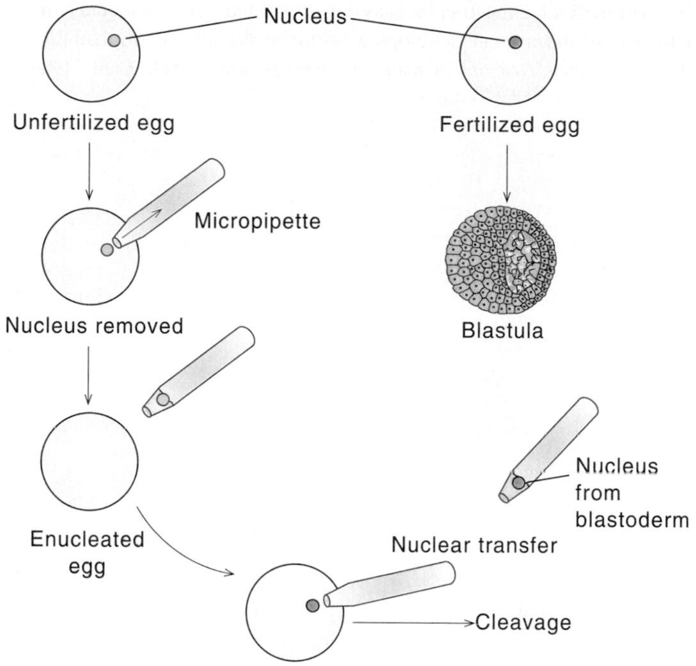

This concludes our discussion of regulatory phenomena in yeast. All of the regulatory processes in yeast are highly reversible, just as they are in *E. coli.* In multicellular eukaryotes additional regulatory mechanisms often show an irreversible character.

Gene Regulation in Multicellular Eukaryotes

We have seen that yeast differentiates into three cell types: The *a* and *α* haploid types and the *a*/*α* diploids. In multicellular eukaryotes we must deal with a much wider range of cell types and a much larger genome. First we consider some aspects of the changing chromosome architecture, which reflects changes in gene activity. Then we examine the underlying changes in the way that expression is controlled at the transcriptional, posttranscriptional, and translational levels. Finally we consider some developmental processes that are triggered by changes in gene expression.

Nuclear Differentiation Starts in Early Development

In the mature adult state, a multicellular eukaryote contains many cells, each with a potential for the constitutive or inducible expression of a unique subset of genes belonging to the total genome.

Whereas most differentiated cells in the adult contain the normal amount of genomic DNA, the expression of only a fraction of the genome in any particular cell type suggests that the average nucleus may have lost its totipotency. The first direct evidence that nuclei irreversibly differentiate during development came from R. Briggs and T. J. King in 1952. Working with frog eggs, they showed that the original nucleus from an unfertilized egg could be replaced by the diploid nucleus from a developing animal (fig. 31.9). If the inserted nucleus came from a frog very early in development (blastula), a mature frog developed. If, however, the nucleus came from a later stage of development, no growth or only limited growth ensued. These nuclear transplantation experiments demonstrated that during development the nucleus assumes a pattern of expression that is difficult or impossible to reverse even when the partially differentiated nucleus is returned to the environment of the egg cytoplasm.

In 1962, J. Gurdon took this type of experiment one step further by showing that adult frogs could be developed by injecting enucleated eggs with nuclei from the intestinal epithelium of tadpoles. The success frequency was much lower than when nuclei from earlier stages of development were used. Nevertheless, such results showed the influence of the cytoplasm on nuclear expression; they also demonstrated that in certain cases nuclei already tentatively com-

Figure 31.10

Manipulation of mouse teratoma cells. The cells from a teratocarcinoma may be dispersed and grown in tissue culture. These cells can be injected into an embryo (*a*), in which case the resulting animal is a chimera in which some cells come from the original parents and others arise from the cells injected into the blastocyst. (*b*) Alternatively, these cells may be implanted subcutaneously, in which case the animal develops a tumor at the site of implantation. (From G. Zubay, *Genetics,* Benjamin/Cummings, Menlo Park, Calif., 1987, p. 301.)

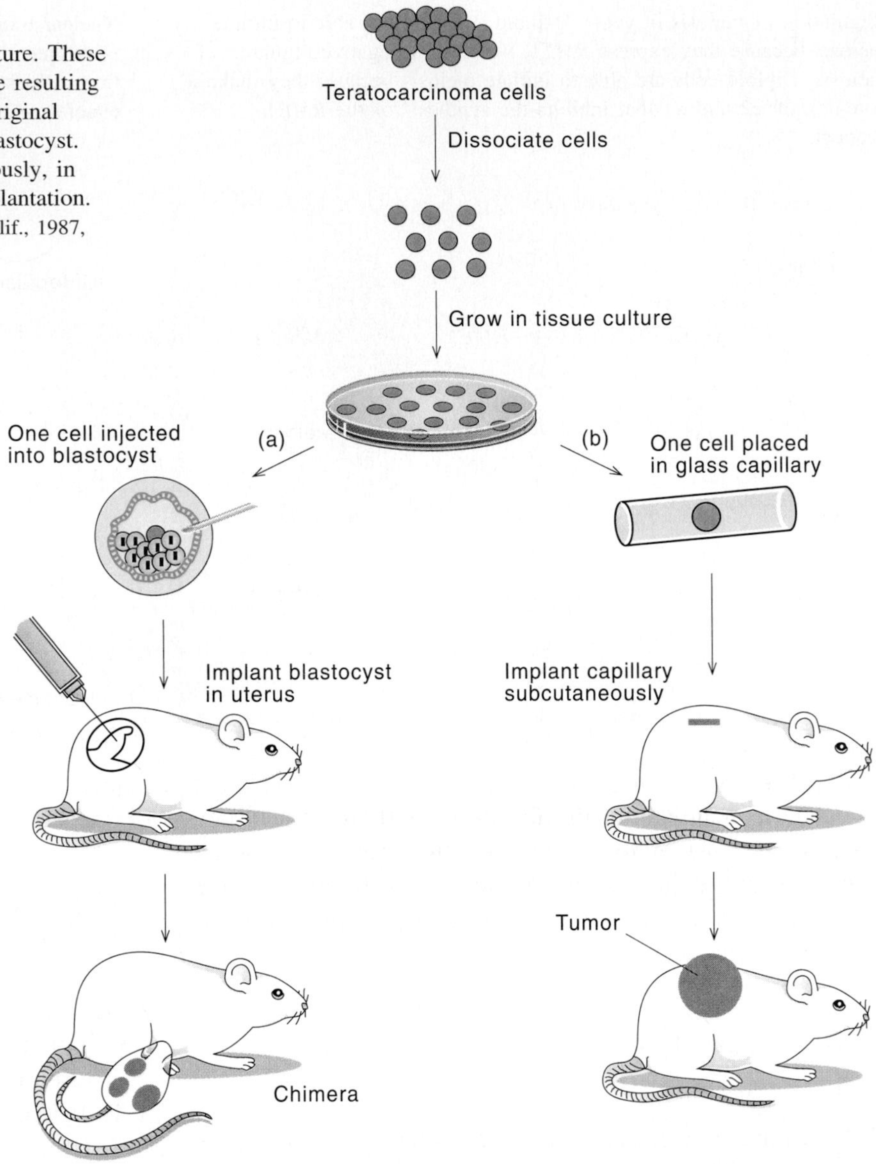

mitted to a specific pathway of development can be reprogrammed by placing them in a different environment. The most important lesson to be gained from such experiments is that the nucleus tends to assume a stable pattern of expression that is ultimately dictated by its surroundings.

The most convincing evidence for the importance of cytoplasm on nuclear development comes from studies on teratoma, a unique type of tumor found in many kinds of mammals. It is composed both of neoplastic cells, like other tumors, and of many kinds of differentiated cells. A typical teratoma may include nerve, muscle, blood, and skin cells, and other differentiated cells all mixed together with undifferentiated neoplastic stem cells. Only the stem cells are neoplastic, producing more stem cells or more differentiated cells, usually both. In many ways the stem cells behave like embryonic cells. They can be dissociated into single cells and cultured *in vitro* like bacterial cells. Such cells can be reintroduced into the animal by subcutaneous implantations, in which case they produce a tumor. Most remarkably, they can be implanted in an early embryo (blastocyst) to produce a hybrid chimera, in which, in favorable cases, all of the tissues possess some cells from the tumor parent (fig. 31.10). Even egg cells have been isolated that are derived from the tumor parent. These egg cells can be fertilized by normal sperm, resulting in progeny that could truly be said to have a tumor for a mother. This result shows that the neoplastic stem cells are capable of reverting to completely normal behavior when subjected to the embryonic environment.

These pioneering studies demonstrating nuclear differentiation and dedifferentiation are supported by a broad range of genetic and biochemical findings that indicate that chromosomal structural differences often can be correlated with changes in the potential for gene expression.

Chromosome Structure Varies with Gene Activity

In interphase cells, chromosomes are in a dispersed form called chromatin. Chromatin exists in a highly condensed form known as heterochromatin or a swollen form known as euchromatin. All nuclei contain both types of chromatin. Some chromosomes or parts of chromosomes are always heterochromatic (constitutive heterochromatin); others are heterochromatic only during certain times of the cell cycle or in certain cell types (facultative heterochromatin). Many studies indicate that euchromatin is relatively active in RNA synthesis, and heterochromatin is relatively inactive. Whereas the euchromatic state seems to be necessary for a high level of transcription, for most genes it is not sufficient. This has been demonstrated by autoradiography of cells grown briefly in the presence of ^{3}H-labeled RNA precursors. Those regions of the genome that are transcriptionally active incorporate the label; the extent of labeling is proportional to the amount of RNA synthetic activity. Some euchromatic regions appear active by this test, whereas others do not.

Giant Chromosomes Permit Direct Visualization of Active Genes

Polytene chromosomes are unusually large chromosomes found in the salivary glands of certain insects. Because of their size, they are convenient vehicles for studying the relationship between chromosome structure and activity. Polytene chromosomes are produced by the repeated replication of interphase chromosomes without separation, resulting in a large number of chromosomes that remain laterally aligned. For example, the DNA content in the giant salivary gland cells of *Drosophila melanogaster* may be as much as a thousand times that of other cells in the fruit fly. Microscopically, the stained chromosomes appear as cross-banded extended bodies (fig. 31.11). About 5,000 bands are in *Drosophila,* and it is tempting to associate single bands with single genes. However, the average DNA content found in each band is 30,000 bp, which is considerably more DNA than necessary to encode the average protein (about 1,000 bp).

Transcription in polytene chromosomes usually is associated with local swellings of the bands, called puffs (see fig. 31.11). Sometimes puffing results in a broadening and lengthening, and sometimes the extended DNA projects laterally into loops that combine to form a large ringlike structure. As a rule, a puff originates from a single band, but it can result from swelling of one or more bands. The extent of incorporation of labeled RNA precursors into a puff is approximately proportional to the size of the puff. A detailed investigation of the puffing patterns as a function of the physiological state of the organism shows that different bands become activated, swollen, and transcriptionally active in a sequentially related, tissue-specific fashion.

Figure 31.11

A segment of giant chromosome from the midge larva of *Rhyncosciara* at different stages during development. The arrows and connecting lines indicate comparable bands. Changes in the extent of swelling of different regions reflect the activity of those regions in transcription. (Source: Drawing based on the results of M. F. Breuer and C. Pavin, *Chromosoma* 7:257–280, 1946.)

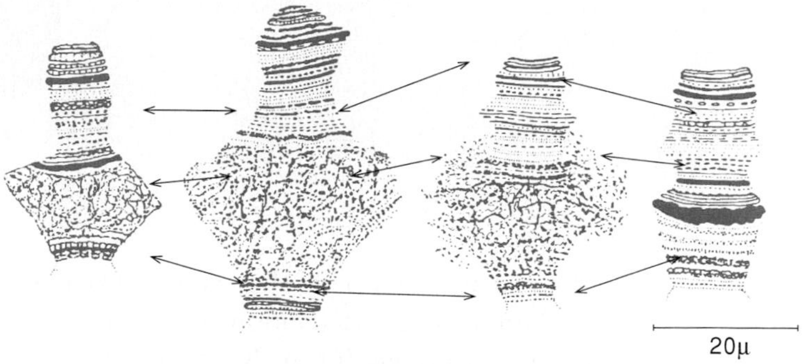

The correlation of puffs with genetic functions has focused on two types of gene products: Secretory proteins and proteins formed in response to heat shock. Certain puffs in *Drosophila* can be artificially induced by the insect hormone, ecdysone, which is instrumental in regulating development. Some puffs are induced directly by exposure to the hormone, whereas others are induced indirectly, depending for puffing on the gene products of the more directly induced puffs. This dependence is shown by adding drugs that inhibit protein synthesis, with the result that secondary puffs do not form.

In Some Cases Entire Chromosomes Are Heterochromatic

In some cases entire chromosomes can be heterochromatized and stay that way through successive cell divisions. Male and female mammals are distinguished by the fact that females carry two X chromosomes, and males carry only one. Invariably, one of the X chromosomes in the somatic tissue of females is inactivated and condensed into a hetero-

Figure 31.12

Diagram of nuclei obtained from cells in the mucous membrane of the human mouth. (*a*) Nucleus from a female, showing one Barr body (arrow). (*b*) Nucleus from a male, with no Barr body. (*c*) Nucleus from an XXX female, showing two Barr bodies. (*d*) Nucleus from an XXXX female, showing three Barr bodies.

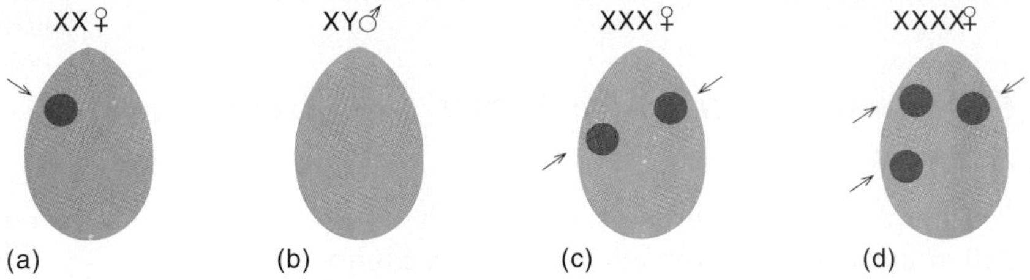

chromatic state known as a Barr body (fig. 31.12). The consequences of X-chromosome inactivation are readily observed in females heterozygous for an X-linked mutation.

For example, the enzyme, glucose-6-phosphate dehydrogenase (G-6-PD), is encoded by an X-linked gene. A female that is heterozygous for this gene may carry two alleles that produce electrophoretically distinct forms of the same enzyme (A and B). When isolated cells from a skin biopsy are cloned, each clone contains either the A or the B form of the G-6-PD enzyme, but never both. If every skin cell of a single organism were to be analyzed for the enzyme, large homogenous patches expressing one or the other of the G-6-PD alleles would be found. This pattern indicates that the decision as to which X chromosome of the female is inactivated is made at the multicellular stage in the embryo. Once the decision is made, that X chromosome remains inactive through successive cell divisions. There is an equal chance that the X chromosome from either parent is inactivated. In genetically abnormal cells that contain three or four X chromosomes, two or three Barr bodies, respectively, can be found (see fig. 31.12).

X-chromosome inactivation provides a simple means of maintaining equal amounts of active X-linked genes in both males and females. This form of so-called dosage compensation is not found in all species. For example, in *D. melanogaster* neither of the X chromosomes in the female cells is condensed into a Barr body. In this species, dosage compensation, which regulates the activity of specific alleles, operates by a different mechanism so that many alleles in the female X are about 50% as active as the corresponding alleles in the male. However, not all alleles in the *Drosophila* X chromosome are regulated in this way, so that for certain genes twice as much gene product is produced in females as in males.

Biochemical Differences between Active and Inactive Chromatin

Much of the preceding discussion on chromatin structure indicates that active chromatin exists in a more swollen state than inactive chromatin. Several biochemical changes accompany the transition from condensed to swollen chromatin. These changes include a redistribution of nucleosomes along the DNA duplex, chemical modification of histones, alteration in the pattern of nonhistone chromosomal protein binding, and chemical modification of the DNA. Currently, most of these changes are discussed in a general, descriptive manner because their causes and consequences are not known.

DNA in Active Chromatin Is More Susceptible to DNase Degradation. The greater accessibility of transcriptionally active chromatin has been elegantly demonstrated by Groudine and Weintraub for the hemoglobin and ovalbumin genes in the chicken. Chromatin was isolated from chicken erythrocytes, in which the hemoglobin genes were recently very active, and from the oviduct, in which the ovalbumin gene is very active. These two chromatins were degraded by DNaseI using an assay that measured gross DNA degradation as well as that of specific genes. In both cases the rate of DNA degradation from the active gene was much greater than that from average DNA. Thus in erythrocyte chromatin the globin gene DNA was rapidly degraded, whereas in oviduct chromatin the ovalbumin gene DNA was rapidly degraded. Extensive regions surrounding the active genes were also quite susceptible to DNaseI hydrolysis.

DNA Methylation Is Correlated with Inactive Chromatin. Methylation of cytosine in CpG sequences may play a regulatory role in the gene expression. In chapter 25 it was noted

Figure 31.13

Hemimethylated DNA. The unmethylated C residue in the indicated sequence is destined to be methylated shortly after DNA replication.

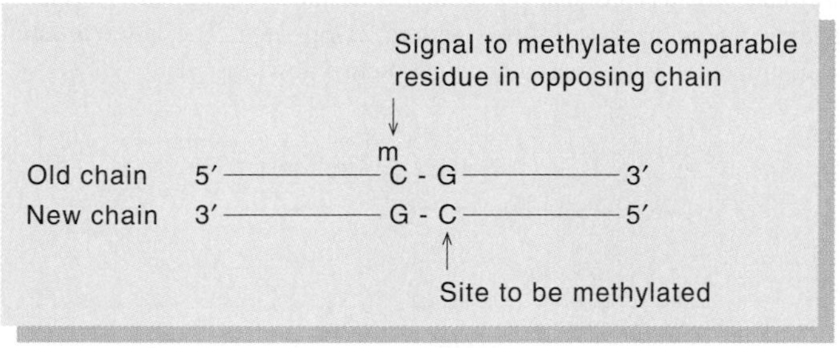

that 5-methylcytosine (5-MC) is the main modified base in vertebrate DNA. Most of the 5-MC occurs in the dinucleotide CpG. Indeed, in mammals and birds approximately 50%–70% of all such dinucleotide sequences are modified. In this connection it is noteworthy that CpG sequences occur much less frequently than would be expected statistically. The mechanism of methylation has been explored by DNA transfection of certain tissue culture grown cells. If the DNA used in transfection is methylated, the pattern of methylation is maintained through many cell duplications. Likewise, when unmethylated DNA is used in transfection, a non-methylated pattern is maintained. These results indicate that under many circumstances methylation is passively maintained by a signal that recognizes hemimethylated DNA. If the C residue in a CpG dinucleotide is methylated, immediately after semi-conservative DNA replication, only the parental DNA chain is methylated; the passive methylation system signals methylation of the corresponding C residue in the new DNA chain (fig. 31.13).

The extent of methylation of a gene is correlated with its ability to transcribe. Given that DNA methylation usually reduces transcription, two important, closely related questions remain unanswered: How is methylation regulated *in vivo?* How does methylation interfere with transcription? Since methylation is known not to interfere with the elongation phase of RNA synthesis, it seems likely that methylation blocks initiation. The binding of polymerase and other regulatory proteins at the initiation locus is sensitive to modification of these nucleotides. The precise inhibition mechanisms, however, await further elucidation.

Before turning to other questions we note that methylation or something like methylation could be at the root of X chromosome inactivation. In this connection it may be relevant that *Drosophila,* which does not show the methylation phenomenon, does not show X-chromosome inactivation either.

Enhancers Are Promoter Elements that Operate over Great Distances. In chapter 28 we saw that several proteins in addition to RNA polymerase II are required for the initiation of transcription at the adenovirus major late promoter; this promoter is prototypical of many eukaryotic promoters. In addition to signals for regulatory protein binding in the immediate vicinity of the polymerase-binding site, other signals function at more distant locations. In yeast, sequences of this type have been called UAS sequences because their action is usually confined to regions located a few hundred bases upstream from the promoters they influence. In vertebrates *cis*-active sequences operating over much greater distances have been discovered. These sequences have been named enhancers because they enhance expression of genes located in their vicinity. Enhancer signals are effective downstream as well as upstream of the promoters they influence. Moreover many enhancers are equally effective in either orientation on the DNA.

The first enhancer was discovered by George Khoury in the SV40 virus (see chapter 26). It should be recalled that the SV40 genome is a circular duplex containing about 5,300 bp. In SV40-infected cells, viral RNA synthesis is divided into early and late phases. Early transcription originates from a block of sites illustrated in figure 31.14. A TATA box, typical of eukaryotic gene promoters (see chapter 28) occurs about 27 bp upstream from the transcription start site. Immediately upstream of the TATA box the SV40 promoter contains a series of three tandemly repeated GC-rich segments that bind a transcription activator protein known as SP1. Further upstream from the transcription start site a tandemly repeated 72-bp sequence occurs (-116 to -188 and -189 to -261 from the 5′ end of the messenger). Removal of one of these sequences has no effect on transcription, but removal of both 72-bp sequences drastically lowers early transcription.

Surprisingly, the precise location or orientation of the 72-bp segment is not critical to the stimulating effect on transcription. Thus, the 72-bp segment remains effective after inversion, or after translocation further upstream or downstream from the transcription start site. Foreign genes inserted into DNA containing the SV40 enhancer are frequently stimulated in the same way as the SV40 early region, demonstrating the general stimulating effect of this enhancer.

Figure 31.14

Region in and around the early transcription start sites of the SV40 genome. The base pair number on the circular genome is indicated. There are several transcription start sites. One cluster of start sites (early), used initially after infection, is located about 27 bp downstream from the TATA box. The other cluster of start sites (late early) is used at later times. This cluster of start sites is located upstream of the TATA box. One of the products of early transcription is the protein known as large T antigen. This protein binds in and around the core *ori*. T antigen inhibits early transcription. The upstream positive control *cis* elements include the tandem 72-bp enhancer sequences and three tandem 21-bp sites. The latter function in conjunction with host-encoded protein known as SP1.

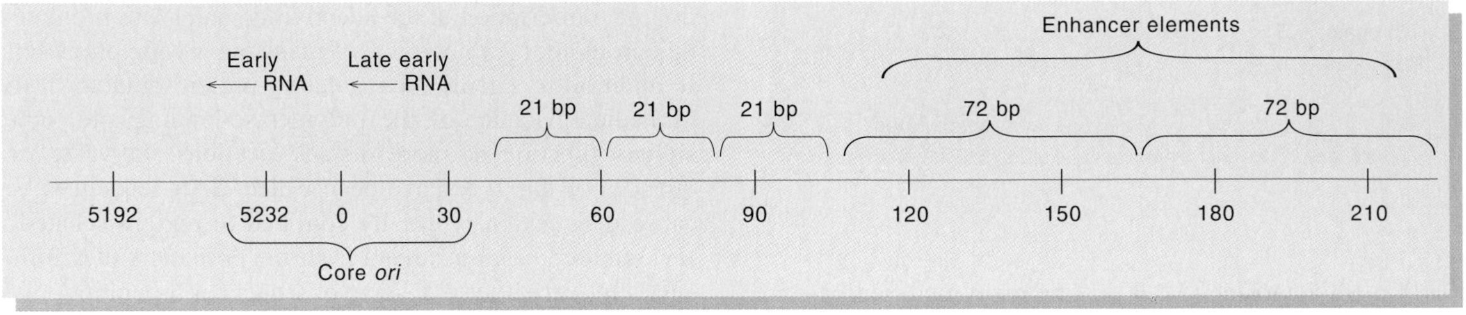

Figure 31.15

Enhancers are frequently composed of two or more components, or enhansons. Each enhanson contains binding sites for two or more proteins, which can interact with each other when bound to the enhanson if not before. An enhancer that contains two enhansons, each of which can bind two different proteins, can be arranged in 2^4, or 16, ways with respect to binding sites. From considerations such as this it is clear that a great deal of variety is possible in enhancer construction with a very limited number of DNA-binding sites for different regulatory proteins. The proteins that bind to the enhancer do not all make direct contact with the DNA. Often one finds pyramids of proteins that bind to each other in which only the base of the pyramid makes direct contact with the DNA.

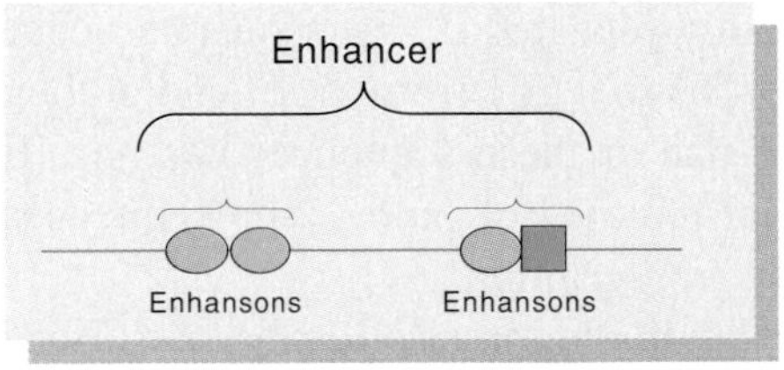

A more detailed analysis of the SV40 enhancer shows that it is composed of elements 15–20 bps in length that bind one or more protein factors specifically. These elements, sometimes called enhansons, are ineffective when separated from one another. It is believed that different enhancers are composed of different combinations of enhansons (fig. 31.15). The emerging picture of the enhancer is of a complex, multicomponent segment of DNA that can bind different combinations of protein activators. This appears to be a major strategy used in multicellular eukaryotes to produce a greater variety of responses. It has the advantage that, given a finite set of regulatory proteins, a much greater variety of combinations can be produced. The use of complexes containing combinations of regulatory proteins has implications for the structures of regulatory proteins that are discussed later on.

It is not certain how proteins that bind to enhancers influence transcription over long distances. However, several proposals have been made: (1) The enhancer element may function as an attachment point to a structural component of the nucleus to stimulate transcription; (2) it may serve as an initial binding site for some factor required for transcription that must subsequently move along the duplex to the initiation point for transcription; or (3) folding or looping of the chromosome may bring the *cis* bound enhancer proteins into close contact with the gene it stimulates (fig. 31.16). Currently, the third possibility is strongly favored, but this does not exclude the other two mechanisms.

DNA-Binding Proteins That Regulate Transcription in Eukaryotes Are Often Asymmetrical

In the previous chapter we discussed DNA-binding proteins that regulate transcription in prokaryotes. The principles that govern recognition between proteins in eukaryotes show some similarities and some differences. In both cases specific recognition is dominated by interactions that take place in the major groove of the DNA. The specific interactions usually involve H bond formation

Figure 31.16

Possible mechanisms for enhancer action. (1) The enhancer draws the associated gene to the nuclear matrix, where it is more accessible to the transcription apparatus. (2) The enhancer is the initial binding site for an element that subsequently moves. (3) The enhancer folds or loops, depending on its polarity, to bind with other promoter elements.

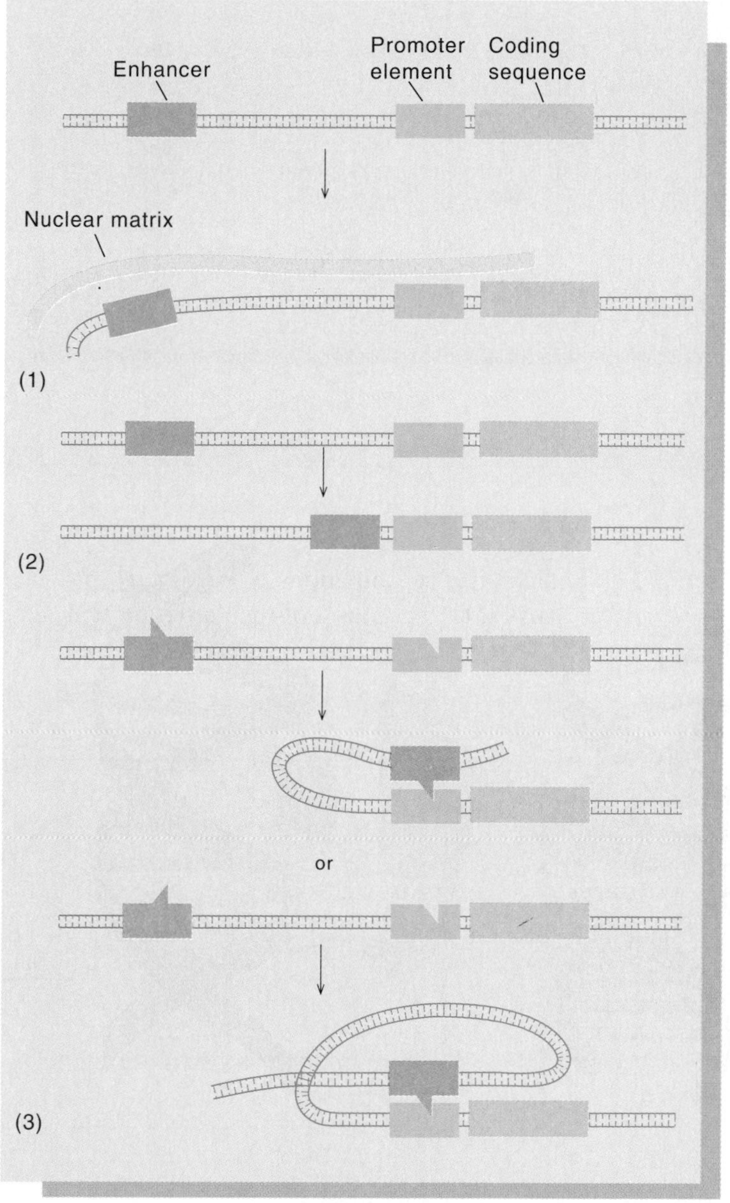

between the base pairs in the DNA and the side chains in the proteins. In both cases the α helix is the most common element used for DNA recognition. The most striking difference between DNA-binding proteins in prokaryotes and eukaryotes has to do with the symmetry of the interaction. In prokaryotes DNA-binding proteins almost always are composed of an equal number of identical subunits that interact in a symmetrical fashion with the DNA. Half-sites in the protein bind to half-sites in the DNA, which are usually spaced one helix turn apart. In eukaryotes most of the cases that have been examined so far involve proteins that interact in an asymmetrical fashion with the DNA. In some cases we find eukaryotic proteins that contain multiple sites for interaction within one polypeptide chain, and in other cases we find two structurally distinct proteins interacting to form heterodimers that interact with the DNA.

The Homeodomain

The homeodomain has a motif that has been recognized in a large family of eukaryotic regulatory proteins. The name "homeodomain" derives from the fact that many of the mutations that affect the body plan in a developing *Drosophila* embryo were referred to as homeotic gene mutations (this is taken up later). Walter Gehring found that many of these homeotic genes and other genes that regulate development in *Drosphila* encode regulatory proteins that possess a common 180-base segment in the 3′ exon. The base homology ranges between 60% and 80%, depending on the specific genes being compared, and the amino acid homology is even higher (up to 87% fig. 31.17). The high basicity of the amino acid sequence in this region—called the homeobox—and other structural features suggest that this region encodes a DNA-binding protein that regulates gene expression by binding to specific sites on the DNA. Indeed, sufficient amino acid homology exists between the homeobox and the helix-turn-helix motif seen in bacterial regulatory proteins to suggest that the gross structures of these very distantly related regulatory proteins are quite similar.

One of the most exciting features relating to the homeobox is that very similar sequences have been identified in many other animals, including frogs, mice, and even humans. This finding suggests that homeobox proteins occur in a wide range of organisms, possibly playing similar roles in regulating developmental processes.

Whereas amino acid sequence comparisons suggested that the homeodomain would contain a helix-turn-helix motif, the 60-residue homeodomain forms a stable structure that can bind to DNA as a monomer. The recognition helix in the homeodomain is longer and makes more contacts with the DNA core than the recognition helices from bacterial regulatory proteins (fig. 31.18).

Although an isolated homeodomain can fold correctly and bind DNA with a specificity similar to that of the intact proteins, it is believed that the precise DNA-binding specificity is modulated by other regions of the protein. Protein–protein interactions may also have a role in modulating many homeodomain–DNA interactions. For example, the

Figure 31.17

The homeobox sequence found in three *Drosophila* regulatory proteins. The sequence of amino acids along the main, continuous line is that found in the *Antp* homeobox. At points where the sequence of ftz proteins is different, the differences are shown above the corresponding Antp proteins, and at points where the sequence of ubx proteins is different, they are shown below the corresponding Antp proteins. Certain proteins isolated from the human and mouse embryos have segments with similar sequences. Note the high basicity of the sequence, which should favor electrostatic binding to nucleic acid.

				5					10					15					20
			Thr																
Arg	Lys	Arg	Gly	Arg	Gln	Thr	Tyr	Thr	Arg	Tyr	Gln	Thr	Leu	Glu	Leu	Glu	Lys	Glu	Phe
Ser			Gly																
				25					30					35					40
					Ile							Asp			Asn			Ser	
His	Phe	Asn	Arg	Tyr	Leu	Thr	Arg	Arg	Arg	Arg	Ile	Glu	Ile	Ala	His	Ala	Leu	Cys	Leu
	Thr		His										Met		Tyr				
				45					50					55					60
Ser															Ser			Asp	Arg
Thr	Glu	Arg	Gln	Ile	Lys	Ile	Trp	Phe	Gln	Asn	Arg	Arg	Met	Lys	Trp	Lys	Lys	Glu	Asn
															Leu				Ile

Figure 31.18

(*a*) Complex formed between the *Drosophila* homeobox protein engrailed and DNA seen from two angles. In addition to the main contacts made by the recognition helix in the major groove, additional contacts are made in the minor groove, which can be seen in (*b*). (From C. R. Kissinger, B. Liu, E. Martin-Blanco, T. B. Kornberg, and C. O. Pabo, Crystal structure of an engrailed homeodomain-DNA complex at 2.8 Å resolution: A framework for understanding homeodomain-DNA interactions, *Cell* 63:579–590, November 2, 1990. Copyright © Cell Press. Reprinted by permission.)

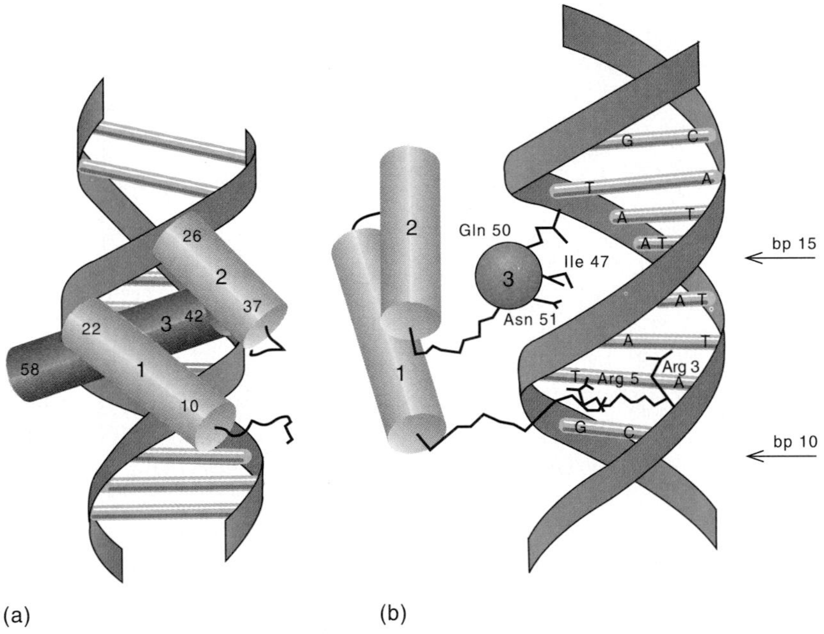

yeast $\alpha 2$ protein forms homodimers but also forms complexes with a related homeodomain protein $a1$ (see fig. 31.8). Each of these complexes has different site preferences, and it is possible that these accessory proteins in addition to adding new DNA contacts actually alter the way that $\alpha 2$ interacts with DNA.

We tentatively conclude that the asymmetrical interaction of the homeodomain protein with DNA gives it a versatility not displayed by the symmetrical helix-turn-helix homeodimers found in bacteria.

Zinc Finger

Another DNA-binding motif very common in eukaryotic regulatory proteins is the zinc finger. This motif was first identified as the DNA-binding structure in the RNA poly-

Figure 31.19

Schematic representation of the C_2-H_2 zinc finger found in Xfin from *Xenopus laevis*. (Adapted from M. S. Lee et al., *Science* 245:645, 1989.) The recognition helix is stabilized by a complex involving zinc. Cysteine sulfurs are in yellow, and histidine nitrogens are blue.

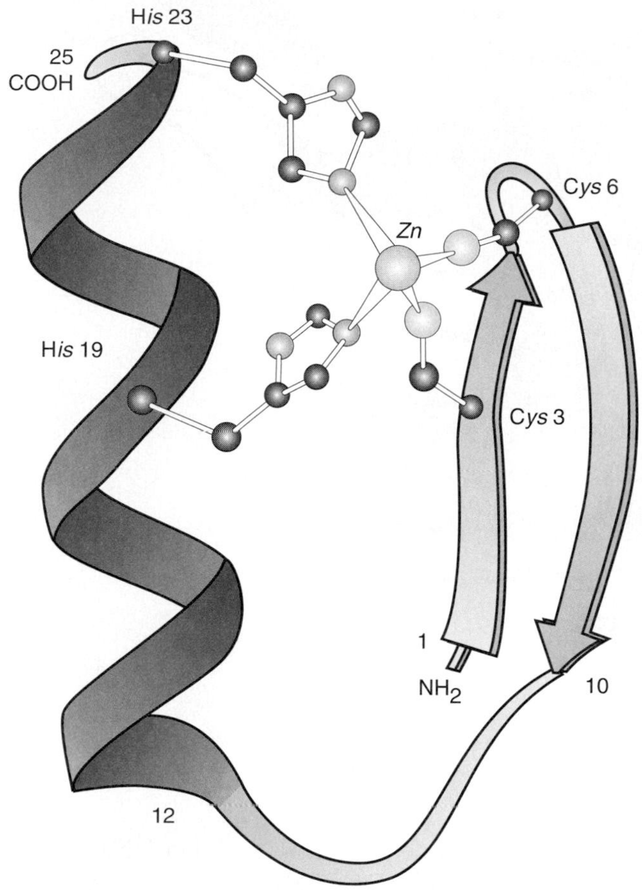

merase III transcription factor TFIIIA which binds to the internal control region of the 5S rRNA gene (discussed in chapter 28). Zinc fingers that resemble TFIIIA are present in the mammalian transcription factor SP1 and in a variety of other regulatory proteins found in eukaryotes. This type of zinc finger motif consists of about 30 amino acids with two cysteine and two histidine residues that stabilize the domain by tetrahedrally coordinating a Zn^{2+} ion (fig. 31–19). A region of about 12 amino acids between the cysteine–histidine pairs is characterized by scattered basic residues and several conserved hydrophobic residues. Proteins in this family usually contain tandem repeats of the 30-residue zinc finger. The crystal structure of a zinc finger–DNA complex containing three fingers from zif268 and a consensus zif-binding site has been reported. The crystal structure shows that the zinc fingers bind in the major groove and wrap part way around the double helix (fig. 31.20). Each finger has a similar way of docking against the DNA and makes base contacts with a 3-bp subsite arranged in a way that reflects the helical pitch of the DNA and the 3-bp periodicity of the subsites.

Leucine Zipper

A third type of DNA-binding domain was first described for the mammalian enhancer-binding protein C-EBP. Proteins of this class show a primary sequence similarity consisting of a highly conserved stretch of about 30 amino acids with a substantial net basic charge immediately followed by a region containing four leucine residues positioned at intervals of seven amino acids. The latter segment, named the leucine zipper by Steve McKnight, is required for dimerization and for DNA binding. It is believed that dimerization of proteins in this group is stabilized by hydrophobic interactions between closely apposed α-helical leucine repeat regions of the two proteins. The two helical cylinders are believed to be oriented in parallel in a coiled-coil fashion (fig. 31.21). The main function of the leucine motifs is to bring two proteins together so they can form homodimers or heterodimers that bind to dissimilar half-sites on the DNA.

Helix-Loop-Helix

The helix-loop-helix motif appears to be another way of creating heterodimers that can bind to asymmetric sites on the DNA. Like the leucine zipper proteins, the helix-loop-helix proteins have a basic region that contacts the DNA and a neighboring region that mediates dimer formation. Based on sequence patterns, it has been proposed that this dimerization region forms an α helix, a loop, and a second α helix. Like the leucine zipper protein, the activity of the helix-loop-helix proteins is modulated by heterodimer formation. For example, the MyoD protein, which appears to be the primary signal for differentiation of muscle cells, binds DNA most tightly when it forms a heterodimer with the ubiquitously expressed E2A protein.

This brief description of three new types of regulatory proteins found primarily if not exclusively in eukaryotes by no means includes all of the known types of DNA-binding proteins. Furthermore, the progress in discovering new regulatory proteins is so rapid that anything we say here is bound to need supplementation if the reader wants to be up to date on this subject.

Transcription Activation Domains of Transcription Factors

Thus far we have focused on the DNA-binding domains of regulatory proteins. Many regulatory proteins have additional domains that are involved in transcription activation.

Figure 31.20

The complex formed between the zif268 zinc finger protein and DNA. The three fingers fit into the major groove and wrap partway around the duplex. The zinc is not shown in this figure. (From N. P. Pavletich and C. O. Pabo, Zinc finger-DNA recognition: Crystal structure of a Zif268 DNA complex at 2.1 Å, *Science* 252:809–817, 10 May 1991. Copyright 1991 by the AAAS. Reprinted by permission.)

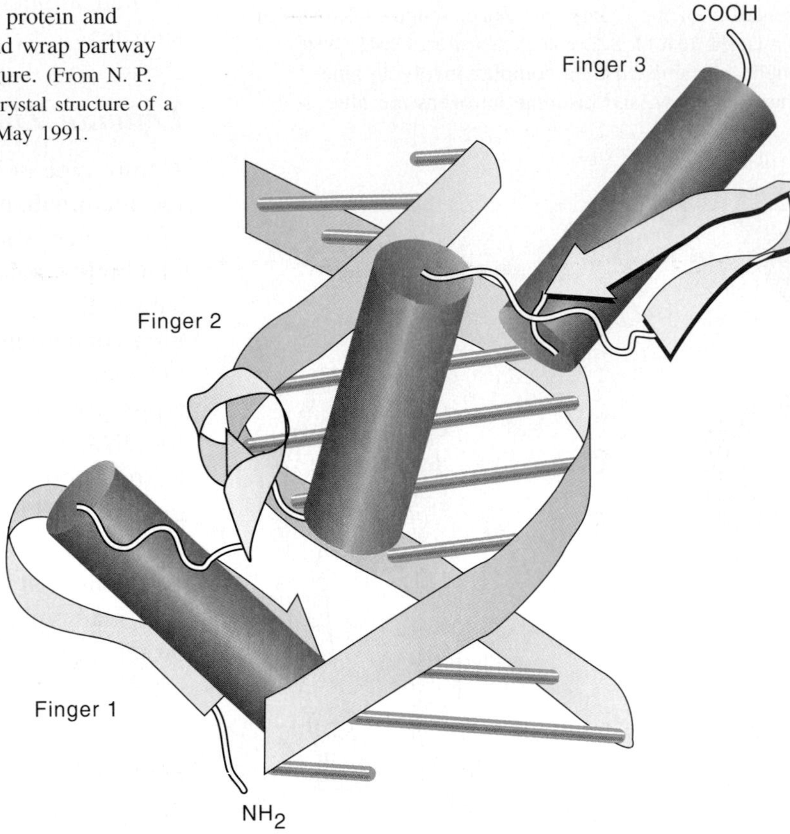

Figure 31.21

Diagrammatic sketch of a leucine zipper dimer. The protein monomers are held together by interaction between leucine side chains (green knobs in the upper part of the structure). The part of the protein monomers that interacts with the major groove of the DNA is shown (in red). (Adapted from C. R. Vinson et al., *Science* 246:911, 1989.)

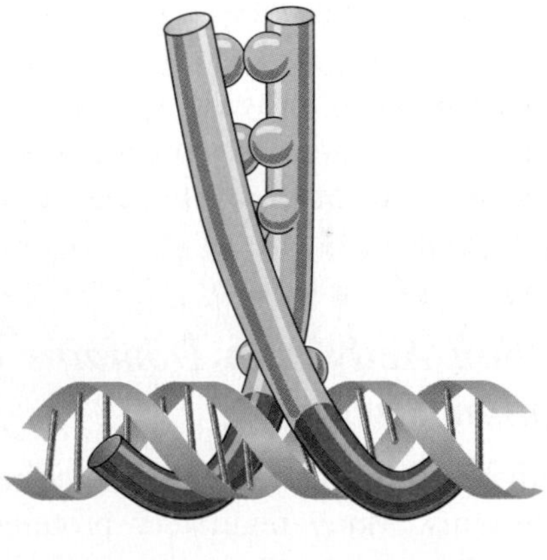

The amino acid sequences of mammalian DNA regulatory proteins suggest the existence of at least three different types of activation domains: Acidic, glutamine-rich, and proline-rich.

Transcriptional activation functions of DNA-binding factors depend on regions of from 30 to 100 amino acids that are separate from the DNA-binding domains. Factors often have more than one activation domain, and several apparently unrelated structural motifs have been identified that confer these functions. The first defined activation regions in eukaryotic transcription factors were identified by studies of the yeast factor, GAL4 (discussed previously). The activation domains of these factors consist of relatively short stretches of amino acids with significant negative charge that can form amphipathic α-helical structures.

Deletion analysis of the transcription factor of SP1 has revealed four separate regions that contribute to transcriptional activation; all lie outside the zinc-finger-binding domain. The two strongest activation domains contain about 25% glutamine and very few charged amino acid residues. Several other transcription factors show glutamine-rich regions in the domain required for transcription activation.

These are but two of the types of domains found to be involved in transcription activation. Undoubtedly, this list will grow in the near future, as will our understanding about how these transcription activation domains work. Thus far they are likely to represent regions that function by contacting other regulatory proteins, transcription factors, and the RNA polymerase itself.

Alternative Modes of mRNA Splicing Present a Potent Mechanism for Posttranscriptional Regulation

The basic mechanisms of RNA splicing were discussed in chapter 28. RNA splicing was first discovered for transcription of adenovirus where different reading frames are connected to the same 5′ end. Thus, we have known about the phenomenon of alternative splicing as long as we have known about splicing itself. Alternative splicing occurs for eukaryotic viruses such as SV40 and polyoma, as well, and it is also a common phenomenon in eukaryotic genes that contain multiple exons in their nascent transcripts. It is clearly a regulatory phenomenon in viruses since we see a shift in the types of splicing as virus infection progresses. For eukaryotic genes it is also a regulatory phenomenon since different modes of splicing are found in different tissues of the same multicellular organism. The type of splicing found usually involves segments from the same transcript, but on occasion transplicing occurs where two independent transcripts participate in a common splicing operation. In the most common splicing situation a promoter is present at one end of the transcript and the combination of exons that is used in the mature mRNA varies according to the pattern of splicing. In some cases one or more exons are excluded from the message by selective splicing. In other cases part of an intron is fused to one of the exons to make the final message. Splicing can also be influenced by the choice of the promoter or the choice of the polyadenylation site. In the latter two situations parts of the upstream or downstream regions of the gene are excluded from the nascent transcript, and so the splicing of the transcript is limited to the RNA remaining.

The possibilities for alternative splicing are enormous. One particularly elaborate example is seen in the gene for tropomyosin in vertebrates. Recall that tropomyosin is a key component of vertebrate striated muscle (see fig. 5.18). The mRNA for tropomyosin found in striated muscle undergoes nine splices in the process of maturation (fig. 31.22). Variants of tropomyosin resulting from alternative splicing are found in other tissues of the same organism. It seems likely that the types of tropomyosins found in different tissues are best suited to the needs of the tissues.

P. Bingham and his coworkers have uncovered another regulatory role for splicing in *Drosophila*. In three different genes they observed that the final splices are subject to regulation. These final splicing operations appear to be regulated by the protein encoded by the mature mRNA. If that protein is present in sufficient concentrations, then it inhibits the final splices, thereby preventing the mRNA from becoming functional and turning out unneeded protein. It is not clear how common this type of feedback inhibition is in *Drosophila* or other species.

Gene Expression Is Also Regulated at the Levels of Translation and Polypeptide Processing

Following messenger formation, the amount and types of proteins can be modulated in additional ways. The initial polypeptide can be processed in various ways so that different polypeptides or proteins are expressed in different tissues. Such a situation exists for processing the precursor polypeptide preproopiomelanocortin (see fig. 24.7). This polypeptide is processed in different ways in the anterior and intermediate lobes of the pituitary gland to give rise to different hormones in these two tissues.

Because of the longer lifetime of eukaryotic messengers, it seems likely that translation level controls should play a greater role in regulation of eukaryotic gene expression. Despite this belief, few mechanisms have been elucidated.

Inactivation of eukaryotic translation factors by covalent modification is one of the few mechanisms known to regulate the rate of translation. Specific protein kinases have been identified that phosphorylate and inactive both eIF-1 and EF-2. The significance of the phosphorylation of EF-2 as a regulatory mechanism of the elongation rate is still not clear, but the phosphorylation of eIF-2 appears to be a general mechanism for controlling translation initiation in many cells.

The regulation of translation through the phosphorylation of eIF-2 is best understood as it operates in the rabbit reticulocyte. Two protein kinases specific for the a subunit of eIF-2 have been purified from reticulocytes. One of these kinases, termed the heme-regulated inhibitor repressor (HRI), serves to coordinate the rate of hemoglobin synthesis (more than 90% of the total protein synthesized in the reticulocyte is hemoglobin) with the availability of hemin (the

Figure 31.22

Alternative modes of splicing of the tropomyosin gene transcript in different tissues. (Source: Adapted from R. E. Breitbart, A. Andreadis, and B. Nadal-Ginard, Alternative splicing: A ubiquitous mechanism for the generation of multiple protein isoforms from single genes, *Ann. Rev. Biochem.* 56:467–495, 1987.)

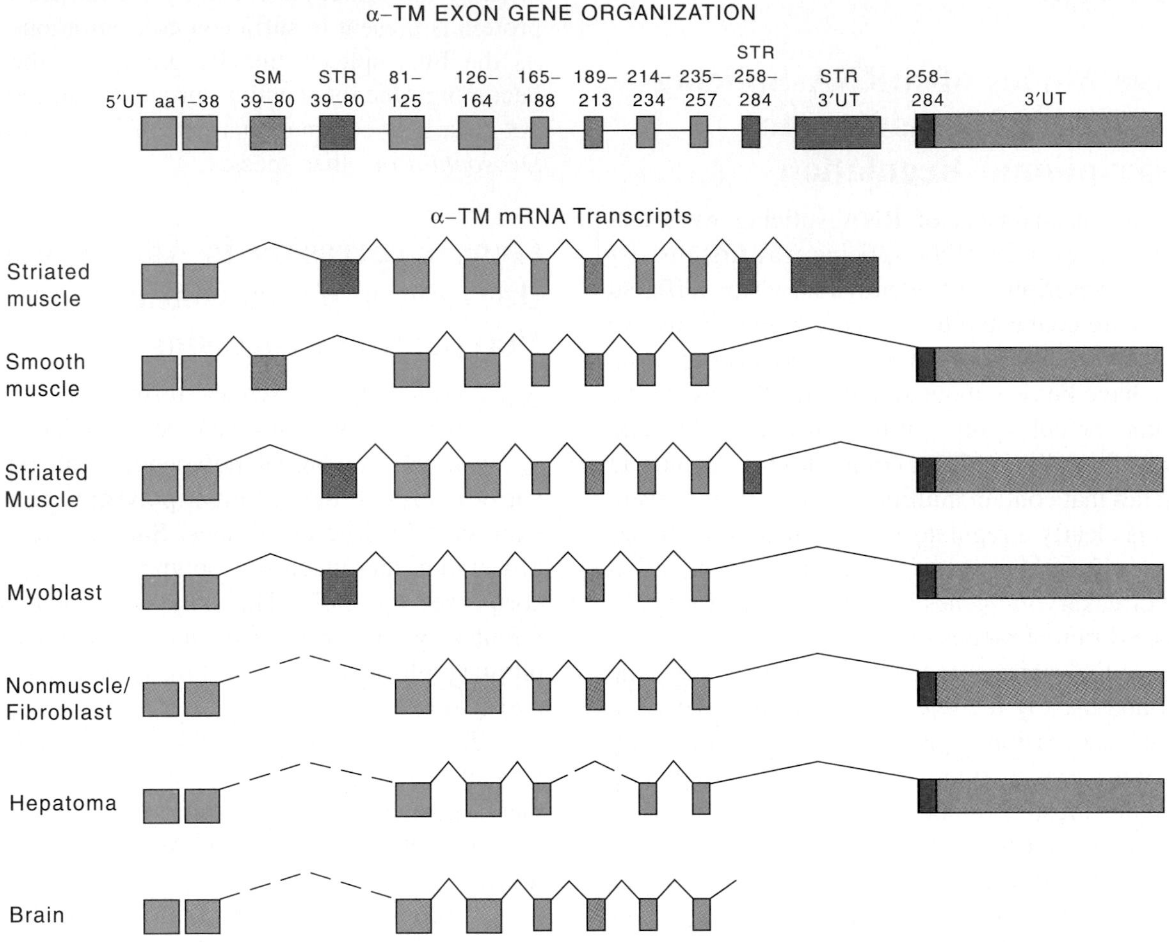

precursor of the heme group in hemoglobin). Hemin binds to and inhibits the activity of HRI, thereby enhancing the rate of globin synthesis (fig. 31.23).

The second eIF-2-specific kinase appears to be present at low levels in most mammalian cells. This kinase is activated by double-stranded RNA and for this reason has been named the double-stranded RNA-activated inhibitor (DAI). DAI may play a role in defending cells against invasion by viruses.

HRI and DAI are different proteins, but both phosphorylate the same amino acid residue in a subunit of eIF-2, and, in consequence, both kinases inhibit protein synthesis by the same mechanism. Phosphorylated eIF-2 is capable of catalyzing the binding of Met-tRNA to the ribosome, but it does so in a stoichiometric rather than a catalytic manner. This mechanism of inhibition results from the fact that eIF-2 dissociates from the ribosome as a complex with GDP, and in order to recycle the bound GDP must exchange for GTP. Phosphorylated eIF-2 binds to, but is unable to dissociate from, the guanine nucleotide exchange factor (GEF) that catalyzes the exchange of GTP for GDP. Because cells contain fewer copies of GEF than eIF-2, all of the GEF can be sequestered by partial phosphorylation of eIF-2, and protein synthesis initiation ceases for lack of GEF. This makes the rate of protein synthesis initiation exquisitely sensitive to the state of phosphorylation of eFI-2.

Figure 31.23

Regulation of protein synthesis in the rabbit reticulocyte. The vast majority of the protein synthesized in the rabbit reticulocyte is hemoglobin. The gross rate of protein synthesis in the reticulocyte is controlled indirectly by the concentration of heme. Heme inactivates a kinase that would otherwise inactivate the initiation complex involving eIF-2 and eIF-2B. The kinase phosphorylates the eIF-2 factor, making it impossible for the eIF-2–eIF-2B complex to exchange GDP for GTP.

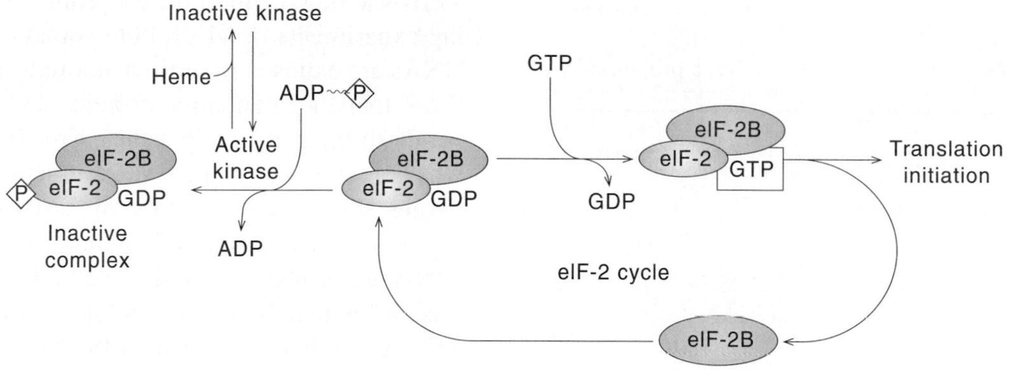

Patterns of Regulation Associated with Developmental Processes

As cells proliferate within the embryo, they assume different properties; changes are evidenced by the genes they express and the proteins they synthesize. Developmental differences can be maintained or altered during subsequent cell duplication. During embryonic growth, cells that are initially capable of following any pathway of development become committed to a particular pathway. As a rule, pathways have many branchpoints, so that when progeny cells of partially committed cells reach a branchpoint they may differentiate further down a more specialized pathway. This process of gradual commitment, repeating itself many times, gives rise to a complex pattern of cell lineages, all arising from the initially fertilized egg.

It seems likely that the pivotal events in the evolution of a differentiated cell reflect changes in gene expression that result from a complex hierarchy of controls. The key to understanding differentiation, therefore, is to identify the regulatory factors responsible for the controls involved in differentiation and to explain how they act.

We focus on regulatory events that occur in development. First we discuss some of the classical studies of gene expression in the embryogenesis of sea urchins and amphibians and then turn to some of the developmental events in fruit flies, the best characterized developmental system.

During Embryonic Development in the Amphibian, Specific Gene Products Are Required in Large Amounts

During early development, when cell divisions are occurring rapidly, demand for certain products associated with the translation apparatus (ribosomes) increases. Special regulatory mechanisms ensure that these gene products are produced in the required amounts and that they are not overproduced in the more mature organism.

Ribosomal RNA in Frog Eggs Is Elevated by DNA Amplification

In eukaryotic organisms ribosomal RNA genes are present in clusters of hundreds to thousands of copies. The nascent transcript, a 45S molecule, undergoes an elaborate series of processing reactions to produce three kinds of ribosomal RNAs: 28S (25–28S), 18S (17–18S), and 5.8S (5.5–5.8S). Multiple copies of ribosomal RNA genes are organized into tandem arrays separated by nontranscribed spacers, which vary in length in different species from about 2,000 to 30,000 bp. *In situ* hybridization studies, using radioactive rRNA and autoradiography, demonstrate that the ribosomal RNA genes are localized around the nucleolus, where rRNA processing and ribosome assembly take place. Germ cells and, in particular, immature eggs (oocytes) often raise the

Figure 31.24

Time course of rDNA amplification in the *Xenopus laevis* oocyte. The amount of rDNA per cell is plotted against time. Note that the ordinate is a logarithmic scale. The sharp drop in rDNA at fertilization is thought to be due to dilution of the amplified rDNA copies by cell division, rather than to their destruction. (From A. P. Bird, Gene reiteration and gene amplification, *Cell Biol.* vol. 3, *Gene Expression: The Production of RNAs,* L. Goldstein and D. M. Prescott (eds.), Copyright © 1980 Academic Press, New York.)

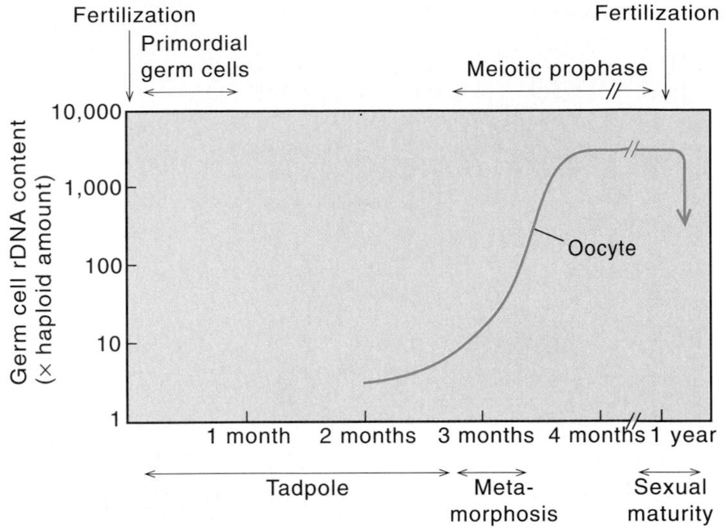

level of rDNA per nucleus by amplification to satisfy the high demand for rRNA during the very rapid early cleavage stages. This type of amplification has been most thoroughly investigated in the frog *Xenopus laevis* (fig. 31.24) in which amplification increases the number of rRNA genes about 1,000-fold. These extrachromosomal genes are transcribed during oogenesis and subsequently discarded.

5S rRNA Synthesis in Frogs Requires a Regulatory Protein

An entirely different mechanism is used to elevate the level of 5S rRNA synthesis during oocyte development. Recall that eukaryotic ribosomes contain the small rRNA components (5S and 5.8S). Whereas the 5.8S rRNA is contained in the unprocessed 45S transcript with the large rRNAs, the 5S exists independently. In amphibians the 5S rRNA genes are organized into the different multigene families the expression of which is under developmental control. For instance, in *X. laevis* the normal haploid genome contains about 20,000 copies of the 5S rRNA genes, organized into three multigenic families. The genes of two major families, comprising about 98% of all the 5S rRNA genes, are expressed only in growing oocytes. The genes of the third family, existing in about 400 copies, are active in most (somatic) cells and growing oocytes. This developmental control enables the oocyte to accumulate 5S rRNA at rates 1,000-fold higher than is possible in somatic cells.

In chapter 28 we noted (see fig. 28.12) that PolIII, which transcribes the 5S genes, requires the binding of three transcription factors to the 5S promoter. These factors are known as TFIIIA, TFIIIB, and TFIIIC. The abundance of TFIIIA is much greater in the oocyte than in most somatic tissue. As a consequence, only in the oocyte is there enough TFIIIA to bind to all of the 5S promoters. Competition binding experiments in which both somatic and oocyte 5S gene DNAs are exposed to limited amounts of the TFIIIA protein show that the regulatory protein binds considerably more firmly to the somatic 5S genes. This fact probably explains why the somatic 5S genes are uniquely active in somatic tissues where the amounts of TFIIIA protein are much lower. Under a variety of conditions, large amounts of 5S rRNA are synthesized in the oocyte but not in somatic cells; this difference is directly related to the concentration of TFIIIA regulatory protein in the two situations.

Early Development in *Drosophila* Leads to a Segmented Structure that Is Preserved to Adulthood

Whereas frogs have proved valuable in the study of isolated embryonic events, they have not been as useful in understanding the overall pattern of early embryonic developments. For this purpose it has been necessary to turn to other organisms to identify the genes involved in regulating development. Many fundamental aspects of developmental biology have been studied using genetic and molecular biology techniques on investigations of *D. melanogaster*.

Some steps in the development of *Drosophila* are shown in figure 31.25. Beginning shortly after fertilization, the new diploid nucleus undergoes a rapid series of divisions with no segregation of nuclei into separate cells. After the eighth division, when there are 256 nuclei (fig. 31.25*c*), the nuclei begin to migrate to the periphery of the egg cytoplasm. After another nuclear doubling, cell membranes form around a group of cells at the posterior end of the egg; these cells are progenitors of germ cells for the subsequent generation. The remaining nuclei continue to divide until there are about 6,000 nuclei at the periphery (fig. 31.25*e*). At this point, about 2 h after fertilization, membranes are formed separating the nuclei into a monolayer of cells. The resulting structure, known as the blastoderm, is essentially a cell monolayer enclosing the yolk.

The blastoderm divides into 14 different segments: Md, Mx, Lb, T1–T3, and A1–A8 (see fig. 31.25*f*). Three of

Figure 31.25

Steps in the development of *Drosophila melanogaster*. The fertilized egg contains a single zygotic nucleus (*a*). This divides (*b*) every 10 min. After eight divisions, when there are 256 nuclei, nuclear migration toward the outer cortex structure begins (*c*). Eventually all the nuclei form a monolayer on the cortex surface. The first nuclei to become enclosed are the pole cells, which become germ cells in the adult organism (*d*). The fully formed blastoderm contains about 6,000 cells, which form a monolayer around the cortex (*e*). Even before a cell membrane has formed around the nuclei, the embryo has become functionally divided into a segmented structure (*f*). During gastrulation (*g*), there is continued cell duplication, a folding of sheets of cells, and mass migration of segments. Eventually this structure hatches into the first larval stage (*h*). The remaining stages between the larva and the adult fly (*i*) are not illustrated. In (*f*) through (*i*), various segments are labeled. Three segments, Md, Mx, and Lb, fuse to make the head structure. Thoracic segments T1–T3 and abdominal segments A1–A8 retain their segmental appearance in the adult.

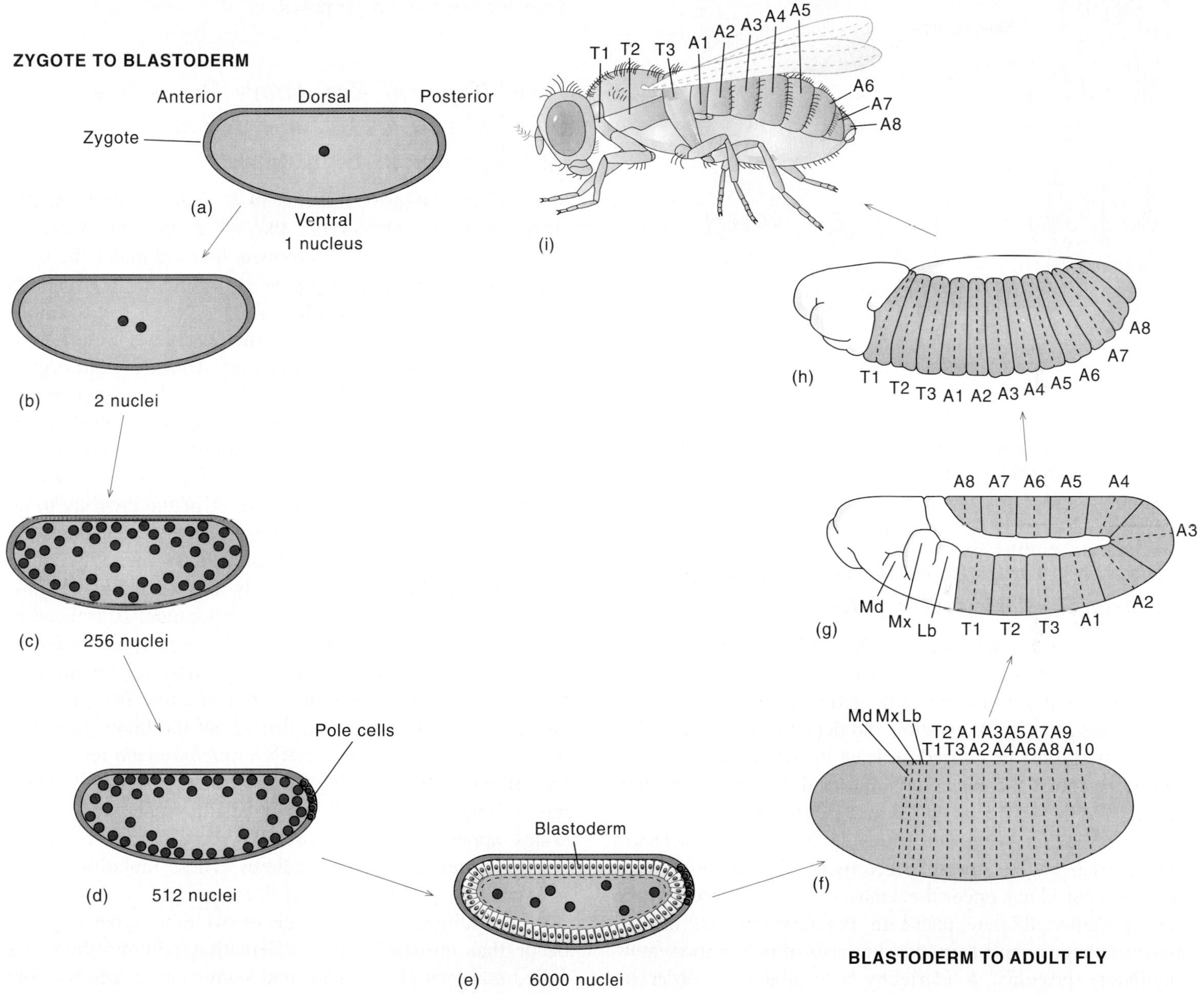

these segments, Md, Mx, and Lb, fuse to become part of the head structure. The remaining segments become subdivided into two compartments, with anterior and posterior parts. During gastrulation there is a continued cell duplication, folding of sheets of cells, and mass migration of segments (see fig. 31.25*g*). The embryo eventually hatches into the first larval stage (see fig. 31.25*h*). Cells within the larva are of two types. About 80% are embryonic precursors to adult tissues rather than larval tissues. These latter cells form packets within the larva, called imaginal disks, that are ar-

Figure 31.26

Imaginal disks in a mature larva and the structures they lead to in the adult fly.

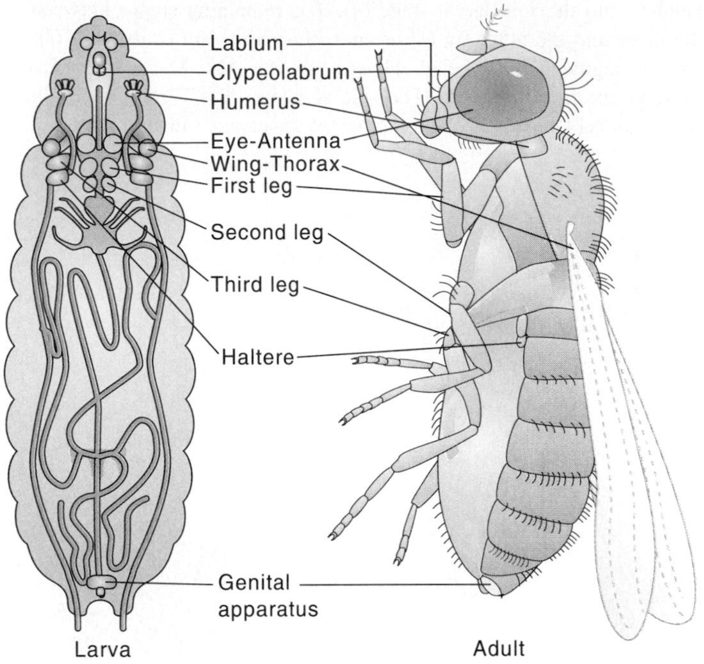

rested in development until pupation (fig. 31.26), when the hormone, ecdysone, signals their differentiation into specific adult structures.

Early Development in Drosophila *Involves a Cascade of Regulatory Events*

There are two distinct phases in Drosophila embryogenesis; the first precedes cellularization of the blastoderm and is associated with a cascade of interacting regulators; the second occurs after cellularization and depends on intercellular signals that must be carried, at least in part, by membrane-bound proteins. We focus on some of the events occurring in the first phase because they are better understood.

Recall that in the infectious cycle of the λ bacteriophage, each stage is characterized by the expression of specific gene products under the control of one or more regulatory proteins. A new phase in the development of the bacteriophage results from the synthesis of one or more new regulatory proteins. A hierarchy determines the order in which the different regulatory proteins make their appearance. Early development in *Drosophila* is very similar except for its greater complexity. In fact, some of the earliest regulatory proteins in the developing oocyte are supplied by surrounding cells. Another difference in *Drosophila* development is that the large size of the oocyte permits the establishment of gradients within a single cell. The regulatory protein encoded by a messenger RNA injected at the anterior end of the oocyte is likely to exist in highest concentration at the anterior end and at lowest concentration at the posterior end. Similarly, mRNAs injected into the oocyte at the posterior end establish regulatory protein gradients in the opposite direction. These gradients are crucial for regulating the expression of genes involved in early development. Some gradients are also established in the perpendicular direction, on the dorsal-ventral axis, but we overlook these for our present purposes.

Three Types of Regulatory Genes Are Involved in Early Segmentation Development in Drosophila

Identifying regulatory mutants in *E. coli* or yeast usually entails detecting phenotypes that affect the amounts of a number of structural gene products that are under the control of the same regulatory gene. In searching for mutants that carry mutations in regulatory genes that affect development, the task of detection is somewhat more complex. In the first place we are always dealing with a diploid organism that carries two alleles for each gene. An organism that carries a mutation in a regulatory gene does not show an abnormal phenotype if the mutant allele is recessive to the wild-type allele. However, on inbreeding two such heterozygotes one would expect one-quarter of the progeny to be homozygous for the mutant allele and therefore show an abnormal phenotype. What does this phenotype look like? It may have misplaced parts or it may have an incorrect number of segments. In many cases development is arrested at the stage when the associated regulatory gene is needed in normal development. The first insight into the genetic system directing *Drosophila* development came from the discovery of bizarre mutations that affect the body plan (fig. 31.27). For example, the mutation *Antennapedia* results in a pair of extra legs sprouting from the head in place of antennae. Another mutation, *Bithorax,* results in an extra pair of wings appearing where normally there should be much smaller appendages called halteres. These mutations occur in regulatory genes known as homeotic genes.

In addition to homeotic genes two other types of genes occur that influence the segmentation pattern; these are called maternal-effect genes and segmentation genes (table 31.2). Maternal-effect genes are so called because they affect the phenotype only according to the information present in the female parent. Maternal-effect mutations occur in genes responsible for establishing the anterior–posterior axes in the young embryo. Segmentation mutations affect the number and polarity of the body segments. Homeotic

Figure 31.27

Abnormal phenotypes resulting from homeotic mutations. *Antennapedia* results in a pair of extra legs sprouting from the head in place of antennae. *Bithorax* results in an extra pair of wings appearing where halteres normally appear.

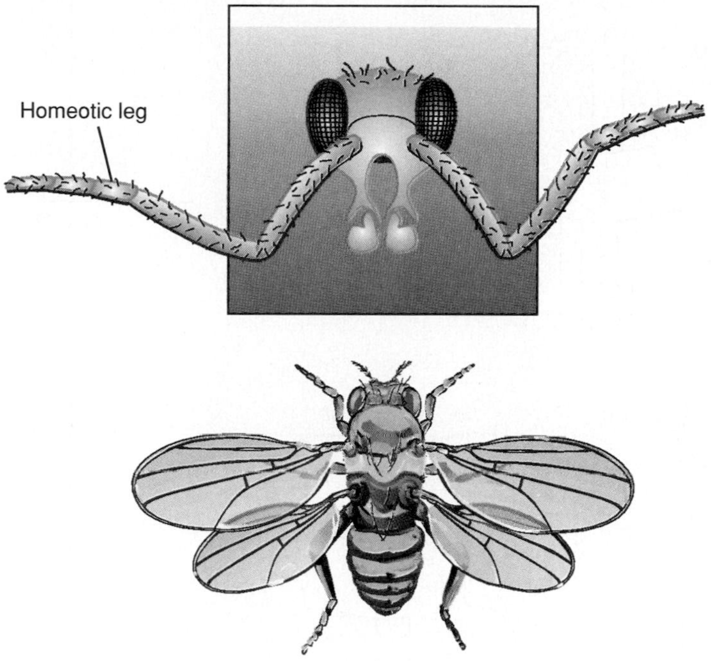

Table 31.2

Regulatory Genes Involved in Segmentation Development

Maternal-effect genes—establish gradients
Anterior
bicoid (*bcd*)
Posterior
nos
Dorsal–ventral
dorsal (*dl*)
Segmentation genes
Gap genes—define four broad regions in the egg
hunchback (*hb*)
Krüppel (*Kr*)
Knirps (*kni*)
Pair-rule genes—define seven bands
runt (*runt*)
hairy (*h*)
fushi tarazu (*ftz*)
even skipped (*eve*)
paired (*prd*)
odd paired (*opa*)
Segment-polarity genes—define 14 bands
engrailed (*en*)
wingless (*wg*)
gooseberry (*gb*)
Homeotic genes—specify structures associated with individual segments
BX-C locus
ANT-C locus

mutations are more specific than segmentation mutations; they result in changes in structures that are uniquely associated with individual segments or subsegments.

Analysis of Genes That Control the Early Events of *Drosophila* Embryogenesis

Once we find a mutation that appears to affect development, we must have a way to detect the presence or absence of the related gene product in the developing embryo. A principal tool for doing this is autoradiography using ^{32}P-labeled specific DNA probes. The embryo is labeled by *in situ* hybridization with the radioactive probe specific for the transcript of a gene. Both the location and intensity of the labeling give an indication of the level of expression of the gene. In some cases the protein product relating to a specific transcript can also be detected by use of a specific antibody-labeling technique. In this way the appearance of many regulatory genes involved in early development has been traced.

Maternal-Effect Gene Products for Oocytes Are Frequently Made in Helper Cells

The first regulatory gene products to play a role in *Drosophila* development are active before fertilization. The bicoid (*bcd*) gene is a major determinant of the anterior–posterior pattern. Like many other maternally expressed genes, *bcd* is transcribed in the ovary in specialized cells that form a cluster around the future anterior pole of the developing oocyte. The *bcd* RNA synthesized in these so-called nurse cells passes into the oocyte through cytoplasmic canals and becomes localized at the anterior pole of the oocyte (fig. 31.28*a*). Similarly the regulatory gene product of *nos* becomes localized at the posterior pole of the egg (see fig. 31.28*a*). Females lacking a functional copy of the *bcd* gene produce eggs that develop into embryos with no head or thorax. Embryos produced by female mutants homozygous for *nos* develop normal head and thoracic segments but lack the entire abdomen.

Figure 31.28

Key events in the expression of developmental regulatory genes in the blastoderm embryo. Events take place in the order shown over a period of a few hours. Measurements are based on the use of probes for detecting the transcripts of various regulatory genes. Description of genes is given in table 31.2. (Source: Adapted from P. W. Ingham, The molecular genetics of embryonic pattern formation in Drosophila, *Nature* 335:25–34, 1988.)

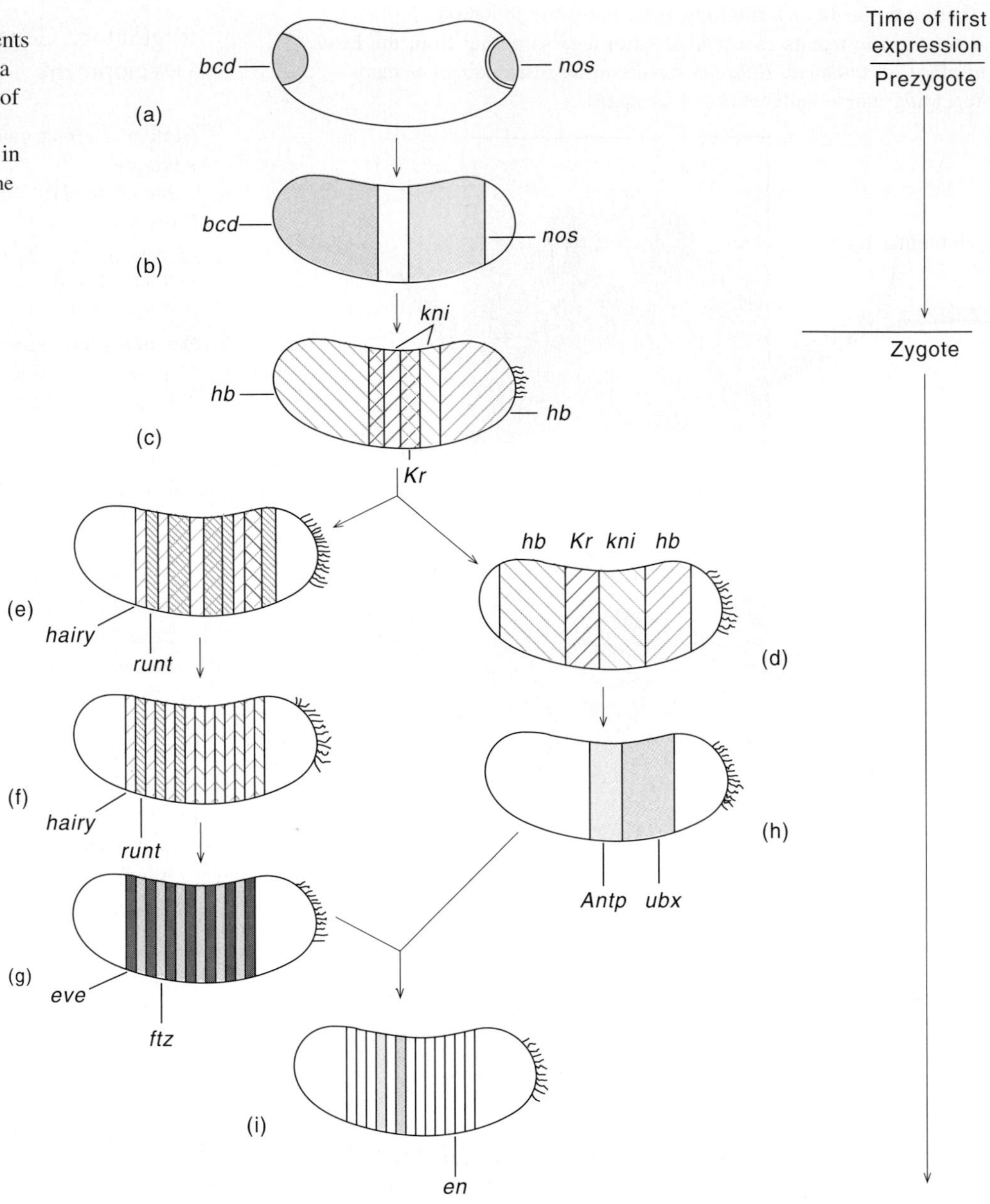

The pattern-forming process continues on fertilization; as the *bcd* mRNA is translated, the protein diffuses from the anterior pole so that it becomes distributed over about half of the length of the egg (see fig. 31.28*b*). Simultaneously, *nos* gene product originally localized at the posterior pole begins to move forward.

By the beginning of the precellular blastoderm stage, gradients for maternally encoded products occur along the anterior–posterior axis of the embryo. This quantitative information is transformed into qualitative differences in the form of region-specific gene expression, by a process requiring interaction between the maternally derived products and the zygotic genome that resulted from fertilization.

Gap Genes Are the First Segmentation Genes to Become Active

The first segmentation genes to become active in the zygote belong to the *gap* class, so called because their mutants lack major regions of the body, thereby creating a gap in the anterior–posterior pattern. The three members of this class are hunchback (*hb*), Krüppel (*Kr*), and Knirps (*kni*). These genes are expressed in two division cycles prior to cellularization. Expression of *hb* is restricted to two regions: One extending from the anterior pole to 50% of the egg length, the other from the posterior pole to about 25% of the egg length (see fig. 31.28*c*). Initially the *Kr* gene is expressed in

Figure 31.29

Influence of developmental regulatory genes on one another. An arrow with a plus sign at the arrowhead ⊕ indicates a positive effect on expression, whereas a minus sign ⊖ at the arrowhead indicates a negative effect on expression. The *gap* genes are essential for the establishment of the *hairy* and *runt* banded patterns, but the precise way in which they do this is unclear.

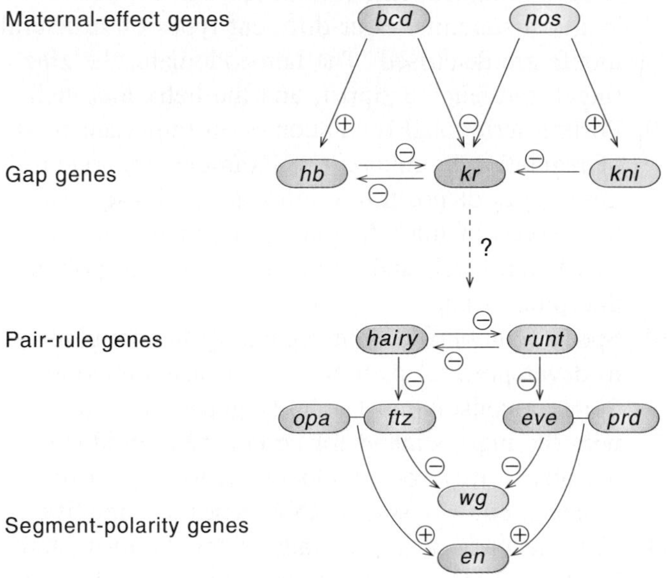

a single broad band in the middle of the embryo, whereas *kni* is expressed in two distinct domains, one anterior and one posterior to the *Kr* band. These transcriptional domains for *hb, Kr,* and *kni* are influenced by the maternally derived information encoded by the *bcd* and *nos* group genes. Thus, mutants that lack *bcd* activity do not express *hb,* whereas *Kr* is extended anteriorly in an unusually broad domain. The absence of the posterior maternal determinants results in the extension of the *Kr* domain posteriorly. These observations suggest that the *bcd* and *nos* group genes both act to repress transcription of *Kr* in the anterior and posterior regions, respectively, thereby restricting its region of expression to the central portion of the embryo. In contrast, *bcd* appears to act as a positive regulator of *hb,* and *nos* appears to be a positive regulator of *kni.* In normal embryos the transcriptional domains of *hb, Kr,* and *kni* narrow with time, giving rise to sharp boundaries of expression. This process is driven by negative effects between the gap genes (see fig. 31.28*d*), *hb* and *Kr,* mutually repressing each other, and *kni* acting as a negative regulator of *Kr.* Thus, the establishment of stable domains of *gap* gene expression is a two-step process: First, a differential response to graded levels of maternal determinants, and second, a mutual repression effect, leading to the generation of stable boundaries between adjacent domains.

The *gap* gene products regulate the position-specific expression of other genes, those belonging to the pair-rule class of segmentation genes.

By scrutinizing the location of different regulatory gene products as a function of time in both normal and mutant embryos, the patterns shown in figure 31.28 were established for all three classes of regulatory proteins that function in early development.

The positive and negative effects established between maternal-effect and pair-rule genes are summarized in figure 31.29. This diagram tells an incomplete story. The most serious omission in our knowledge is how the gradients of maternal-effect and *gap* genes help to deliver the signal for the establishment of the strict periodic patterns of expression seen for most of the segmentation genes (see fig. 31.28). It seems unlikely that the signals in the form of gradients could ever institute such a pattern without other factors being involved. Possibly some of the signals for expression of pair-rule genes are established in the oocyte before fertilization.

Summary

Some mechanisms of gene expression that are found in eukaryotes are rarely, if ever, seen in prokaryotes. Still, *trans*-acting regulatory proteins that bind to *cis* effector sites on the genome are present in eukaryotic systems as well as in *E. coli*. The best understood unicellular eukaryote is the budding yeast *Saccharomyces cerevisiae*. Gene regulation, particularly of development, can be quite complex in multicellular eukaryotes. Our discussion in this chapter focused on the following points.

1. In eukaryotes, individual genes encode transcripts for a single polypeptide chain. Although functionally related genes are clustered, each gene has its own promoter.
2. In yeast, genes of the *GAL* system are under the joint control of two *trans*-acting genes that encode regulatory proteins: *GAL4* and *GAL80*. GAL4 protein is an activator that binds at an upstream site, and GAL80 is a repressor that inhibits GAL4 action by binding to it. The ability of the GAL80 protein to bind to GAL4 is lost in the presence of galactose, which binds to the GAL80 protein, causing a reversible allosteric change in its structure.

3. Yeast has two haploid cells of opposite mating types and one nonmating diploid cell that results from the fusion of haploid cells of opposite mating type. The mating type is determined by the *MAT* locus. Information for the mating type is stored at other, silent loci and is expressed only if it is transposed to the *MAT* locus. Mating-type information is not expressed at the storage loci because of a complex repressor system of proteins interacting in *cis* fashion over a considerable distance at the control centers of the storage loci.
4. The greatest difference between the regulatory systems in yeast and *E. coli* is that yeast regulatory proteins can bind at a long distance from the RNA polymerase-binding site and still be effective.
5. Complex multicellular eukaryotes differentiate irreversibly so that different cell types express a different profile of genes. Genes that are expressed are usually associated with swollen chromatin. Proteins found in active regions of the genome show characteristic modifications.
6. Enhancers are elements of the genome that generally stimulate transcription. They resemble yeast UAS sequences in yeast in that they can function over long distances. In fact, they can function over even greater distances than UAS sequences, and they are effective in either orientation: Either upstream or downstream from the promoter. A possible mechanism for enhancer and UAS function over long distances is that the chromosome folds to bring the proteins bound at the enhancer site in close proximity to other regulatory proteins or RNA polymerase bound at the promoter.
7. A typical enhancer is composed of a cluster (usually two or three) of *cis* sites called enhansons; each enhanson contains binding sites for a unique combination of regulatory proteins. Enhansons must be clustered to be effective.
8. In chapter 30 we discussed DNA-binding proteins that regulate transcription in prokaryotes. In prokaryotes and eukaryotes specific recognition is dominated by H bond interactions that take place in the major groove of the DNA. In both cases the α helix is the most common element used for DNA recognition. The most striking difference between DNA-binding proteins in prokaryotes and eukaryotes has to do with the symmetry of the interaction. In prokaryotes the binding proteins almost always interact in a symmetrical fashion with the DNA. In eukaryotes most of the cases that have been examined so far involve proteins that interact in an asymmetrical fashion with the DNA. In many cases the regulatory proteins interact in multisubunit complexes that contain nonidentical subunits. Four different types of structural motifs are discussed: The homeodomain, the zinc finger, the leucine zipper, and the helix-loop-helix.
9. Posttranscriptional regulation is an important mode of regulation in eukaryotes. Examples are given of three types of posttranscriptional regulation: Alternative modes of mRNA splicing, regulation at the translation level, and alternative modes of polypeptide processing.
10. Special combinations of regulatory factors give rise to developmental patterns of cell differentiation. Some organisms must amplify genes to keep pace with the high demand for certain gene products, especially in early development. Histones, ribosomal proteins, and ribosomal RNA genes are amplified.
11. The existence of many kinds of regulatory mutants has helped to advance our understanding of early development in the fruit fly, *Drosophila melanogaster*. Regulatory gene products are proteins that activate or repress other genes. Early development in *Drosophila* is a sequence of events in which different regulatory proteins gradually come into play in cascade fashion, controlling a wide range of enzymes and structural proteins and also influencing each other. Until the blastoderm stage the nuclei in a developing *Drosophila* embryo are not separated by cellular membranes. As a result, the regulatory proteins and other gene products may diffuse freely from their site of synthesis to other nuclei in the embryo. At the late blastoderm stage, the nuclei become cellularized. From this point on, the influence of regulatory proteins made in one cell must be exerted on another cell at the level of the cell membrane.

Selected Readings

Achneider, R. J., and T. Shenk, Impact of virus infection on host cell protein synthesis. *Ann. Rev. Biochem.* 56:317–332, 1987.

Atchison, M. L., Enhancers: Mechanisms of action and cell specificity. *Ann. Rev. Cell. Biol.* 4:127–153, 1988.

Binétruy, B., T. Smeal, and M. Karin, Ha-Ras augments c-Jun activity and stimulates phosphorylation of its activation domain. *Nature* 351:122–127, 1991.

Bingham, P. M., T. Chou, I. Mims, and Z. Zachari, On/off regulation of gene expression at the level of splicing. *Trends Genet.* 4:134, 1988.

Brietbart, R. E., A. Andreadis, and B. Nadal-Ginard, Alternative splicing: A ubiquitous mechanism for the generation of multiple protein isoforms from single genes. *Ann. Rev. Biochem.* 56:467–495, 1987.

Brown, D. D., How a simple animal gene works. In *The Harvey Lectures,* Series 76, pp. 27–44. New York: Academic Press, 1982.

Chandler, V. L., B. A. Maler, and K. R. Yamamoto, DNA sequences bound specifically by glucocorticoid receptor *in vitro* render a heterologous promoter responsive *in vivo:* A steroid specific enhancer. *Cell* 33:489–499, 1983.

Chen, H-Z., T. Hoey, and G. Zubay, Purification and properties of the *Drosophila* zen protein. *Mol. Cell. Biochem.* 79:181–189, 1988. First evidence that a homeobox protein binds to DNA as a monomer.

Comai, L., N. Tanese, and R. Tjian, The TATA-binding protein and associated factors are integral components of the RNA polymerase I transcription factor, SL1. *Cell* 68:965–976, 1992.

Duncan, I., The bithorax complex. *Ann. Rev. Genet.* 21:285–320, 1987.

Evans, R. M., The steroid and thyroid hormone receptor superfamily. *Science* 240:889–895, 1988.

Ferre-D'Amare, A. R., G. C. Prendergast, E. B. Ziff, and S. K. Burley, Recognition by Max of its cognate DNA through a dimeric b/HLH/Z domain. *Nature* 363:38–44, 1993.

Forsburg, S. L., and L. Guarente, Communication between mitochondria and the nucleus in regulation of cytochrome genes in the yeast *Saccharomyces cerevisiae. Ann. Rev. Cell. Biol.* 5:153–180, 1989.

Funder, J. W., Mineralocorticoids, glucocorticoids, receptors and response elements. *Science* 259:1132–1133, 1993.

Gabrielsen, O. S., and A. Sentenac, RNA polymerase III (C) and its transcription factors. *Trends Biochem. Sci.* 16:412–416, 1991.

Gehring, U., Steroid hormone receptors: Biochemistry, genetics and molecular biology. *Trends Biochem. Sci.* 12:399–402, 1987.

Gehring, W., Homeo boxes in the study of development. *Science* 236:1245–1252, 1987.

Gehring, W. J., The molecular basis of development. *Sci. Am.* 253(4):152–162, 1985.

Gimeno, C. J., and G. R. Fink, The logic of cell division in the life cycle of yeast. *Science* 257:626, 1992.

Giniger, E., and M. Ptashne, Cooperative binding of the yeast transcriptional activator *GAL4. Proc. Natl. Acad. Sci.* 85:382–386, 1988.

Gruenberg, D. A., S. Natesan, C. Alexandre, M. Z. Gilman. Human and *Drosophila* homeo domain proteins that enhance the DNA-binding activity of serum response factor. *Science* 257:1089–1095, 1992.

Grunstein, M., Histones as regulators of genes. *Sci. Am.* 267:68–74, 1992.

Guarente, L., UASs and enhancers: Common mechanism of transcriptional activation in yeast and mammals. *Cell* 52:303–305, 1988.

Guarente, L. P., Regulatory proteins in yeast. *Ann. Rev. Genet.* 21:425–452, 1987.

Gurdon, J., Egg cytoplasm and gene control in development. The Croonian Lecture, 1976. *Proc. R. Lond. B.* 198:211–247, 1977.

Hahn, S., Structure and function of acidic transcription activators. *Cell* 72:481–483, 1993.

Hanes, S. D., and R. Brent, A genetic model for interaction of the homeodomain recognition helix with DNA. *Science* 251:426–430, 1991.

Harrison, S. C., A structural taxonomy of DNA-binding domains. *Nature* 353:715–719, 1991.

Herskowitz, I., Life cycle of the budding yeast *Saccharomyces cerevisiae. Microbiol. Rev.* 53:536–553, 1988.

Hinnebusch, A. G., Involvement of an initiation factor and protein phosphorylation in translational control of GCN4 mRNA. *Trends Biochem. Sci.* 15:148–152, 1990.

Horvitz, H. R., and P. W. Sternberg, Multiple intercellular signalling systems control development of the caenorhabditis elegans vulva. *Nature* 351:535–541, 1991.

Johnson, P. F., and S. L. McKnight, Eukaryotic transcriptional regulatory protein. *Ann. Rev. Biochem.* 58:799–839, 1989.

Karin, M., and T. Smeal, Control of transcription factors by signal transduction pathways: The beginning of the end. *Trends Biochem. Sci.* 17:418–422, 1992.

Karlsson, S., and A. W. Nienhius, Developmental regulation of human globin genes. *Ann. Rev. Biochem.* 54:1071–1108, 1985.

King, T., and R. Briggs, Serial transplantation of embryonic nuclei. *Cold Spring Harb. Symp. Quant. Biol.* 21:271–290, 1956.

Koleske, J., and R. A. Young, An RNA polymerase II holoenzyme responsive to activators. *Nature* 368:466–469, 1994.

Klevit, R. E., Recognition of DNA by Cys_2His_2 zinc fingers. *Science* 253:1367–1393, 1991.

Lai, E., and J. E. Darnell, Jr., Transcriptional control in hepatocytes: A window on development. *Trends Biochem. Sci.* 16:427–429, 1991.

Lamb, P., and S. L. McKnight, Diversity and specificity in transcriptional regulation: The benefits of heterotypic dimerization. *Trends Biochem. Sci.* 16:417–433, 1991.

Laybourn, P. J., and J. T. Kadonaga, Role of nucleosomal cores and histone H1 in regulation of transcription by RNA polymerase II. *Science* 254:238–245, 1991.

Lee, M. S., S. A. Kliewer, J. Provencal, P. E. Wright and R. M. Evans, Structure of the retinoid X receptor α DNA binding domain: A helix required for homodimeric DNA binding. *Science* 260:1117–1121, 1993.

Leuther, K. K., and S. A. Johnston, Nondissociation of GAL4 and GAL80 *in vivo* after Galactose Induction. *Science* 256:1333–1336, 1992.

Li, E., C. B. Beard, and R. Jaenisch, Role for DNA methylation in genomic imprinting. *Nature* 366:362–365, 1993.

Marmorstein, R., M. Carey, M. Ptashne, and S. C. Harrison, DNA recognition by GAL4: Structure of a protein-DNA complex. *Nature* 356:408–414, 1992.

Marzluff, W. F., and N. B. Pandey, Multiple regulatory steps control histone messenger-RNA concentrations. *Trends Biochem. Sci.* 12:49–51, 1988.

McClintock, B., Controlling elements and the gene. *Cold Spring Harb. Symp. Quant. Biol.* 21:197–216, 1956.

McKinney, J. D., and N. Heintz, Transcriptional regulation in the eukaryotic cell cycle. *Trends Biochem. Sci.* 16:430–434, 1991.

Melton, D. A., Pattern formation during animal development. *Science* 252:234–241, 1991.

Miner, J. N., and K. R. Yamamoto, Regulatory crosstalk at composite response elements. *Trends Biochem. Sci.* 16:423–426, 1991.

Mitchell, P. J., and R. Tjian, Transcriptional regulation in mammalian cells by sequence-specific DNA binding proteins. *Science* 245:371–378, 1989.

Nevins, J. R., Transcriptional activation by viral regulatory proteins. *Trends Biochem. Sci.* 16:435–439, 1991.

Parkhurst, S. M., D. Bopp, and D. Ish-Horowicz, X:A ratio, the primary sex-determining signal in *Drosophila,* is transduced by helix-loop-helix proteins. *Cell* 63:1179–1191, 1990.

Pearce, D., and K. R. Yamamoto, Mineralocorticoid and glucocorticoid receptor activities distinguished by nonreceptor factors at a composite response element. *Science* 259:1161–1164, 1993.

Prendergast, G. C., and E. B. Ziff, A new bind of Myc. *Trends Genet.* 8:92–96, 1992.

Ptashne, M., How gene activators work. *Sci. Am.* 260(1):41–47, 1989.

Raghow, R., Regulation of messenger RNA turnover in eukaryotes. *Trends Biochem. Sci.* 12:3358–3360, 1987.

Roeder, R. G., The complexities of eukaryotic transcription initiation: Regulation of preinitiation complex assembly. *Trends Biochem. Sci.* 16:402–407, 1991.

Ruvkun, G., and M. Finney, Regulation of transcription and cell identity by POU domain proteins. *Cell* 64:475–478, 1991.

Schler, A. F., and W. J. Gehring, Direct-homeo domain-DNA interaction in the autoregulation of the *fushi tarazu* gene. *Nature* 356:804–806, 1992.

Schwabe, J. W. R., and D. Rhodes, Beyond zinc fingers: Steroid hormone receptors have a novel structural motif for DNA recognition. *Trends Biochem. Sci.* 16:291–296, 1991.

Sherman, A., M. Shefer, S. Sagee, and Y. Kassir, Post-transcriptional regulation of IME1 determines initiation of meiosis in *Saccharomyces cerevisiae. Mol. Gen. Genet.* 237:375–384, 1993.

Singer, S. J., Intercellular communication and cell-cell adhesion. *Science* 255:1671–1677, 1992. Considers aspects of regulation we didn't have time to cover.

Stark, G., M. Debatisse, E. Giulotto, and G. M. Wahl, Recent progress in understanding mechanisms of mammalian gene amplification. *Cell* 57:901–908, 1989.

Struhl, K., Molecular mechanisms of transcriptional regulation in yeast. *Ann. Rev. Biochem.* 58:1051–1077, 1989.

Struhl, K., Helix-turn-helix, zinc-finger and leucine-zipper motifs for eukaryotic transcriptional regulatory proteins. *Trends Biochem. Sci.* 14:137–140, 1989.

Thompson, C. C., and S. I. McKnight, Anatomy of an enhancer. *Trends Genet.* 8:232–236, 1992.

Weinberg, R. A., Finding the anti-oncogene. *Sci. Am.* 259(3):44–51, 1988.

Weinzierl, R. O. J., B. D. Dynlacht, and R. Tjian, Largest subunit of *Drosophila* transcription factor IID directs assembly of a complex containing TBP and a coactivator. *Nature* 362:511–517, 1993.

Problems

1. On the basis of what you know about genes in prokaryotic and eukaryotic cells, define a gene. Make sure that your definition is brief and concise. Does your definition have any limitations or problems?
2. Why is attenuation control in eukaryotes unlikely?
3. The domain-swap experiment illustrated in figure 31.4 demonstrated that the *lexA* DNA-binding domain from *E. coli* is functionally interchangeable with the equivalent domain from the GAL4 protein if and only if the *lexA* DNA recognition site replaces the GAL4 equivalent. Is the *GAL1* gene with a *lexA* binding site upstream transcribed in the presence of intact lexA protein (with no GAL4 activation domain)? Why or why not?
4. A *GAL4* mutation ($GAL4^c$) leads to constitutive synthesis of the *GAL1* gene product in haploid yeast. Propose an explanation for the effect of this mutation.
5. How do you expect a deletion of *HMLE* to affect the expression of mating-type genes in yeast? Compare this effect with the deletion of the α_2 gene from MAT_α. Consider both homothallic and heterothallic backgrounds.
6. Color blindness is X-chromosome linked. Bearing in mind the phenomenon of X-chromosome inactivation, suggest an explanation for the observation that females who are heterozygous for the defective gene show no signs of color blindness.
7. Briggs and King were able to grow a differentiating embryo from an egg of *Rana pipiens,* the chromosomes of which were replaced with a single diploid nucleus from another embryo. What does this tell us about the state of the nucleus in the developing embryo?
8. There is no large difference in the frequency of histones in transcribed regions of the genome compared with untranscribed regions. Why don't nucleosomes interfere with transcription in eukaryotes?
9. High salt concentrations weaken the interaction of histones with DNA but have little effect on the binding of many regulatory proteins. Explain this observation in terms of how these molecules interact with DNA.
10. List some characteristics that distinguish active from inactive chromatin.
11. The restriction endonuclease *Hpa*II cleaves the sequence CCGG only if the second C is unmethylated. The enzyme *Msp*I cleaves the same sequence, whether or not it is methylated. How do the globin-specific sequences in erythroblast DNA (erythroblasts are red blood cell precursors) differ from those of other tissues in their susceptibility to these two enzymes?
12. Explain how a DNA sequence (enhancer sequence) located 5,000 bp from a gene transcription start site can stimulate transcription even if its orientation is reversed.
13. Discuss the types of structural motifs found in eukaryotic transcription factors.
14. What kind of changes have to be made in a typical eukaryotic structural gene for its protein product to be expressed in bacteria?
15. In the *Xenopus* oocyte a large number of ribosomes are made in a short time to handle the rapid demand for cell growth during cleavage stages. How is this large amount of rRNA made in such a short time?
16. Based only on the definition of maternal-effect genes, segmentation genes, and homeotic genes, which do you predict act earliest in development of the *Drosophila* embryo, and which act latest?
17. Given that specific subsets of homeotic genes are required for the development of specific segments of the *Drosophila* embryo, suggest a possible mechanism whereby the necessary spatially restricted pattern of homeotic gene expression might be achieved. Incorporate the observed effect of homeotic mutations on segment morphology into your model.
18. Given a cloned fragment of a *Drosophila* gene, how could you determine which chromosomal band(s) contain the gene?

SUPPLEMENT

3

Principles of Physiology and Biochemistry: Immunobiology

Supplement Outline

When vertebrates are invaded by foreign agents, they can mobilize a versatile set of adaptive immune responses to form specifically reactive cells and proteins. These responses constitute the principal means of defense against pathogenic microorganisms and viruses and probably also against host cells that undergo transformation into cancer cells.

Throughout most of this century, the subject of immunology has attracted some of the keenest minds in biology. As a result, the intricacies of this fascinating subject have been largely unraveled, and its understanding is providing a strong bridge between the fields of biochemistry and physiology. In this chapter we give a brief introduction to some of the major findings in immunobiology and the experiments that have led to our current level of understanding.

Overview of the Immune System

The immune system was first studied in humans, but mice became a popular subject for immune system studies when researchers began to appreciate how close the mouse and

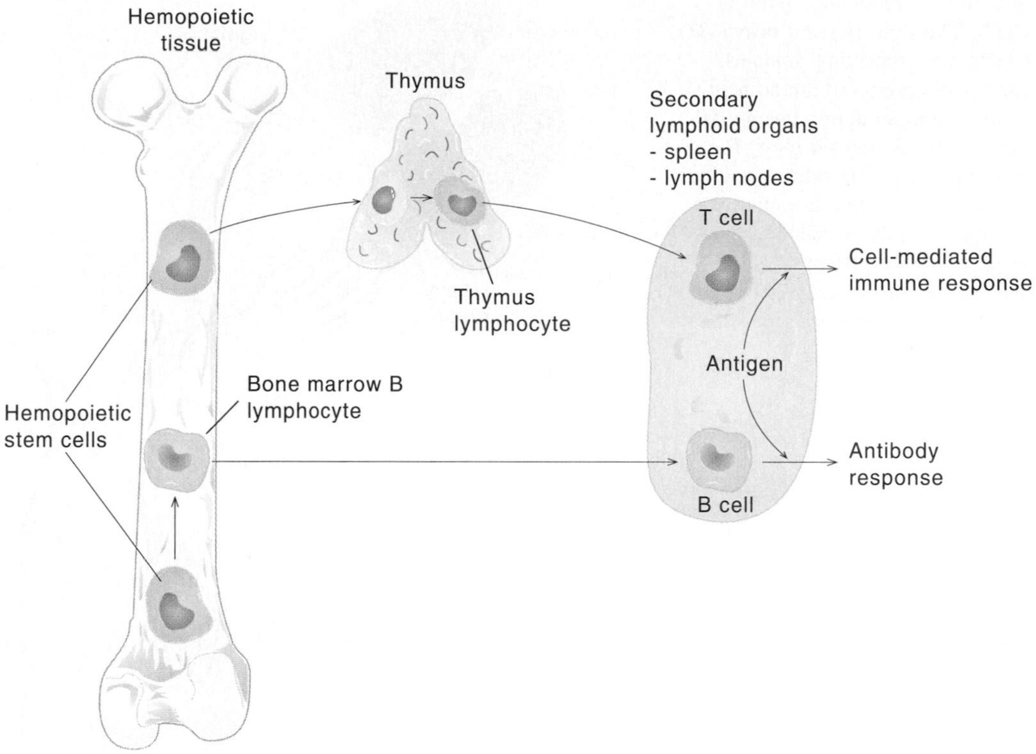

Figure S3.1

B cells and T cells follow different pathways of development. In mammals, B lymphocytes mature in the bone marrow and then migrate to secondary lymphoid organs. On exposure to foreign substances known as antigens, they proliferate to produce immunoglobulins. T lymphocytes mature in the thymus gland. They also can be stimulated to proliferate by exposure to an appropriate antigen.

human systems were in their organization and action. Currently, both systems are under study in many different laboratories. Although we focus here on mice, we often refer to parallel observations on humans.

The immune system is an example of a developmental process that takes place in the mature organism. In the new cells that are constantly being generated, changes arise in the genes of the immune system. This variability results from DNA splicing and point mutations. It benefits the organism by providing the cells of the immune system with the widest possible range of specificities. As soon as the organism is invaded by foreign agents, usually viruses or bacteria, the immune system is activated. Those immune system cells that carry the specific immune receptors for interacting with the foreign agent are stimulated to proliferate. In a matter of days, clones of immune system cells with the appropriate specificity have been produced, and the organism fends off the invader with the specific immunologic tools provided by those cells. The clones of cells tend to persist for some time, usually months to years, which accounts for the fact that the second time the same foreign invader makes its presence known, it is usually rejected promptly and without crisis.

Two different classes of white cells, or lymphocytes are associated with the immune response: The *B cells* and the *T cells*. In mammals, B lymphocytes mature in the bone marrow and then migrate to secondary lymphoid organs (fig. S3.1). On exposure to foreign substances known as antigens, they proliferate and produce immunoglobulin proteins known as antibodies, which they secrete into the bloodstream. T lymphocytes, by contrast, mature in the thymus gland before migrating to the secondary lymphoid organs. They also can be stimulated to proliferate by exposure to an appropriate antigen, but their specific effector molecules remain firmly bound to the cellular membrane, as opposed to being secreted. Three main types of T cells have been recognized; T killer cells, which specifically destroy target cells; T helper cells, which promote the maturation of antigen-stimulated B and T cells; and T suppressor cells, which block the effects of T helper cells.

The Humoral Response: B Cells and T Cells Working Together

There are two basically different types of immune response. The first, called the humoral, or antibody, response, involves the concerted action of both B cells and T cells, and the active agents are the antibody, or immunoglobulin, proteins secreted by the B cells into the bloodstream. The combined B–T immune response is characteristic of most vertebrates. The second type of immune response, called the cell-mediated response, involves only T cells, and the active agent is the circulating T cell itself, which attacks the foreign agent. This type of immune re-

Figure S3.2

Structure of immunoglobulin G (IgG). The light (L) and heavy (H) chains have repeating domains, each with about 110 amino acid residues and an approximately 60-member S—S bonded loop. The subscripts L and H refer to regions of the sequence that are relatively constant or quite variable, respectively, in different IgG species. (Illustration copyright by Irving Geis. Reprinted by permission.)

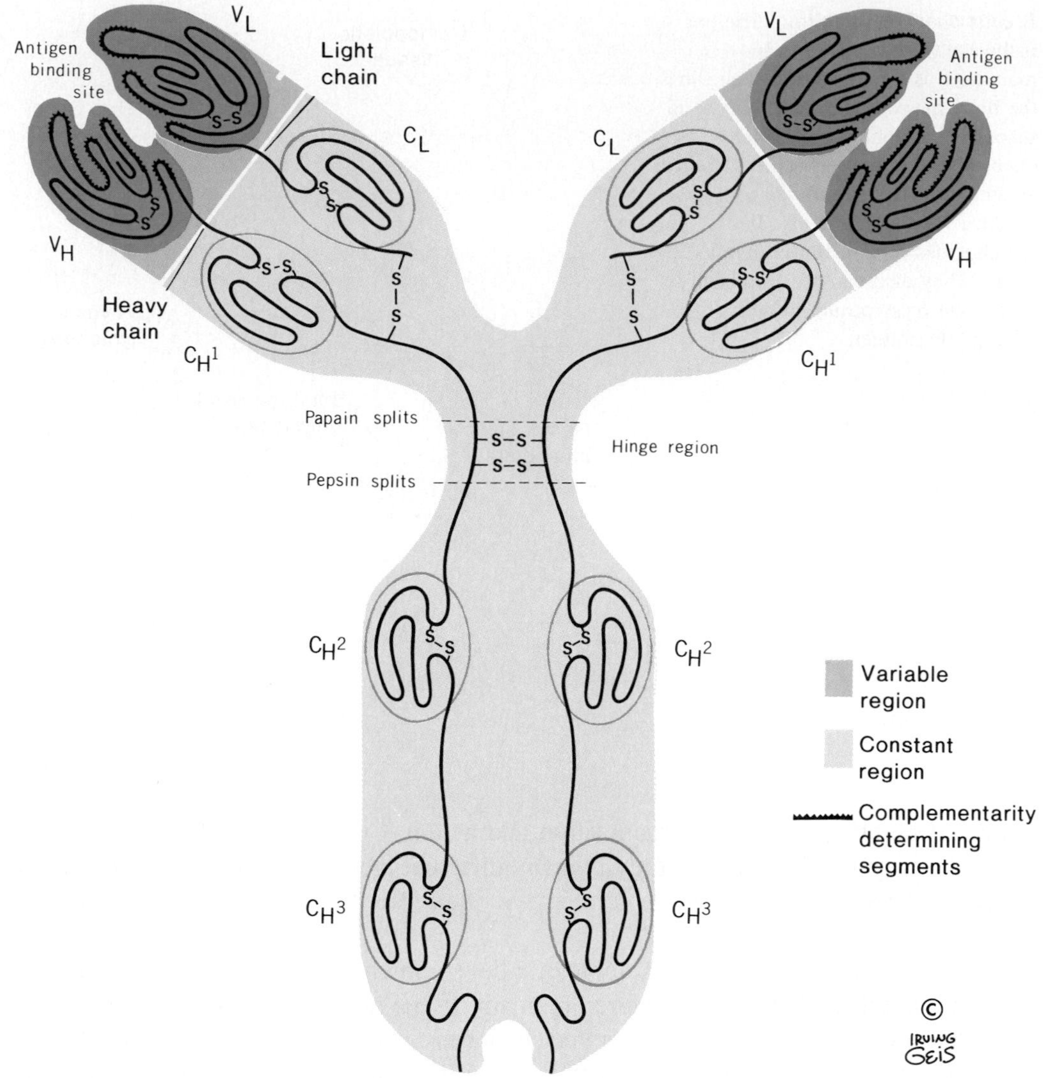

sponse is limited to certain groups of vertebrates, including mammals. We first discuss the B–T cell response mediated by antibodies.

Immunoglobulins Are Extremely Varied in Their Specificities

A detailed investigation of immunoglobulin structure provided the first leads about how immunoglobulins of such a wide variety are synthesized. The most common type of immunoglobulin is the 7S molecule, known as immunoglobulin G (IgG). This immunoglobulin has a molecular weight of about 150,000 daltons. It is composed of four polypeptide chains: Two heavy (H) chains with molecular weights of about 50,000 and two light (L) chains with molecular weights of about 25,000 (fig. S3.2).

To obtain a detailed understanding of immunoglobulin structure, it was necessary to isolate pure antibody proteins for sequencing. Fortunately, it was discovered that certain plasma cell tumors known as myelomas produce enormous amounts of pure immunoglobulins, which have the same gross structure as the mixed immunoglobulins isolated from serum. Each myeloma appears to originate from a single cell turned cancerous. As a result, each myeloma serves as the source of a pure antibody protein, the sequence and other properties of which can be determined. In some cases myelomas synthesize both heavy and light chains, like normal antibody-forming cells, but in other cases they synthesize only one or the other type of chain. Comparison of the sequences from a number of different immunoglobulins derived from myelomas has shown that both the heavy and the light chains for immunoglobulins of the same class are di-

Figure S3.3

Approximate locations of the hypervariable regions in the heavy and light chains of IgG. Each hypervariable segment is believed to make up part of the site that binds to antigen, known as the complementarity-determining region (CDR). The CDR regions are located in the loop regions of the variable domains where they can make close contact with antigen.

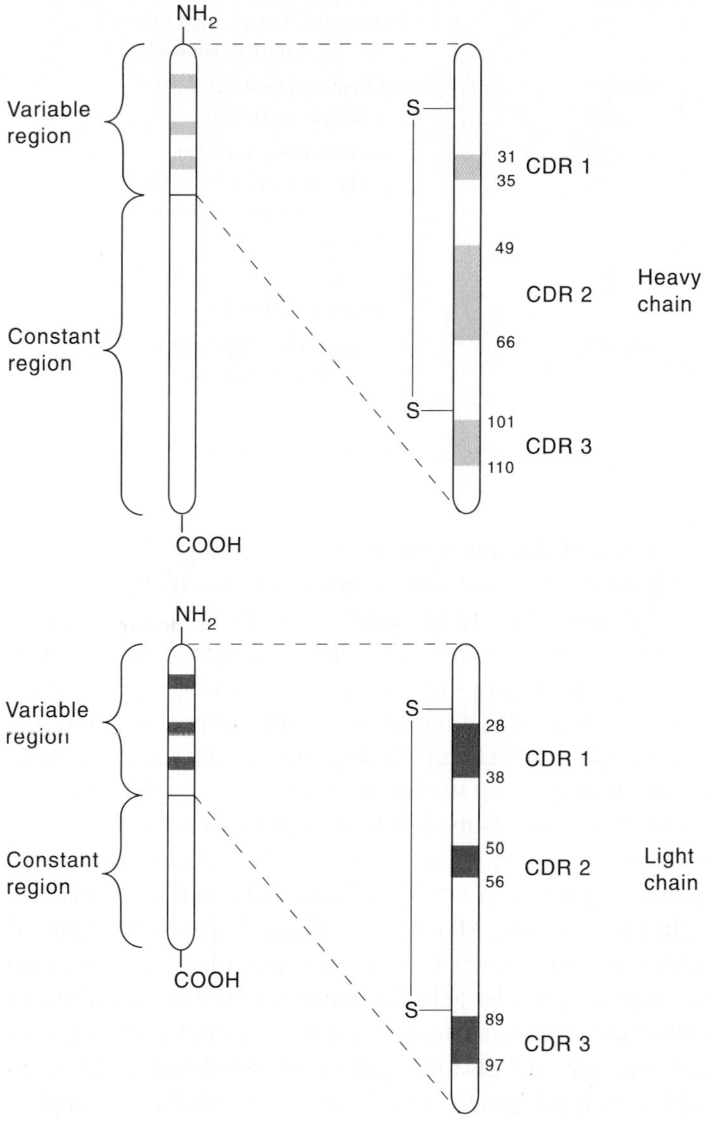

vided into regions of relatively constant sequence (the *C* segments) and regions of relatively variable sequence (the *V* segments). In the intact immunoglobulins the antigen-binding domain is composed exclusively of *V* segments originating from the H and L chains (see fig. S3.2), each tetramer containing two equivalent sites for the binding of a specific antigen.

As information on antibody sequences has accumulated, it has become increasingly clear that within the *V* regions of both the heavy and light chains, three segments account for most of the variability (fig. S3.3). These regions are called the hypervariable regions, and they are believed to be the parts of the antibody molecule in most direct contact with the antigen in the antigen–antibody complex.

Immunoglobulin G (IgG), the tetrameric species we have been discussing thus far, is not the only type of immunoglobulin found (table S3.1). Most of the other known antibodies—IgM, IgA, IgD, and IgE—also involve a closely related tetrameric structure, sometimes forming larger aggregates and always associated with different functions. For instance, IgM is a 19S antibody accounting for 5%–10% of the serum Ig. It is an aggregate of five tetramers that is formed early during the immune reaction, soon to be diminished in quantity and overshadowed by large amounts of IgG. IgA occurs in various polymeric forms. It normally accounts for about 15% of the total immunoglobulin found in serum, and in addition it is the principal immunoglobulin in exocrine secretions. Each of the different classes of immunoglobulin possesses distinct, heavy chains. Indeed, even within the human IgG class, antisera tests reveal four different types of IgG (IgG1 through IgG4), each with a distinctive H chain, comprising about 70, 19, 8, and 3%, respectively, of the total IgG proteins. The light chains, of which there are two types, are common to all classes of immunoglobulins.

Immunoglobulins that are produced in response to antigens are themselves highly immunogenic (that is, they stimulate antibody formation) when injected into genetically nonidentical organisms. The serological responses induced by using immunoglobulins as antigens or immunogens have been useful in classifying them according to their antigenic determinants.

So-called isotypic determinants are shared by all immunoglobulin molecules of a given class. For example, the human IgG molecules are classified into four isotypes because they are recognized by heterologous antisera produced in one species against the immunoglobulins of another species. Within a given isotype the different immunoglobulins can usually be separated into sets, called allotypes, that are distinguished by minor antigenic differences. Allotypic differences usually reflect alternative amino acid substitutions within otherwise quite similar amino acid sequences of the *C* region of the antibody. Allotypes show typical Mendelian inheritance patterns. Antisera useful for discriminating between allotypes are usually made by injecting an individual who lacks a specific allotype with immunoglobulins from an individual who carries the allotype. Finally, idiotype refers to the specific antigenic determinant of the antibody. Some idiotypic determinants are limited to a single immunoglobulin; others are shared by a small number of immunoglobulins. Whereas isotypic and allotypic differences usually result from differences in the

Table S3.1

Different Isotypes Found in Humans

Class	Heavy Chain	Light Chain	Molecular Formula	Molecular Weight (daltons)	Physiological Functions
IgG	γ	κ or λ	$\gamma_2\kappa_2$ $\gamma_2\lambda_2$	150,000	Complement fixation; placental transfer; stimulation of ingestion by macrophages
IgA	α	κ or λ	$\alpha_2\kappa_2$ $\alpha_2\lambda_2$	160,000 320,000	Localized protection of external secretions
IgM	μ	κ or λ	$\mu_2\kappa_2$ $\mu_2\lambda_2$	900,000	Complement fixation; early immune response; stimulation of ingestion by macrophages
IgD	δ	κ or λ	$\delta_2\kappa_2$ $\delta_2\lambda_2$	185,000	Found on cell surfaces; function unknown
IgE	ϵ	κ or λ	$\epsilon_2\kappa_2$ $\epsilon_2\lambda_2$	200,000	Stimulates mast cells to release histamines

constant portion, or *C* segments, of the immunoglobulin polypeptide chain, idiotypic differences usually result from differences in the variable portion, or *V* segment. As we know already, this is the region that contains the binding site for the foreign agent or antigen.

Antibody Diversity Is Augmented by Unique Genetic Mechanisms

Antibodies have been studied most extensively in the mouse, an ideal vertebrate for both genetic and biochemical manipulations. It is believed that mice can synthesize more than a million antibodies with different antigenic specificities. This enormous diversity of proteins is generated from a limited amount of genetic information with the help of two mechanisms: Somatic recombination and somatic mutation.

DNA Splicing Brings Different Parts of the Antibody Gene Together. The involvement of somatic recombination was first demonstrated by examining the structure of a specific antibody gene in embryonic and adult immunoglobulin-forming tissue. The adult tissues favored for many studies are myelomas. These tumorous tissues, which we mentioned earlier, produce a homogeneous population of polypeptide chains. They have served as a convenient source of pure immunoglobulin polypeptide chains as well as their mRNAs.

Recall that the generalized structure of the predominant serum antibody consists of two identical heavy chains and two identical light chains (see fig. S3.2). In the mouse there are two classes of light chains, which differ appreciably in the constant regions of the polypeptide chains. These are referred to as the kappa (κ) and the lambda (λ) light chains. Using highly inbred, genetically identical (isogenic) strains of mice, S. Tonegawa and his co-workers isolated the DNA from embryonic cells and two different myeloma tumor cells: One that produces homogeneous λ light chains (strain H2020) and one that produces κ light chains (strain MOPC321). These DNAs were digested with the *Eco*R1 restriction enzyme and electrophoresed on an agarose gel. The gels contained an enormous variety of restriction fragments representing total nuclear DNA, and consequently no discrete pattern of bands could be seen with a stain for nucleic acid. However, when the electrophoresed gels containing the DNA were denatured and hybridized with ^{32}P-labeled DNA containing the sequences found in the RNA for the λ chain, a specific pattern of bands showed up in autoradiographs (fig. S3.4). The R1 digest of DNA from the λ-containing myeloma (H2020) showed four bands; the DNAs of the embryo and the κ-containing myeloma (MOPC321) showed three bands in common with the first DNA but were missing the fourth band.

These results strongly suggested that at some point during development from the embryonic state, a rearrange-

Figure S3.4

Analysis of DNA fragments containing λ_1 gene sequences from mouse embryo and myeloma cells. High-molecular-weight DNAs extracted from myeloma H2020, a λ-chain producer (A), from a 13-day-old BALB/c embryo (B), and from myeloma MOPC321, a κ-chain producer (C) were digested to completion with *Eco*R1, electrophoresed on agarose gel, and transferred to nitrocellulose membrane filters and hybridized with a nick-translated *Hha*I fragment of the plasmid B1 DNA. (Source: After C. Brack, M. Hirama, R. Lemhard-Schuller, and S. Tonegawa, A complete immunoglobulin gene is created by somatic recombination, *Cell* 15:1–14, 1978.)

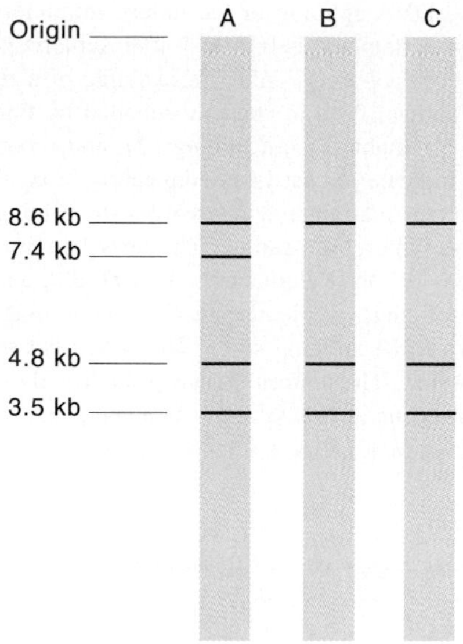

Figure S3.5

Arrangement of mouse λ_1 gene sequences in embryos and λ_1 chain-producing plasma cells. The vertical arrows point to *Eco*R1 restriction sites. The blue dashed diagonal lines point to hypothesized splice points that explain the difference in structure of the region in the two cell types. The boxed regions represent coding regions, and I_1 and I_2 between boxed regions indicate first and second introns in the spliced gene.

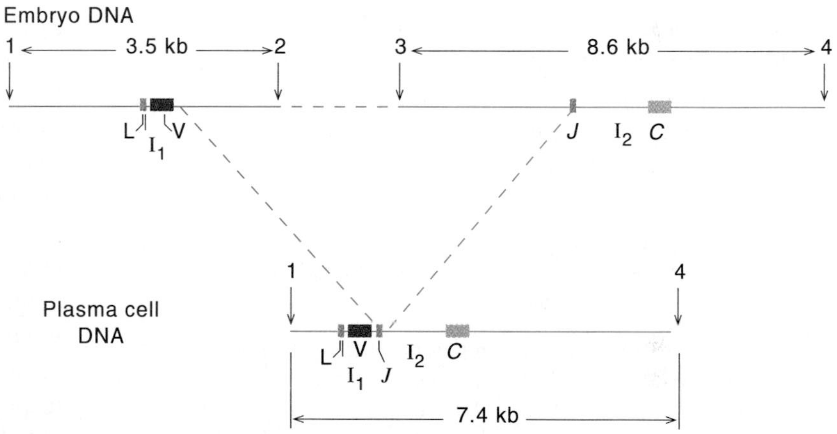

ment had taken place in the H2020 myeloma cells on one of the homologous pairs of chromosomes carrying the λ-chain gene. Further analyses with more specific radioactive probes, containing sequences from either C_λ (the constant region of the λ chain) or V_λ (the variable region of the λ chain), were done. Only the 7.4-kb fragment originating from the λ myeloma cell hybridized to both probes. Thus, the *Eco*R1 fragment must contain both V_λ and C_λ DNA. The 8.6-kb fragment hybridized to the C_λ but not the V_λ probe. Therefore this fragment must contain C_λ sequences but no V_λ sequences. Both the 3.5-kb fragment and the 4.8-kb fragment hybridized exclusively to the V_λ probe, so they must contain only V_λ sequences. Still further analyses indicated that the 7.4-kb fragment arose from a recombinational event between the 3.5-kb fragment and the 8.6-kb fragment (fig. S3.5). The 4.8-kb fragment originates from another V_λ sequence located in the same chromosome. The mouse has only two V_λ sequences in the embryo. As a rule, only one of the pairs of homologous chromosomes recombines to yield a productive antibody gene. This explains why the 3.5-kb and the 8.6-kb bands are still visible in the H2020 myeloma DNA preparation. The κ myeloma cell producer shows the same pattern as the embryonic cell. Had it been probed with a DNA carrying the V_κ antibody sequence, it would have shown differences from the embryonic pattern.

Figure S3.6

Organization and DNA splicing of the mouse immunoglobulin genes. The organization is depicted before and after somatic cell rearrangement. For each class only one example of a rearranged chromosome is shown. A light chain is encoded by three distinct DNA elements, a variable (*V*), a joining (*J*), and a constant (*C*) element. Boxes indicate exons; lines connecting boxes indicate introns. Somatic cell rearrangement involves the joining of a *V* and a *J* segment on the same chromosome. The heavy chain is encoded by four distinct types of DNA elements: a *V*, a *D*, a *J*, and a *C* element. In heavy-chain splicing a *V* element, a *D* element, and a *J* element are joined in two DNA-splicing steps. The splice between *D* and *J* elements occurs first. The splices shown provide only one example of many that might occur in this system. (From G. Zubay, *Genetics*, Benjamin/Cummings, Menlo Park, Calif. 1987, p. 808.)

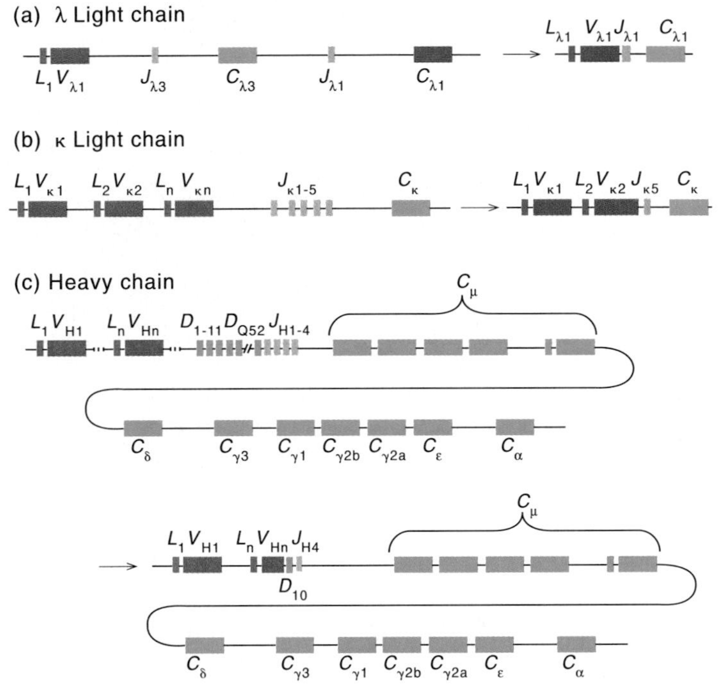

The success achieved by Tonegawa and others in this type of sequence detection of the immunoglobulin genes led to a massive research effort and a general picture of antibody gene organization. All antibody polypeptide chains are derived from split genes. Antibody genes are encoded by three unlinked gene families located on different chromosomes. In the human, chromosomes 2, 22, and 14 encode those gene clusters associated with the kappa (κ) light chains, the lambda (λ) light chains, and the heavy chains, respectively.

The light chain is encoded by three distinct DNA elements—a variable (V_L), a joining (J_L), and a constant (C_L) element (fig. S3.6). During the differentiation of a B cell, a V_L element (encoding the first 95 amino-terminal residues) and a J_L segment (encoding about 13 amino acids) are joined by DNA splicing. A complete light-chain gene transcript consists of three exons and two introns, arranged in the following order; 5′L, I_1, *VJ*, I_2, *C*, 3′. Here L refers to sequences in the leader, or signal, peptide, which is removed in the mature immunoglobulin, and I_1 and I_2 refer to the first and second introns, respectively, which are removed by RNA splicing. The heavy chain is encoded by four distinct DNA elements: a V_H, a *D*, a J_H, and a C_H element. In heavy-chain variable region formation, a V_H element (encoding about 99 amino acids), a *D* element (encoding 1–15 amino acids), and a J_H element (encoding about 15 amino acids) are joined in two DNA-splicing steps. The heavy-chain gene family has several closely linked C_H genes, which determine the immunoglobulin class.

Class switching, involving an additional DNA-splicing operation, frequently occurs with heavy-chain genes. Thus, an immature B cell, which initially expresses a μ chain, results in IgM antibodies. Subsequently, the same cell can be induced to differentiate further so that the same V_H region of the genome becomes relocated next to a C_γ gene, causing the cell to produce an IgG antibody with the same V_H region associated with a different C_H region. In cases of class switching, the second DNA splicing involves removal of the intervening genes. Thus, in the case of C_μ to C_γ switching, the region containing C_δ must be removed by the second splice reaction (fig. S3.7). In the case of C_μ to C_α class switching, all of the *C* genes between C_μ and C_α must be removed by the second splice reaction.

Class switching of the heavy-chain genes illustrates a temporally regulated process associated with the DNA rearrangements that can occur in the antibody-forming genes. Because any given cell normally synthesizes only one type of antibody, a special type of regulatory mechanism must respond in a systematic way to antigenic stimulation. In B-cell development it is known that the V_H-D-J_H joining occurs first, leading to the formation of μ heavy chains. Subsequently, V_L-J_L joining occurs, with the production of functional light chains.

As a rule, the DNA-splicing reactions leading to the joining process occur in only one of the alleles for each of these chains. The mechanism for this phenomenon, known by the name allelic exclusion, is not understood. Another type of exclusion process results in the production of only κ or λ light chains in a given B cell. This process is called isotypic exclusion. Indirect evidence that the κ gene family is expressed before the λ gene family comes from the examination of human B-cell lines that are active κ- or λ-chain producers. In all B cells producing λ chains, the C_κ genes are either deleted or rearranged. By contrast, in all cells producing κ chains, the C_λ genes are in

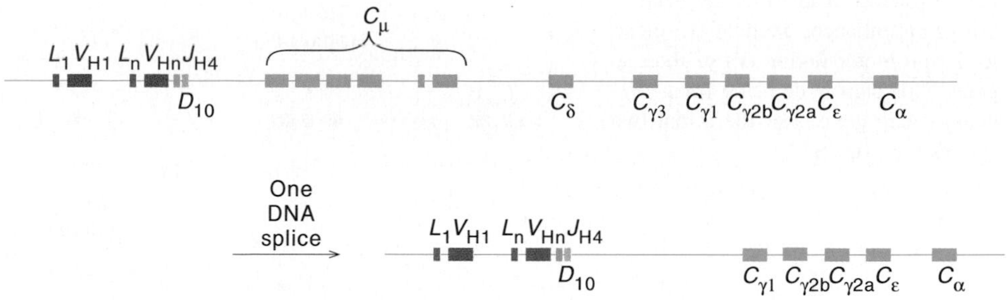

Figure S3.7

Class switching of heavy-chain genes involves additional DNA splicing. Switching form C_μ to $C_{\gamma 1}$ is illustrated. The additional splicing involves removal of the continuous segment carrying the C_μ through the $C_{\gamma 3}$ genes. (From G. Zubay, *Genetics*, Benjamin/Cummings, Menlo Park, Calif. 1987, p. 809.)

the germ-line configuration. The obvious inference is that a B cell becomes a κ producer first and then becomes a λ producer. Other evidence exists for regulatory signals that may control isotypic exclusion. When a pre-B cell is fused with a myeloma cell that has been producing a functional heavy chain but no functional light chain, the hybrid cell is capable of expressing a new light chain as a result of a *V-J* joining of gene segments in the pre-B cell. If, however, the myeloma cell was already producing a functional light chain, no new light chains or DNA rearrangements result from the cell fusion process.

Alternative Pathways Exist for RNA Splicing of Heavy-Chain mRNAs. The heavy-chain transcript contains numerous exons and introns. In addition to the introns between the L_H and V_H exons and between the V_H-J_H and C_H exons, each C_H region contains several introns. The hinge regions of γ chains (see fig. S3.2) and the intramembrane and cytoplasmic portions of all C_H chains are encoded by independent exons. Most of the RNA-splicing operations are presumed to occur by standard mechanisms similar to those described in chapter 28. In the case of the heavy-chain genes, it has been shown that alternative RNA-splicing patterns involving the C_H region can yield molecules that function as membrane-bound or secreted forms of the antibody. Thus, two membrane exons have been detected in the 3′ flanking regions of both the C_μ and C_γ genes. Alternative pathways of RNA splicing give rise to two mRNAs, which encode separately the membrane-bound and the secreted forms of the heavy chains. It seems likely that the alternative splicing pattern is temporally regulated, because the membrane-bound forms are favored before and the secreted forms are favored after antigenic stimulation. Amino acid as well as nucleotide sequencing of the μ_M chain (μ_M stands for the membrane-bound form of the μ chain) has revealed a 26-residue hydrophobic segment at the COOH terminus, which anchors in the lipid bilayer of the membrane.

Somatic Mutation Contributes to Antibody Diversity. Various estimates, some exceeding a million, have been put forward for the number of different antibodies an organism can make. A good deal of this diversity results from the different combinations of heavy and light chains that can be generated by making alternative splicings between the different gene segments that make up the light and heavy chains (table S3.2). Superimposed on the variation from this source, additional diversity is provided by somatic mutation within the coding regions. Several subsets of κ and heavy chains and their germ-line *V* segments have been analyzed by cloning and sequencing. The results confirm that somatic mutations amplify the diversity encoded in the germ-line genome.

A V_κ probe prepared for the κ genes expressed in a particular myeloma, MOPC167, detected only one major band in the Southern gel blot analysis of total cellular DNA. This unusual situation permitted a relatively simple sequence comparison to be made between this segment, found in the germ-line, and its rearranged counterparts in two myeloma lines: MOPC167 and MOPC511. Four and five nucleotide differences were found, respectively, in the V_κ regions of these two lines. The changes were completely

Table S3.2

Estimated Number of Germ-Line Gene Segments for Different Mouse Antibody Genes

	Light Chains		Heavy Chains
	κ	λ	*H*
C	1	4	8
V	90–300	2	100–200
J	5	4	4
D	—	—	12

From Geoffrey Zubay, *Genetics* (p. 810). Copyright © 1987 Benjamin/Cummings Publishing Company Inc., Menlo Park, Calif. Reprinted by permission of the author.

Figure S3.8

DNA splicing of an antibody gene brings an enhancer element (E) close to the promoter region (P) of the gene. This step is depicted for a heavy-chain gene after the initial two splices (see fig. S3.6).

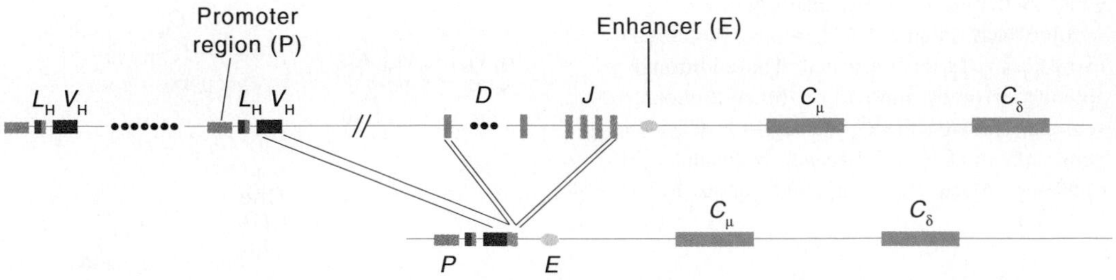

Figure S3.9

Pluripotent stem cells in the marrow give rise to more stem cells and to various progenitors. The lymphocyte progenitors give rise to B-cell and T-cell lineages. The final differentiation step in both B-cell and T-cell development requires antigenic stimulation.

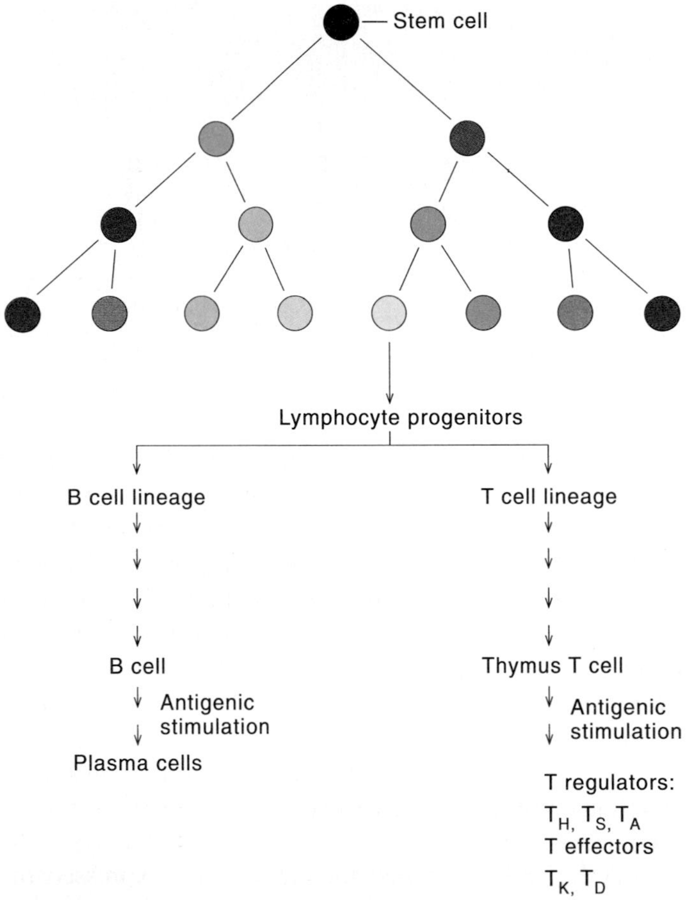

different in both lines. This investigation and other similarly directed investigations show that the regions in which mutations are introduced during somatic differentiation are highly restricted to the V_1-J_1 or V_H-D-J_H segments. The density of mutations observed is about three times higher in the complementarity-determining regions (CDRs) of the variable segments than in the connecting framework regions (FRs) of the variable segments. (See fig. S3.3 for the approximate location of the CDR regions.) The CDRs are the regions in the immunoglobulin that play a critical role in the interaction with antigen. Mutations produced in these regions must occur either during or after DNA splicing, as they are not found in unrearranged (unspliced) DNA. Somatic mutation clearly increases the number of possible antibodies almost without limit. Most importantly, it leads to selectable changes in just those regions of the immunoglobulins that are responsible for antigen recognition. Several mechanisms for hypermutation have been proposed; none have been verified.

DNA Splicing Brings an Enhancer Element Close to the Promoter. DNA sequences derived from the germ-line J_H-C_μ region are required for accurate and efficient transcription from a functionally rearranged V_H promoter (fig. S3.8). Similar to viral transcriptional enhancer elements (see chapter 28 and supplement 4), these cellular sequences stimulate transcription from either the homologous V_H gene segment promoter or a heterologous SV40 promoter. They are active when placed on either the 5′ or the 3′ side of the rearranged V_H gene segment, and they function when their orientation is reversed. However, unlike viral enhancers, the immunoglobulin gene enhancer appears to act in a tissue-specific manner because it is active in mouse B cells but not in mouse fibroblasts.

The discovery of this enhancer suggests another benefit that results from splicing; it brings an enhancer in close proximity to the promoter just when that promoter activity becomes an important part of the cell's function.

Interaction of B Cells and T Cells Is Required for Antibody Formation

Thus far, our discussion of antibody formation has centered on those reactions that occur at the gene level and on the structure of the antibody. However, since the immune response *in vivo* involves reactions between whole cells, we

Figure S3.10

Kinetics of IgM and IgG appearance in the serum following a first and a second exposure to the same antigen. On first exposure a delay of several days occurs before any antibody appears in the serum. IgM appears before IgG. After many days the serum level falls. If the system is exposed to a second equivalent dose of antigen, the appearance of IgM and IgG is much faster, and the maximum level of IgG is much higher. (From G. Zubay, *Genetics,* Benjamin/Cummings, Menlo Park, Calif. 1987, p. 812.)

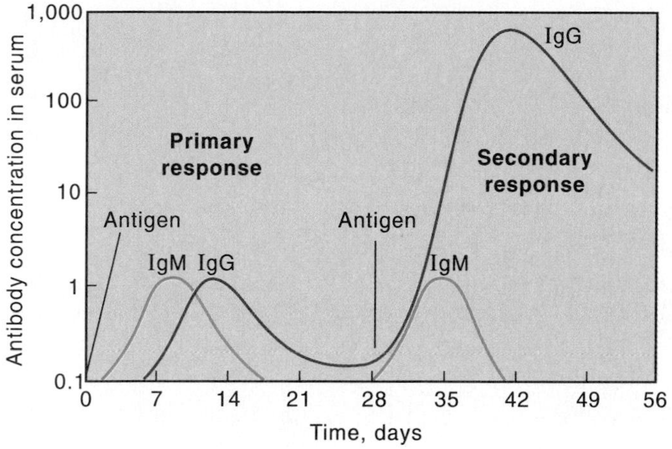

Figure S3.11

An immature B cell (*a*) and a fully developed antibody-producing B cell (*b*). Most notable in (*b*) is the expanded cytoplasm with densely packed endoplasmic reticulum. (Electron micrographs courtesy of Dorthea Zucker-Franklin, New York University Medical Center.)

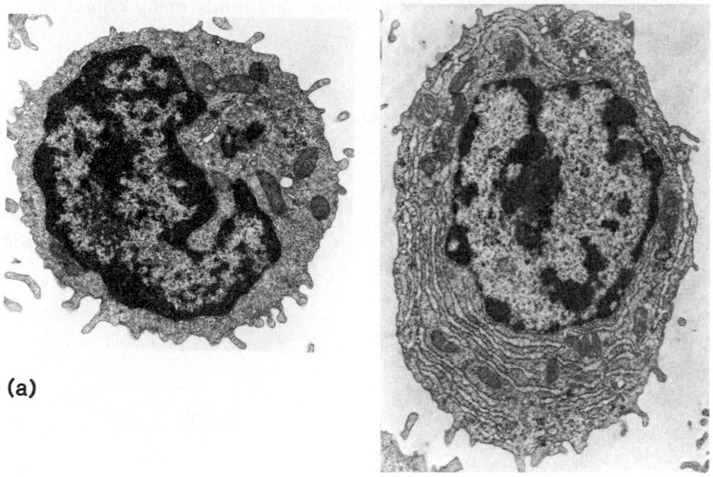

(a)

(b)

must also consider what happens at the level of cell-to-cell interactions. Some of the pluripotent stem cells of the bone marrow replicate in an undifferentiated state, maintaining their pluripotency, whereas others replicate and differentiate at the same time. The latter cells can differentiate along various pathways (fig. S3.9). Some of those that become committed to the lymphocyte pathway differentiate further into B-cell and T-cell lineages. T cells mature in the thymus, whereas B cells mature in the marrow in most vertebrates. Subsequently, both cell types relocate in secondary lymphatic organs, the spleen and the lymph nodes, and also more diffusely in all tissues. Here they remain dormant until they encounter specific antigens that trigger them to complete their differentiation into fully mature effector B cells and effector and regulator T cells. Each B cell contains immunoglobulins of a specific idiotype bound to its surface, usually of the IgM type and to a lesser extent of the IgD and IgG types.

Antigens Stimulate the Formation of B-Cell Clones. The first time a foreign antigen is injected into the bloodstream, there is a long delay before specific antibody appears, and the response is fairly weak. The second time the same antigen is injected, the response is both more rapid and more intense (fig. S3.10). A great deal of research effort has gone into seeking an explanation for this difference between primary and secondary responses. Specific labeling studies show that secondary lymphatic organs contain a very heterogeneous mixture of B cells, carrying membrane-bound antibodies of different idiotypes. When sufficient amounts of antigen bind to the surface antibody, the B cell becomes triggered to divide and make more cells of the same type. Some of these continue to divide, making clones of memory cells in the lymphatic organs, whereas others terminally differentiate into plasma cells richly laced with endoplasmic reticulum that is geared to the production of the antibody of the same idiotype as that which is membrane-bound (fig. S3.11). Whereas the idiotype and consequently the V_L and V_H parts of the antibody are the same as in the original, unstimulated B cell, the *C* regions change. The first secreted antibody is of the secretory IgM type. This is gradually replaced by the secretory IgG type, which dominates in the secondary response. As we have noted (see fig. S3.7), the replacement of IgM by IgG is indicative of an additional DNA splicing.

Helper T Cells Trigger B-Cell Division and Differentiation. The importance of T cells in mediating the B-cell response to most antigens is demonstrated by the fact that thymectomized animals, which are depleted in T cells, usually show a greatly attenuated reaction to foreign antigen. A normal response can be restored by transplantation of thymus tissue from a genetically identical animal. Activated T helper cells interact with B cells whose antibodies can recognize the same antigen that stimulated the helper T cell. The complex between the T and B cell results in the clustering of a large number of transmembrane proteins causing a

Figure S3.12

Stimulation of antibody-forming cells involves three types of cells. Typically, antigen is phagocytosed by a macrophage. The phagocyte partially digests the antigen and "presents" the processed antigen on its outer plasma membrane. A specific helper T cell binds the antigen to a receptor bound to its outer plasma membrane. The T cell usually has many of the same types of receptors and binds many copies of the antigen in a similar way. This stimulates T-cell proliferation. These T cells interact with B cells that display similar processed antigen. Several B-cell receptors involved in a similar way become focused in a local region of the outer plasma membrane, a process known as CAP formation. CAP formation triggers the proliferation and differentiation of the B cells involved.

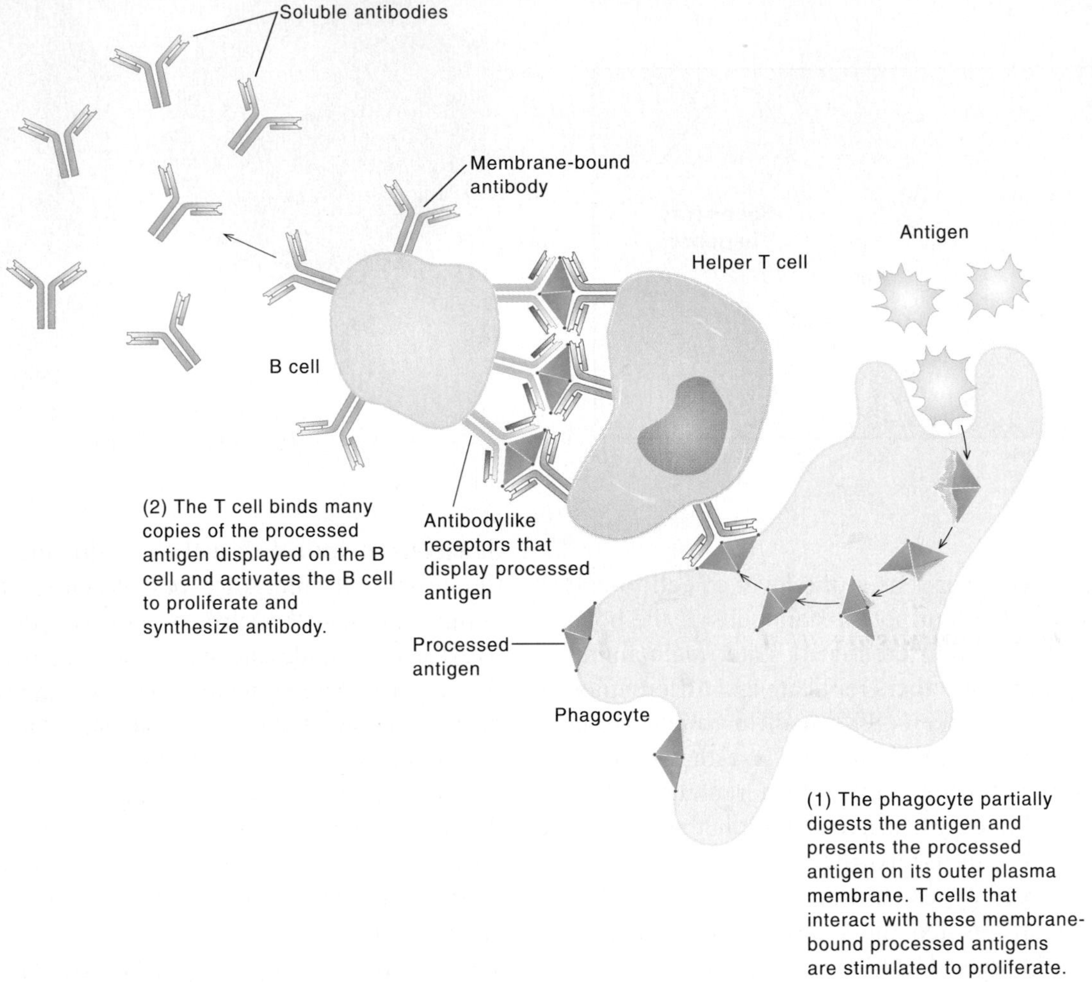
Soluble antibodies
Membrane-bound antibody
Antigen
Helper T cell
B cell
(2) The T cell binds many copies of the processed antigen displayed on the B cell and activates the B cell to proliferate and synthesize antibody.
Antibodylike receptors that display processed antigen
Processed antigen
Phagocyte
(1) The phagocyte partially digests the antigen and presents the processed antigen on its outer plasma membrane. T cells that interact with these membrane-bound processed antigens are stimulated to proliferate.

process known as CAP formation. CAP formation by some unknown means triggers the proliferation and differentiation of the B cells involved.

Support for the role of T helper cells in focusing the antigen comes from the observation that T-independent stimulation of B cells is frequently elicited by polymeric antigens. Such polymeric antigens constitute another way of presenting a large amount of antigen at one location on the B-cell membrane, thereby focusing the B-cell receptors.

Before they can help other lymphocytes respond to antigen, helper T cells must be activated themselves. This activation occurs when a helper T cell recognizes a foreign antigen bound on the surface of a specialized antigen-presenting cell. The latter cells are found in most tissues; they are derived from bone marrow and constitute a heterogeneous group, including dendritic cells in lymphoid organs and certain types of macrophages. Together with B cells, which can present antigen to helper T cells, these are the main cell types that react with helper T cells. The sequence of events involving antigen-processing cell, T helper cell, and antibody-forming B cell is depicted in figure S3.12.

Note that the mechanism proposed for helper T-cell function requires that T cells have surface receptors that recognize antigen-processing cells, antigen itself, and the appropriate B cell. We will shortly discuss the nature of T-cell receptors, as well as the surface structures they recognize, in a broader context.

Figure S3.13

The principal stages in complement activation. Complement activation occurs exclusively on the microbial cell membrane, where it is triggered by bound antibody or microbial envelope polysaccharides, both of which activate early complement components. Two sets of early components belong to two distinct pathways of complement activation. Activation of each complement system involves a cascade of proteolytic reactions. Each component of the complement system is a proenzyme that is activated by the preceding component of the chain by a limited proteolytic cleavage. The ultimate result of this chain reaction is the development of a complex that attacks the cell membrane.

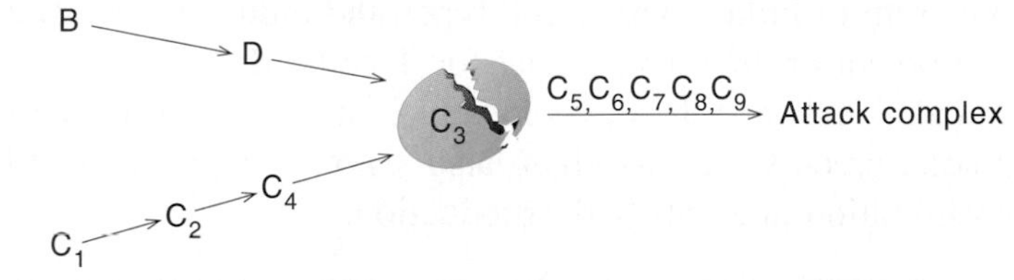

The Complement System Facilitates Removal of Microorganisms and Antigen–Antibody Complexes

Many antigen–antibody complexes are eliminated by phagocytosis. Others are attacked by complement, a group of serum proteins that aids in the defense against microorganisms. Individuals with a deficiency in their complement system are subject to repeated bacterial infections, just as are individuals deficient in antibodies themselves.

About 20 different proteins are included in the complement system: Proteins C1–C9, factors B and D, and a series of regulatory proteins. All these proteins are made in the liver, and they circulate freely in the blood and extracellular fluid. Activation of the complement system involves a cascade of proteolytic reactions. In addition to forming membrane attack complexes, the proteolytic fragments released during the activation process promote dilation of blood vessels and the accumulation of phagocytes at the site of infection.

The activation of complement occurs exclusively on the microbial cell membrane, where it is triggered either by bound antibody or microbial envelope polysaccharides, both of which activate early complement components. Two sets of early components belong to two distinct pathways of complement activation: C1, C2, and C4 belong to the pathway that is triggered by antibody binding; factors B and D belong to the alternative pathway that is triggered by microbial polysaccharides (fig. S3.13). The early components of both pathways ultimately act on C3. The early components and C3 are proenzymes that are activated sequentially by limited proteolytic cleavage. As each proenzyme in the sequence is cleaved, it is activated to generate a serine protease, which cleaves the next proenzyme in the sequence. Many of these cleavages liberate small peptide fragments and expose a membrane-binding site on the larger fragment. The larger fragment binds tightly to the target cell membrane by its newly exposed membrane-binding site and helps to carry out the next reaction in the sequence. In this way complement activation is confined largely to the cell surface, where it began.

The components of the complement attack complex have a very short half-life if released from the complex. This limitation confines their destructive action to the point of assembly, the surface membrane of the microorganism.

The Cell-Mediated Response: A Separate Response by T Cells

Most B cells are fixed in the secondary lymphatic organs: The spleen and the lymph nodes. They are stimulated by antigen only if the antigen is transported to the lymph nodes. In many cases, however, antigens are poorly transported. To rid the organism of antigens that are not readily transported, a second type of immune system exists that is independent of the B cells. This second type of immunity is not found in all vertebrates; it is restricted to mammals, birds, certain amphibians, and bony fishes. Long before the distinction between T and B cells was appreciated, it was known that certain types of immunity, which manifest a delayed-type hypersensitivity response, could be transferred from immunized animals to nonimmunized animals with leukocytes but not with antibody-containing serum. Subsequently, other forms of leukocyte-mediated immunity were discovered that involved increased phagocytic and bacteriocidal activity of macrophages, lysis of virus-infected cells and tumor cells, and the rejection of skin grafts from genetically dissimilar organisms.

The two types of cell-mediated immunity involve basically different types of T cells. In the delayed-type hypersensitivity response, the T cell, T_D, that reacts specifically with antigens secretes interleukins (see Box S3A) that attract and activate macrophages or other leukocytes, thereby causing a slowly developing inflammatory response. A second type of T cell, known as the T killer cell, T_K, reacts specifically with antigen that is bound to target cells, causing their lysis.

BOX S3A The Role of Interleukins

The interactions between immune and inflammatory cells are mediated in large part by certain proteins, called lymphokines, or interleukins (IL), that are able to promote cell growth, differentiation, and functional activation. Interleukins resemble hormones, which also function as intercellular messengers. In contrast to hormones, however, they are secreted by isolated cells rather than discrete glands. Several interleukins have been described; each has unique biological activities as well as some that overlap with the others. For example, macrophages produce IL-1 and IL-6, whereas T cells produce Il-2–IL-6, and bone marrow stromal cells produce IL-7. IL-1 and IL-6 not only play important roles in immune cell function, they also stimulate a spectrum of inflammatory cell types and induce fever. IL-2 is a potent proliferative signal for T cells. IL-1, Il-3, IL-4, and IL-7 enhance the development of a variety of hematopoietic precursors. IL-4–IL-6 also serve to enhance B-cell proliferation and anti-body production.

Tolerance Prevents the Immune System from Attacking Self-Antigens

A thorough understanding of the immune system requires an explanation for how the immune system is able to discriminate between foreign antigens and its own antigens (self-antigens). The ability to recognize self-antigens and thus to avoid making an antagonistic response to them is called tolerance. Some serious diseases are believed to be due to a breakdown in the self-tolerance mechanism. Conversely, conditions occur under which a foreign antigen is able to establish itself so that it becomes tolerated, for either a short or a long time. Tolerance and rejection are so closely related that it seems likely that a full understanding of one entails a full understanding of the other.

Several conditions favor the establishment of tolerance: (1) Tolerance is much easier to establish in the fetus or the newborn than in the adult. Thus, if an organism is exposed as a fetus to a soluble foreign antigen or a foreign tissue graft, it may become indefinitely tolerant to the antigen. This situation is illustrated for two genetically nonidentical strains of mice in figure S3.14. Cells from the Y mouse are injected into a newborn X mouse. The same X mouse as an adult is the recipient of a skin graft (transplant) from a Y-type mouse. Normally such a graft would be rapidly rejected. However, because of the early exposure to the Y cells, the graft is tolerated indefinitely. This result indicates that the mouse that was exposed to the Y cells just after birth recognizes Y cells later in life as self-antigens. (2) Very high levels of antigen frequently favor the development of tolerance, whereas intermediate levels of antigen favor the development of an immune response. (3) Frequently, the route of antigen administration is critical in the development of tolerance. An intravenously injected antigen is more likely to promote a tolerant condition than antigen injected subcutaneously.

Two cellular mechanisms may account for most forms of tolerance. In one of these the clones of T or B cells that could otherwise be activated by the antigen are destroyed or inactivated. In the other mechanism, the clones in question are still present but do not respond because they are blocked by T_S cells. In general, B cells are more difficult to make tolerant than T cells. This difference is evident from observations that longer and higher doses of antigen are required to establish tolerance to the B-dependent system. Moreover, the period of tolerance is generally shorter when B cells are involved.

Adult mice may be made tolerant to foreign antigens by destroying their immune system with whole-body irradiation and then supplying them with transplants of bone marrow and thymus from a foreign donor, thus refurbishing their system with competent B and T cells, respectively. With the new immune system the animal is now tolerant to any tissue from a mouse that is isogenic with the mouse used to supply the marrow and the thymus.

The irradiated host organism also supplies a useful means of testing the state of B and T cells from a nonirradiated isogenic mouse. B cells may be effectively transferred

Figure S3.14

Establishment of tolerance to tissue grafts. If a newborn X mouse is injected with Y cells when it is young, it is tolerant to Y tissue transplants when it becomes an adult. If the X mouse is not exposed to Y cells at an early age, it readily rejects the tissue graft.

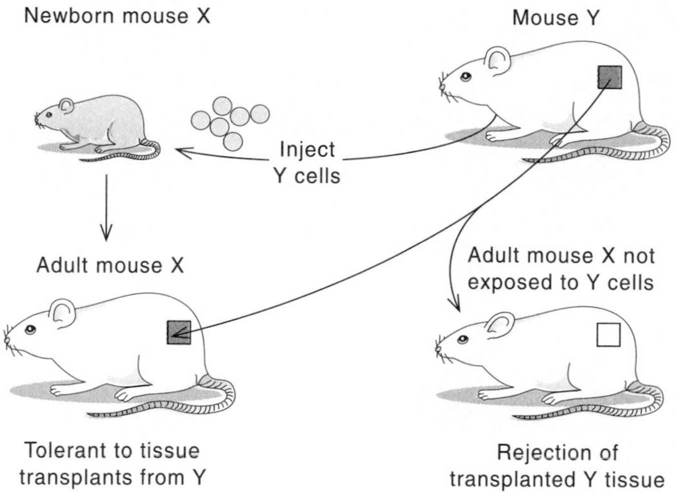

by bone marrow transplants, and T cells may be transferred by thymus tissue transplants. The donor cells may come from a normal, untreated isogenic strain. In this event the irradiated recipient shows normal immunological behavior. Alternatively, the donor cells may come from a donor that has been made either sensitive or tolerant to a particular antigen. By doing various combinations of experiments using different isogenic donors, it is possible to determine the relative importance of B-cell and T-cell involvement with different antigens.

T Cells Recognize a Combination of Self and Nonself

All T cells, whether they fight infection directly or regulate the activity of other effector cells, recognize antigen only in combination with a cell membrane surface. Of greatest significance, the antigen must be associated with a cell from an organism that is genetically similar or identical to that of the T cell. This requirement was first demonstrated in an experiment where T cells were removed from a mouse that had been immunized with a virus. Immunization meant that the extracted T cells had been activated to direct themselves against cells infected with this particular virus. Thus, these T cells could kill fibroblasts infected with the virus in tissue culture. However, they could not kill fibroblasts infected with the virus if the fibroblasts came from a mouse with a different genetic background. Thus, sensitized killer T cells do not recognize foreign antigens alone; they recognize foreign antigens only in combination with determinants present on their host's own cells (fig. S3.15). These common determinants are a subset of the proteins encoded by the so-called major histocompatibility complex (MHC).

MHC Molecules Account for Graft Rejection

MHC molecules were recognized long before their major biological function was appreciated. They were initially defined as the main target antigens in transplantation reactions, and because of this they became known as major histocompatibility antigens. Experiments on mice demonstrated that graft rejection is an immune response to the surface antigens of the grafted cells. Subsequently, it was shown that these reactions are mediated by T cells and that they are directed against a family of glycoproteins encoded by a group of genes called the major histocompatibility complex. MHC molecules are transmembrane proteins found on the cell membranes of all higher vertebrates.

A vertebrate does not normally need to be protected against invasion by foreign vertebrate cells, so the antagonistic reaction of its T cells to foreign MHC molecules was both an obstacle to transplant operations and an enigma to immunologists. The enigma was solved when it was discovered that MHC molecules contributed to the binding of T lymphocytes on those host cells that have foreign antigen on their surface, as in the case of the virus-infected cells we have just described (see fig. S3.15).

There Are Two Major Types of MHC Proteins: Class I and Class II

There are two major classes of MHC proteins that have similar types of structures (fig. S3.16). Class I molecules contain three external domains, each about 90 residues in length; a transmembrane region; and a cytoplasmic domain. The third external domain is noncovalently associated with a small polypeptide known as the β_2 microglobulin. Class II molecules are composed of two noncovalently associated polypeptide chains: α and β, whose overall structure resembles that of the class I complex.

Several genes encode each type of MHC protein, and several alleles represent different MHC proteins for each of these genes. This gives rise to a tremendous variety of MHC proteins, with the consequence that it is extremely unlikely that any two members of the same species will possess the same assortment of MHC proteins. Thus, two humans are not likely to carry the same MHC proteins on their cell membranes unless they are identical twins. This is the reason why transplants between two individuals are almost

Figure S3.15

Activated T cells bind and kill only those cells that display a foreign antigen and a familiar cell surface antigen.

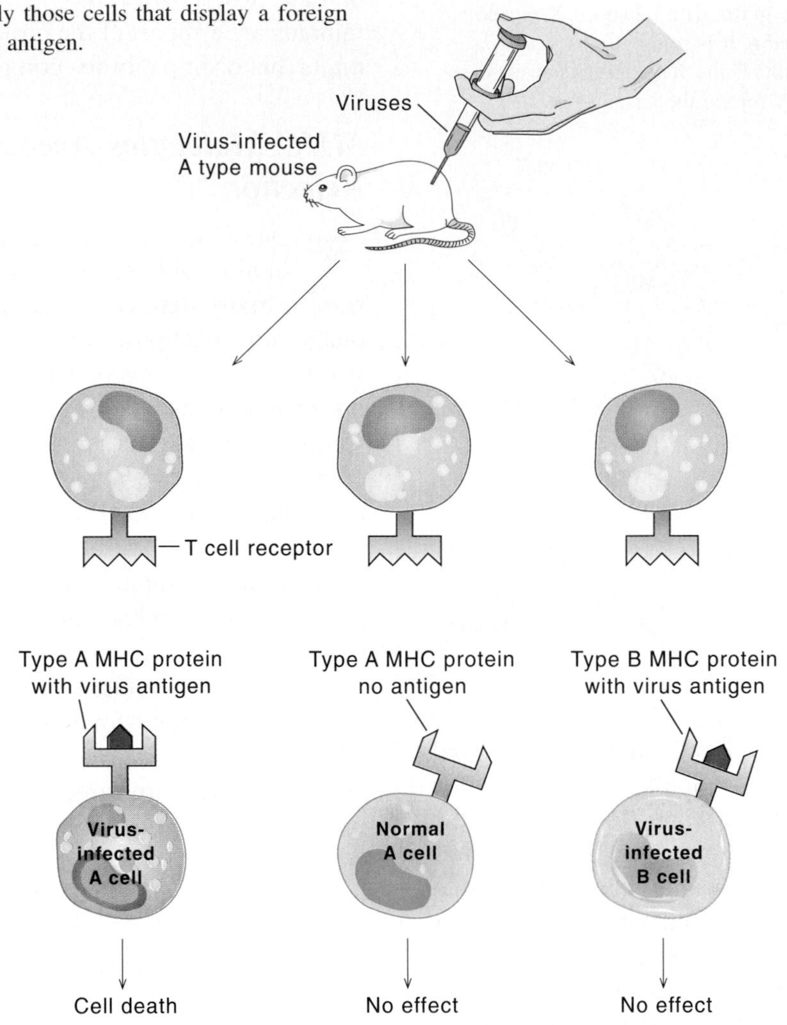

always rejected—it is usually just a matter of time. Highly inbred strains of mice, which are genetically identical, display the same MHC proteins and accordingly take grafts from each other without any rejection.

The two classes of MHC proteins are displayed on different cell types. Class I MHC proteins are found on almost all nucleated cells, including killer T cells. Class II MHC proteins are found mainly on cells involved in the immune response, including antigen-presenting cells, B cells, and T helper cells, but not T killer cells.

T-Cell Receptors Resemble Membrane-Bound Antibodies

The specific receptors found on the membranes of T cells resemble the highly specific antibodies found on the membranes of B cells. Specific T-cell receptors (TCRs) are composed of two disulfide-linked peptide chains (called α and β). Each of these chains share with antibodies the distinctive property of a variable amino-terminal region and a constant carboxyl-terminal region.

With one exception, all the mechanisms used by B cells to generate antibody diversity are also used by T cells to generate T-cell receptor diversity. The one mechanism that does not appear to operate in T-cell receptor diversification is somatic hypermutation. This is presumably because mutation would be likely to generate killer T cells that would wantonly attack self-molecules. This is much less of a problem for B cells, since most self-reactive B cells could not be activated without the aid of specific helper T cells.

Additional Cell Adhesion Proteins Are Required to Mediate the Immune Response

In addition to membrane-bound antibodies, TCR proteins and MHC proteins, other transmembrane proteins are required to obtain the specific cell–cell interactions necessary to mediate the immune response. These additional proteins

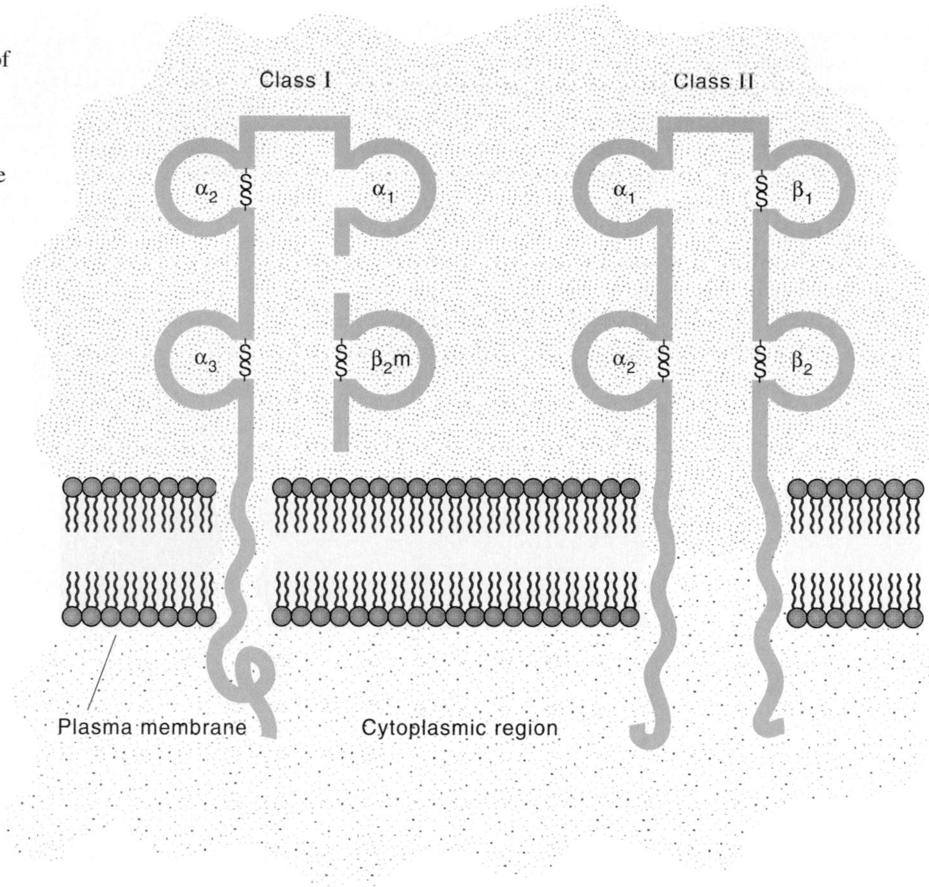

Figure S3.16

Structures of class I and class II molecules as they would appear in the lipid bilayers of the cell membrane. Models show striking similarities between the two types of molecules. S—S indicates disulfide bridges in the class I and class II molecules and the β_2 microglobulin.

are also important in a number of nonimmune reactions in which cell–cell adhesion is important to the organism. We only consider a limited number of these additional proteins because of the enormous complexity of this subject. Three transmembrane proteins directly involved in the T-cell reaction with other cells are members of a family of cell–cell adhesion proteins (the CD family of proteins). The CD3 protein is found on both killer and helper T cells; it forms a complex with the membrane-bound TCR proteins. The CD8 and CD4 proteins interact with this complex, but unlike the CD3 proteins, CD8 is only found on killer T cells, whereas CD4 is only found on helper T cells. The presence of CD4 and CD8 proteins limits the range of binary complexes formed by T cells with other cells. This is because CD4 and CD8 have selective affinities for MHC class II and MHC class I proteins, respectively. As a result, killer T cells only form complexes with cells that display a processed foreign antigen in a complex with the MHC I protein; this includes most nucleated cells other than those that belong to the immune system. Similarly, helper T cells are limited to reactions with cells that display the appropriate processed antigen in a complex with the MHC II protein; this includes B cells or phagocytes (fig. S3.17).

From this discussion we can see that a productive interaction between T cells and other cells requires a highly specific reaction between a TCR protein and a processed antigen that is supplemented by less specific interactions between other transmembrane proteins on the two cell types.

Left out of this discussion is an explanation for the source of the processed antigen on the antigen-presenting B cell. Immature B cells display a membrane-bound antibody, which interacts with circulating complementary antigen. Once bound to the antibody, the antigen is internalized, processed, and ultimately displayed by the MHC II protein on the cell surface of the same B cell. It is the interaction between such a B cell and the appropriate helper T cell that triggers the B cell to divide and differentiate.

Immune Recognition Molecules Are Evolutionarily Related

The striking structural similarity between immune recognition proteins almost certainly reflects their evolution from a common ancestor. All of them contain one or more immunoglobulinlike domains, each with about 90 amino acids stabilized by a conserved disulfide linkage. The simplest member of this family of genes is β_2-microglobulin. Possibly this or a similar protein was the ancestor for the remaining immune recognition proteins.

Figure S3.17

Antigen-specific reactions involving T cells. (*a*) Killer T cells interact and destroy target cells that display complexes of processed antigen with an MHC I protein. Effective interaction requires that the TCR protein interact specifically with the processed antigen. All of the remaining reactions between the two cells are general reactions that occur between other killer T cells and other target cells. (*b*) Helper T cells interact specifically with antigen-presenting B cells, provided a specific complex is made between the TCR protein and the processed antigen complexed to the MHC II protein and the processed antigen complexed to the MHC II protein on the B cell. The unlabeled membrane proteins in this figure represent other nonspecific protein–protein interactions that strengthen the binary cell reaction. Processed antigen is indicated in red.

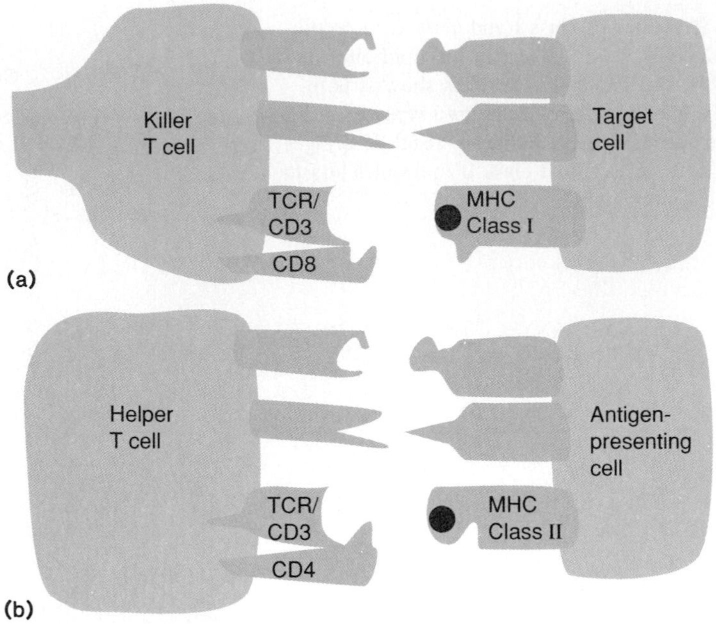

Summary

In this supplement we considered the major findings that led to our understanding of the immune system in vertebrates, especially mice and humans. The following points are the highlights of our discussion.

1. Two different classes of white cells (lymphocytes) are associated with the immune response: The B cells and the T cells. B cells produce antibodies that are secreted and aggregated foreign substances known as antigens. T cells are of different types: Some of them assist B cells to become antibody-forming cells; others mount an attack on antigens of their own.
2. The most common type of antibody is a tetramer composed of two identical heavy (H) chains and two identical light (L) chains. Each chain is divided into a relatively constant sequence (the *C* segments) and a relatively variable sequence (the *V* segments). Within the variable regions there are subsegments known as the hypervariable regions. These subsegments are the regions that specifically bind different antigens.
3. According to some estimates, each organism is capable of making more than a million different antibodies from a limited amount of genetic information. This feat is accomplished with the help of somatic recombination and somatic mutation. Somatic recombination involves splicing at the DNA level. It brings specific *C* and *V* regions together to make a unit that can be transcribed. Somatic mutation is focused on those parts of the gene that encode the hypervariable regions.
4. T helper cells focus the antigen on an immature B cell. In response, some B cells turn into antibody-forming cells, and some B cells turn into memory cells for production of specific antibodies at a later date. The first time an organism is exposed to a specific antigen it takes longer to mount an immunological response. That is because of the lack of memory cells for making a specific antibody.
5. Complement is a group of serum proteins that aids in the defense against microorganisms and removal of antibody–antigen complexes.
6. T cells give rise to an immune response of their own. The two types of cell-mediated immunity involve basically different types of T cells. In the delayed-type response, the T cell reacts specifically with antigens and secretes lymphokines. These are substances that attract macrophages or other leukocytes, thus produc-

ing a slowly developing inflammatory response. A second type of T cell reacts specifically with antigen bound to target cells and causes their lysis.

7. Tolerance prevents the immune system from attacking self-antigens. To understand tolerance we must appreciate how T cells work. T cells recognize a combination of self and nonself. The cell surface antigens recognized by T cells are known as the major histocompatibility complex (MHC). If two organisms carry the same histocompatibility antigens, the tissues from the two organisms are completely compatible. The histocompatibility antigens resemble antibodies in structure.

Selected Readings

Ada, G. L., and G. Nossal, The clonal selection theory. *Sci. Am.* 257(2):62–69, 1987.

Darnell, J., H. Lodish, and D. Baltimore, *Molecular Cell Biology.* New York: W.H. Freeman Co., Scientific American Books, 1990. For a more comprehensive textbook treatment of this subject, see chapter 25.

Golde, D. W., The stem cell. *Sci. Am.* 265(6)86–93 (December) 1991.

Honjo, T., and S. Habu, Origin of immune diversity: Genetic variation and selection. *Ann. Rev. Biochem.* 54:803–830, 1985.

Hood, L., M. Kronenberg, and T. Hunkapiller, T cell antigen receptors and the immunoglobulin super gene family. *Cell* 40:225–229, 1985.

Hunkapiller, T., and L. Hood, The growing immunoglobulin gene superfamily. *Nature* 323:15–17, 1986.

Kaappes, D., and J. L. Strominger, Human class II major histocompatibility genes and proteins. *Ann. Rev. Biochem.* 57: 991–1028, 1988.

Klein, J., *Immunology.* London: Blackwell Scientific Publications, 1990.

Leder, P., The genetics of antibody diversity. *Sci. Am.* 246(5):102–115, 1982.

Lieber, M. R., Site-specific recombination in the immune system. *FASEB J.* 5:2934–2944, 1991.

Milstein, C., Monoclonal antibodies. *Sci. Am.* 243(4):66–74, 1980. An exciting, tissue-culture technique for obtaining pure antibodies specific for an antigen of choice.

Mizel, S. B., The interleukins. *FASEB J.* 3:2379–2388, 1989.

Nowak, M. A., R. M. Anderson, A. R. McLean, T. F. W. Wolks, J. Goudsmit, and R. M. May, Antigenic diversity thresholds and the development of AIDS. *Science* 254:963–969, 1991.

Rini, J. M., U. Schulze-Gahmen, and I. A. Wilson, Structural evidence for induced fit as a mechanism for antibody–antigen recognition. *Science* 255:959–965, 1992.

Silver, M. L., H-C. Guo, J. L. Strominger, and D. C. Wiley, Atomic structure of a human MHC molecule presenting an influenza virus particle. *Nature* 360:367–372, 1992.

Singer, S. J., Intercellular communication and cell–cell adhesion. *Science* 255:1671–1677, 1992. A comprehensive review of the major cell–cell interactions important in immune reactions.

Springer, T. A., Adhesion receptors of the immune system. *Nature* 346:425–434, 1990. An excellent review.

Sutton, B. J., and H. J. Gould, The human IgE network. *Nature* 366:421–428, 1993. An understanding of the IgE system holds the key to understanding and intervening in the aetiology of allergic diseases.

Tonegawa, S., Somatic generation of antibody diversity. *Nature* 302: 801–803, 1983. A classic paper.

Tonegawa, S., The molecules of the immune system. *Sci. Am.* (December) 1985.

Weiss, R. A., How does HIV cause AIDS? *Science* 260:1273–1278, 1993. An exciting up-to-date account of the various routes whereby the HIV virus attacks and destroys the immune system.

SUPPLEMENT

4

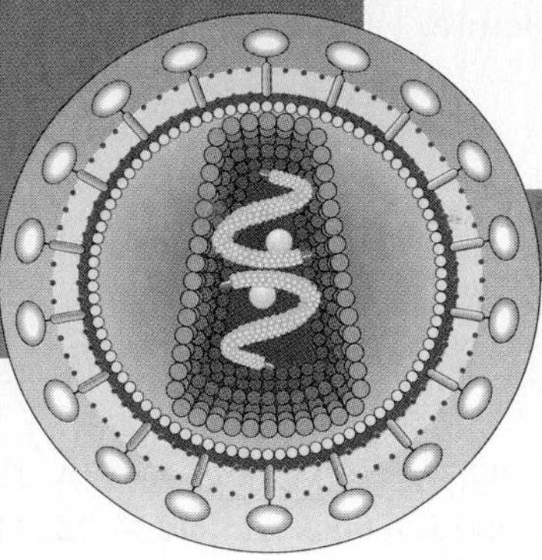

Principles of Physiology and Biochemistry: Carcinogenesis and Oncogenes

Supplement Outline

In the complex developmental process that leads to a mature human being from a fertilized egg, each cell takes up a role in a specialized organ or tissue. Each group of cells proliferates to a certain point and then stops. In the case of a solid organ such as the liver, we see cells of several different types that grow until the organ takes on a fixed size and shape and then cease to grow. Normally, further growth occurs only at the very slow rate required for replacement of aging cells. However, if part of the liver is amputated, a rapid regenerative process ensues so that the liver regrows to approximately the same size it had before the amputation. Other organs or tissues may show more or less growth than the liver, but as a rule, new cells appear only as old and damaged cells are replaced. Furthermore, the cells of normal tissues do not as a rule mix.

Normally you do not find liver cells mixing with intestinal cells or any other types of cells, suggesting that cells

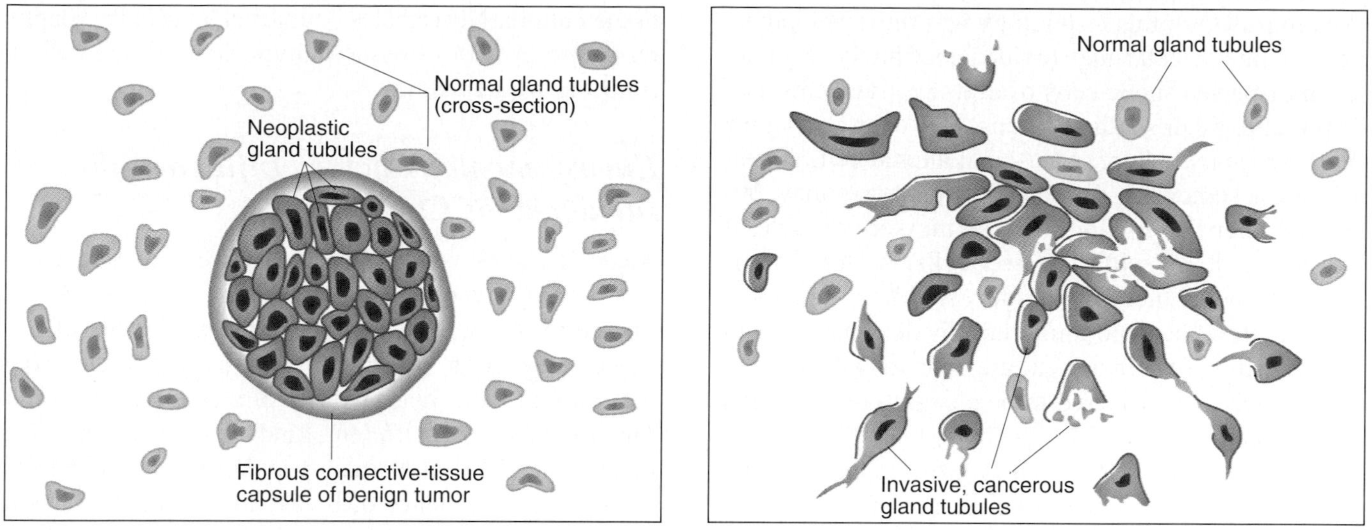

Figure S4.1

A benign glandular adenoma (*a*) and a malignant glandular adenocarcinoma (*b*). In both cases the tumor cells differ morphologically from the surrounding tissue and show a pattern of disorganized growth.

have surface properties that favor interaction of like cells. This was demonstrated in the 1940s by experiments in which embryonic tissues were dissociated into single cells and allowed to reassociate in the presence of other cell types; they usually reassociated with cells of a similar type. It is now known that adhesion molecules are located on cell surfaces that encourage such associations. This was alluded to in supplement S3 when we described the interaction of T cells with other cell types of the immune system. Cell adhesion molecules that favor the association of like cells are of two types: Some require Ca^{2+} for adhesion, the cadherins, and others do not, the cell adhesion molecules (CAMs). When these molecules are not present, as in culture-transformed cell lines, like cells adhere poorly to each other or to other cultured cells. It seems likely that when tumor cells become invasive the character of their cell adhesion molecules has been altered. We mention this here because it is probably an important property of malignant cells, but we have little more to say about it in this supplement because this aspect of malignancy is not well characterized.

Cancer cells are relatively easy to recognize when they occur in groups in the organism (fig. S4.1). A cluster of cancer cells divides more rapidly, and it usually clashes with the architecture of normal cells in its immediate surroundings. From cytological inspection alone, we might speculate that cancer cells are breaking the rules for normal cell growth and associations. It appears that the processes that regulate cells are disrupted; cell growth is uncontrolled, and patches of morphologically distinct cells show by their invasive character that they do not recognize normal territorial boundaries between cells of different types. Cancer cells are deadly because ultimately they invade and disrupt a vital area. In this supplement we present a brief overview of some aspects of carcinogenesis and oncogenes.

Cancers: Cells Out of Control

Cancer is not a stage in the orderly evolution of a complex organism; it is an organismic catastrophe. Perhaps it is only reasonable to expect that multicellular eukaryotes, with their many dividing cells, occasionally produce a mutant cell that gets out of control and destroys the system. The notion that cancer cells are out of control was proposed more than half a century ago. In the intervening time we have learned a great deal about the biochemistry of growth regulation. This knowledge has helped us to understand the molecular basis of cancer in many cases. Conversely, the study of cancer has added to our knowledge of normal growth regulation.

Transformed Cells Are Closely Related to Cancer Cells

In the 1950s, systematic methods were developed for taking cells directly from the tissues of whole animals and inducing them to grow as single cells in liquid culture.

An outgrowth of this technology was the development of pure cell lines—cells that divide indefinitely. Primary cultures of dispersed single cells from tissue fragments usually die out after 50 or so duplications. However, during the growth of a primary culture, occasional altered cells appear that possess a somewhat different morphology; they frequently take over the culture, because they are capable of initiating new cultures with far fewer cells. A clone that is derived from one of these indefinitely dividing variants—which has in effect become immortal—is designated a cell line. The transition of primary culture cells to cells that can grow indefinitely is considered to be a step in the direction of becoming cancerous.

Typically, continuous cell lines grow in culture only on a solid support or ''anchor'' (the surface of a petri dish, for example) and in the presence of relatively high concentrations of nutrients. Even then, they divide only as long as the culture is sparse. When the cell density increases beyond a critical point, the growth rate decreases sharply; in fact, the cells of some continuous lines stop dividing altogether once they have formed a confluent monolayer.

At the density that normally halts growth, rare transformed cells continue to multiply and may reach cell densities up to 20 times higher than those of untransformed cells. By picking and subculturing such transformed cells, it is possible to establish clonal lines of transformants and to ask in what ways such cells differ from their untransformed progenitors.

The differences between transformed and untransformed cells can be classified in three ways: Changes in cell growth, changes in cell surface properties, and genetic alterations. These changes are too great in number and too diverse in quality to be caused directly by one or a few mutations. Clearly, then, most of the observed alterations must be secondary and tertiary events that have occurred as a consequence of some primary event. It is the aim of much current work with tumor-causing agents, especially certain viruses, to discover the molecular nature of this primary event.

Transformed cells sometimes, but not always, give rise to tumors when injected into an isogenic animal. There may be numerous reasons for the failure to produce tumors on all occasions. One of the best known reasons for lack of tumor production is immunological rejection. The transformed cell resulting from exposure to a tumor virus frequently displays viral antigens which are recognized as foreign to the animal host. This recognition excites a positive immune response that can lead to rejection of the transformed cells.

Just as transformed cells frequently do not develop into tumors when injected into whole animals, so is it that tumor cells do not always grow well when dispersed into tissue culture. Nevertheless, tumor cells usually adapt more readily to growth in tissue culture than normal cells taken from the whole animal.

Environmental Factors Influence the Incidence of Cancers

It seems very unlikely that we will find cures for many types of cancer in the foreseeable future. Therefore it is a good idea to put considerable effort into cancer prevention, especially because there are reasons for believing that the vast majority of cancers are preventable. The frequency of different kinds of cancer varies enormously in different human populations (table S4.1). For example, the incidence of breast and colon cancer is much higher in the United States than in Japan, whereas that of stomach cancer is much higher in Japan. In these cases, diet is most likely the culprit. Liver cancer is highest in third world countries. Once again, diet seems the most likely cause. Lung cancer is more prevalent in countries where smoking is popular, and indeed, it is well documented that heavy smokers are much more likely to get lung cancer than nonsmokers. Overall, it is clear that much of the variation in cancer incidence is environmental rather than genetic, since these differences tend to disappear from one generation to the next in migrant populations. For example, the high incidence of stomach cancer in Japan is not matched by a similar figure among Japanese-Americans in the United States.

The conviction is growing that the carcinogenic agents in the environment are active as cancer-causing agents because they produce mutations. The hunt for carcinogens has been facilitated by Bruce Ames, who developed a test for carcinogens based on the mutagenic action of a compound on bacteria.

Ames tests have shown that many carcinogens originate from food and chemical pollutants. In addition to chemicals, ionizing radiation is carcinogenic. For example, the indications are strong that the ionizing radiation we get from the sun is the major cause of skin cancer. Thus, the incidence of skin cancer in the United States is much higher in the South than in the North. High dosages of radioactive materials or x-rays are also strongly correlated with the incidence of numerous forms of cancer.

In many cases, exposure to certain viruses is correlated with specific types of cancers. For example, liver cancer is 300 times more common in individuals who have previ-

Table S4.1

Variation in the Incidence of Cancers by Country

Type of Cancer	Country of High Incidence	Country of Low Incidence	Ratio (High/Low)
Skin	Australia	India	200
Esophagus	Iran	Nigeria	200
Lung	England	Nigeria	35
Prostate	United States	Japan	15
Liver	Mozambique	England	40
Breast	Canada	Israel	7
Rectum	Denmark	Nigeria	20
Colon	United States	Nigeria	10
Ovary	Denmark	Japan	6
Stomach	Japan	Uganda	25

ously been infected with hepatitis B virus. Although we can only speculate on the precise reasons for this, such numbers indicate that liver disease resulting from hepatitis B virus infection is almost certainly the major cause of human liver cancer.

A cancer of the B lymphocytes known as Burkitt's lymphoma is strongly correlated with infection by the Epstein-Barr virus. Once again, we do not know the precise chain of events that lead to cancer. Hepatitis B and Epstein-Barr are DNA viruses. Certain RNA viruses have also been shown to be correlated with cancer. The best known is the human immunodeficiency virus (HIV-1, the virus that causes AIDS.; fig. S4.2). The AIDS virus is believed to cause cancer indirectly by incapacitating the immune system. Various types of cancer follow infection by the AIDS virus, a fact testifying to the key role the immune system plays in protecting the body from cancer. The RNA virus known as human T-cell leukemia virus type I (HTLV-I) causes T-cell leukemia. It is believed that this virus carries a cancer-causing gene (an oncogene), the continuous expression of which upsets the normal metabolism, causing infected cells to become transformed to cancer cells.

Cancerous Cells Are Genetically Abnormal

In the previous section we stated that a strong correlation exists between the mutagenic properties of a chemical and its potency as a carcinogen. This is consistent with other evidence that mutation often plays a causal role in carcinogenesis.

The most graphic illustrations of the correlation between genetic abnormality and cancer come from the frequent association of certain types of tumors with highly specific chromosomal translocations. Chromosomal translocations involve the movement of a segment of chromosomes from one location to another. Frequently, translocations occur between different chromosomes in a reciprocal fashion. This is the case for translocations between human chromosome 8 and the chromosomes that carry the antibody genes: 14, 2, and 22.

A reciprocal exchange of segments of chromosome from the ends of chromosomes 8 and 14 is frequently found in the genome of cancer cells of the Burkitt's lymphoma (fig. S4.3). Burkitt's lymphoma is a tumor of B lymphocytes; for unknown reasons, its occurrence is most common in Africa. The resulting tumor cells usually secrete antibodies. The fact that antibody genes, which are so active in B cells, map to the same chromosomal regions involved in the specific translocations in Burkitt's tumors led to the suggestion that a specific oncogene on chromosome 8 might be activated if it were placed near one of these antibody genes. This suggestion proved to be correct; the oncogene involved is *c-myc.* We have more to say about *c-myc* later.

The first chromosome abnormality found to be associated with a cancer was the Philadelphia chromosome, which arises as a result of a reciprocal translocation between chromosomes 9 and 22 (fig. S4.4). The tumor associated with this translocation is chronic myelogenous leukemia and results from the activation of the *c-abl* oncogene, which is normally located on chromosome 9.

Whereas it is a common belief that tumor formation involves mutation, certain properties associated with the etiology of teratocarcinoma do not fit with this concept. Tera-

Figure S4.2

AIDS virion structure. Numbers associated with proteins are kilodalton masses (e.g., gp120 is a 120-kd protein). (© Scientific American, Inc., George V. Kelvin. Reprinted by permission.)

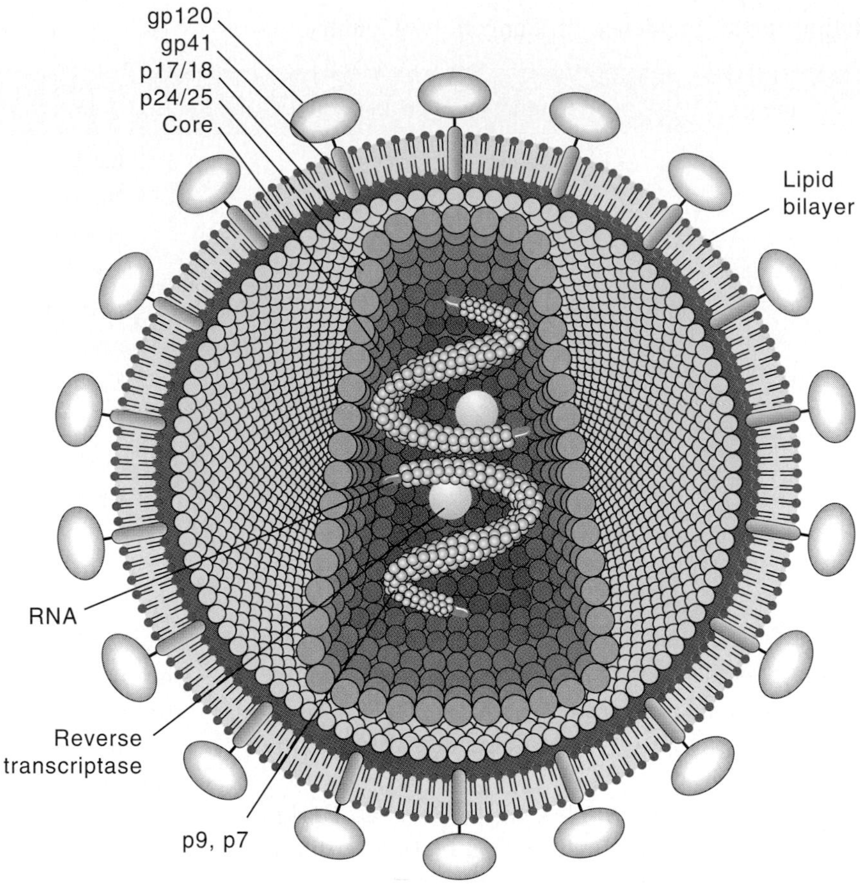

tocarcinoma is a unique type of tumor in which the tumor is associated with a cluster of cells of different types, including stem cells and differentiated cells of various types. Only the stem cells are oncogenic. These stem cells may be replicated *in vitro.* On subcutaneous injection these cells produce tumors, but on injection into a blastocyst embryo these stem cells revert to normal cells, which can be identified in the adult animal by the genetic markers they carry. If these stem cells bear a mutation that accounts for their oncogenicity, then it is hard to see how this mutation could revert to wild type by placing it in the cytoplasmic environment of the embryo. Possibly the teratoma is an exceptional situation in which the tumor is caused by abnormal development not necessitating a mutation.

Some Tumors Arise from Genetically Recessive Mutations

The oncogenes *c-myc* and *c-abl* are dominant to their wild-type alleles. Many tumors arise from oncogenes that show recessive behavior when compared with their wild-type alleles. Such is the case with the genetic locus implicated in the malady known as bilateral retinoblastoma, in which tumors develop in the retinas of both eyes. Humans with this condition usually inherit a genetically recessive mutation on chromosome 13. Sometimes the mutation is microscopically visible as a small chromosomal deletion. In general, recessive mutations are not apparent unless they are present in the homozygous state. However, in the course of the very extensive cell duplication that ensues in retinal development, there is a high probability that at least one cell in each retina will mutate in the wild-type allele of the genetic locus in question. This change results in a homozygous state and in the conversion of the normal cell to a tumorous cell that proliferates rapidly. As a result of this somatic mutation, both eyes usually develop tumors. It seems likely that genes that give rise to tumors when they are defective or lacking, as in retinoblastoma, are of the regulatory type. Some other tumors that result from recessive genetic lesions are listed in table S4.2).

Many Tumors Arise by Mutational Events in Cellular Protooncogenes

A growing number of cellular genes have been implicated in a wide variety of cancers. These genes do not lead to cancers in their normal state but only after a mutational event

Figure S4.3

Translocation associated with Burkitt's lymphoma. The protooncogene *c-myc,* on the end of chromosome 8q, is translocated to the end of chromosome 14q, adjacent to the immunoglobulin constant gene, *Ig-Cμ.* (From J. Yunis, The chromosomal basis of human neoplasia, *Science* 221:227–236, Jan. 1, 1983. Copyright 1983 by the AAAS. Reprinted by permission.)

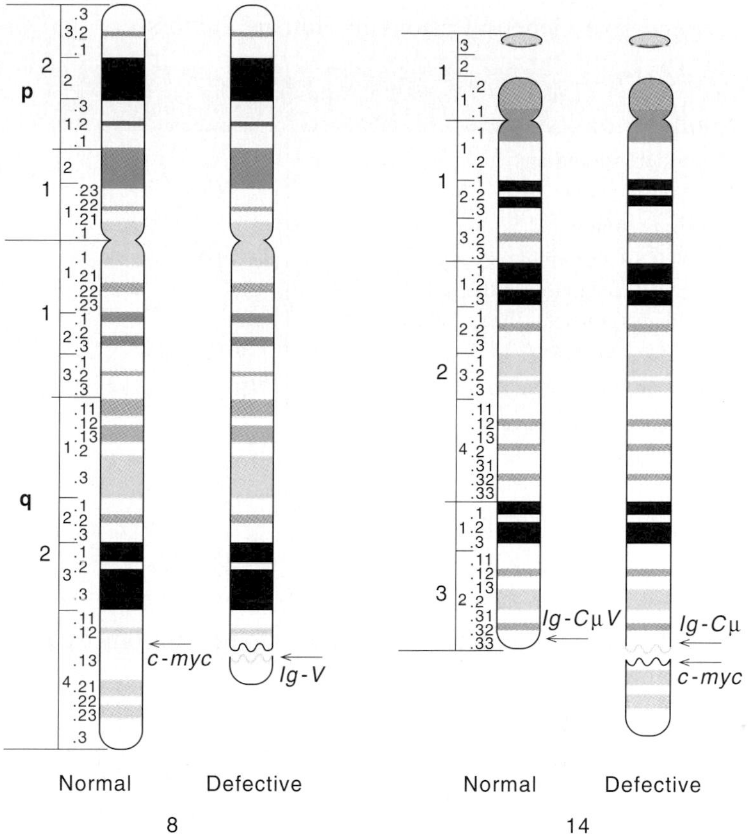

Figure S4.4

The Philadelphia chromosome arises by a reciprocal translocation between chromosomes 9 and 22. Note that the *abl* protooncogene is moved in the translocation process.

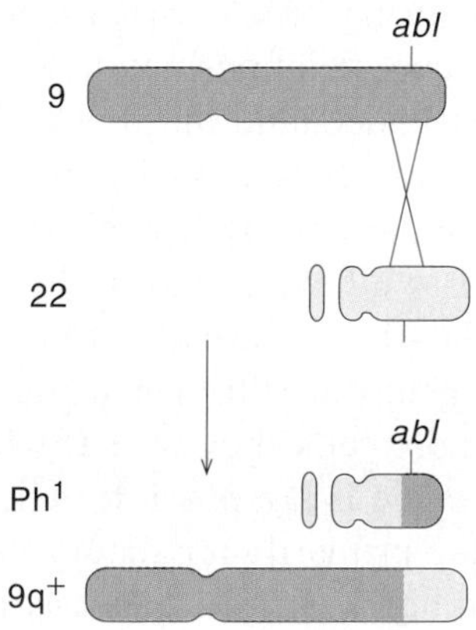

has somehow resulted in their altered expression. For that reason such genes are called protooncogenes in their normal state and oncogenes in their mutated, cancer-causing state. Most protooncogenes, including *c-abl* and *c-myc,* were not known until very similar genes were found as a result of studies of tumor-causing viruses.

Oncogenes Are Frequently Associated with Tumor-Causing Viruses

The first virus linked to tumor production was discovered in 1908 by Ellerman and Bang. Their finding was followed in 1911 by the better known discovery by Peyton Rous of a virus that produces sarcomas in chickens (now known as Rous sarcoma virus, or RSV). Later (1932) Shope showed that a papilloma virus produced cutaneous tumors in rabbits, and Lucke (1934) showed that adenoviruses produced renal adenocarcinoma in the frog. These and other discoveries made in the 1940s and 1950s indicated that exposure to certain viruses could result in tumor production. However, viral oncogenes had not yet been identified, and skeptics

Table S4.2

Recessive Genetic Lesions in Human Tumors

Tumor	Chromosomal Locus
Retinoblastoma and osteosarcoma	13 (q14)
Wilm's tumor (nephroblastoma)	11 (p13)
Embryonal tumor of the kidneys, muscle, liver, and adrenal glands	11p—
Bladder carcinoma	11p
Acoustic neuroma or meningioma	22

still classified cancer-causing viruses along with other carcinogens.

The development of tissue culture techniques in the 1950s permitted a systematic investigation of the causal link between tumor viruses and cellular transformation in culture. In many ways the properties of cells transformed in culture resembled those of tumor cells *in vivo,* and so the tissue culture system was considered to be a convenient way to explore the oncogenic properties of viruses. It was noted with some tumor viruses that the response of cells of different species was quite different. On cells that were normally permissive for virus replication, no tumors arose. This was hardly surprising, in view of the fact that the virus replication cycle, particularly for the DNA viruses, usually results in cell death. What was surprising was the finding that the same virus on another cell type, in which the virus could not replicate, sometimes resulted in cell transformation. Observations of this sort led Renato Dulbecco to hypothesize that a parallel existed between lysogeny of *Escherichia coli* by λ and transformation of animal cells by oncogenic viruses. In λ, the viral genome can exist either as an independent entity, leading to virus replication and cell death, or as an integrated passive state, leading to a lysogenic cell (see chapter 30). With λ the two states—lytic, or active, and lysogenic, or passive—are attainable in the same cell type. Tumor-causing viruses are capable of existing in two states also. However, in this case the two states, actively replicating and passively integrated, are not normally attainable in the same cell type. Usually two cell types of different species are required. In one cell type the viral genome invariably exists in the actively replicating form, causing immediate cell death; in another cell type the viral genome can exist only in a passive integrated form. Dulbecco suggested that in the integrated state the viral genome expresses a limited number of gene products that lead to cellular transformation.

So far so good. Dulbecco's suggestion appeared to constitute a reasonable hypothesis for explaining the oncogenicity of certain DNA viruses. But it was also known that certain RNA viruses cause tumors. Indeed, one of the first known cancer-causing viruses, the Rous sarcoma virus, was an RNA virus. To explain the oncogenicity of RNA viruses, Dulbecco proposed that cancer-causing RNA viruses form a DNA copy of their genome that can be integrated into the host cellular genome. This bold proposal—that an RNA genome could be converted into a DNA copy—was confirmed by experiments in both David Baltimore's and Howard Temin's laboratories. Both groups found an enzyme, reverse transcriptase, uniquely associated with RNA tumor-causing viruses that could synthesize DNA from an RNA template.

The finding that many viruses cause cancer only after inserting their genome into the host genome has spurred new efforts in cancer research. More and more attention has been paid to viruses. With the help of genetic manipulations, the cancer-causing genes within the viruses have been identified and their gene products have been studied. The results of these studies are providing a broad platform of understanding about the kinds of biochemical processes that result in the conversion of normal cells to cancer cells.

The Role of DNA Viral Genes in Transformation Reflects Their Role in the Permissive Infectious Cycle

Usually, tumor-causing DNA viruses possess oncogenes that bear no sequence relationships to any host genes. We can probably best understand the role of DNA viral oncogenes in cellular transformation by appreciating the role these genes play in the virus lytic cycle (table S4.3). First of all, the oncogenes we have found in the adenovirus, polyoma virus, and SV40 virus are all expressed early in virus infection. These genes also trigger a transient transformation process in most cells they infect, where a permissive virus replication cycle is not possible. Thus, it seems likely that these genes are primarily regulatory. In SV40 and polyoma the so-called big-T genes are clearly regulatory in productive infection because they are involved in regulating

Table S4.3

Properties of DNA Virus Oncogenes

Viral Gene and/or Gene Product	Function in Lytic Infection	Function in Transformation	Other Properties
Adenovirus			
E1A 9 kd, 26 kd, 32 kd	Required for expression of other early genes	Immortalizes cells; partial transformation	General transcription activator
E1B 9 kd, 55 kd	Required for lytic infection	Required for full transformation and tumorigenic cells Either 19 kd or 55 kd suffices	Associates with p53 55-kd protein found primarily in nucleus; 19-kd protein in endoplasmic reticulum and in nucleus
SV40			
Big T 90 kd	Binds to specific sites; necessary for DNA synthesis; inhibition of early RNA synthesis and activation of late RNA synthesis	Immortalizes and transforms	Stimulates host DNA synthesis on microinjection; associates with p53
Small t	None (?)	Required for full transformation on some cell lines	—
Polyoma			
Big T 100 kd	Binds to DNA; necessary for DNA synthesis and regulation of early and late RNA synthesis	Immortalizes cells; only amino terminal 40% fragment is required	Associates with p53; ATPase activity; gyrase activity
Middle T 56 kd	None (?)	Transforms cells and makes them tumorigenic	Associates with $pp60^{c\text{-}src}$
Small t	None (?)	Transformed cells adhere less firmly when small t is present	Found in nucleus

both viral DNA and RNA synthesis. In adenovirus there is no comparable gene, but the early genes *E1A* and *E1B* may play a similar, albeit indirect role. *E1A* functions as a transcription activator.

Another lead as to how a DNA virus transforms cells is to look at the cellular proteins they influence. There is a striking parallel here between the functions of adenovirus *E1B* and SV40 big T in this regard. These two genes are required for immortalization or transformation, and they both form a tight association with the cellular protein p53, which is considered to be a good candidate for a protooncogene-related protein.* Finally, the polyoma middle-T protein, which is required for transformation by polyoma virus, forms a strong complex with the cellular protein encoded by the protooncogene *c-src*. It seems likely that these associations between viral oncogenes and host proteins result in changes in regulatory functions, and that if we knew what regulatory functions were being affected we would be on the threshold of understanding the biochemical mechanism(s) involved in transformation. We will have more to say about this after we have considered the retroviruses.

* In its normal protooncogenic form, p53 inhibits SV40 replication; in its mutated oncogenic state, it does not inhibit SV40 replication, nor does it bind to T antigen.

p53 Is the Most Common Gene Associated with Human Cancers

The p53 protein acts as a negative regulator of cell growth; it was first detected through its association with the SV40 big-T oncoprotein in virus-transformed cells. Viral oncoproteins like SV40 T antigen and adenovirus *E1B* sequester p53 protein in inactive complexes. Whereas this results in an increase in the steady-state concentration of p53, the p53 is no longer able to reach its normal site of action in the nucleus.

Many mutant *p53* alleles favor growth and transformation in cells that continue to carry intact, wild-type *p53* gene copies; accordingly, such mutant *p53* alleles act in a dominant fashion. The dominant behavior of mutant *p53* could be explained by the normal active form of the protein. In its active form p53 protein assembles into homotetramers and higher order homooligomeric structures. Defective subunits of such an oligomerizing protein (for example, mutant p53 molecules) may participate in forming a multisubunit complex together with wild-type monomers and, in so doing, poison the function of the complex as a whole.

Normal p53 protein binds DNA in a sequence-specific manner and thus most likely regulates gene transcription. Co-transfection experiments show that wild-type p53 activates the expression of genes adjacent to a p53 DNA-binding site. Cells bearing oncogenic forms of p53 have lost this activity.

The best demonstration that the loss of active normal p53 explains the oncogenic behavior of mutant p53 comes from studies in which a null mutation was introduced into the gene by homologous recombination in murine embryonic stem cells. Mice homozygous for the null allele appear to be normal but are prone to the development of a variety of neoplasms by 6 months of age. These observations suggest that a normal *p53* gene is dispensable for embryonic development but that its absence predisposes the animal to neoplastic disease.

Somatic mutation of *p53* has been implicated as a causal event in the formation of a large and ever-increasing number of common tumors, including those involving the hematopoietic organs, bladder, liver, brain, breast, lung, and colon. In view of the fact that many other oncogenes exist, we must ask why *p53* is so often implicated in commonly occurring tumors. This appears to stem from its genetic and biochemical traits. Point mutations create carcinogenic *p53,* and such simple genetic changes occur readily. Unlike *ras,* for example, in which point mutations productive for cancer are limited to specific changes in two or three codons, the cancer-favoring missense mutations of *p53* can occur in at least 30 distinct codons in its reading frame. In addition, these point mutations often create dominant alleles that produce shifts in cell phenotype even without a reduction to homozygosity.

Retroviral-Associated Oncogenes that Are Involved in Growth Regulation

The first intensively studied oncogene associated with a retrovirus was the RSV *src* gene. It is clear that the *src* gene found in RSV, *v-src,* is structurally quite similar to a normal host gene found in the chicken genome called *c-src.* Subsequent work has shown that at least 30 other animal retroviruses exist that have acquired a cellular oncogene during their evolution. This uncanny association of retroviruses with cancer-causing genes has made retroviruses useful devices for scanning the cellular genome for the presence of protooncogenes (fig. S4.5).

A listing of some of the better characterized retrovirus-associated cellular oncogenes appears in table S4.4. In this table the name of the oncogene is indicated in the left-most column. Proceeding to the right in the table are indicated: (1) The virus of origin, (2) the viral gene product, (3) the cellular homolog of the viral product; (4) the activity associated with the viral gene product, and (5) the subcellular location of the viral gene product, which in most cases is similar to the subcellular location of the cellular homolog.

The src *Gene Product*

The protein encoded by the *src* gene has a molecular weight of 60 kd. Like many other oncogenic proteins, it is bound to the inner plasma membrane, and it possesses a tyrosine phosphokinase activity; that is, it catalyzes the addition of a phosphate group to the tyrosine hydroxyl groups of various proteins. It is considered highly likely that the kinase activity of the src protein is associated with its transforming activity because its loss in mutant *src* genes leads to loss of transforming activity. However, since many different proteins are phosphorylated by the src kinase, it is not clear which phosphorylations are crucial to the complex physiological response that triggers transformation.

The sis *Gene Product*

Many oncogenes are associated with tumors produced from a restricted class of differentiated cells. Such is the case with the *sis* oncogene, found in the simian sarcoma virus. This oncogene is believed to be closely related to the gene for platelet-derived growth factor, PDGF. PDGF is a small protein, synthesized in platelets, that stimulates the growth and division of target cells that carry a membrane-bound

Table S4.4

Some Retroviral Oncogenes and Their Cellular Homologs

Oncogene	Viral Origin[a]	Viral Gene Product	Cellular Homolog	Activity	Subcellular Location
sis	Simian sarcoma virus	$p28^{sis}$	PDGF B-chain	PDGF agonist	Cytoplasm
src	Rous sarcoma virus	$p60^{v\text{-}src}$	$p60^{c\text{-}src}$	Tyrosine kinase	Plasma membrane
fps[b]	Fujinami sarcoma virus	$p140^{gag\text{-}fps}$			
abl	Abelson murine virus	$P120^{gag\text{-}abl}$	$p150^{c\text{-}abl}$	Tyrosine kinase	Plasma membrane
erbB	Avian crythroblastosis virus	$gp72^{erbB}$	Truncated EGF receptor	Tyrosine kinase	Plasma membrane
myc	Avian myelocytomatosis virus MC29	$P110^{gag\text{-}myc}$	$p58^{c\text{-}myc}$	Binds DNA	Nucleus
H-ras	Harvey murine sarcoma virus	$p21^{v\text{-}Hras}$	$p21^{c\text{-}H\text{-}ras}$	Threonine kinase binds GDP or GTP	Plasma membrane
K-ras	Kirsten murine sarcoma virus	$p21^{v\text{-}Kras}$	$p21^{v\text{-}Kras}$		
N-ras	See footnote *c*.		$p21^{c\text{-}N\text{-}ras}$		

[a] Only one example of a virus strain is given for each oncogene.
[b] *fps* and *fes* are homologous genes from chicken and cat, respectively.
[c] The transforming gene product, $p21^{N\text{-}ras}$, identified by transfection experiments.

Figure S4-5

Hypothesis for how a retrovirus picks up a cellular protooncogene. The RNA–DNA cycle for retroviruses is shown in figure 26.25. The proviral DNA is believed to recombine the cellular DNA in such a way that it picks up a region of the cellular DNA containing a protooncogene. Red segments represent coding regions of C-*onc*.

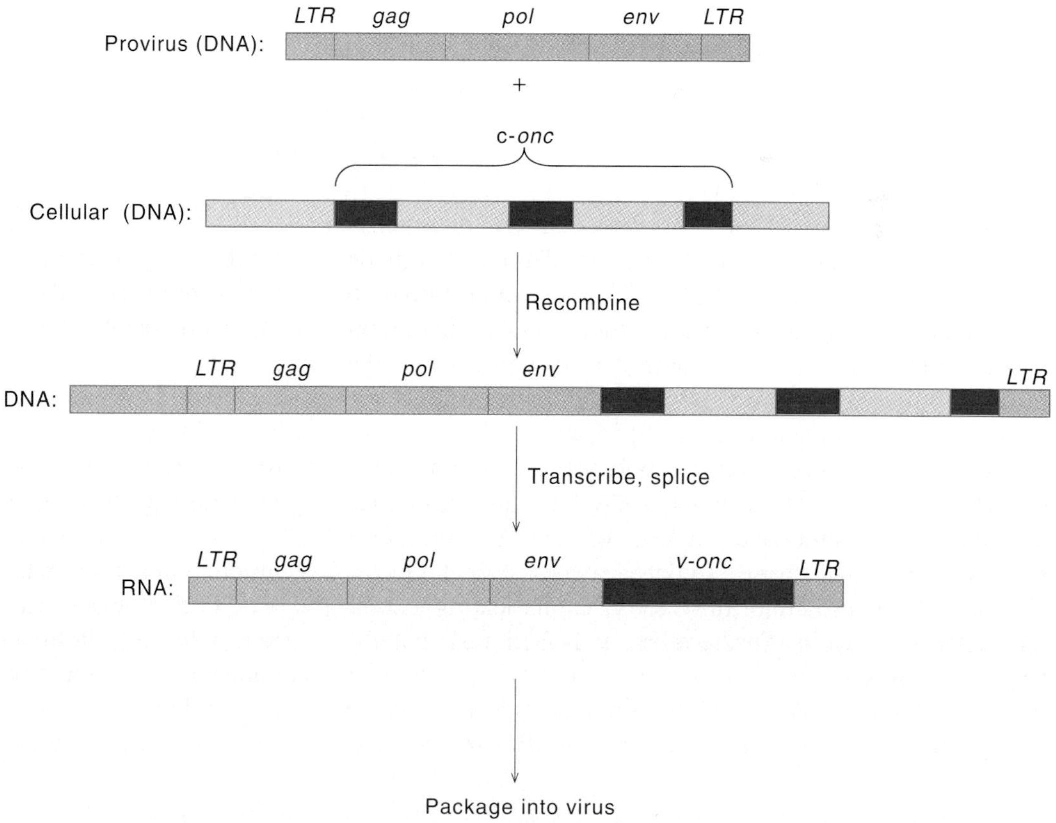

Figure S4.6

Mechanism of mitogenesis in normal and transformed cells. (*a*) Schematic representation of the growth-factor-related mitogenic pathway in normal cells. Here, (1) represents the growth factor, (2) the growth factor receptor, and (3) the intracellular messenger system that transmits the mitogenic signal from the receptor to the nucleus. (*b*) Schematic representation of a possible perturbation of the growth-factor-related mitogenic pathway in transformed cells. Here, (1) represents endogenous production of growth factor that may stimulate the cell. The endogenously produced factor may be secreted and interact with growth factor receptors at the cell surface (as shown) or, alternatively, activate the receptor in an intracellular compartment.

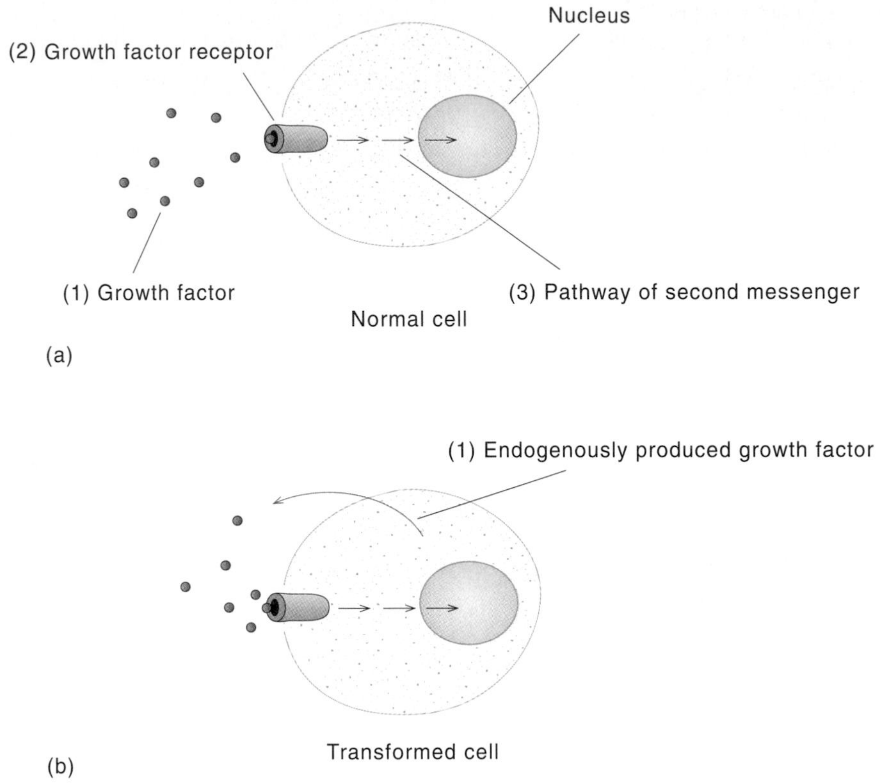

specific receptor for PDGF—in particular, the mesenchymal cells involved in wound healing. Tumors originate when the target cells carrying the PDGF receptor mutate to cells capable of synthesizing their own PDGF. This situation is believed to lead to uncontrolled growth because the factor continuously stimulates cell proliferation in the very cells in which it is synthesized (fig. S4.6).

The PDGF protein contains two polypeptide chains: A and B, with molecular weights of 28,000 and 32,000, respectively. The amino acid sequence of the PDGF protein is very similar to that of the predicted sequence of the transforming p23 sis protein of the SSV virus, indicating a common ancestral origin. Furthermore, it is believed that the SSV virus causes tumors by synthesizing large amounts of the viral p23 sis protein, which stimulates the unregulated proliferation of target cells that carry the PDGF receptor.

The erbB *Gene Product*

Epidermal growth factor (EGF) is a small mitogenic protein that stimulates the proliferation of cells carrying specific membrane-associated EGF receptors. The EGF receptor has a strong amino acid sequence homology with $gp65^{erbB}$, the transforming protein of avian erythroblastosis virus (AEV). The EGF receptor has tyrosine kinase activity, like the src protein. In addition, the EGF receptor contains an extracellular domain that binds the EGF growth factor.

The *v-erbB* oncogene acts to expand a pool of highly mitotic, undifferentiated erythroid precursor cells, but these are poorly tumorigenic, because they differentiate at high rates into postmitotic, end-stage red cells. Another potential oncogene, *v-erbA,* blocks differentiation of erythroid precursors but creates no tumors because it is unable to provide the mitogenic impetus needed to expand the pool of stem cells. The two genes, *erbA* and *erbB,* carried into erythroid precursors together by avian erythroblastosis virus, act in concert to create an aggressive erythroleukemia; *v-erbB* drives expansion of the pool of undifferentiated precursor cells, whereas *v-erb4* blocks their conversion by the differentiation pathway. The *v-erbA* allele that participates in formation of chicken erythroleukemias is a mutant version of a transcriptional regulatory protein, the chicken thyroid hormone (triiodothyronine) receptor. Function of the wild-type receptor protein is blocked in the presence of *v-erbA,* because the latter occupies critical DNA-binding sites in a way that precludes association by the wild-type receptor protein and inhibits transcription.

Figure S4.7

Interaction between the *ras* protooncogene and GDP. The protein is shown as a green ribbon, interacting with GDP (dark purple, yellow, and orange). The dark purple rectangle represents the guanosine; the dark yellow pentagon, the ribose; and the orange circles, the phosphates of GDP. The domains of the protein interacting with each of these parts of GDP are represented by sleeves on the protein ribbon, color-coded to match the corresponding part of GDP. The P domain of the protein contains glycine 12; this amino acid is in a critical position next to the phosphates of the nucleotide. The N and C termini of the protein are labeled. (From L. Tong et al., Structure of the ras protein, *Science* 245:243, July 21, 1989. Copyright 1989 by the AAAS. Reprinted by permission.)

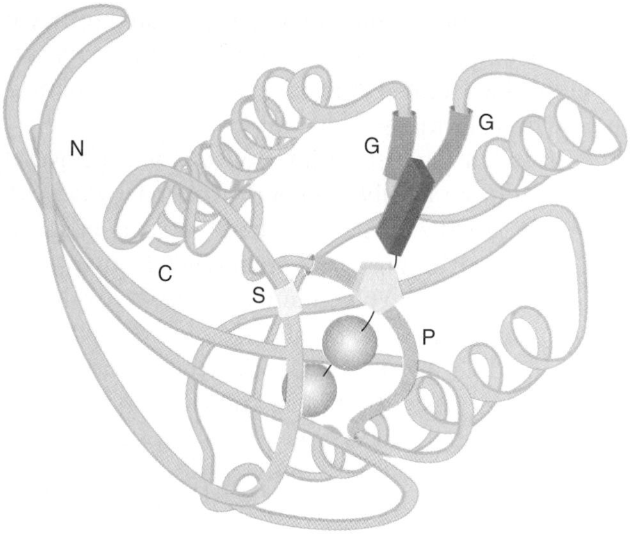

The ras *Gene Product*

One of the best understood protooncogenes is *ras.* The *ras* gene found in normal mammalian cells is closely related to the *ras* oncogenes of Harvey and Kirsten murine sarcoma viruses, *H-ras* and *K-ras,* respectively. *Ras* oncogenes have also been isolated with the help of DNA transfection techniques. Activated *ras* genes have been isolated from a wide variety of dissimilar neoplasms, including carcinomas, sarcomas, neuroblastomas, lymphomas, and leukemias. Some of these *ras* genes are associated with the murine sarcoma viruses (e.g., *N-ras* in table S4.4). All members of the *ras* gene family encode closely related proteins of approximately 21 kd, which have been designated p21. Unlike the oncogenes associated with many retroviruses, the level of p21 expression is similar in normal cells and in many different human tumor-cell lines. Nucleotide sequence analysis of the *H-ras* transforming gene isolated from a human bladder carcinoma cell line has shown that the transforming activity of this gene is the consequence of a point mutation altering amino acid 12 of wild-type p21 from glycine to valine (fig. S4.7).

Subcellular fractionation and immunofluorescence of normal cellular and viral ras proteins have indicated that p21 is localized on the inner face of the plasma membrane. Both the normal and the mutant ras proteins bind GTP specifically and strongly, but only the normal protein has GTPase activity.

We understand normal *ras* gene function better than that of many other protooncogenes because the yeast *Saccharomyces cerevisiae* carries *ras* genes, and many researchers have taken advantage of the powerful genetic tools available in yeast to study them. *S. cerevisiae* contains two closely related but distinct genes, *RAS1* and *RAS2,* that encode proteins that are highly homologous to the mammalian ras proteins. Although neither *RAS1* or *RAS2* by itself is an essential gene, *RAS* function is required for the continued growth and viability of yeast cells. Thus, $ras1^-$, $ras2^-$ double mutants are nonviable unless they carry a suppressor mutation *bcyL.* The *bcyL* mutation suppresses lethality in adenylate-cyclase-deficient yeast. We will see the significance of this fact presently. Yeast strains that are $ras2^-$ are noticeably depressed in cAMP levels. In $ras1^-$, $ras2^-$, *bcyL* triple mutants, the levels of cAMP are virtually undetectable. But cells containing *RAS2 val19,* a *RAS2* allele with a missense mutation analogous to the one that activates the transforming potential of mammalian *ras* genes, have cAMP levels significantly higher than those observed in wild-type cells. Membranes from $ras1^-$, $ras2^-$, *bcyL* triple mutants lack the GTP-stimulated adenylate cyclase activity present in membranes from wild-type cells, and membranes from the *RAS2 val19* yeast strain have elevated levels of a GTP-independent adenylate cyclase activity. Mixing membranes from *ras1*$^-$, *ras2*$^-$ yeast with membranes from an adenylate-cyclase-deficient yeast leads to the reconstitution of a GTP-dependent adenylate cyclase. It appears, then, that *RAS* encodes a protein or proteins that regulates activity of membrane-bound adenylate cyclase in a GTP-dependent manner. In the yeast mutant *RAS2 val19,* the regulatory properties of this protein appear to be disrupted so that the adenylate cyclase is overactive and no longer requires GTP for activation. Further experiments have been done to show that the *ras* genes from yeast cells can function in mammalian cells and *vice versa.* Thus, it has been shown that yeast carrying *ras1*$^-$, *ras2*$^-$ mutations remain viable if they carry the mammalian *H-ras* gene. Conversely, a mutant yeast *RAS* gene has been isolated that can bring about tumorigenic transformation in mammalian cells.

From what has been said, you might expect that ras proteins would be involved in activating adenylate cyclase in mammalian cells. However, there is no indication that this is the case. In mammals, other heterotrimeric GTP-binding proteins are involved in the adenylate cyclase pathway (see table 24.5). Apparently, the *ras* system is a useful signaling module that has been adapted to various uses in different organisms; it could be likened to an electronic switching device, that serves different functions depending on the devices it is attached to. In this connection we note that a structural analysis indicates a similarity between $p21^{ras}$ structure and the G-domain of the protein synthesis elongation factor Tu(EF-Tu) from *E. coli.*

The myc *Gene Product*

The *c-myc* gene, identified originally as the cellular homolog of the transforming determinant carried by avian myelocytomatosis virus, is altered in association with a broad spectrum of neoplasms. Consistent with the observation that altered *c-myc* is associated with tumors of diverse origin, it has been observed that normal *c-myc* is expressed in a variety of tissues. Thus, *c-myc* appears to encode a function associated with a ubiquitous metabolic pathway.

The *c-myc* gene encodes a 49-kd protein that is highly concentrated in the nucleus of the cell. The concentration of c-myc protein normally varies appreciably with the metabolic state of the cell, increasing by more than an order of magnitude in cells just prior to chromosome duplication and cell division.

In most human tumors associated with the *c-myc* gene, the concentration of c-myc protein is greatly amplified. Burkitt's lymphoma, which we discussed earlier, is a notable exception. In this case the c-myc transcript is marginally, and in some cases not at all, increased by comparison with control lymphoblastoid cell lines. Recall that in Burkitt's lymphoma, reciprocal chromosomal translocations are found that involve a chromosome carrying *c-myc* proto-oncogene (chromosome 8) and one of the chromosome segments (usually chromosome 14) bearing immunoglobulin genes.

In at least one Burkitt's lymphoma cell line, the structure of the amino-acid-coding portion of the translocated *c-myc* gene, and hence its predicted protein product, has not been altered, a fact suggesting that activation of *c-myc* must be mediated via a regulatory disturbance. Normally, c-myc protein synthesis is strongly regulated with respect to the cell cycle and tightly repressed in quiescent cells. The translocated *c-myc* gene appears to be somewhat deregulated with respect to its expression during various phases of the cell cycle.

Members of the *myc* gene family have been implicated in the control of normal cell proliferation as well as in neoplasia. A more direct role for *myc* genes in transformation is indicated by their ability to transform primary rat embryo fibroblasts in association with the *c-H-ras* oncogene.

An important clue to *c-myc* function was the discovery in the conserved carboxy-terminal regions of three structural motifs, the leucine zipper (LZ), helix-loop-helix (HLH) and basic region (B). These motifs were originally defined in a number of other sequence-specific DNA-binding proteins but had not previously been found within a single protein.

The comparatively weak homooligomerization efficiency of the c-myc HLH/LZ suggested that a partner protein(s) might exist that heterooligomerizes with c-myc to form a specific DNA-binding complex. A protein termed max (for myc-associated "X" factor) was shown to interact with c-myc in a manner that required the integrity of the c-myc HLH/LZ motif. Max was also demonstrated to associate with N- and L-myc proteins, but not with the HLH/LZ proteins USF or AP-4, nor with several other HLH or leucine zipper proteins. In DNA-binding assays, c-myc-max specifically recognized a c-myc binding site (CACGTG) in a manner that required both an intact max basic region and the HLH/LZ motifs.

In yeast model systems it has been observed that myc is a transcription activator but only when present in a heterodimer with max. Max appears to be essential for DNA binding. Max dimer can bind to DNA on its own but it does not activate transcription on its own. Mammalian genes that are normally activated by myc are still not known.

It has been reported that the myc amino-terminal and central regions can specifically interact with the retinoblastoma (RB) tumor suppressor protein *in vitro.* This finding prompts the suggestion that RB may directly facilitate myc function, but it is not yet known whether the observed interaction reflects a physiologically relevant association *in vivo.*

The jun *and* fos *Gene Products*

The *v-jun* oncogene was discovered as a 0.93-kb insert in the genome of a replication-defective retrovirus, avian sarcoma virus 17 (ASV17), isolated from chicken sarcoma. ASV17 causes fibrosarcomas in chickens and oncogenic transformation in cultured avian embyronic fibroblasts. The oncogenic potential of ASV17 is due to the presence of the *v-jun* gene, which is derived from the cellular *c-jun* gene. The gene is believed to be oncogenic in the virus because it is expressed there in higher amounts.

A great deal of excitement was generated by the discovery that the jun protein can substitute for the yeast transcription activator GCN4. Comparative analysis of the sequences of the two proteins showed that a strong homology occurs over about one-third of their length in the DNA-binding parts. Both of these proteins bind to the same consensus sequence, TGACTCA.

The oncogene of the FBJ murine osteosarcoma virus (*fos*) codes for a related nuclear protein that participates in transcriptional regulation. In human fibroblasts the fos protein is mostly associated with *c-jun.* The fos–jun complex binds specifically to DNA. Since fos alone does not show specific DNA binding, it is believed that jun is responsible for this affinity. Although jun can form homodimers that bind to DNA, the heterodimers formed between fos and jun show a greater affinity. The heterodimers are also more effective in transcription activation; therefore the heterodimer is probably the functionally relevant state of the jun and fos proteins.

Structural analysis indicates that the fos and jun proteins belong to a class of DNA-binding proteins that share the conserved structural motif known as the leucine zipper (see fig. 31.21). Thus, the dimerization of these two proteins is mediated by hydrophobic interaction between the leucine side chains of two leucine zipper domains.

The jun–fos protein complex interacts with regulatory regions of numerous genes. We have yet to find out which of these target genes are involved in aberrant cellular growth.

The Transition from Protooncogene to Oncogene

All of the oncogenes thus far discovered are associated with a cellular homolog that is required for normal growth. The transition of the protooncogene to an oncogene is accompanied by abnormal expression of the gene products. In many cases, such as *src, jun,* or *sis,* excessive amounts of the gene product are synthesized. In some other cases, such as *H-ras* or *K-ras,* normal amounts of the oncogenic product are synthesized, but the protein encoded by the oncogene is altered so that it behaves differently. In still other cases such as *myc* the oncogene product is similar to the cellular homolog in amount as well as structure, but the time during the cell cycle when it is produced is altered.

Oncogene products assume specific locations within the cell. Usually they are associated with one of two locations: The plasma membrane or the nucleus. This specificity is consistent with the hypothesis that oncogene products are associated with normal cellular metabolism relating to regulation of cell proliferation; it seems likely that many of the elements regulating cell proliferation would be found at the membrane and nucleus of the cell.

It is not surprising to find oncogenes with varying specificity for producing tumors. Some components of the cell proliferation regulatory apparatus are probably quite general. An oncogene like *myc* is probably associated with one of these and is therefore associated with tumors of widely varying origins. By contrast, an oncogene like *sis* is specifically associated with cells that are designed to be triggered into proliferation by PDGF. Hence, tumors associated with this oncogene are limited to cell types possessing the PDGF membrane receptor.

Carcinogenesis Is a Multistep Process

A wide range of observations indicate that tumorigenesis is a multistep process involving several mutations, each of which results in discrete changes in the cellular metabolism. If this is the case, then we might expect that any particular oncogene would have the capacity for affecting only one step in the overall process. We have seen that a multistep biochemical pathway ordinarily requires a separate enzyme for each step. Enzymes that function in the different steps of a pathway are sometimes said to complement one another. The question is, in carcinogenesis do different oncogenes show a similar complementation? For example, can purely cellular oncogenes such as *N-ras* complement one or the other of the polyoma oncogenes? In fact, middle-T antigen and *N-ras* oncogene produced no new phenotypes when they were cotransfected into rat embryo fibroblasts (REFs), but large-T antigen and *ras* together achieved dramatic results, producing rapidly expanding foci. This study shows that the conversion of a normal cell to a tumor cell can be achieved by the complementary action of two distinct oncogenes; in this case, one is cellular and one is viral.

Similar studies have shown that the *H-ras* and the *myc* oncogenes can cooperate to produce dense foci of morphologically transformed cells from REF cultures (fig. S4.8). Thus, two cellularly derived oncogenes have been shown to complement each other to produce a fully transformed phenotype. Parallel experiments have been done on whole mice to show that the combination of *ras* and *myc* is highly tumorigenic. Experiments of this sort suggest that at a mini-

Figure S4.8

Rat embryo fibroblasts after several days of growth on plates. Normally these cells stop growing as they approach confluency (top frame). Overgrowth of the monolayer by rounded cells is indicative of cellular transformation. Two transformed foci are seen in cells pretreated with both *ras*- and *myc*-containing DNA (bottom frame). The transformed cells produced similar transformed cells when repleated and when injected into whole animals produce tumors with a much higher frequency than normal cells.

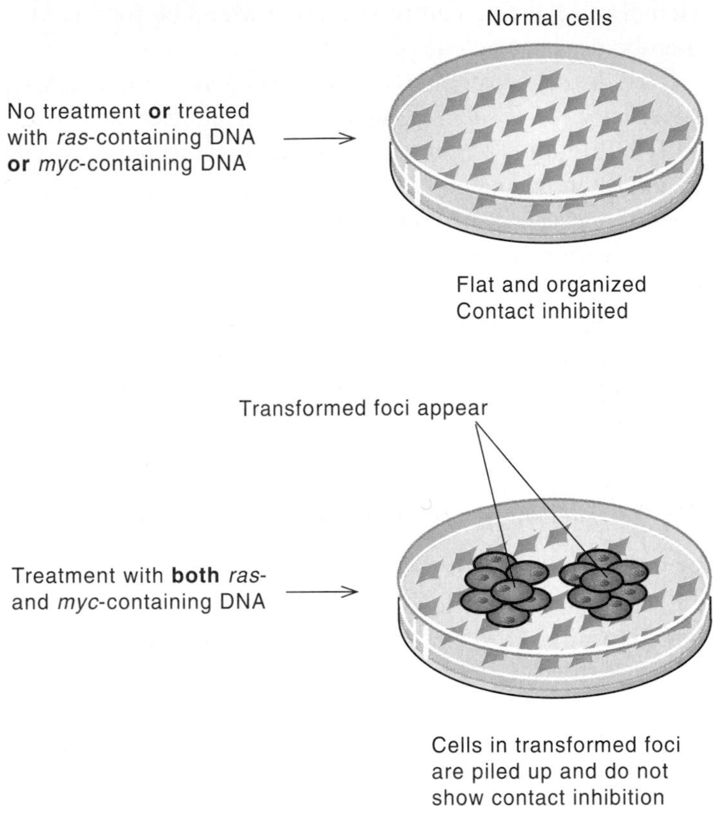

mum, two different oncogene functions are required to convert a normal cell into a tumorigenic one, but it is too early to say that only two are required in general. Nevertheless, these results are very exciting because they are the beginning of genetic complementation assays that should help us to classify oncogenes into different complementary functions.

One further comment on the multistep nature of carcinogenesis: In our discussion here, we focused on some of the early steps in the process that lead to uncontrolled growth but said nothing about those transitions that convert transformed cells into invasive cells. We also concentrated on oncogenes related to cancer-causing viruses because these are most likely to be the first understood in terms of their biochemical function.

Summary

From the time of their discovery, cancer cells have always appeared to be unruly. We have reached that stage in our understanding of cancer cells when we can point to specific aberrations in the genome. In this chapter we took the view that an understanding of cancer can be achieved by analyzing the regulatory pathways involved in growth control because most cancers appear to originate from mutations in specific genes involved in growth regulation. The following points are the highlights of our discussion.

1. Many properties of transformed cells grown in tissue culture resemble cancer cells. The factors that lead to uncontrolled growth *in vivo* can be studied *in vitro* by the effects they have on tissue culture cells.
2. To judge by the frequency of occurrence of cancers in different countries, environmental factors have more influence on the incidence of cancers than inherited genetic factors do. This conclusion is reinforced by studies of migrant populations.
3. Chromosomal translocations are frequently associated with specific types of cancer. This is direct evidence that genetic abnormalities can lead to cancer.
4. A number of tumors arise from recessive mutations in which the mutations appear to be in growth-control genes.
5. Growth-control genes that lead to cancers when they are altered in some way are referred to as proto-oncogenes. They become oncogenes, that is, cancer-causing genes, by mutation. Host protooncogenes are frequently very similar in structure to oncogenes carried by tumor causing viruses.
6. Dulbecco proposed that cancer-causing viruses insert their oncogenes into the host genome. It appears that

cancer-causing viruses are associated with a limited number of DNA viruses and RNA viruses known as retroviruses, which replicate through a DNA intermediate. It is this DNA intermediate that usually gets inserted into the host genome.

7. Abnormal expression accompanies the transition from protooncogene to oncogene. Three types of abnormalities occur: (1) Excessive production of the gene product; (2) altered behavior of the gene product, such as a change in its regulatory properties; and (3) expression at a time during the cell cycle when the gene is not normally expressed.
8. A fully developed cancer appears to arise in steps, each step showing a breakdown in normal regulation.

Selected Readings

Aaronson, S. A., Growth factors and cancer. *Science* 254:1146–1152, 1991.

Amati, B., S. Dalton, M. W. Brooks, T. D. Littlewood, G. L. Evans, and H. Land, Transcriptional activation by the human c-myc oncoprotein in yeast requires interaction with max. *Nature* 359:423–426, 1992.

Cho, Y., Gorina, S., Jeffrey, P. D., and Pavletich, N. P., Crystal structure of a p53 tumor suppressor-DNA complex: Understanding tumorigenic mutations. *Science* 265:346–358, 1994.

Cobrinik, E., S. F. Dowdy, P. W. Hinds, S. Mittnacht, and T. A. Weinberg, The retinoblastoma protein and the regulation of cell cycling. *Trends Biochem. Sci.* 17:312–315, 1992.

Cooper, G. M., *Elements of Human Cancer.* Boston: Jones and Bartlett Publishers, 1991.

Donehower, L. A., M. Harvey, B. L. Slagle, M. J. McArthur, C. A. Montgomery, J. S. Butel, and A. Bradley, Mice deficient for p53 are developmentally normal but susceptible to spontaneous tumours. *Nature* 356:215–221, 1992. A remarkable new technique for obtaining null mutations demonstrates that embryogenesis is normal in the absence of p53. However, animals develop a variety of neoplasms in the first 6 months when p53 is lacking.

Downward, J., The ras superfamily of small GTP-binding proteins. *Trends Biochem. Sci.* 15:469–472, 1990.

Feig, L. A., The many roads that lead to ras. *Science* 260:757–758, 1993. Ras can be activated by a number of different transduction pathways.

Gallo, R. C., The AIDS virus. *Sci. Am.* 256(1):46–56, 1987.

Halauska, F. G., Y. Tsujimoto, and C. M. Croce, Oncogene activation by chromosome translocation in human malignancy. *Ann Rev. Genet.* 21:321–345, 1987.

Hausen, H., Viruses in human cancers. *Science* 254:1167–1173, 1991.

Henderson, B. E., R. K. Ross, and M. C. Pike, Toward the primary prevention of cancer. *Science* 254:1131–1144, 1991.

Jacks, T., A. Fazeli, E. M. Schmitt, R. T. Bronson, M. A. Goodell, and R. A. Weinberg, Effects of an Rb mutation in the mouse. *Nature* 359:295–300, 1992.

Lane, D. P., p53, guardian of the genome. *Nature* 358:15–16, 1992. Proposes that p53 monitors the integrity of the genome.

Levine, A. J., The p53 protein and its interaction with the oncogene products of the small DNA tumor viruses. *Virology* 177:419–426, 1990.

Linzer, D. I. H., The marriage of oncogenes and anti-oncogenes. *Trends Genet.* 4:245–247, 1988.

Liotta, L. A., Cancer cell invasion and metastasis. *Sci. Am.* 266:54–63, 1992. A most important aspect of carcinogenesis that we did not deal with in our short supplement.

Mack, D. H., J. Vartikar, J. M. Pipas, and L. A. Laimins, Specific repression of TATA-mediated but not initiator-mediated transcription by wild-type p53. *Nature* 363:281–283, 1993. The p53 protein may repress the activity of certain promoters by direct interaction with TATA box-dependent transcription machinery.

Makela, T. P., P. J. Koskinen, I. Vastrik, and K. Alitalo, Alternative forms of max as enhancers or suppressors of myc-ras cotransformation. *Science* 256:373–376, 1992.

Malkin, D., F. P. Li, L. C. Strong, J. F. Fraumeni, C. E. Nelson, D. H. Kim, J. Kassel, M. A. Gryka, F. Z. Bischoff, M. A. Tainsky, and S. H. Friend, Germ line p53 mutations in a familial syndrome of breast cancer, sarcomas, and other neoplasms. *Science* 250:1233–1238, 1990.

Milburn, M. V., L. Tong, A. M. deVos, A. Brunger, Z. Yamaizumi, S. Nishimura, and S. H. Kim, Molecular switch for signal transduction: Structural differences between active and inactive forms of protooncogenic ras proteins. *Science* 247:939–945, 1990.

Paparassiliou, A. G., M. Trier, C. Chavrier, and D. Bohmann, Targeted degradation of *c-fos,* but not *v-fos,* by a phosphorylation-dependent signal on *c-jun. Science* 258:1941–1949, 1992.

Solomon, E., J. Borrow, and A. D. Goddard, Chromosome aberrations and cancer. *Science* 254:1153–1160, 1991.

Varmus, H. E., Reverse transcription. *Sci. Am.* 257(3):56–64, 1987.

Weinberg, R. A., Finding the anti-oncogene. *Sci. Am.* 259(3):44–51, 1988.

Weinberg, R. A., The retinoblastoma gene and cell growth control. *Trends Biochem. Sci.* 15:199–202, 1990.

Weinberg, R. A., Tumor suppressor genes. *Science* 254:1138–1146, 1991. Provides an update on the mechanisms of action of a number of better known oncogenes.

Willinghofer, A., and E. F. Pai, The structure of ras protein: A model for a universal molecular switch. *Trends Biochem. Sci.* 16:382–387, 1991.

Appendices

Comparative Sizes of Biomolecules, Viruses and Cells

Frame 1

Water, Amino Acids, DNA and Protein Structure

Starting at the far left, we see a water molecule, two common amino acids, alanine and tryptophan, a segment of a DNA double helix, a segment of a protein single helix, and the folded polypeptide chain of the enzyme copper, zinc superoxide dismutase or SOD. With respect to the relative sizes of some of these molecules and structures, the water molecule is roughly half a nanometer (nm) across, the DNA and protein helices are about 2 nm and 1 nm in diameter, respectively, and the SOD, a small, globular protein of about 150 amino acids, is about 6 nm in width. SOD catalyzes the breakdown of harmful, negatively charged oxygen radicals, thereby protecting people against neurodegenerative diseases such as Lou Gehrig's disease.

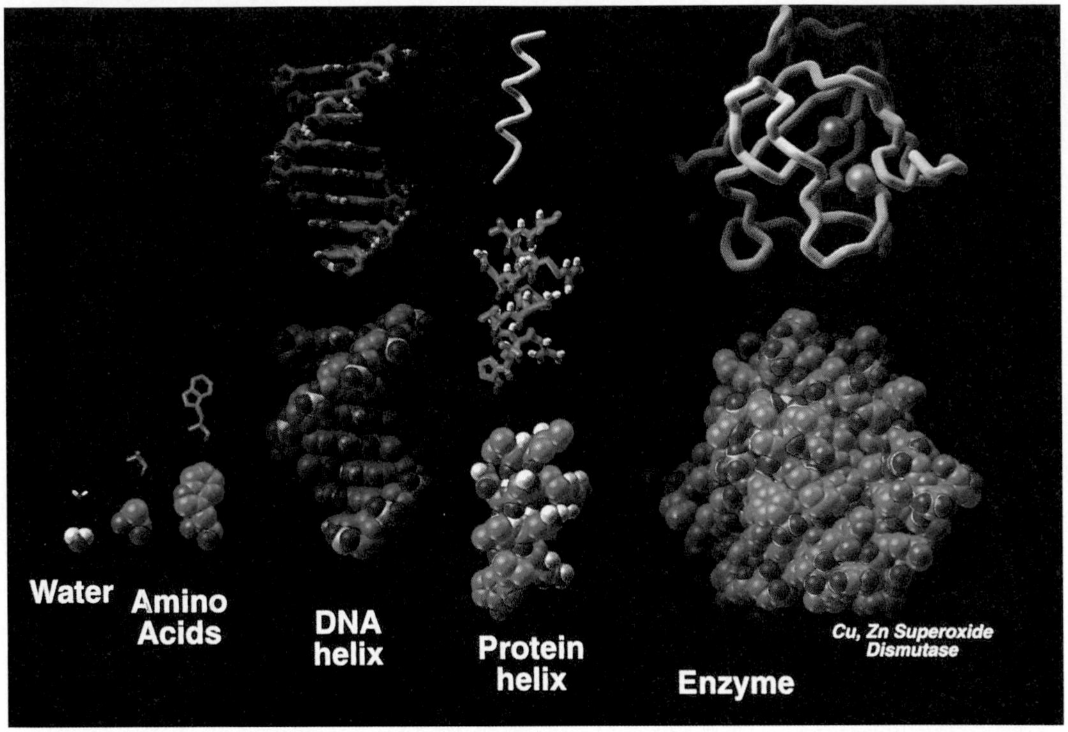

Frame 2

Proteins and Viruses

At the far left, we can see the nucleic acid and protein structures shown in frame 1. In addition, we show a much larger protein, the immunoglobulin G antibody molecule. Four separate polypeptide chains join to make up an antibody molecule: two heavy chains (blue) of about 400 amino acids and two light chains (purple) of about 200 amino acids. The antibody is about 16 nm in width. Finally, at the far right, we show the core particle from a small plant virus, the reovirus. Only the icosahedral protein coat of the virus can be seen. The reovirus particle is about 60 nm across. The nucleic acids of the virus are sequestered inside the virus core. The reovirus family is unusual in that its nucleic acids are all double-stranded RNA molecules.

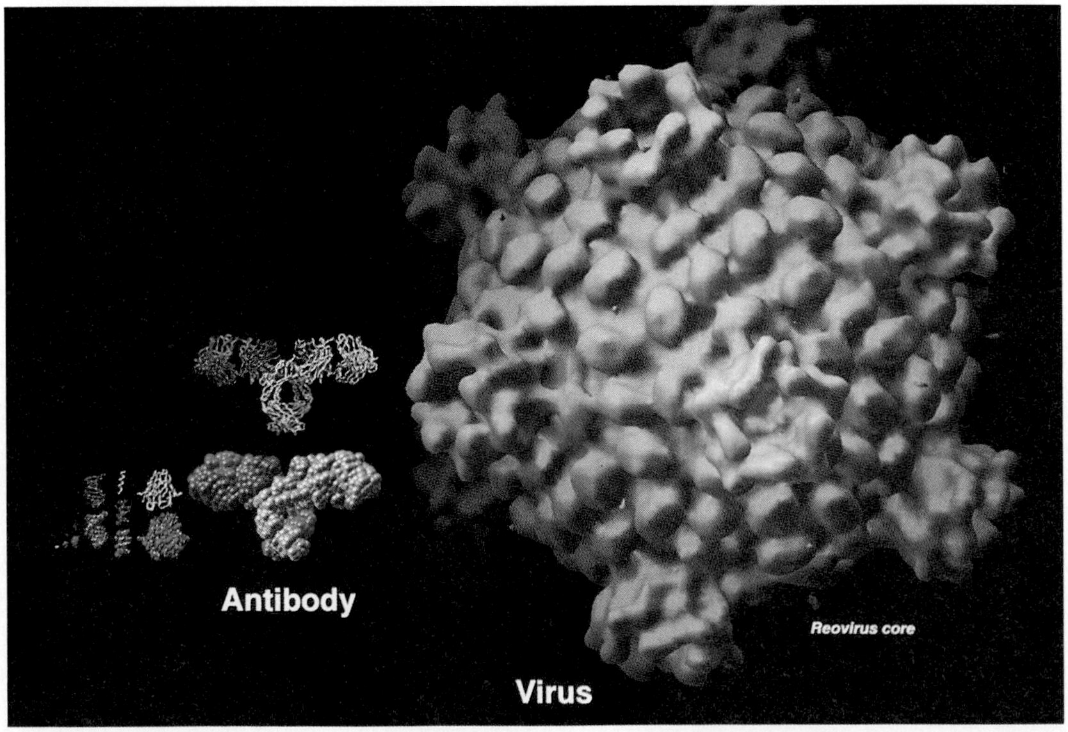

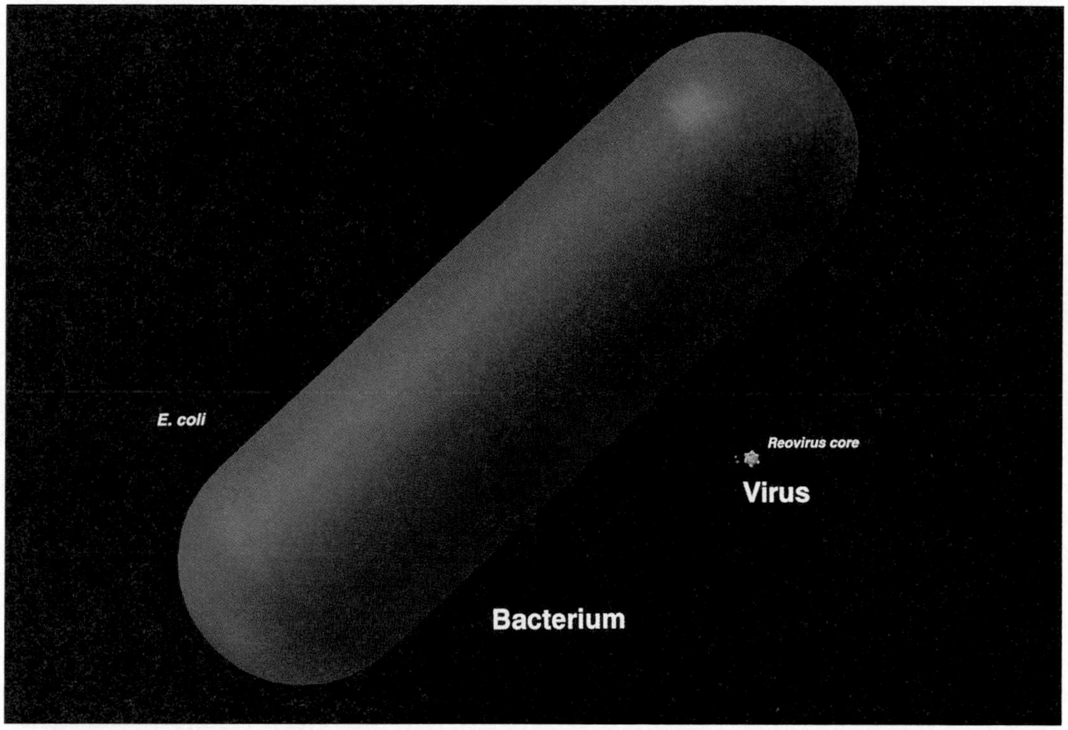

Frame 3

A Bacterial Cell

The common bacterium *E. coli,* shown here in a highly stylized form, is about 2 μm long. The vertebrate gut contains enormous numbers of *E. coli* cells which aid in digestive processes. To the right of the bacterial cell, the same reovirus core particle shown in frame 2 is displayed for size comparison.

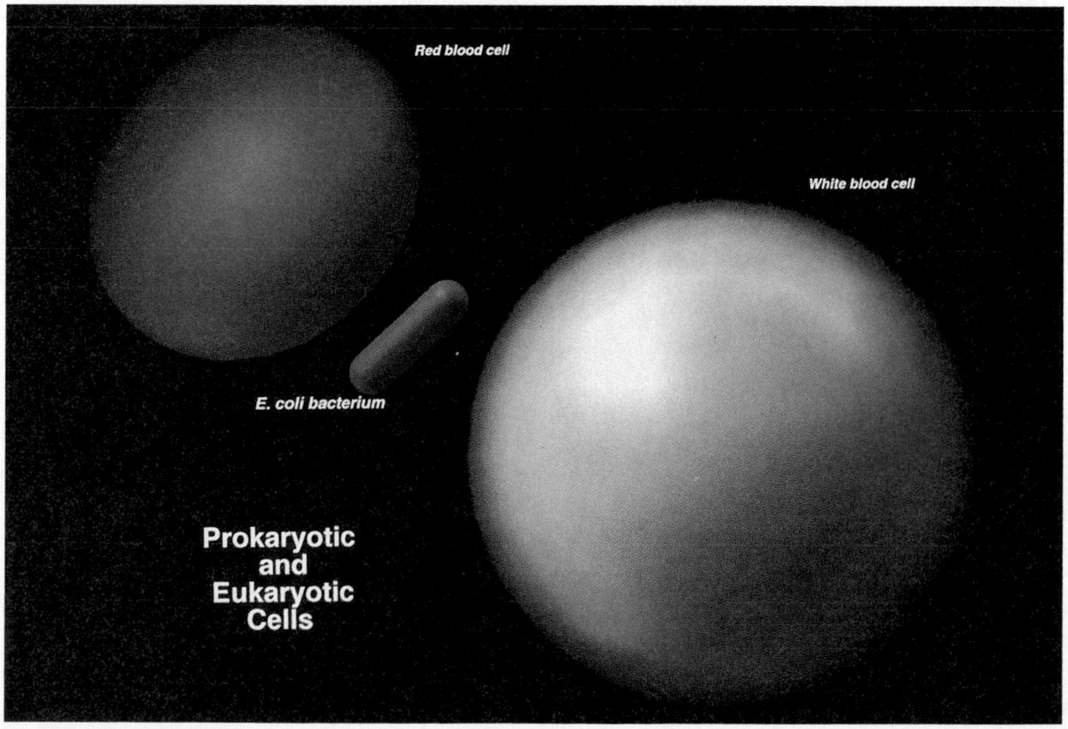

Frame 4

Two Human Cells

Two human blood cells, a white cell about 10 μm across and a red cell about 7 μm across, are shown here in highly stylized representations. The *E. coli* cell from frame 3 is shown between the two human blood cells for size comparison. (Frames 1–4: SOD enzyme by Parge, Hallewell, Getzoff, and Tainer. Antibody by Silverton, Navia, and Davies. Reovirus by Dryden and Yeager. Graphics by Michael Pique using AVS software. Concept by Arthur Olson. Visualization advice by David Goodsell. Images copyright/cW/1994 by The Scripps Research Institute.)

Common Abbreviations in Biochemistry

A	adenine
ACh	acetylcholine
ACTH	adrenocorticotropic hormone
ADP	adenosine-5′-diphosphate
AIDS	acquired immune deficiency syndrome
Ala (A)	alanine
AMP	adenosine-5′-monophosphate
Asn (N)	asparagine
Asp (D)	aspartic acid
ATP	adenosine-5′-triphosphate
BChl	bacteriochlorophyll
bp	base pair
BPG	D-2,3-bisphosphoglycerate
C	cytosine
Cys (C)	cysteine
CAP	catabolite gene activator protein
CDP	cytidine-5′-diphosphate
CMP	cytidine-5′-monophosphate
CTP	cytidine-5′-triphosphate
CoA or CoASH	coenzyme A
CoQ	coenzyme Q (ubiquinone)
cAMP	adenosine 3′,5′-cyclic monophosphate
cGMP	guanosine 3′,5′-cyclic monophosphate
cyt	cytochrome
d	2′-deoxy-
DHAP	dihydroxyacetone phosphate
DHF	dihydrofolate
DHFR	dihydrofolate reductase
DMS	dimethyl sulfate
DNA	deoxyribonucleic acid
cDNA	complementary DNA
DNase	deoxyribonuclease
DNP	2,4-dinitrophenol
ER	endoplasmic reticulum
FAD	flavin adenine dinucleotide (oxidized form)
$FADH_2$	flavin adenine dinucleotide (reduced form)
FBP	fructose-1,6-bisphosphate
fMet	*N*-formylmethionine
FMN	flavin mononucleotide (oxidized form)
$FMNH_2$	flavin mononucleotide (reduced form)
F1P	fructose-1-phosphate
F6P	fructose-6-phosphate
G	guanine
G protein	guanine-nucleotide binding protein
GDP	guanosine-5′-diphosphate
Gly (G)	glycine
GMP	guanosine-5′-monophosphate
Gln (Q)	glutamine
Glu (E)	glutamic acid
GSH	glutathione
GSSG	glutathionine disulfide
GTP	guanosine-5′-triphosphate
Hb	hemoglobin
HDL	high-density lipoprotein
HETPP	hydroxyethylthiamine pyrophosphate
HGPRT	hypoxanthine-guanosine phosphoribosyl transferase
His (H)	histidine
HIV	human immunodeficiency virus
HMG-CoA	β-hydroxy-β-methylglutaryl-CoA
HPLC	high-performance liquid chromatography
IDL	intermediate-density lipoprotein
IF	initiation factor
IgG	immunoglobulin G
Ile (I)	isoleucine
IMP	inosine-5′-monophosphate
IP_1	inositol-1-phosphate
IP_3	inositol-1,4,5-trisphosphate
IPTG	isopropylthiogalactoside
K_m	Michaelis constant

kb	kilobase pair
kDa	kilodaltons
LDL	low-density lipoprotein
Leu (L)	leucine
Lys (K)	lysine
Man	mannose
MHC	major histocompatibility complex
Met (M)	methionine
NAD^+	nicotinamide-adenine dinucleotide (oxidized form)
NADH	nicotinamide-adenine dinucleotide (reduced form)
$NADP^+$	nicotinamide-adenine dinucleotide phosphate (oxidized form)
NADPH	nicotinamide-adenine dinucleotide phosphate (reduced form)
NDP	nucleoside-5′-diphosphate
NAM	*N*-acetylmuramic acid
NMR	nuclear magnetic resonance
NTP	nucleoside-5′-triphosphate
Phe (F)	phenylalanine
P_i	inorganic orthophosphate
PEP	phosphoenolpyruvate
PFK	phosphofructokinase
PG	prostaglandin
2PG	2-phosphoglycerate
3PG	3-phosphoglycerate
PIP_2	phosphatidylinositol-4,5-bisphosphate
PK	pyruvate kinase
PLP	pyridoxal-5-phosphate
PP_i	inorganic pyrophosphate
Pro (P)	proline
PRPP	phosphoribosylpyrophosphate
PS	photosystem
Q	ubiquinone or plastoquinone
QH_2	ubiquinol or plastoquinol
RER	rough endoplasmic reticulum
RF	release factor or replicative form
RFLP	restriction-fragment length polymorphism
RNA	ribonucleic acid
hnRNA	heterogeneous nuclear RNA
mRNA	messenger RNA
rRNA	ribosomal RNA
snRNA	small nuclear RNA
tRNA	transfer RNA
snRNP	small ribonucleoprotein
RNase	ribonuclease
Ru1,5P	ribulose-1,5-bisphosphate
Ru5P	ribulose-5-phosphate
R5P	ribose-5′-phosphate
RSV	Rous sarcoma virus
s	Svedberg constant
SAM	*S*-adenosylmethionine
SDS	sodium dodecyl sulfate
Ser (S)	serine
S7P	sedoheptulose-7-phosphate
SRP	signal recognition particle
T	thymine
THF	tetrahydrofolate
Thr (T)	threonine
TLC	thin-layer chromatography
TMV	tobacco mosaic virus
TPP	thiamine pyrophosphate
Trp (W)	tryptophan
TTP	thymidine-5′-triphosphate
Tyr (Y)	tyrosine
U	uracil
UDP	uridine-5′-diphosphate
UDPG	UDP-glucose
UMP	uridine-5′-monophosphate
UQ	ubiquinone
Val (V)	valine
VLDL	very-low-density lipoprotein
XMP	xanthosine-5′-monophosphate
Xu5P	xylulose-5′-phosphate

Organic Chemistry and Its Relationship to Biochemistry

By definition organic chemistry deals with the chemistry of carbon regardless of its origin. Since biochemistry deals with carbon chemistry only insofar as it concerns living processes, it comprises a distinct subdivision of organic chemistry. In addition to this major difference between organic chemistry and biochemistry, there are two others. The range of conditions used in organic chemistry far exceeds that of biochemistry. Organic chemistry is the study of carbon chemistry in both organic and aqueous solvents, whereas biochemistry is the study of reactions that take place in aqueous solvents. Organic chemistry often involves the study of reaction conditions that are devised by the chemist in the laboratory; biochemistry is concerned exclusively with reactions that occur in living systems. Finally, in organic chemistry the reactions and the reaction conditions often serve no purpose except the ones intended by the chemist. In biochemistry the reactions are functionally related to the needs of the organism.

Regardless of these differences, organic chemistry and biochemistry overlap considerably. As a result, the principles that govern organic chemistry provide a strong foundation for exploring biochemical phenomena. Here we review some of the regions of overlap between the two disciplines. We start with the fundamental properties that affect the chemistry of atoms and molecules. We then consider some of the molecules, their functional groups, their reactions, and the catalysts that accelerate their reactions.

Atoms Are Composed of Protons, Neutrons and Electrons

All atoms contain a centrally located nucleus composed of a mixture of protons and neutrons (except for the hydrogen nucleus, which contains only a single proton). Electrons surround the nucleus in a series of shells. Electrons in the innermost shell are the hardest ones to remove from the atom, and electrons in the outermost shell are the easiest to remove. The outermost shell is called the valence shell because, in most chemical reactions between atoms, electrons are either added to or removed from this shell.

Electrons rotate around the nucleus at high speeds, and determining their exact location at any given time is impossible. Nevertheless, for simplicity, electrons often are depicted as small spheres occupying discrete orbits (fig. 1). A more realistic depiction, the electron orbital model, indi-

Figure A.1

Bohr models of the six most common atoms found in living organisms.

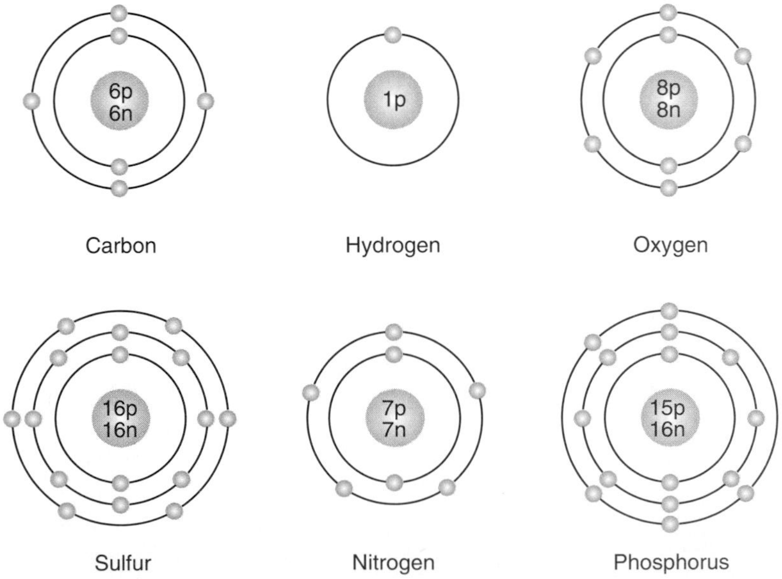

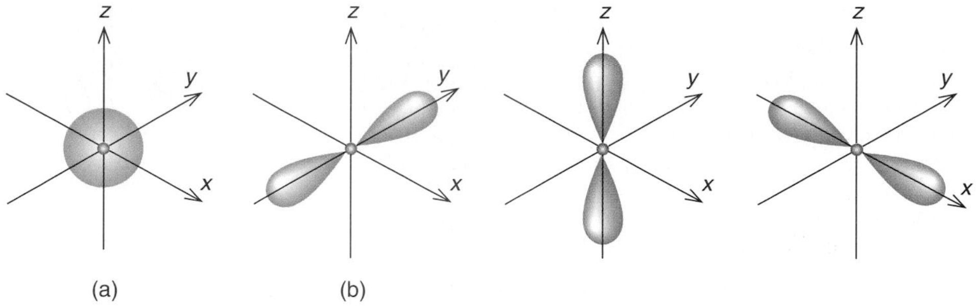

Figure A.2

Electron orbitals. Model depicting the volume of space in which an electron is likely to be found 90% of the time. (*a*) The first energy level consists of one spherical orbital containing up to two electrons. The second energy level has four orbitals, each describing the distribution of up to two electrons. One of the orbitals of the second energy level is spherical; the other three are dumbbell-shaped and arranged perpendicular to one another, as the axis lines indicate (*b*). The nucleus is at the center, where the axes intersect.

cates the volume of space in which a particular electron is likely to be found most of the time (fig. 2). The shell closest to the nucleus consists of a single orbital containing one or two electrons. Both the second and third shells consist of four orbitals, each containing one or two electrons.

Subatomic particles from which all atoms are composed are distinguished by mass and electrical charge. A proton has a mass of one unit and a positive electrical charge of one unit. A neutron also has a mass of one unit, but it has no net charge. An electron has a mass only about 1/1800 that of a nuclear particle, but it has a negative charge equivalent in magnitude to the positive charge of the proton. The net charge and mass of an atom are the sums of the charges and masses of its constituent particles. In calculating mass we can usually ignore the weight of the electrons. Thus an atom's weight approximately equals the total number of protons and neutrons. Another term, atomic number, equals the number of protons in an atom's nucleus.

Atoms Combine to Form Molecules

Except for the inert gases, atoms tend to interact with other atoms to form molecules. Hydrogen, oxygen, and nitrogen each readily form simple diatomic molecules. Invariably, molecules have properties that are quite different from those of the constituent elements. For example, a molecule of sodium chloride contains one atom of sodium (Na) and one atom of chlorine (Cl). Sodium is a highly reactive silvery metal, whereas chlorine is a corrosive yellow gas. When equal numbers of Na and Cl atoms interact, vigorous reaction occurs and white crystalline solid sodium chloride is formed.

Molecules are described by writing the symbols of the constituent elements and indicating the numbers of atoms of each element in the molecule as subscripts. For example, the sugar molecule glucose is represented as $C_6H_{12}O_6$, which indicates that it contains 6 atoms of carbon, 12 atoms of hydrogen, and 6 atoms of oxygen.

Atoms react with one another by gaining, losing, or sharing electrons to produce molecules. The type of chemical bond that forms between atoms depends on the number of electrons the atoms have in their outermost valence shells.

Ionic Bonds Are Formed Between Oppositely Charged Atoms (Ions)

Atoms of the elements prevalent in living things (carbon, nitrogen, oxygen, phosphorus, and sulfur) are most stable chemically when they have eight electrons in their valence shells. This chemical tendency to fill a valence shell is called the octet rule. One way that this is done is for an atom with one, two, or three electrons in this shell to lose them to an atom with, correspondingly, seven, six, or five electrons in the valence shell—a kind of chemical give-and-take. A sodium atom, for example, has a single electron in its outermost shell. As a result, sodium has a strong tendency to lose this single electron because then it will have 8 electrons in its outermost shell. By contrast chlorine has a strong tendency to gain a single electron to fill its valence shell, which contains only seven electrons.

The attraction between oppositely charged ions results in an ionic bond, such as the one that holds NaCl together. The oppositely charged ions Na^+ and Cl^-, attract each other in such an ordered manner that a crystal results (fig. 3).

Covalent Bonds Are Formed Between Atoms That Share Electron Pairs

Atoms that have three, four, or five electrons in their valence shells are more likely to share electrons in a covalent bond than to swap them in the electronic give-and-take of an ionic bond. In both organic chemistry and biochemistry covalent bonds are much more common than ionic bonds. Carbon, with four electrons in its outer shell, can attain the stable eight-electron configuration in its outer shell by sharing electrons with four hydrogen atoms, each of which has one electron in its only shell. The resulting compound containing four single covalent bonds is methane, CH_4 (fig. 4).

Two or three electron pairs can also be shared in covalent bonds called double and triple bonds, respectively. A single atom of carbon forms a double bond with the oxygen

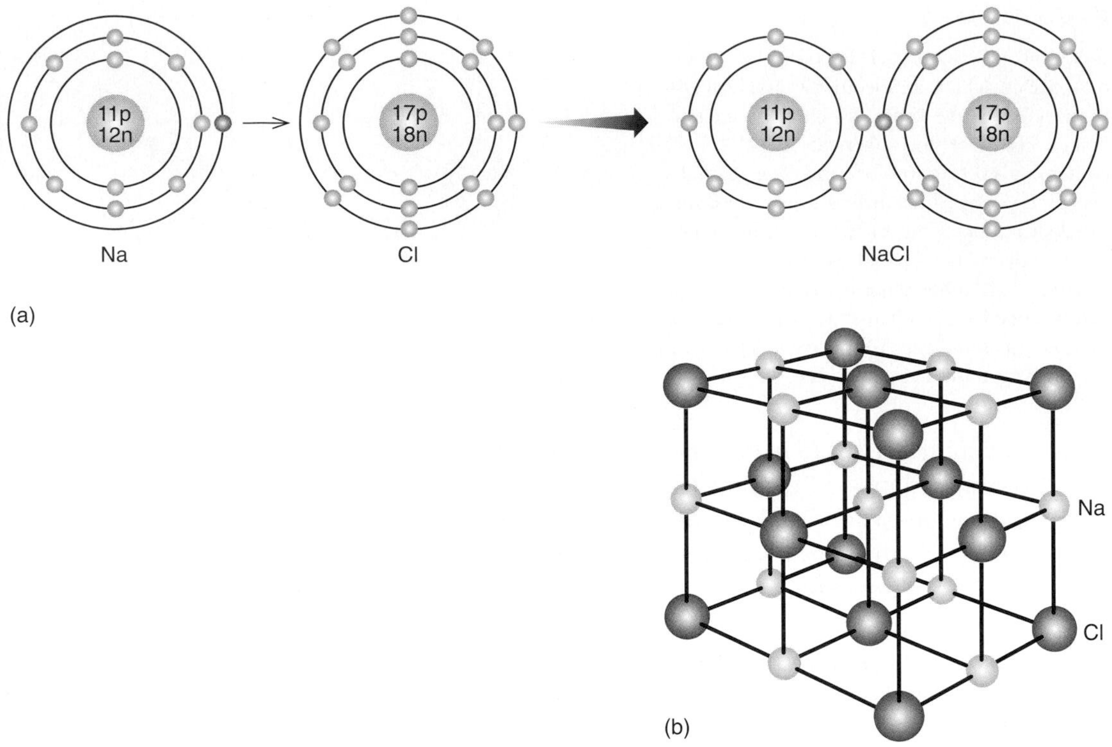

Figure A.3

An ionically bonded molecule (NaCl). (*a*) A sodium atom (Na) can donate the one electron in its valence shell to a chlorine atom (Cl), which has seven electrons in its outermost shell. The resulting ions (Na^+ and Cl^-) bond to form the compound sodium chloride (NaCl). The octet rule has been satisfied. (*b*) The ions that constitute NaCl form a regular crystalline structure in the solid state.

in formaldehyde and a triple bond with the nitrogen in hydrogen cyanide (fig. 5).

Weak Forces Lead to Intramolecular or Intermolecular Interactions

The strong interactions that result in either ionic or covalent bonds determine the primary structure of molecules. Molecules often interact, not by forming more bonds of this type, but rather by forming weaker bonds by polar and apolar interactions. Polar interactions result in polar molecules when the electrons in a covalent bond are not equally shared. In a water molecule (H_2O), for example, the single oxygen atom has a greater affinity for the shared electrons than do the two hydrogen atoms. Thus most of the time the shared electrons are closer to the oxygen atom than to the hydrogen atoms. Because of this unequal sharing, the hy-

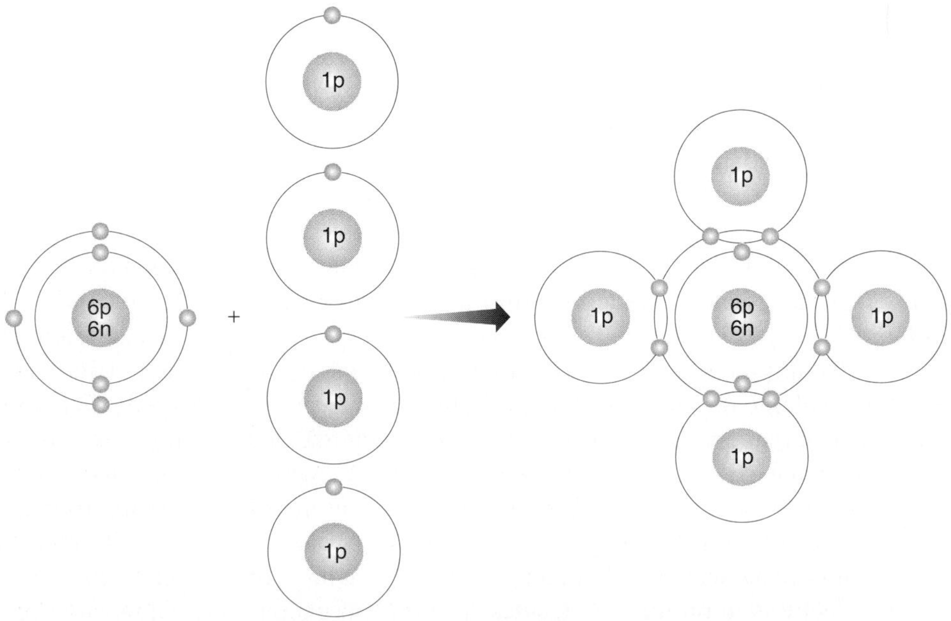

Figure A.4

A covalently bonded molecule. By sharing electrons, one carbon and four hydrogen atoms complete their outermost shells to form methane. In carbon, the outermost shell contains four electrons whereas in hydrogen the outermost shell contains only one electron.

Figure A.5

Covalent linkages formed by carbon with hydrogen, oxygen and nitrogen. Each shared electron pair is represented by two dots or a single straight line.

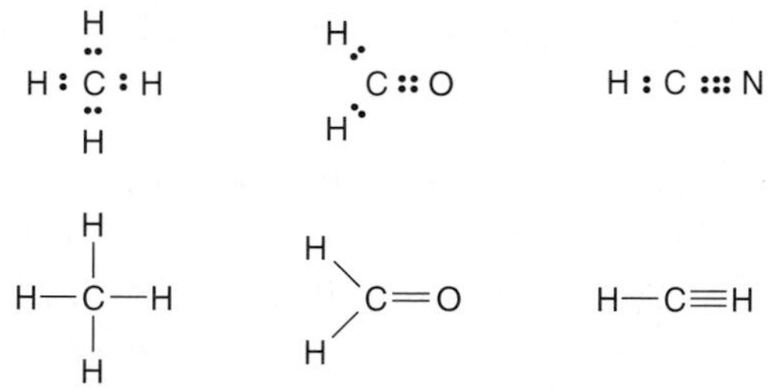

Figure A.6

Two commonly occurring fatty acids (*a*) and comparable hydrocarbons (*b*).

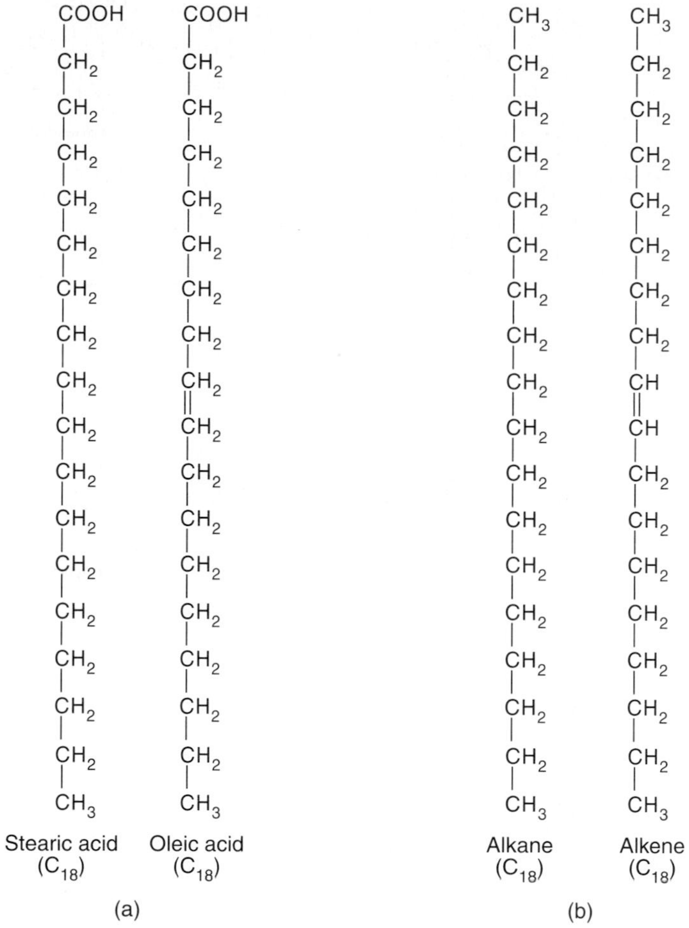

drogen atoms have a partial positive charge and the oxygen atoms have a partial negative charge. This polarity within individual water molecules results in electrostatic interactions between nonbonded H's and O's in liquid water. Apolar molecules are attracted to each other by so-called van der Waals forces. Whereas polar and apolar bonds are much weaker than ionic bonds and covalent bonds, they frequently outnumber the stronger bonds in the liquid state when molecules are nearly close packed. As a result, these weak forces exert a strong influence on the structure of water and the conformation of polymers in an aqueous solvent. Since water itself is polar, a polymer in water tends to fold in a way that exposes its polar groups to the aqueous solvent while burying its apolar groups so that they can interact with each other. In an apolar solvent the exact opposite occurs. The apolar groups on the polymer are exposed for interaction with the apolar organic solvent, whereas the polar groups are buried within the polymer structure so that they can interact with each other.

Four Types of Organic Compounds Are Most Commonly Found in Biochemical Systems

Chemically, the simplest organic compounds found in biochemical systems are the fatty acids (fig. 6). For the most part fatty acids consist of long zig-zag chains with carbon backbones and hydrogen substituents. At one end they usually contain a carboxyl group. In organic chemistry the same compound without the carboxyl group is called an alkane. If two hydrogen atoms are removed from adjacent carbons in a fatty acid, the carbons form a double bond with one another. The comparable organic compound without the carboxyl group is called an alkene. Alkanes and alkenes are highly insoluble in aqueous solvents because of their uniformly apolar character. Fatty acids are partially soluble in aqueous media because of their polar carboxyl group. The carboxyl group also is an important functional group that participates in chemical reactions.

Sugars also possess a carbon backbone but they differ in major respects from fatty acids in that one of the hydrogens attached to each carbon usually is replaced by a hydroxyl group. This change has a dramatic effect on the solubility properties of sugars, making them highly soluble in aqueous solvents. Most sugars are much shorter than fatty acids because of the tendency of sugars to form five- or six-membered rings (fig. 7).

Amino acids are more varied in composition than either fatty acids or sugars (fig. 8). Most amino acids important to biochemistry contain a central carbon with four different substituents: a hydrogen, an amino group, a carboxyl group, and an R group that varies for different types of amino acids. These R groups may be positively charged, contain sulfur, be negatively charged, or be neutral. Neutral R groups are further divided into polar and apolar types. The water solubility of amino acids varies by R group.

At the highest level of complexity in biochemical systems are the nucleotides (fig. 9). These molecules contain three components: a five-carbon sugar to which are attached

Figure A.7

Most sugars contain only 5 or 6 carbons. In aqueous solution, the circular hemiacetal form is strongly favored over the linear form.

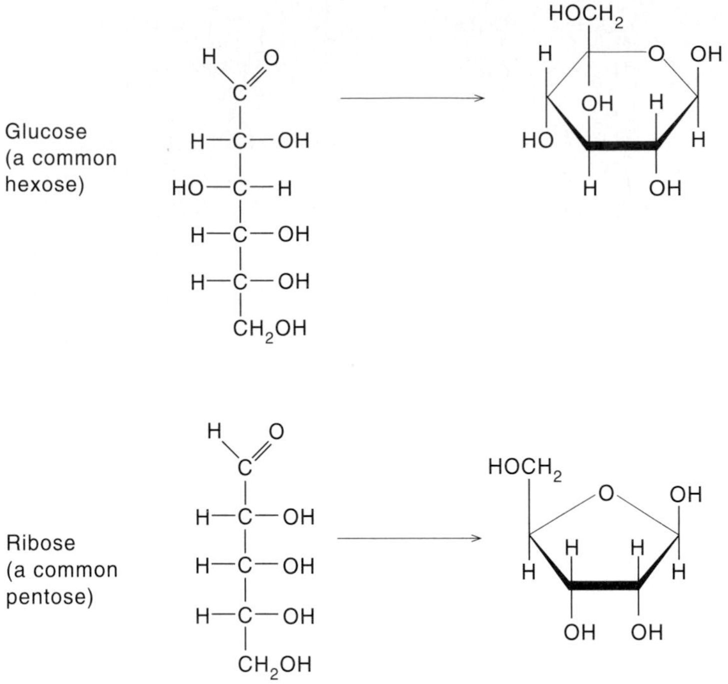

a phosphate group on one side and a heterocyclic base on the other side. Because of the phosphate group and the hydroxyl groups on the sugar, nucleotides tend to be highly soluble in water and poorly soluble in apolar organic solvents. Whereas twenty different types of amino acid are commonly found in biochemical systems, only five types of heterocyclic nitrogen bases are commonly found in nucleotides.

Each of the molecules described in this section represents a component of a considerably larger structure in living cells. Fatty acids are major components of lipids and membranes. Sugars are major components of polysaccharides and cell walls. Amino acids are the substance from which proteins are synthesized. Nucleotides are the building blocks of nucleic acids and chromosomes.

Most Molecules Found in Cells Are Chiral

When you look into a mirror, you see a reflection that is very similar in appearance to yourself—the main difference being that your right and left sides are switched. No matter which way you turn, you cannot duplicate your mirror image. The same is true of most amino acids, which have four different substituents attached to their α carbon atoms. Amino acids exist in two chiral forms just like you and your mirror image. We can take this analogy a step further. Just as you know that you will never see anybody who looks just like your mirror image, so in biomolecules only one of two chiral forms is usually found. This characteristic is true of most of the biomolecules found in cells.

Functional Groups Are Important in Both Organic Chemistry and Biochemistry

Biomolecules contain a limited number of functional groups that are the reactive centers of the molecules (table 1). Biochemical reactions involving these functional groups are closely related to reactions studied in organic chemistry. Here we describe some of the best known functional groups and the reactions in which they participate.

Alcohols, which contain a hydroxyl functional group, can undergo dehydration reactions with either carboxylic or phosphoric acid to form esters. The double arrows used in this and subsequent equations indicate that the reaction can go in either direction.

Figure A.8

The structure of α-amino acids found in proteins. A central carbon atom is linked to four different substituents (*a*). The R group substituents are of 20 different types.

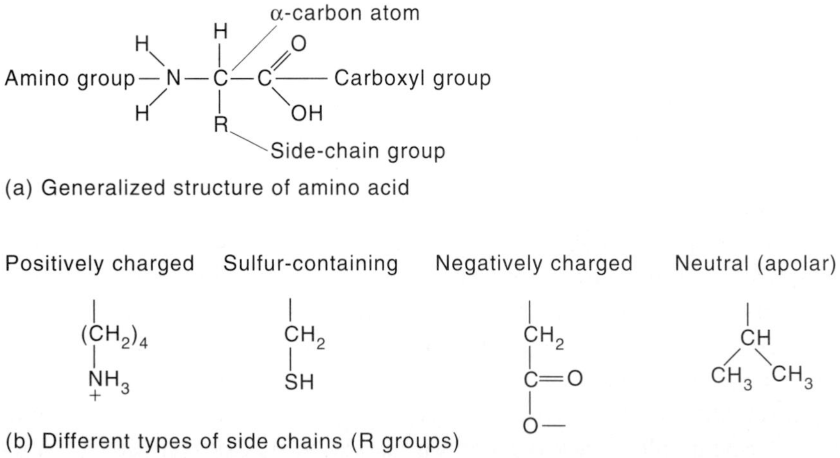

Figure A.9

The nucleotide is a three component structure (*a*). There are five different bases commonly found in nucleotides (*b*).

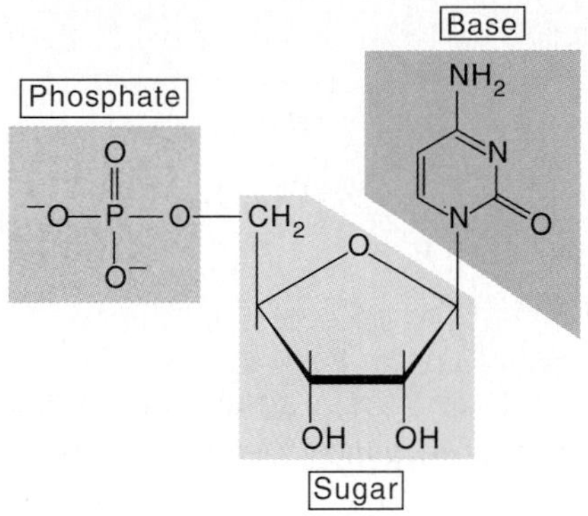

(a) Generalized structure of a nucleotide

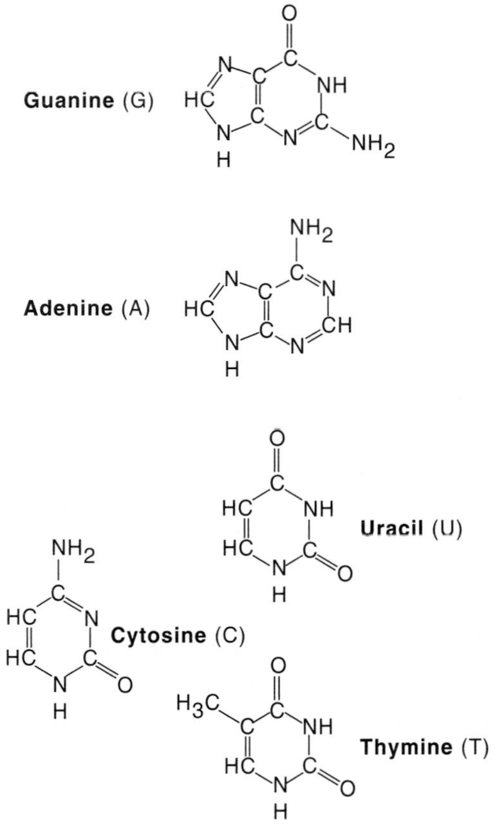

(b) Different bases found in nucleotides

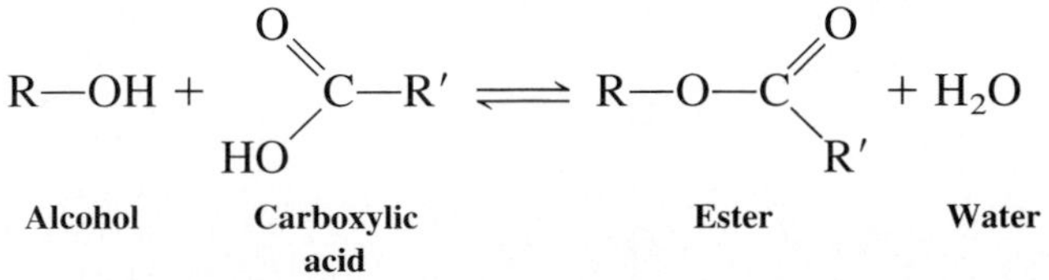

$$R{-}OH + \underset{HO}{\overset{O}{C}}{-}R' \rightleftharpoons R{-}O{-}\overset{O}{\underset{R'}{C}} + H_2O \quad (1)$$

Alcohol   **Carboxylic acid**   **Ester**   **Water**

$$R{-}OH + HO{-}\overset{O}{\underset{O}{P}}{-}O^- \rightleftharpoons R{-}O{-}\overset{O}{\underset{O}{P}}{-}O^- + H_2O \quad (2)$$

Alcohol   **Phosphoric acid**   **Phosphoric acid ester**   **Water**

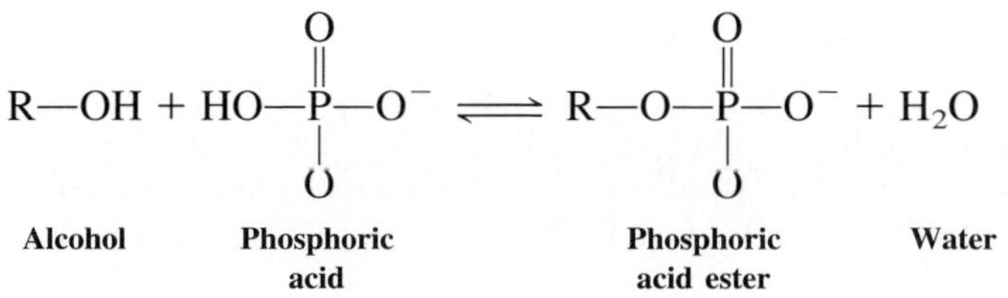

Thiols contain sulfhydryl groups (—SH) and can substitute for alcohols in some reactions, leading to the formation of thiol esters.

$$R{-}SH + \underset{HO}{\overset{O}{C}}{-}R' \rightleftharpoons R{-}S{-}\overset{O}{\underset{R'}{C}} + H_2O \quad (3)$$

Thiol   **Carboxylic acid**   **Thiol ester**   **Water**

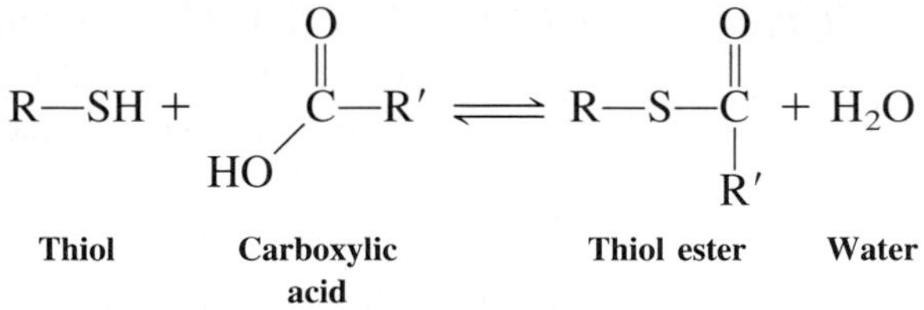

Two alcohols can react with each other to form an ether.

$$R{-}OH + HO{-}R' \longrightarrow R{-}O{-}R' + H_2O \quad (4)$$

Alcohol₁   **Alcohol₂**   **Ether**

Alcohols may undergo a dehydrogenation reaction to form a carbonyl derivative (aldehyde or ketone).

$$R{-}\overset{H}{\underset{R}{C}}{-}OH \underset{+2H}{\overset{-2H}{\rightleftharpoons}} \underset{R}{\overset{R}{C}}{=}O \quad (5)$$

Alcohol   **Aldehyde or ketone**

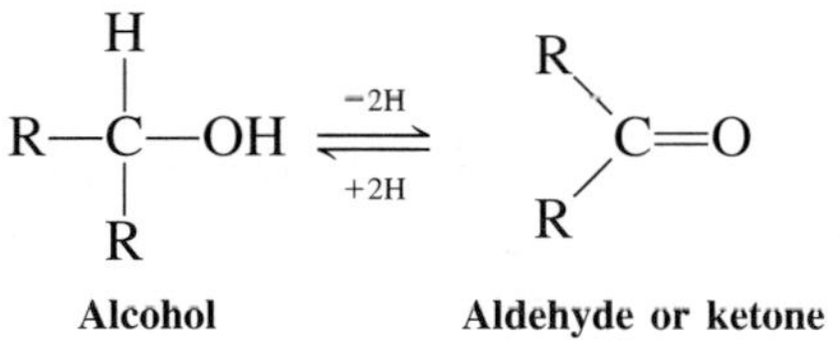

Amines undergo reactions with carboxylic acids comparable to the formation of esters from alcohols. The product is known as an amide.

$$R{-}\overset{H}{\underset{H}{N}} + \underset{HO}{\overset{O}{C}}{-}R' \rightleftharpoons$$

Primary amine   **Carboxylic acid**

$$R{-}\overset{H}{\underset{}{N}}{-}\overset{O}{\underset{}{C}}{-}R' + H_2O \quad (6)$$

Amide

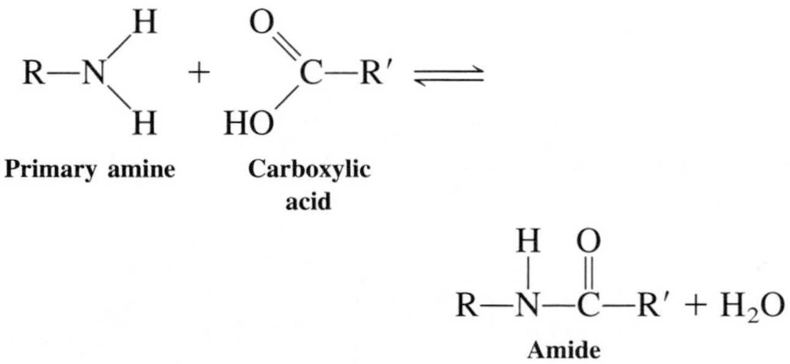

Amines may undergo dehydrogenation reactions leading to the formation of imines, which are frequently unstable in water and hydrolyze to ketones, or, in cases where one of the R groups is an H, to aldehydes.

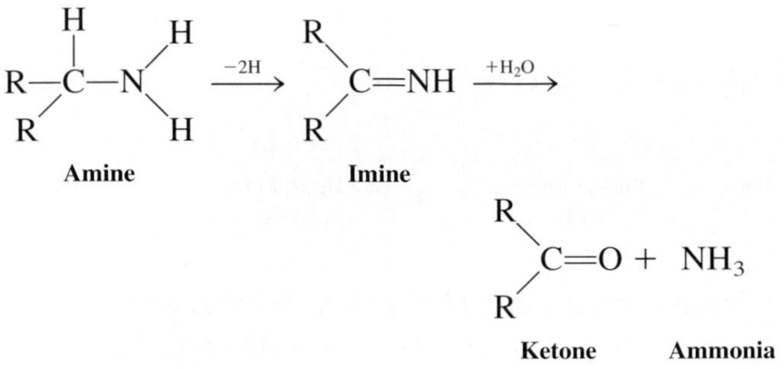

(7)

Aldehydes and ketones both may be reduced to alcohols by hydrogenation (see the alcohol dehydrogenation reaction, equation 5). Aldehydes may react with either water or alcohol to form aldehyde hydrates or hemiacetals, respectively (also see figure 7 for intramolecular hemiacetals formed by sugars). Reaction of an aldehyde with two molecules of alcohol leads to acetal formation.

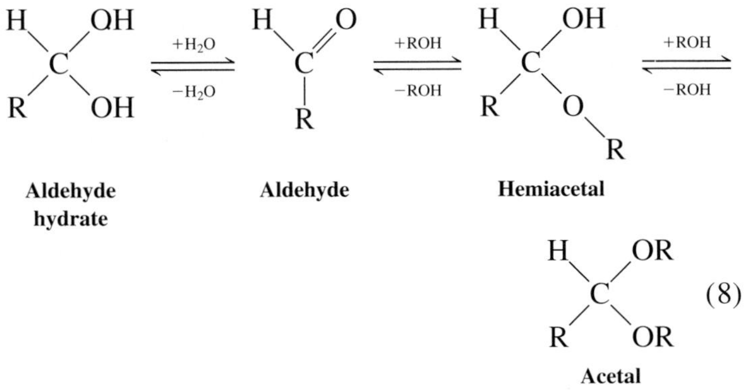

(8)

Dehydrogenation of an aldehyde hydrate leads to carboxylic acid formation.

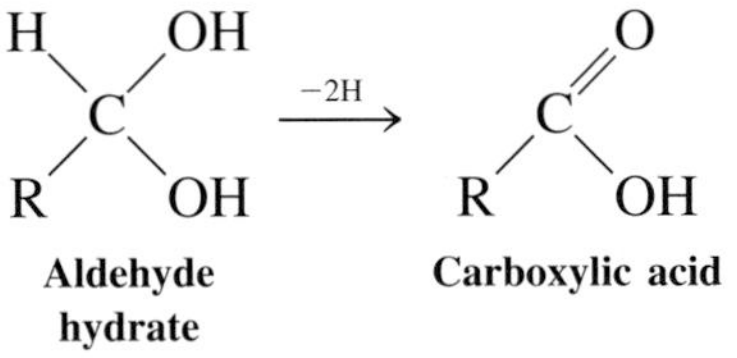

(9)

Aldehydes and ketones may also isomerize to the enol form as long as the adjacent carbon atom is bonded to at least one hydrogen atom. In the reaction a hydrogen migrates and the double bond shifts. At equilibrium the keto form is strongly preferred.

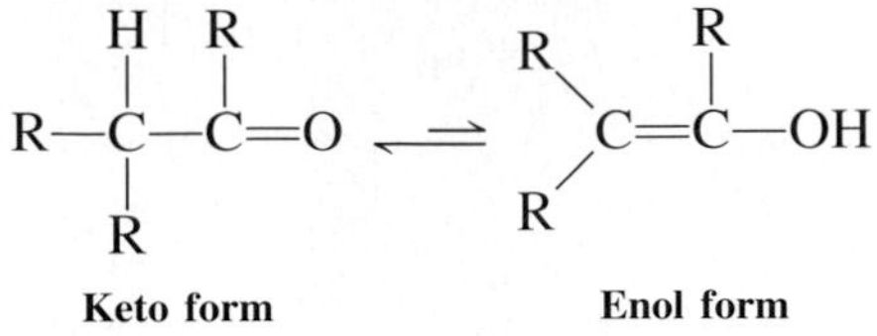

(10)

Pyrophosphates may hydrolyze to inorganic phosphoric acid (phosphate) and an organophosphoric acid.

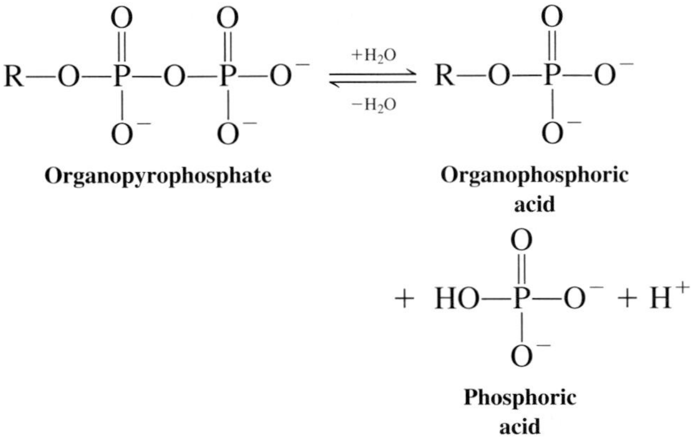

(11)

Such hydrolysis reactions yield considerable energy, which can be utilized in biosynthesis.

All the functional groups described are electrostatically neutral in organic solvents. However, in water many of these functional groups either lose or gain protons to become charged species. Such ionization reactions are extremely important in biochemical systems because they frequently influence solubility and reactivity.

Carboxylic and phosphoric acids lose one or more protons in water to become negatively charged. The ionized forms are stabilized by resonance as shown:

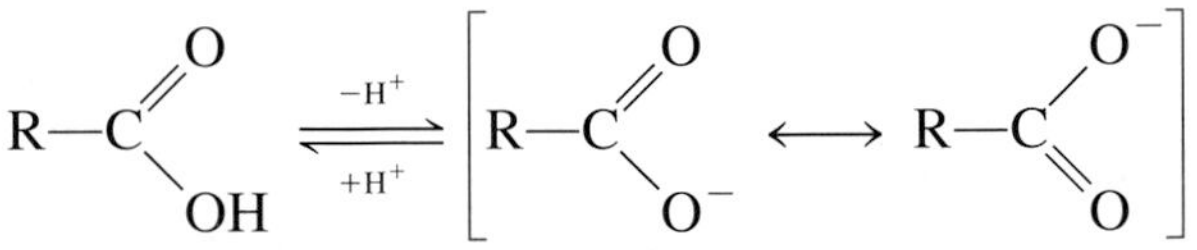

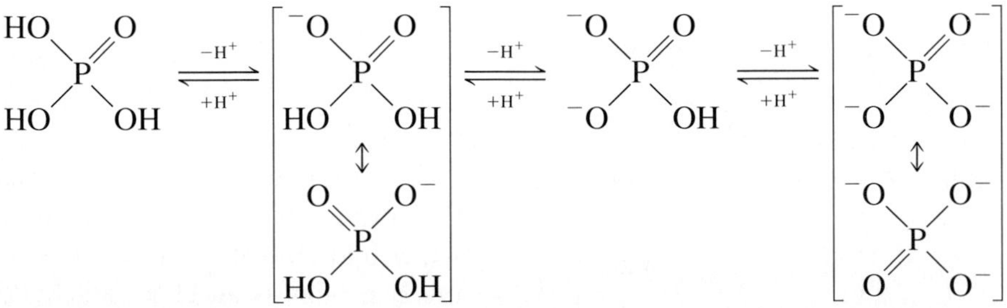
(12)

Table 1

Organic Functional Groups of Biochemical Importance

Type of Compound	General Structural Formula	Characteristic Functional Group	Name of Functional Group	Example Biomolecule with Functional Group	Chapter Reference
Alcohols	R—O—H	—OH	Hydroxyl group	Glycerol (polyhydroxyl) 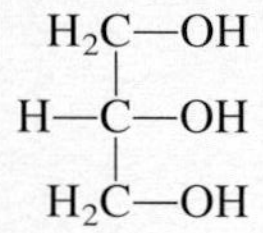	17
Ethers	R—O—R′	—O—	Ether	Thromboxane A_2 (cyclic ether) 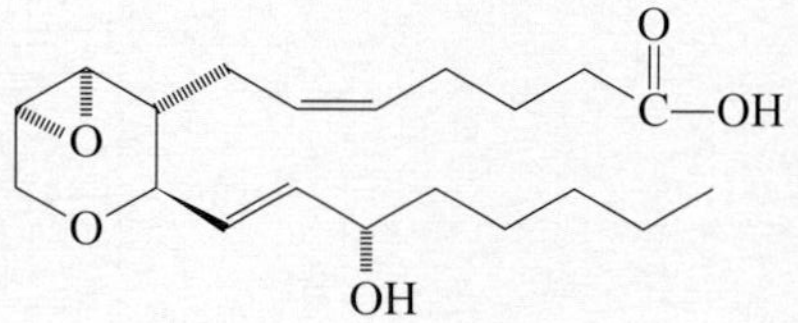	19
Thiols	R—S—H	—SH	Sulfhydryl	Coenzyme A CoASH	10
Aldehydes	R—C(=O)—H	—C(=O)—H	Aldehyde	Glyceraldehyde-3-phosphate 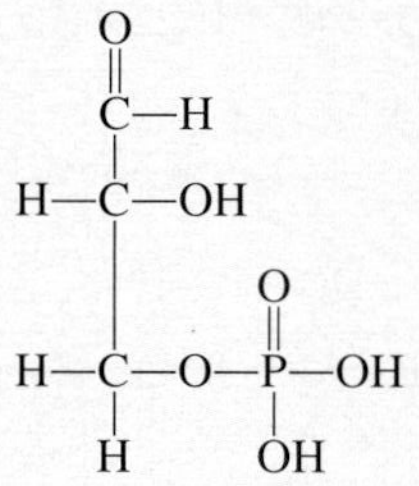	12

Ketones	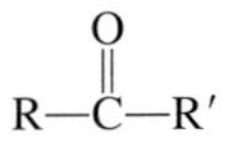	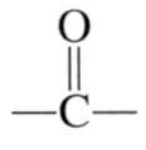	Keto	Fructose-6-phosphate 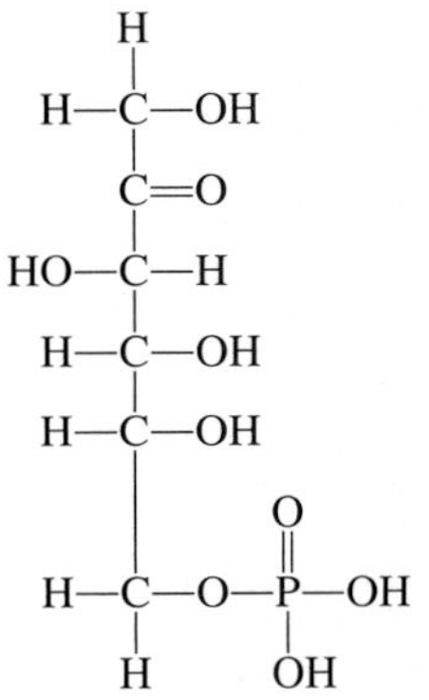	12
Carboxylic acids	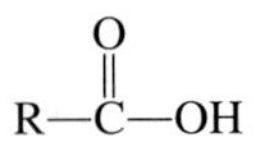	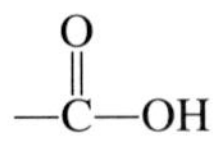	Carboxy	Stearic acid 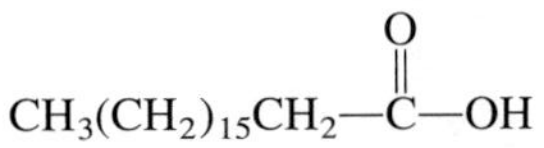	17
Mixed acid anhydride	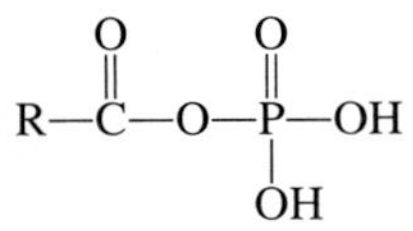	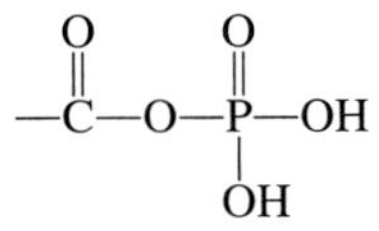	Phosphoric acid anhydride	Glycerate-1,3-bisphosphate 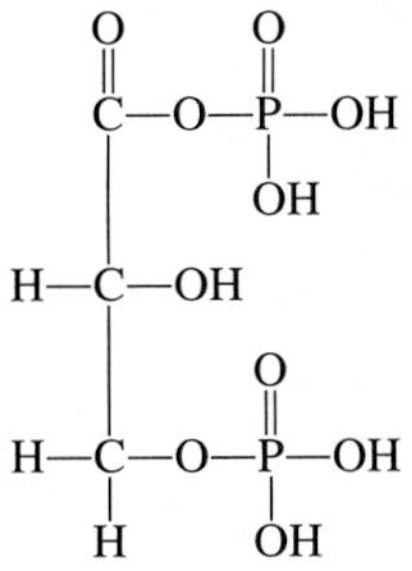	12
Esters	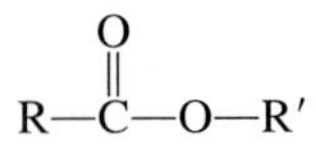	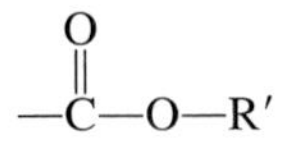	Ester	Cholesterol ester 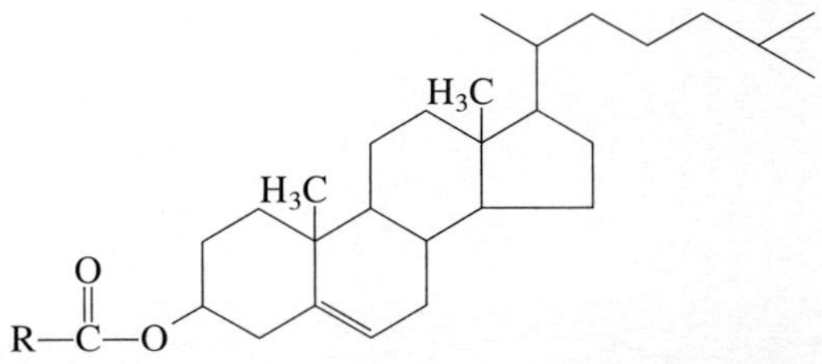	20
Hemiacetals	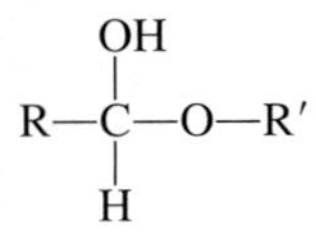	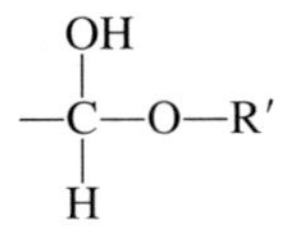	Hemiacetal	Glucose 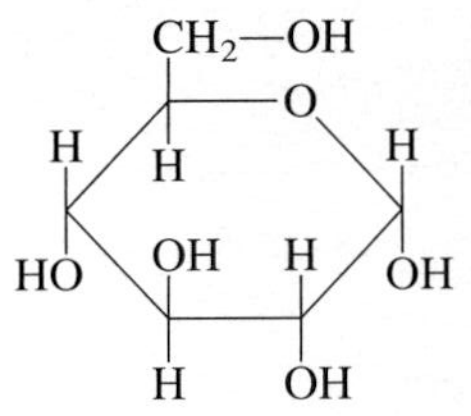	12

Table 1

Organic Functional Groups of Biochemical Importance (continued)

Type of Compound	General Structural Formula	Characteristic Functional Group	Name of Functional Group	Example Biomolecule with Functional Group	Chapter Reference
Acetals	R—CH(O—R″)—O—R′	—CH(O—R″)—O—R′	Acetal	Sucrose 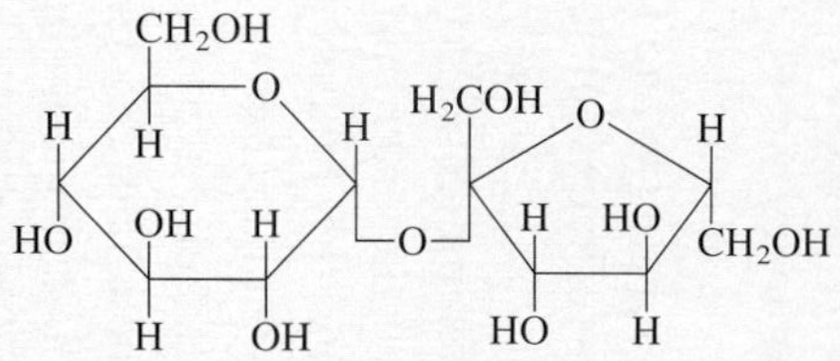	16
Lactones	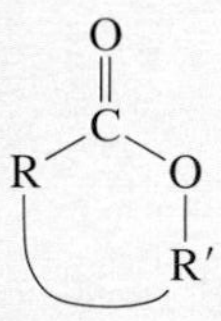	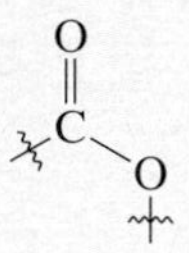	Lactone	6-Phospho-D-gluconolactone 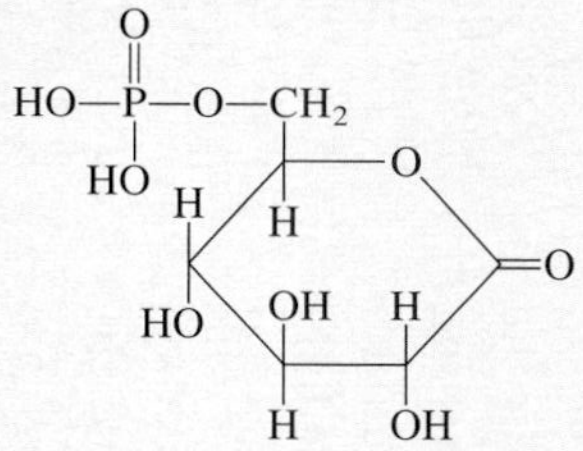	12
Amines	R—NH_2	—NH_2	Amino	Lysine $H_2N—(CH_2)_4—CH(NH_2)—C(=O)—OH$	3
Amides	R—C(=O)—N(R‴)(R‴)	—C(=O)—N(R‴)(R‴)	Amido	*N*-Acetyl-D-glucosamine 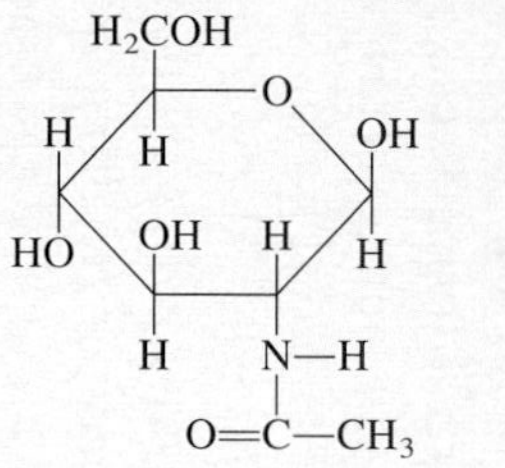	16
Alkenes	(R‴)(R‴)C=C(R‴)(R‴)	>C=C<	Alkenyl	Palmitoleic acid $CH_3(CH_2)_5—CH=CH—(CH_2)_7—C(=O)—OH$	17

Note: R, R′, R″ are abbreviations for any alkyl or aryl group.
R‴ is the abbreviation for any alkyl group or hydrogen.

Amines usually add a proton to become positively charged.

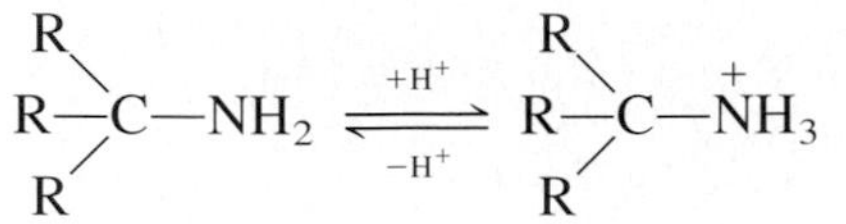

(13)

Near neutrality (10^{-7} M H^+), where most biochemical systems function, the carboxyl group exists mainly in the negatively charged form, phosphoric acid exists mainly in the diionized form, and amino groups exist mainly in the positively charged form. These conditions have interesting consequences for amino acids which contain one amino group and one carboxyl group. The amino acids are usually neutral overall, even though they contain two charged groups, one resulting from the deprotonation of the carboxyl group and the other resulting from the protonation of the amino group. Amino acids existing as dipolar ions are called zwitterions.

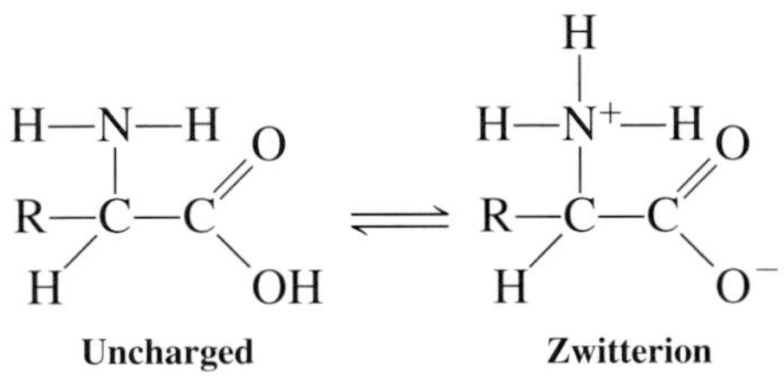

(14)

The preceding are some of the more important reactions involving covalent bond breakage or formation in biochemistry. By now two things should be apparent about biochemical reactions: (1) the number of reactions in biochemistry is much more limited than in ordinary chemistry; and (2) as far as the reactants and products are concerned, biochemical reactions may be understood in the same terms as ordinary chemical reactions.

Reaction Mechanisms, Arrow Formalism, and Catalysis

So far we have described reactions strictly in terms of their reactants and products. That doesn't tell us much about how or why a reaction proceeds from reactants to products. The reaction mechanism sometimes is quite a complex process, and for many well-known reactions we still don't understand the mechanism. We confine this discussion to consideration of the relatively simple mechanism of ester hydrolysis. Ester hydrolysis entails the reaction of a water molecule with an ester, leading to the production of an alcohol and a carboxylic acid. It is depicted in equation 1, going from right to left.

In attempting to find a mechanism for this or any other reaction we first must recall that all covalent bonds are formed by electrons; the rearrangement or breakage of such bonds starts with the migration of electrons. In the most general sense, reactive groups function either as electrophiles or nucleophiles. The former are electron-deficient substances that are attacked by electron-rich substances. The latter are electron-rich substances that attack electron-deficient substances.

In the case of ester hydrolysis the O of the water is an electron-rich substance and the carbonyl carbon of the ester is an electron-deficient substance. Ester hydrolysis occurs in three stages: (1) the initial stage in which electrons flow from the water molecule to the ester; (2) the intermediate stage in which the ester carbonyl forms a tetrahedral complex involving the hydroxyl group originating from the water molecule that releases a proton; and (3) the final stage in which carboxylic acid and alcohol are formed.

This reaction is kinetically favored because the oxygen of water is a nucleophile, whereas the carbonyl carbon is an electrophile. In the initial step the curved, colored arrows that represent the flow of electron pairs suggest three electron pair migrations that happen in quick succession (figure 10a). An electron pair migrates from the O—H of the water molecule to the O. An electron pair from the O attacks the carbonyl carbon, and an electron pair between the C and the O in the carbonyl migrates to the O of the carbonyl. The intermediate step involves the results of these and two additional electron pair migrations from the relatively unstable intermediate products that lead to the final products.

Normally, ester hydrolysis at ambient temperatures in water at neutral pH will occur very slowly—over a period of many days—unless something is done to accelerate the reaction. Usually, reactions are accelerated by raising the temperature because molecules react more vigorously as the temperature rises. However, a better way to accelerate a reaction is to add a catalyst, that is, a molecule that accelerates the reaction without undergoing any net consumption. Frequently, the main job of the catalyst is to make a potentially reactive center more attractive for reaction, i.e., to potentiate the electrophilic or nucleophilic character of the reactive centers of the reacting species.

In aqueous solution, protons (actually present as hydronium ions) or hydroxide ions are the catalysts most commonly used for nonenzymatic reactions. The way in which acid or base catalysts work in ester hydrolysis is illustrated in Figure 10b and c. As a result of the electronegativity of the oxygen atom in the ester >C=O group, the oxygen has a fractional negative charge δ^- and the carbon has a fractional positive charge δ^+. Either acid or base catalysts may be used to accelerate hydrolysis of the ester. In acid catały-

Figure A.10

Reaction mechanisms for ester hydrolysis in the absence (*a*) and presence of catalysts (*b–e*).

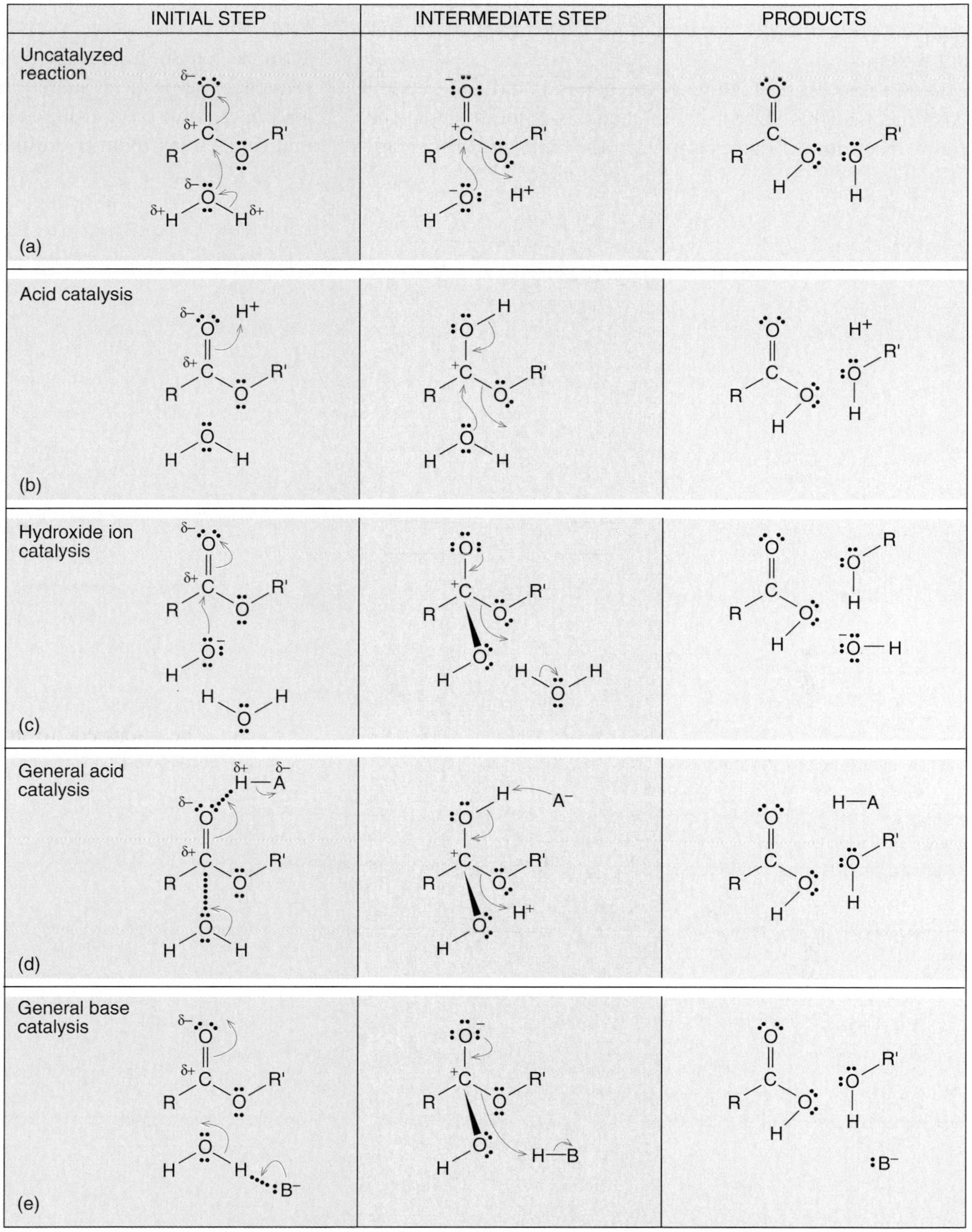

sis, a proton acting as an electrophile is attracted to the oxygen. That leads to an intermediate stage that accentuates the positive charge on the carbon atom, making it a more attractive electrophile to an attacking nucleophile—in this case, water. Water is a poor nucleophile and would attack the carbon slowly without such an inducement, so this step is the key to the catalysis. The remaining reactions leading to ester hydrolysis and regeneration of the catalyst occur rapidly and spontaneously. In hydroxide ion catalysis of the ester, the carbon atom is attacked directly by a stronger nucleophile, OH^-. Again after hydrolysis, the catalyst is regenerated.

To avoid the harshness of pH extremes, generalized acids or bases frequently are used to replace protons or hydroxide ions as catalysts. These compounds are capable of yielding protons or absorbing protons, respectively, during the course of a reaction. Enzymes almost always utilize generalized acids or bases because of the delicacy of biochemical systems.

An interesting feature of the general acid and general base mechanisms shown is the catalysis of both steps (see figure 10d and e). For example, in the first step of general acid catalysis, HA adds a proton and thus is acting as an acid, but in the second step A^- removes a proton and is acting as a base. In the general base catalysis mechanism the sequence is reversed. Such sequential catalysis by a general acid or general base group is much more common in enzymatic reactions than in ordinary chemical reactions.

Appendix A
Some Landmark Discoveries in Biochemistry

In this appendix we list, in chronological order, some of the most important discoveries made in biochemistry during the past two centuries. It is impossible, for reasons of space, to give credit to every worker who has made a significant contribution, but it is possible to identify certain events as milestones, and thus to show how progress in this field has accelerated with the passage of time.

1770–1774
Priestly showed that oxygen is produced by plants and consumed by animals.

1773
Rouelle isolated urea from urine.

1828
Wohler synthesized the first organic compound, urea, from inorganic components.

1838
Schleiden and Schwann proposed that all living things are composed of cells.

1854–1864
Pasteur proved that fermentation is caused by microorganisms.

1864
Hoppe-Seyler crystallized hemoglobin.

1866
Mendel demonstrated the segregation and independent assortment of alleles in pea plants.

1893
Ostwald showed that enzymes are catalysts.

1898
Camillio Golgi described the Golgi apparatus.

1905
Knoop deduced the β oxidation mechanisms for fatty acid degradation.

1907
Fletcher and Hopkins showed that lactic acid is formed quantitatively from glucose during anaerobic muscle contraction.

1910
Morgan discovered sex-limited inheritance in *Drosophila.*

1912
Warburg postulated a respiratory enzyme for the activation of oxygen.

1913
Michaelis and Menten developed a kinetic theory of enzyme action.

1922
McCollum showed that lack of vitamin D causes rickets.

1926
Sumner crystallized the first enzyme urease.

1926
Jansen and Donath isolated vitamin B_1 (thiamine) from rice polishings.

1926–1930
Svedberg invented the ultracentrifuge and used it to demonstrate the existence of macromolecules.

1928
Levene showed that nucleotides are the building blocks of nucleic acids.

1928
Szent-Gyorgyi isolated ascorbic acid (Vitamin C).

1928–1933
Warburg deduced the iron-prophyrin presence in the respiratory enzyme.

1929
Burr and Burr discovered that linoleic acid is an essential fatty acid for animals.

1931
Englehardt discovered that phosphorylation is coupled to respiration.

1932
Warburg and Christian discovered the "yellow enzyme," a flavoprotein.

1933
Krebs and Henseleit discovered the urea cycle.

1933
Embden and Meyerhof demonstrated the intermediates in the glycolytic pathway.

1935
Schoenheimer and Rittenberg first used isotopes as tracers in the study of intermediary metabolism.

1935
Stanley first crystallized a virus, tobacco mosaic virus.

1937
Krebs discovered the citric acid cycle.

1937
Warburg showed how ATP formation is coupled to the dehydrogenation of glyceraldehyde-3-phosphate.

1938
Hill found that cell-free suspensions of chloroplasts yield oxygen when illuminated in the presence of an electron acceptor.

1939
C. Cori and G. Cori demonstrated the reversible action of glycogen phosphorylase.

1939
Lipmann postulated the central role of ATP in the energy-transfer cycle.

1939–1946
Szent-Gyorgyi discovered actin and the actin-myosin complex.

1940
Beadle and Tatum deduced the one gene–one enzyme relationship.

1942
Bloch and Rittenberg discovered that acetate is the precursor of cholesterol.

1943
Chance applied spectrophotometric methods to the study of enzyme–substrate interactions.

1943
Martin and Synge developed partition chromatography.

1944
Avery, MacLeod, and McCarty demonstrated that bacterial transformation is caused by DNA.

1947–1950
Lipmann and Kaplan isolated and characterized coenzyme A.

1948
Leloir discovered the role of uridine nucleotides in carbohydrate metabolism.

1948
Hogeboom, Schneider, and Palade refined the differential centrifugation method for fractionation of cell parts.

1948
Kennedy and Lehninger discovered that the tricarboxylic acid cycle, fatty acid oxidation, and oxidative phosphorylation all take place in mitochondria.

1949
Christian deDuve discovered lysosomes.

1950–1953
Chargaff discovered the base equivalences in DNA.

1951
Pauling and Corey proposed the α-helix structure for α-keratins.

1951
Lynen postulated the role of coenzyme A in fatty acid oxidation.

1952
Palade, Porter, and Sjostrand perfected thin sectioning and fixation methods for electron microscopy of intracellular structures.

1952–1954
Zamecnik and his colleagues developed the first cell-free systems for the study of protein synthesis.

1953
Vincent du Vigneaud synthesized the first biologically active peptide hormone, ocytocin.

1953
Woodward and Bloch postulated a cyclization scheme for squalene, leading to cholesterol.

1953
Sanger and Thompson determined the complete amino acid sequence of insulin.

1953
Hokin and Hokin showed that acetylcholine induces the rapid biosynthesis of phosphatidylinositol in pigeon pancreas.

1953
Horecker, Dickens, and Racker elucidated the 6-phosphogluconate pathway of glucose catabolism.

1953
Watson and Crick and Wilkins determined the double-helix structure of DNA.

1954
Hugh Huxley proposed the sliding filament model for muscular contraction.

1955
Ochoa and Grunberg-Manago discovered polynucleotide phosphorylase.

1955
Kennedy and Weiss described the role of CTP in the biosynthesis of phosphatidylcholine.

1956
Kornberg discovered the first DNA polymerase.

1956
Umbarger reported that the end product isoleucine inhibits the first enzyme in its biosynthesis from threonine.

1956
Dorothy Crawfoot Hodgkin determined the structure of coenzyme B12.

1956
Ingram showed that normal and sickle-cell hemoglobin differ in a single amino acid residue.

1956
Anfinsen and White concluded that the three-dimensional conformation of proteins is specified by their amino acid sequence.

1956
Leloir determined the pathway to uridine diphosphate glucose (UDPG).

1957
Hoagland, Zamecnik, and Stephenson isolated tRNA and determined its function.

1957
Sutherland discovered cyclic AMP.

1958
Weiss, Hurwitz, and Stevens discovered DNA-directed RNA polymerase.

1958
Meselson and Stahl demonstrated that DNA is replicated by a semiconservative mechanism.

1959
Wakil and Ganguly reported that malonyl-CoA is a key intermediate in fatty acid biosynthesis.

1960
Kendrew reported the x-ray analysis of the structure of myoglobin.

1961
Jacob and Monod proposed the operon hypothesis.

1961
Jacob, Monod, and Changeux proposed a theory of the function and action of allosteric enzymes.

1961
Mitchell postulated the chemiosmotic hypothesis for the mechanism of oxidative phosphorylation.

1961
Nirenberg and Matthaei reported that polyuridylic acid codes for polyphenylalanine.

1961
Marmur and Doty discovered DNA renaturation.

1962
Racker isolated F_1 ATPase from mitochondria and reconstituted oxidative phosphorylation in submitochondrial vesicles.

1966
Maizel introduced the use of sodium dodecylsulfate (SDS) for high-resolution electrophoresis of protein mixtures.

1966
Crick proposed the wobble hypothesis.

1966
Gilbert and Muller-Hill isolated the lac repressor.

1968
Glomset proposed the theory of reverse cholesterol transport in which HDL is involved in the return of cholesterol to the liver.

1968
Meselson and Yuan discovered the first DNA restriction enzyme. Shortly thereafter Smith and Wilcox discovered the first restriction enzyme that cuts DNA at a specific sequence.

1969
Zubay and Lederman developed the first cell-free system for studying the regulation of gene expression.

1970
Howard Temin and David Baltimore discovered reverse transcriptase.

1971
Vane discovered that aspirin blocks the biosynthesis of prostaglandins.

1972
Jon Singer and Garth Nicolson proposed the fluid mosaic model for membrane structure.

1973
Cohen, Chang, Boyer, and Helling reported the first DNA cloning experiments.

1975
Brown and Goldstein described the low-density lipoprotein receptor pathway.

1975
Sanger and Barrell developed rapid DNA-sequencing methods.

1976
Michael Bishop and Harold Varmus discovered the c-src gene in uninfected cells, which is homologous to the v-src gene in the Rous sarcoma virus.

1977
Starlinger discovered the first DNA insertion element.

1977
McGarry, Mannaerts, and Foster discovered that malonyl-CoA is a potent inhibitor of β oxidation.

1977
Splicing of RNA simultaneously discovered in Broker's and Sharp's laboratories.

1977
Nishizuka and coworkers reported the existence of protein kinase C.

1978
Shortles and Nathans did the first experiments in directed mutagenesis.

1978
Tonegawa demonstrated DNA splicing for an immunoglobulin gene.

1981
Cech discovered RNA self-splicing.

1981
Steitz determined the structure of CAP protein.

1981–1982
Palmiter and Brinster produced transgenic mice.

1983
Mullis amplified DNA by the polymerase chain reaction (PCR) method.

1984
Schwartz and Cantor developed pulsed field gel electrophoresis for the separation of very large DNA molecules.

1984
Michel, Deisenhofer, and Huber determined the structure of the photosynthetic reaction center.

1984
Blobel discovered the mechanism for protein translocation across the endoplasmic reticulum membrane—the signal hypothesis.

1988
Elion and Hitchings shared the Nobel Prize for design and synthesis of therapeutic purines and pyrimidines.

1989
Synder and colleagues purified and reconstituted the inositol-1,3,4-P_3 receptor.

Appendix B

A Guide to Career Paths in the Biological Sciences

After You Receive Your Bachelor's Degree, What's Next?

When you receive your bachelor's degree in the biological sciences (e.g., biology, biochemistry, microbiology, etc.), you may choose to continue your education by attending graduate school, medical school, or some other professional school, or you may want to look for a job. If you are interested in going on to graduate school for a masters (MS) or doctorate (Ph.D.) degree in the biological sciences or to medical school for an MD or MD/Ph.D. degree, you can usually obtain information on specific schools and programs from your faculty advisor, on-campus career center, or library.

Job opportunities are available for bachelor's level biological sciences graduates if you are interested in finding employment directly after graduation. Because majors and job titles don't always match and colleges and universities don't typically have employment services, finding employment most often requires quite a bit of research on your part. Extensive job descriptions for virtually every type of position available are provided in *The Dictionary of Occupational Titles,* published by the U.S. Department of Labor. Also, research and reference firms such as Peterson's Guides, Inc., publish guides to job opportunities for science graduates.

Job Opportunities for Bachelor's Level Biological Science Graduates

As indicated in *Peterson's Job Opportunities for Engineering, Science, and Computer Graduates 1993,** job titles listed in employment advertisements can be categorized by functional occupational area. They are accounting/finance, administration, information systems/processing, marketing/sales, production/operations, research/development, and technical/professional services. All these functional areas are not directly applicable to people with degrees in the biological sciences, but thinking about jobs in terms of these areas can help broaden your career options.

The most applicable functional occupational areas for people with a degree in the biological sciences are marketing/sales, production/operations, research/development, and technical/professional services. A large number of companies and other organizations have entry-level positions available for bachelor's level biological sciences graduates, including the following.

Marketing / Sales

American Cyanamid Company
1 Cyanamid Plaza
Wayne, NJ 07470
Contact: Office of Technical Recruiting

Eli Lilly and Company
Lilly Corporate Center
Indianapolis, IN 46285
Contact: Office of Corporate Recruiting

Johnson & Johnson
1 Johnson & Johnson Plaza
New Brunswick, NJ 08933
Contact: College Relations Coordinators in the Personnel Office of each Johnson & Johnson Division

Marion Merrell Dow, Inc.
10236 Marion Park Drive
Kansas City, MO 64137
Contact: Staffing Manager

Parke-Davis, Warner-Lambert Company
201 Tabor Road
Morris Plains, NJ 07950
Contact: Director, Human Resources, Sales & Marketing

Production / Operations

Calgon Vestal Laboratories
Calgon Corporation
St. Louis, MO 63166
Contact: Personnel Department

Ciba-Geigy Corporation
444 Saw Mill River Road
Ardsley, NY 10502
Contact: Manager, College Relations and Staffing

* *Peterson's Job Opportunities for Engineering, Science, and Computer Graduates 1993,* Copyright © 1992 Peterson's Guides, Inc., Princeton, New Jersey.

Genentech, Inc.
460 Point San Bruno Boulevard
South San Francisco, CA 94080
Contact: Human Resources Department

Merck & Co., Inc.
126 East Lincoln Avenue
Rahway, NJ 07065
Contact: Office of College Relations

Millipore Corporation
80 Ashby Road
Bedford, MA 01730
Contact: Manager of Employment

Research / Development

Argonne National Laboratory, University of Chicago
9700 South Cass Avenue
Argonne, IL 60439
Contact: Recruiting Coordinator

Coca-Cola Foods
2000 St. James Place
Houston, TX 77056
Contact: Manager, Professional Staffing

General Mills, Inc.
One General Mills Boulevard
Minneapolis, MN 55446
Contact: Director of Recruitment and College Relations

Memorial Sloan-Kettering Cancer Center
1275 York Avenue
New York, NY 10021
Contact: Administrator, College Relations

Centers for Disease Control
Department of Health and Human Services
1600 Clifton Road, NE
Atlanta, GA 30333
Contact: Employment Office

National Cancer Institute, National Institutes of Health
9000 Rockville Pike
Bethesda, MD 20892
Contact: Personnel Staffing Specialist

Technical / Professional

American Software USA, Inc.
470 East Paces Ferry Road, NE
Atlanta, GA 30305
Contact: Manager, Corporate Recruiting

Energy and Environment Analysis, Inc.
1655 North Fort Meyer Drive
Arlington, VA 22209
Contact: Recruiting Coordinator

General Electric Company
3135 Easton Turnpike
Fairfield, CT 06431
Contact: Recruiting Support Services

Patent and Trademark Office
U.S. Department of Commerce
2011 Crystal Drive, One Crystal Park
Arlington, VA 20231
Contact: Office of Personnel

Consumer Product Safety Commission
5401 Westband Avenue
Bethesda, MD 20816
Contact: Director, Division of Personnel Management

Appendix C

Answers to Selected Problems

Chapter 2

1. Chemists can often make a thermodynamically unfavorable reaction proceed to some extent by manipulating experimental reaction parameters such as pressure, temperature, or concentrations of reactants. Biochemists generally cannot alter such reaction parameters because organisms function within limited concentration ranges and at essentially constant pressure and temperature.
3. A state function describes the thermodynamic parameters of the system under consideration at a particular moment. Only the difference between initial and final states, not the path taken to achieve these states, is important in most thermodynamic considerations. Enthalpic contributions defining the thermodynamic state are considered only at the initial and final states. Enthalpy is independent of pathway and is therefore a state function.
5. In thermodynamics the total system is the universe, consisting of a particular system and its environment. If the particular system under study is a closed system, it can exchange internal energy as heat and work with its surroundings, but no exchange of matter can occur. An open system can also exchange matter with its surroundings. Therefore, although any energy lost by a particular system is gained by its environment (and vice versa), the energy of the universe remains constant. This is the first law of thermodynamics. Distinguishing between the system and the universe is important to differentiating between a situation in which energy can change and a situation in which energy is constant.
7. Entropy decreases because of hydration effects. The ionized species will order much of the water during hydration, decreasing the total number of free molecules.
9. The reaction will proceed toward oxaloacetate formation in the cell if low product concentration is maintained. Oxidation of NADH by the mitochondrial electron transport system and utilization of oxaloacetate in the formation of citrate shifts the malate-oxaloacetate reaction toward oxaloacetate production.
11. (a) For reaction (P1), $\Delta G^{\circ\prime} = -2.4$ kcal/mole;
    (b) $\Delta G^{\circ\prime}$ of ATP hydrolysis is -7.9 kcal/mole.
13. (a) $\Delta G^{\circ\prime} - 1.5$ kcal/mole; thermodynamically favorable as written.
    (b) $\Delta G^{\circ\prime} = +1.7$ kcal/mole; thermodynamically unfavorable as written.
    (c) $\Delta G^{\circ\prime} = -7.3$ kcal/mole; thermodynamically favorable as written.
15. In the first case, where the repressor protein is cut in half, the binding enthalpy for each part would be essentially half the enthalpy value for the intact repressor. The entropy would be less favorable (less positive) because of the chelation effect. As a result, the free energy would be less favorable (less negative).

    In the second case, where one of the binding sites on the DNA is eliminated, the binding enthalpy for the repressor would be approximately half that for repressor binding to the unmodified DNA sequence. The entropy would depend on the extent of hydration and the extent of mobility of the unbound portion of the repressor. Again, we would expect the free energy for binding to be less favorable.

Chapter 3

1. Berzelius's proposal of $C_{40}H_{62}N_{10}O_{12}$ has a molecular mass of 874 g/mole, of which 140 g/mole is nitrogen (or 16.0% nitrogen by mass).
3. $$\text{pH} = 6.39 + \log\frac{0.0133\ \text{mole}/0.25\ \text{L}}{0.0060\ \text{mole}/0.25\ \text{L}} = 6.39 + 0.35 = 6.74$$
   $[H^+] = 10^{-6.74} = 1.82 \times 10^{-7}$ M
5. pI histidine $= [(pK_{\text{amino}}) + (pK_{\text{imidazole}})]/2$
   $\text{pI} = 7.69$
   pI aspartic acid $= [(pK_{\text{carboxyl}}) + (pK_{\text{R}})]/2$
   $\text{pI} = 2.95$
   pI arginine $= (pK_{\text{amino}} + pK_{\text{R}})/2$
   $\text{pI} = 10.74$
   The sum of positive and negative charge contributions is

α-amino group	α-carboxyl	β-carboxyl	
(+1)	(−0.9)	(0.1)	= 0

These data demonstrate that the net charge on aspartic acid is zero at pH 2.95, which verifies 2.95 as the isoelectric pH.

7. Aspartic acid: pH 2, 4, 10. (The pH range 1 to 5 will be buffered by the α-carboxyl and the β-carboxyl groups.)
   Histidine: pH 2, 9, 6 (imidazole side chain).
   Serine: pH 2, 9. (The pK_a of the alcohol is outside the range of pH normally considered for buffers.)
9. Threonine and isoleucine
11. NH_2-Ala-Ala-Lys-Ala-Ala-Phe-Ala

Chapter 4

1. Consider the entropic effect of decreasing water organization by moving the hydrophobic residue side chains from an aqueous to a nonaqueous environment.
3. The α helix is a rodlike element that cannot easily change direction. Loops, β bends, and ''random'' structure break the helical structure and allow these directional changes.
5. An α helix broken at the Pro-Asn-Ala region with the hydrophobic residues on the exterior should insert into the membrane.
7. The right- or left-handedness of a helix is the same as a conventional screw or bolt. When turned clockwise, a right-handed screw advances. The same is true of a helix or a helical spring.
9. These substances can cause the loss of a protein's shape by disrupting hydrogen bonding and electrostatic interactions.
11. The detergent replaces the membrane, producing a soluble enzyme, and allows the enzymes to be purified free of the membrane.

Chapter 5

1. The early chemists had no way of knowing that each hemoglobin contains four Fe^{2+}.

   4.9 mg of Fe_2O_3

3. (a) $CO_2 + H_2N\text{-hemoglobin} \longrightarrow {}^{-}O\overset{\overset{\displaystyle O}{\|}}{C}HN\text{-hemoglobin} + H^+$

   (b) $HOCO_2^{-} + H_2N\text{-hemoglobin} \longrightarrow {}^{-}O\overset{\overset{\displaystyle O}{\|}}{C}HN\text{-hemoglobin} + H_2O$

5. Red blood cells pass single file through the capillaries. In a sickle cell crisis the red blood cells ''jam together,'' clogging the capillaries. The associated tissues become starved for oxygen, producing the pain of a sickle cell crisis.
7. Individuals with sickle cell anemia have a functional gene for the gamma chain. If the production of fetal hemoglobin could be ''turned back on'' the affected individuals could function normally except during pregnancy.
9. Unprotonated. The protonated form of histidyl F8 is positively charged and is less likely to interact favorably with the Fe^{2+} of the heme.
11. 457 Å. Because the two 457-Å-long strands are twisted into a helix, the resulting TM is shorter than 457 Å.
13. Rigor is probably caused by the depletion of ATP and a considerable discharge of calcium from the sarcoplasmic reticulum.

Chapter 6

1. It is often a rapid effective way to reduce the volume of crude extracts and at the same time eliminate a major portion of the total protein.
3. Difference in charge allows the separation of phosphorylase a from phosphorylase b with the use of DEAE-cellulose. Phosphorylase a and phosphorylase b should elute as a single peak upon gel filtration.
5. Heat treatment of protein solutions denatures and precipitates some of the proteins, while others remain both soluble and stable. Thermal lability is determined empirically for each enzyme or protein of interest.
7. The protein is positively charged at a pH more acidic than the pI. Therefore the protein will probably adhere to (bind to) the CM-cellulose if the pH is between 4 and 6.
9. The student's ''pure'' protein contains at least two components separable by the criterion of mass/charge but which share a common subunit molecular weight. The multiple protein bands appearing in the nondenaturing gel possibly arose through deamidation of glutamine or asparagine residue side chains.
11. (a) Proteins 1 and 3 should elute in the initial wash buffer, but proteins 2 and 4 are predicted to bind to the column. Based solely on isoelectric point, we might predict that protein 4, then protein 2, would be eluted in the salt gradient.
    (b) Proteins 2 and 4 would be eluted in the initial wash buffer from the column, whereas proteins 1 and 3 would be predicted to adhere to the column. Based solely on pI values, protein 3 would be predicted to elute prior to protein 1 in the KCl gradient.
    (c) Proteins with molecular weight greater than the limit (protein 62,000 M_r) are excluded from entry

into the gel and elute in the void volume (V_o). The other proteins in the solution will elute in the order protein 3, protein 1, and protein 4.

Chapter 7

1. Reaction order is the power to which a reactant concentration is raised in defining the rate equation. The example is first order in A and B, second order overall, and second order in C.
3. Let $v = V_{max}/2$ and substitute into equation 25:

$$\frac{V_{max}}{2} = \frac{V_{max}[S]}{[S] + K_M}$$

Solving yields $K_M = [S]$.

5. No
7. Steady-state approximation is based on the concept that the formation of [ES] complex by binding of substrate to free enzyme and breakdown of [ES] to form product plus free enzyme occur at equal rates. A graphical representation of the relative concentrations of free enzyme, substrate, enzyme-substrate complex, and product is shown in figure 7.8 in the text. Derivation of the Michaelis-Menten expression is based on the steady-state assumption. Steady-state approximation may be assumed until the substrate concentration is depleted, with a concomitant decrease in the concentration of [ES].
9. $[S] = 1.0 \times 10^{-3}$ M, $[I] = 9.1 \times 10^{-4}$ M
   $[S] = 1.0 \times 10^{-5}$ M, $[I] = 1.8 \times 10^{-5}$ M
   $[S] = 1.0 \times 10^{-6}$ M, $[I] = 9.9 \times 10^{-6}$ M
11. An enzyme is a catalyst for a chemical reaction, so the rate of an enzyme catalyzed reaction increases with increased temperature. However, the catalyst, a protein, is structurally labile and is inactivated (denatured) at elevated temperatures. The precise temperature at which the enzyme is inactivated varies with the specific enzyme. There is no "temperature optimum" for a catalyst (enzyme).
13. $V_{max} = 3.3 \times 10^{-7}$ Ms^{-1}; $K_m = 2.4 \times 10^{-4}$ M; $k_{cat} = 3.3 \times 10^4$ s^{-1}; specificity constant $= 1.4 \times 10^8$ M^{-1}s^{-1}

Chapter 8

1. The substrate recognition site is a pocket on hexokinase into which glucose, then ATP, bind. Glucose and hexokinase bind together, changing the shape of both and forming a binding site for ATP.
3. Specificity of bond cleavage would most certainly change. The favorable electrostatic interaction between the Lys or Arg side chain of the peptide substrate and the aspartate in the binding pocket would be replaced by electrostatic repulsion upon substitution of the lysine residue. Thus selective binding of Arg or Lys to the substrate pocket would be precluded. The modified trypsin might therefore (a) cleave peptide bonds randomly, but exclude bonds on the carboxyl side of Lys or Arg residues, or (b) exhibit specificity for peptide cleavage on the carboxyl side of acidic amino acids (Asp, Glu). Side chains of these amino acids may fit well into the substrate binding pocket and have favorable electrostatic interaction with the lysine therein.
5. We expect iodoacetate and para-mercuribenzoate to inhibit plant proteases and diisopropylfluorophosphate to inhibit serine proteases.
7. Histidyl 12 accepts the 2′-OH hydrogen to start the reaction. The protonated histidyl 12 ultimately provides the hydrogen to the cyclic ester to produce the 3′-phosphate product.
9. Renaturation of denatured protein is dictated by the primary structure of the protein. The trypsin family of enzymes and carboxypeptidase A are synthesized as proenzymes that are proteolytically activated. The proteolyzed, active enzymes have primary structures different from the gene product and are not active upon renaturation. In addition, zinc is a cofactor required for carboxypeptidase A activity.
11. The structure of the transition-state analog is complementary to the structure of the active site. The analog thus binds tightly to the active site.
13. (a) Substitution of Asp for Lys 86 markedly decreases activity and several explanations are possible. The lysine is a critical residue either at or near the active site or is essential for maintaining a catalytically competent conformation of the enzyme. (b) Lysines 21 and 101 are probably outside the catalytic site and may not be evolutionarily conserved. Their replacement with aspartate yielded no great change in enzymatic activity. (c) Lysine 86 is essential for enzymatic activity and would be conserved.

Chapter 9

1. The Ile in chymotrypsinogen is the 16th amino acid. During the conversion to chymotrypsin a 15 amino acid peptide is cleaved from chymotrypsinogen. The original amino acid number system was retained in chymotrypsin, making the N-terminus amino acid number 16.
3. Covalent modification requires an enzyme to control an enzyme. Allosterism requires only a binding site on an enzyme that interacts (via an equilibrium) with a particular small molecule.

5. Phosphorylation of serine or threonine (and possibly tyrosine) on the target protein may be largely influenced by the amino acid sequence around these residues. These amino acid sequences may define a specific motif or recognition site for the protein kinase.
7. The substrate concentration required for half-maximal activity ($S_{0.5}$) of an allosterically regulated enzyme will depend on the cumulative effects of allosteric activators and/or inhibitors also present. Hence $S_{0.5}$ may be decreased with allosteric activators and may be increased with allosteric inhibitors.
9. Allosteric regulation of an enzyme having a binding site for a regulatory molecule and an active site on the same subunit is not uncommon. Regulatory molecules bind at sites separate from the active site and induce a conformational change that affects substrate binding to the active site on the same as well as adjacent subunits.
11. The methylene group prevents the elimination of a phosphate, which is required to convert the postulated intermediate into carbamoyl aspartate. The suggested oxygen analog might eliminate phosphate (i.e., be a substrate for aspartate carbamoyltransferase).

Chapter 10

1. The following coenzymes contain the AMP moiety: NAD^+, NADH, $NADP^+$, NADPH, FAD, $FADH_2$, and CoASH.
3. (a) Refer to figure 10.5b in the text for the structures of each intermediate step.
   Steps 1. & 2. Form Schiff base
   3. Remove proton α-C
   4. Protonate C-4′
   5. & 6. Hydrolyze Schiff base
   Release α-keto acid and pyridoxamine phosphate
   Transfer of amino group from pyridoxamine phosphate to pyruvate, forming alanine occurs by reversal of these steps. Other transaminases use other α-amino acids and α-keto acids.
   (b) Figure 10.5b in the text gives the structures of intermediates.
   Steps 1. Decarboxylation of α-COO^-
   2. Protonate α-Carbon
   3. & 4. Hydrolyze Schiff base
   (c) β-Decarboxylation. Form Schiff base steps 1–3, solution 3a.
5. Refer to figure 10.2 in the text.
   (a) Steps 1. Deprotonation of TPP to ylid form, nucleophilic addition of ylid to α-ketogroup
   2. Decarboxylation
   3. Resonance stabilization
   4. Protonation, elimination of TPP
   (b) Steps 1. Deprotonation to ylid form, nucleophilic addition of ylid to fructose-6-phosphate
   2. Oxidation of C-3, release of erythrose-4-phosphate
   3. Resonance stabilization of intermediate
   4. Elimination of −OH from C-2
   5. Transfer of acyl group to phosphate
7. (a) Thiamine pyrophosphate, lipoic acid, FAD, NAD^+, CoASH (an α-ketoacid dehydrogenase); (b) biotin; (c) FAD.
9. (a) Reduced flavins in solution rapidly reduce O_2 to superoxide and H_2O_2, metabolites that are toxic to the cell. Enzyme-bound flavins are usually shielded from rapid oxidation by O_2. (b) Tightly bound NAD(P) is an advantage to enzymes catalyzing rapid $H:^-$ removal and readdition in a stereospecific fashion. Freely diffusing NADH is an advantage in transferring reducing equivalents among various enzyme-catalyzed reactions.
11. (a) Redox agents: FAD, FMN, NAD^+, $NADP^+$, and lipoyl; (b) acyl carriers: CoASH, lipoyl, and thiamine pyrophosphate; (c) both acyl carriers and redox agents: lipoyl.
13. The hydroxyl hydrogen on C-3 is the most acidic because the resulting anion is resonance stabilized.

Chapter 11

1. The transition from left to right involves dehydration reactions; the transition from right to left involves hydrolysis reactions.
3. Catabolism involves pathways composed of enzymes and chemical intermediates that are involved primarily in the breakdown of large molecules into small molecules, often by oxidation processes. Anabolism is the collection of the enzymes and chemical intermediates involved in the biological synthesis of larger molecules from smaller molecules. Anabolism often involves reduction processes.
5. The advantage of subcellular compartments is that a specific pathway can be isolated from comparable pathways that might use similar or identical chemical intermediates. Compartmentalization simplifies regulation of the various processes.
7. (a) The concentration of most metabolites measured under steady-state conditions in the cell usually does not exceed the K_m value. For an enzyme whose reaction can be described by simple Michaelis-Menten kinetics, the observed velocity, v, is $0.5V_{max}$ if substrate concentration equals the

K_m value. The velocity of most enzymes is likely significantly less than V_{max} *in vivo.*

(b) End-product inhibition usually occurs at the committed step in a metabolic pathway or at a branch point in the pathway.

(c) Catabolic pathways tend to be convergent rather than divergent. Metabolic convergence of precursors into common intermediates of a metabolic pathway provides an efficient route for the metabolism of a variety of metabolites by a limited number of enzymes.

(d) Enzymes that are regulated in metabolic pathways most frequently exhibit cooperative kinetic responses rather than hyperbolic responses with respect to substrate concentration and are frequently responsive to allosteric regulation by products, energy charge, or concentration ratio of $NAD(P)H/NAD(P)^+$. The rate of enzymatic activity over a narrow range of substrate concentration can be changed dramatically by allosteric activators or inhibitors. Such dramatic changes in velocity in response to small changes in substrate concentration is not observed with enzymes exhibiting Michaelis-Menten kinetics.

(e) Low energy charge signals the cell that a need for ATP formation exists, and pathways (glycolysis and Krebs cycle) leading to ATP formation are activated. Anabolic pathways that demand high ATP concentrations are inhibited at low energy charge. In the latter case, the ATP required to drive biosynthesis is in low supply. Conversely, increased energy charge inhibits pathways leading to ATP formation and activates anabolic pathways.

(f) Formation of multienzyme complexes is a strategy frequently used for efficient catalysis and control of metabolic pathways. Substrates diffuse shorter distances between active sites in multienzyme complexes than if the enzymes were not organized. Frequently, intermediates covalently bound to cofactors (e.g., lipoamide, biocytin) are moved among active sites within the complex, effectively trapping intermediates in the complex and increasing the concentration of substrates at the enzyme active site.

(g) Separation of catabolic and anabolic pathways diminishes the likelihood of futile cycling of metabolites. Enzymes catalyzing the β-oxidation of fatty acids are located in the mitochondrial matrix, whereas enzymes catalyzing the synthesis of palmitate are located in the cytosol.

9. The ''committed step'' in a reaction sequence steers the metabolite to a sequence of reactions whose intermediates have no other function in the cell. Control of the committed step prevents wasteful accumulation of these single-purpose intermediates and obviates the necessity of controlling each enzyme in a pathway.

Chapter 12

1. The number of sugars is 2^n where n is the number of chiral centers but does not include the chiral carbon involved in producing the D series of sugars. Because C-2 of the ketoses is a carbonyl group, only the geometry of C-3 and C-4 remain to produce the four D-ketohexoses ($2^2 = 4$). The carbonyl group on C-1 of the aldoses allow C-2, C-3, and C-4 to produce eight D-aldohexoses ($2^3 = 8$).
3. (a) Trehalose is two α-D-glucoses linked through C-1 of each sugar. However, the correct answer to this question would also include (b) two β-D-glucoses linked through C-1 and (c) one α-D-glucose linked through C-1 to C-1 of a β-D-glucose.
5. Maltose, lactose, and cellobiose are reducing sugars; sucrose and trehalose (problem 3) are nonreducing sugars.
7. Glucose C-3 and C-4 are lost as CO_2 during ethanol production.
9. By assessing the rate of $^{14}CO_2$ production from ^{14}C-1-glucose versus ^{14}C-6-glucose, we can determine the relative significance of the different pathways in a tissue.
11. Glycerate-2,3-bisphosphate was encountered as an allosteric effector of hemoglobin.
13. In the first example the actual chemistry is identical: an aldose-ketose interconversion. Note that the geometry of the alcohol on C-2 is identical. The second example has essentially the same chemistry. Logically, to do the same chemistry, nature could have stumbled upon the same mechanism twice. More likely, however, the mechanism evolved once and, following gene duplication, the segments of the gene controlling the substrate specificity changed on one copy of the gene.
15. The carboxyl group of pyruvate is lost as CO_2 during the pyruvate decarboxylase–catalyzed formation of acetaldehyde. The pyruvate carboxyl group is formed by the oxidation of glyceraldehyde-3-phosphate. The aldehyde (C-1) carbon is derived directly from the aldolase-dependent cleavage of the fructose-1,6-bisphosphate. In this cleavage, C-4 of glucose becomes C-1 of glyceraldehyde-3-P_i and C-3 of glucose becomes C-3 of dihydroxyacetone phosphate. DHAP is isomerized to Ga3P_i. In this isomerization, C-3 of DHAP (originally C-3 of glucose) becomes C-1 of Ga3P. Thus, labeling either C-3 or C-4 of

glucose will ensure that label is released as CO_2 upon fermentation to ethanol.

17. (a) Triose phosphate isomerase deficiency would inhibit conversion of DHAP to Ga3P and would cause accumulation of DHAP, preventing half of the glucose molecule (C1-C3) from being metabolized through the remainder of the glycolytic pathway. There would be a recovery of only 2 of the possible 4 moles of ATP from glucose, resulting in no net formation of ATP. In addition, DHAP, a product of the aldolase reaction, would likely reverse the aldolase (reaction) and eventually inhibit glycolysis. Either result would be lethal to a cell whose only energy source was glycolysis.
    (b) The small amount of TPI activity would likely allow glycolysis to proceed slowly, but low energy (ATP) level will limit the growth rate under anaerobiosis. However, the yield of ATP is significantly greater when the pyruvate, formed during glycolysis, is oxidized to CO_2 and H_2O. Hence, the growth rate of the mutant should be correspondingly greater under aerobic growth conditions but not as great as the wild type.
    (c) Cells expressing DHAP phosphatase would likely not grow anaerobically if glycolysis of glucose to lactate were the only pathway for ATP formation. The combined activities of TPI and DHAP phosphatase would be predicted to deplete the pool of triosephosphate and the yield of ATP per glucose would likely be less than 1.
19. (a) $K'_{eq} = 30$
    (b) Hexoses brought into the glycolytic pathway must be phosphorylated to provide the appropriate substrate for the glycolytic enzymes and to trap the sugar within the cell. Phosphorylation at the expense of ATP or group translocation at the expense of PEP are common methods to activate the sugar molecules. Sucrose phosphorylase uses the exergonic lysis of the glycosidic bond between the hemiacetyl OH group of glucose and the hemiketal OH group of fructose to drive the endergonic phosphorylation of the hemiacetal C-1 OH group of glucose. Transfer of the phosphate to C-6, catalyzed by phosphoglucomutase, provides substrate for entry into the glycolytic pathway without addition of ATP. The net ATP yield will therefore be 3 rather than 2 moles of ATP per mole of glucose derived from sucrose. The fructose can be phosphorylated by ATP and used in the glycolytic pathway.
21. The free energy available from a reaction depends on the energies of the products compared with the substrates. Dehydration of 2-phosphoglycerate ''traps'' phospho-(enol)pyruvate in the enolate form. The hydrolysis products of PEP are phosphate and the enol form of pyruvate, but (enol) pyruvate is significantly less stable than (keto) pyruvate and rapidly tautomerizes to the more stable keto form. The tautomerization drives the reaction strongly toward products, resulting in a larger free energy difference between substrate and product.
23. Given the ratio of ATP/ADP and the K'_{eq} of 10^6, the equilibrium ratio of [Pyry]/[PEP] would be about 10^5. This calculation supports the metabolic irreversibility of the pyruvate kinase reaction.
25. 6-phosphogluconate + $NADP^+ \longrightarrow$
    $NADPH + H^+ + CO_2$ + Ribulose-5-phosphate
27. (a) Unregulated hepatic PK theoretically could become part of a futile cycle. Net: GTP $\longrightarrow$ GDP + P_i plus formation of cytosolic NADH at the expense of mitochondrial NADH.
    (b) Activation by fructose-1,6-bisphosphate decreases $S_{0.5}$ for PEP and increases PK activity at a given PEP concentration. During gluconeogenesis, the fructose-1,6-bisphosphate concentration should diminish, owing to the hydrolytic activity of the FBPase-1. The low FBP concentration, coupled with the elevated ATP levels, could inhibit the hepatic pyruvate kinase.

Chapter 13

1. The oxidation occurs first, creating a β-keto-carboxylate, which readily loses CO_2 via a resonance-stabilized carbanion.
3. Citrate is a tertiary alcohol, which does not oxidize readily.
5. The following sequence will do the task: α-ketoglutarate, the tricarboxylic acid cycle to oxaloacetate, to phosphoenolpyruvate, to pyruvate, to acetyl-CoA, into the tricarboxylic acid cycle.
7. Both of the oxidative-decarboxylation steps in the tricarboxylic acid cycle are bypassed by the glyoxylate cycle.
9. (a) Increased concentrations of acetyl-CoA slow the activity of pyruvate dehydrogenase.
    (b) The removal of succinyl-CoA for heme synthesis removes any control it has over lowering the activity of α-ketoglutarate dehydrogenase and citrate synthase.
11. (a) False. Lipoamide transacetylase catalyzes reduction of the disulfide on lipoamide concomitantly with oxidation and transfer of the hydroxyethyl group from thiamine pyrophosphate. Dihydrolipoamide dehydrogenase catalyzes oxidation of dihydrolipoamide and reduction of NAD^+. (b) False. Hydrolysis of

acetyl-CoA thioester should yield as much free energy as succinyl-CoA hydrolysis, a process coupled to ADP phosphorylation (succinate thiokinase).
(c) False. The methyl group of acetyl-CoA could be derived from pyruvate, from β oxidation of long-chain fatty acids, or from amino acid metabolism.
(d) False. If aconitase failed to discriminate between the ($—CH_2—COO^-$) groups, half the CO_2 would arise from oxaloacetate and half from the acetate carboxylate. The two CO_2 molecules released by oxidative decarboxylation of isocitrate and α-ketoglutarate are derived from the carboxyl groups of oxaloacetate with which the acetyl-CoA was condensed. (e) False. Malate can easily be dehydrated to fumarate by reversal of the fumarase reaction.

13. (a) Without malate synthase, the yeast would be unable to grow on 2-carbon precursors as the sole carbon source because TCA cycle intermediates could not be synthesized.
    (b) TCA cycle activity would markedly diminish if the cycle intermediates were being used in biosynthetic pathways. In addition, the biosynthetic pathways (lipids, amino acids, and carbohydrates) dependent on TCA cycle intermediates would also be inhibited by the lack of metabolites.
    (c) If the PDH were inhibited more strongly than usual by acetyl-CoA, we might suspect that acetyl-CoA concentration in the mitochondrial matrix would markedly decrease, in turn limiting activity of citrate synthase and diminishing TCA cycle activity. Hence growth of the organism may be limited by lowered energy production and by diminished concentrations of biosynthetic precursors supplied by the TCA cycle.
15. (a) Fumarate + NADH + H^+ $\longrightarrow$ succinate + NAD^+.
    (b) Phosphoenolpyruvate + CO_2 $\longrightarrow$ oxaloacetate + P_i
    (PEP carboxylase)
    Oxaloacetate + NADH + H^+ $\longrightarrow$ L-malate + NAD^+
    (malate dehydrogenase)
    L-Malate $\longrightarrow$ fumarate + HOH
    (fumarase)
    Fumarate + NADH + H^+ $\longrightarrow$ succinate + NAD^+
    (fumarate reductase)
    (c) In the reactions shown in part (b), four reducing equivalents (two hydride groups) are transferred to carbon acceptors. Reduction of oxaloacetate to malate by malate dehydrogenase (MDH) and reduction of fumarate to succinate by fumarate reductase each requires hydride (or equivalent) transfer from NADH to the organic substrate. In the reduction of pyruvate to lactate via LDH only one hydride is used. Thus, two equivalents of NAD^+ are resupplied to glycolysis by the activities of MDH and fumarate reductase, whereas only one equivalent of NAD^+ is regenerated by LDH. Fumarate is one of the terminal electron acceptors used during anaerobic respiration in *E. coli.* However, each mole of PEP carboxylated is at the expense of 1 mole equivalent of ATP that could have been formed as a product of pyruvate kinase.
17. The committed step in a metabolic pathway is usually under metabolic control. Inhibition of the committed step in a metabolic sequence or pathway prevents the accumulation of unneeded intermediates and effectively precludes activity of the enzymes using those intermediates as substrates. The decarboxylation of pyruvate and the oxidative transfer of the hydroxyethyl group by pyruvate dehydrogenase constitutes the committed step in the pyruvate dehydrogenase catalytic sequence and is a logical control point.

Chapter 14

1. In hemoglobin and myoglobin the heme serves as a carrier with oxygen–heme iron (Fe^{2+}) interacting in a ligand-metal coordination relationship. In the cytochromes heme plays a redox role, with the iron interconverting between Fe^{2+} and Fe^{3+}.
3. (a) 12.5
   (b) 16
   (c) 10
5. Both iron and copper have two common stable ions; the transition between Fe^{2+} and Fe^{3+} or Cu^{1+} and Cu^{2+} are both redox reactions.
7. (a) Heme of the mitochondrial *b*-type cytochrome interacts hydrophobically with adjacent hydrophobic residues from the membrane-spanning α helices. The heme iron is fully coordinated through two imidazole groups from histidines in the protein. Heme in the *c*-type cytochromes is covalently bound to the protein through thioethers formed by the addition of cysteinyl sulfhydryl groups to the vinyl substituents on the heme ring. The iron is also fully liganded to a nitrogen from the imidazole group of histidine and a sulfur from the thioether linkage of methionine providing the fifth and sixth ligands. Heme *a* differs from the protoheme (heme) of the *b*- and *c*-type cytochromes by substitution of a formyl group at ring position 8 and a 17-carbon isoprenoid chain at position 2.

(b) Neither CO nor CN^- at low concentration interact with the cytochrome *b* or *c* heme iron because no open ligand position is available to the iron in these heme proteins.

9. (a) In biological systems, the iron-sulfur centers are obligatorily single electron donors/acceptors, regardless of the number of Fe atoms in the center or their initial oxidation state.
(b) The 4Fe-4S cluster is found in iron-sulfur proteins that transfer electrons at low and at high potential. The reduction potential is a measure of the ease of addition of an electron to the couple, compared to the standard hydrogen electrode. Thus the protein component of the iron-sulfur protein affects the reduction potential.

11. $E°$ (pH 6) = +170 mV; $E°$ (pH 8) = 50 mV.

13. (a) The large amount of UQ is necessary to ensure efficient transfer of electrons from the mitochondrial dehydrogenases to complex III.
(b) (i) If UQ (ubiquinone) were limiting, the rate of oxidation of NADH by the mitochondrial electron-transfer system would also be limited. NADH concentration would increase, the NAD^+ supply would decrease, and the NAD^+-dependent dehydrogenases would be inhibited. Subsequently, the rate of oxidation of NADH-producing substrates would decrease.
(ii) Electrons from succinate oxidation are also transferred to ubiquinone from the succinate dehydrogenase, so a deficiency of the quinone would limit succinate oxidation.
(iii) Ascorbate plus a redox mediator reduces cytochrome *c* but does not transfer electrons to ubiquinone. Limiting amounts of the quinone should not affect ascorbate-dependent reduction of cytochrome *c*.
(c) The deficiency of UQ will decrease the rate of extramitochondrial NADH oxidation resulting in an increase in the amount of pyruvate reduced to lactate. The lactate content would be expected to increase rapidly upon mild exercise.

15. (a) The uncoupler 2,4-dinitrophenol circumvents respiratory control in the mitochondria by short-circuiting the proton gradient. The lipophilic weak acid transports H^+ across the membrane, bypassing the F_1-F_0 complex. Substrates will be oxidized independently of ADP or ATP concentrations, and O_2 reduction will be more rapid than in state 3 respiration.
(b) Uncoupled mitochondria oxidize NADH and succinate but fail to phosphorylate ADP. Cellular processes will continue to utilize ATP, causing an accumulation of ADP and AMP. Increased levels of NAD^+ and ADP or AMP activate both the TCA cycle and glycolysis. The rate of oxidation of carbohydrates and fatty acids would markedly increase.
(c) The uncoupled mitochondria use little, if any, energy to phosphorylate ADP. The energy is dissipated as heat, leading to elevated body temperature (hyperthermia) and profuse perspiration in an effort to decrease body temperature.
(d) 2,4-Dinitrophenol is a lipid soluble weakly acidic compound thought to allow equilibration of protons across the inner mitochondrial membrane. Although protons are translocated from the matrix across the inner membrane during electron transfer, the proton gradient would immediately be depleted without passing through the F_0-F_1–dependent ADP phosphorylation system. Respiratory (ADP) control would be lost.

17. The transport process is electrogenic if the export of one molecule coupled with the import of another molecule yields a net charge difference across the membrane. In general terms, transfer of A^{3-} from the matrix and A^{3-} into the matrix yields a net negative charge on the cytoplasmic side of the membrane. Electrogenic processes are driven by the membrane potential ($\Delta\Psi$).
Neutral transport processes exchange molecules of net identical charge (sign and magnitude) in the opposite vectorial direction or oppositely charged molecules in the same vectorial direction. Processes coupled to the export of OH^- are equivalent to H^+ import, that is, to an energetically favorable decrease in the proton gradient.

Chapter 15

1. The word fixed refers to converting a gas into a liquid or solid.
3. Chlorophyll molecules are flat, completely conjugated ring compounds that obviously have a resonance form. Chlorophylls have a ring containing 9π bonds or 18π electrons, which fits the $n = 4$ situation in the Huckel $(4n + 2)$ rule. The $(cis, trans, trans)_3$ double bond pattern of the chlorophylls also fits that of [18]annulene, which is aromatic.
5. Upon illumination, the chromatophore P870 is activated by absorption of a photon of light. The absorbance at 870 nm decreases because the π-cation radical of the oxidized chromatophore has a lower absorbance at that wavelength. Thus the trace monitoring 870 nm decreases upon illumination of the

chromatophore. The c-type cytochrome is added initially in the reduced form (cyt c^{2+}) and absorbs at 550 nm. Oxidized cytochrome c(cyt c^{3+}) has only a small absorbance at 550 nm. Electron transfer from reduced cytochrome c to the π-cation radical ($P870^+$) regenerates the ground state P870 and forms oxidized cytochrome c. The absorbance of the P870 increases to the initial level and the absorbance at 550 nm of the cytochrome c pool decreases.

7. (a) The absorbance at 275 nm decreases because the concentration of ubiquinone is diminished and because the semiquinone radical does not absorb 275 nm light. The absorbance increase at 450 nm is consistent with the formation of semiquinone radical that absorbs 450 nm light. The second flash activates transfer of a second electron from the photocenter to reduce the bound semiquinone to dihydroquinone. The dihydroquinone absorbs at neither 275 nm nor 450 nm. Reduction of the semiquinone form abolishes the absorbance at 450 nm. The decrease in absorbance at 275 nm is consistent with the decrease in oxidized (ubiquinone) concentration.
   (b) In this experiment, reduced cytochrome c is the electron donor to $P870^+$. Were the reductant omitted, the $P870^+$ might oxidize the reduced quinone, or the activated reaction center P870* might return to ground state by emission of fluorescence.
9. Singlet-state oxygen causes cumulative oxidative damage to the chloroplast. Carotenoids compete with the oxygen for the triplet-state chlorophyll and inhibit singlet oxygen production. Plants deficient in carotenoids risk photooxidative damage because of an increased flux of singlet oxygen.
11. The mechanisms are different, but the energetic outcome is the same.
13. The enolate intermediate (figure 15.26) reacts with oxygen to produce a cycloperoxide intermediate that decomposes into glycerate-3-phosphate and glycolate-2-phosphate (figure 15.27).
15. Some of the many differences include the involvement of erythrose-4-phosphate and dihydroxyacetone phosphate in the production of sedoheptulose-1,7-bisphosphate, phosphorylation of ribulose-5-phosphate to ribulose-1,5-bisphosphate, the addition of CO_2 to ribulose-1,5-bisphosphate, and the production of two glyceraldehyde-3-phosphates.

Chapter 16

1. The sequence, bond geometry, and linkage type of monosaccharides in an oligosaccharide or polysaccharide is determined by the specificity of the glycosyltransferases involved.
3. In general, amino groups are added to biological molecules in locations once occupied by keto groups. Because ketoses have C-2 as the carbonyl group the majority of the amino sugars in nature are 2-amino sugars.
5. Sugars have a large number of hydroxyl groups that form glycosidic bonds with the anomeric carbons of other sugars; these anomeric hydroxyl groups can be either α or β conformation. More than eighty different glycosidic linkages have been identified.
7. Patients with I-cell disease do not phosphorylate the mannose residues on the glycoproteins that are lysosome-bound. These ''lysosomal hydrolases'' are therefore secreted.
9. Gluconic acid cannot form a cyclic hemiacetal and cannot form the glucosidic bonds required for participation in oligo- or polysaccharides.
11. The undecaprenol phosphate functions as a carrier and is acted upon by the series of reactions.
13. The major problem in the synthesis of complex carbohydrates outside the cell is the lack of an external energy source such as ATP. The biosynthesis of a bacterial cell wall such as the peptidoglycan occurs mainly inside the cell, the final step being the cross-linking of the peptidoglycan strands outside the cell. This reaction is a transpeptidation, which does not require any energy source.

Chapter 17

1. The lecithin serves as an emulsifying agent that allows the aqueous and lipid phases to be dispersed in each other and increases the time required for phase separation.
3. The fatty acyl groups are placed in these two different positions by different enzymes.
5. The hydrophobic amino acid side chains on the exterior of the integral membrane protein interact with the hydrophobic lipid of the membrane exterior and are stable in the nonaqueous environment. These residues pack in the interior, hydrophobic environment of globular proteins.
7. Peripheral proteins are bound to the inner or outer aspects of the membrane through weak ionic interactions that include association with phospholipid head groups, by electrostatic or ionic interaction with a hydrophilic region of an integral membrane protein or through divalent metal ion bridging to the membrane surface. Peripherally bound proteins may be released without disrupting the membrane. Thus in-

creased salt concentration shields ionic charges and weakens the charge–charge interactions between the peripheral protein and the membrane components.

Integral proteins are dissolved into the lipid bilayer of the membrane through interactions of the hydrophobic amino acid side chains and fatty acyl groups of phospholipids. In order to remove integral membrane proteins, the membrane must be disrupted by addition of detergents or other chaotropic reagents to solubilize the protein and to prevent aggregation and precipitation of the hydrophobic proteins upon their removal from the membrane.

9. In principle, placing a hydrophilic residue in a nonaqueous environment is energetically unfavorable. In integral proteins with multiple α helices that span the membrane, hydrophilic side chains from different helical segments may interact and in some cases form a channel through which ions may diffuse. Portions of the helical segments exposed to the lipid will contain primarily hydrophobic amino acid residues.
11. Visualizing a protein actually revolving in a controlled manner within the membrane is difficult.
13. (a) $\Delta\Psi = +18$ mV; (b) from side 1 to side 2; (c) Side 1: 64.3 mM K^+, 50 mM Na^+, 144.3 mM Cl^-. Side 2: 85.7 mM K^+, 85.7 mM Cl^-. $\Delta\Psi = 7.5$ mV.
15. Sucrose uptake should be inhibited by a proton ionophore if uptake is by a proton symport. If a protein-binding system was operational, membrane vesicles or cells subjected to osmotic shock would be defective in uptake. If a Na^+ symport was involved, uptake would be dependent on extracellular Na^+. If a PTS was operational, sucrose phosphorylation would be dependent on PEP and not ATP in a crude cell extract.

Chapter 18

1. Arachidic acid produces 134 moles of ATP/mole, whereas arachidonic acid gives 126 moles of ATP/mole of fatty acid. This difference in ATP quantities is rather minor. Therefore any biological difference caused by dietary saturated versus unsaturated fats being due to their ATP yields is unlikely.
3. (a) 29.5 or 30 moles ATP
   (b) 64 moles ATP
   (c) 33% and 32%, respectively
   (d) 2.25 versus 2.13
5. (a) Oxidation of 1 mole of glucose yields 32 moles of ATP. 350 kcal of energy are stored as ATP or 1.9 kcal/g. (b) Oxidation of 1 mole of palmitate yields 106 moles of ATP. 1200 kcal of energy are stored as ATP, or 4.7 kcal/g. (c) Lipids are more highly reduced than are carbohydrates and supply more reducing equivalents to the electron-transport system than do carbohydrates. (d) Lipids have approximately 2.5 times greater energy storage capacity per gram than do carbohydrates and are stored as compact, hydrophobic globules. Storage of an equivalent energy as carbohydrate would require at least 2.5 times the mass, not considering the water of hydration that would accompany the carbohydrate.
7. The availability of citrate has no relationship to the flow of metabolites through the tricarboxylic acid cycle. Increased citrate concentrations result in increased cytoplasmic acetyl-CoA concentrations, which in turn increases fatty acid biosynthesis.
9. From a reaction sequence viewpoint it is a cyclic process. From a substrate viewpoint it is a decreasing spiral because the substrate's length decreases with each turn around the spiral.
11. Many mechanisms are conceivable. One possibility is the approach of a phosphate on succinyl-CoA to produce succinyl phosphate (a mixed anhydride).
13. (a) Ketone body formation in liver supplies an easily transported, water-soluble, energy-rich metabolite that can be used in lieu of glucose in many nonhepatic tissues. (b) β-hydroxybutyrate supplies an additional hydride (2 reducing equivalents) compared with acetoacetate. (c) Consider the reactions catalyzed by β-hydroxybutyrate dehydrogenase, 3-ketoacyl-CoA transferase, and thiolase.
15. (a) The ^{14}C-labeled methyl group of acetyl-CoA will be C-16 of palmitate. (b) Only one deuterium atom from each labeled malonyl-CoA will remain in the reduced lipid chain. Carbons 2, 4, 6, 8, 10, 12, and 14 of palmitic acid will each have one deuterium label. (c) The ^{14}C label will be lost by decarboxylation and no label will remain in the palmitate.
17. Glucose-6-phosphate dehydrogenase, 6-phosphogluconate dehydrogenase, and the $NADP^+$-malic enzyme are sources of the 14 moles of NADPH required for biosynthesis of palmitate.
19. The thioesterase activity of the fatty acid synthase prefers the palmitoyl acyl carrier protein thioester as substrate.

Chapter 19

1. Phosphatidylserine decarboxylase is a pyridoxal phosphate enzyme.
3. The synthesis of phosphatidylcholine is thermodynamically feasible because of the ATP used to phosphorylate choline and the CTP used to form CDP-choline.

5. One possibility is to remove the fatty acyl groups that have been damaged by oxidation.
7. Carriers of a defective gene for the hexosaminidase A enzyme still have a functional copy of the gene.
9. The activated component is different. In the first case (figure 19.4) the minor component is activated, whereas in the production of phosphatidylinositol (figure 19.6) the diacylglycerol is activated.
11. Arachidonic acid is stored in membranes as phospholipids with C_{20} polyunsaturated fatty acids in the SN-2 position. Phospholipase A_2 releases arachidonic acid, which is then used to synthesize prostaglandins, which induce inflammation.

Chapter 20

1. No
3. These frequencies are consistent if one assumes that the gene has only two alleles: $A + B = 1$ and $A^2 + 2AB + B^2 = 1$ describes the population where A^2 are "normal," 2AB are heterozygous and B^2 are homozygous for familial hypercholesterolemia. $2AB = 1/500$, $AB = 0.001$, $A = 0.999$ and $B = 0.001$. Therefore, $B^2 = (0.001)^2 = 0.000001$, which is the one in a million indicated in the text. These frequencies would be inconsistent if the gene has more than two alleles.
5. If a normal person is given an inhibitor for HMG-CoA reductase, cholesterol synthesis is inhibited in the liver. Lower levels of cholesterol then signal the synthesis of increased levels of LDL receptors. This increases the uptake of LDL into the liver and reduces serum LDL. In a patient with FH, this has little effect because there are no LDL receptors. The only effect is that the liver does not make as much cholesterol and does not contribute as much to serum LDL levels. The new liver will make normal amounts of LDL receptors and have normal uptake of LDL from the blood. This result will dramatically lower serum LDL levels and prevent the new heart from developing coronary artery disease. If the liver transplant had not been done, the heart transplant would have been to no avail.
7. The ^{14}C in HMG-CoA derived from 2-[^{14}-C]-acetate is marked in the structure.

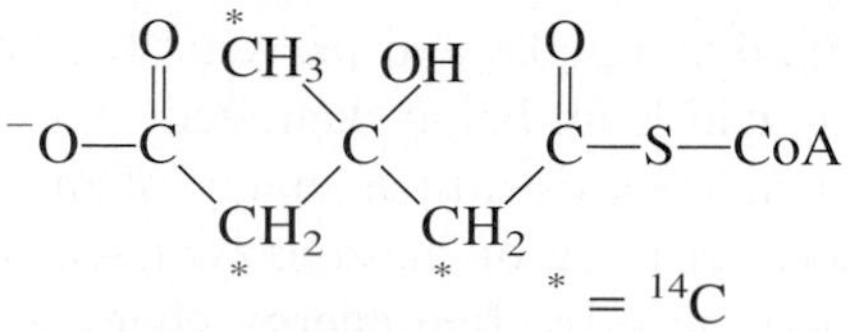

9. The evolution of membranes, organelles, and organs combined with the evolutionary development of new metabolic sequences from previously developed enzymatic mechanisms probably best explains this phenomenon.

Chapter 21

1. The oxide ion is a poor leaving group.
3. The two sequences differ by an ATP.
5. Presumably the availability of valine inhibited an enzyme, such as acetohydroxy acid synthase, in the early stages of valine, isoleucine and leucine biosynthesis.
7. Only when the environmental tryptophan is depleted will material flow through the pathway, or in this case, because of the auxotroph, only part way through the pathway.
9. In this two-step pathway the acetyl group is added in the first reaction and an acetate leaves, and an HS^- is added in the second step. The acetyl group facilitates the removal of the serine oxygen. The OH group is a poor leaving group; acetate is better.
11. Glutamine, the product of glutamine synthase, is the source of nitrogen required in the synthesis of a number of diverse, structurally unrelated compounds synthesized by different pathways. Total inhibition of the glutamine synthase by a single product would in turn inhibit the synthesis of all compounds requiring glutamine.
13. Pyruvate, from glycolysis of glucose, is carboxylated to oxaloacetate or oxidized to acetyl-CoA. These metabolites enter the Krebs cycle, are metabolized to α-ketoglutarate and oxaloacetate, then transaminated to aspartate or glutamate. Asn, Gln, and Pro are synthesized from Asp or Glu. The cycle replenishes intermediates via the anaplerotic reactions (e.g., carboxylation of pyruvate to form oxaloacetate).
15. Hydroxyproline is formed by a posttranslational modification of proline residues in the protein. The ^{14}C-labeled hydroxyproline is not incorporated directly into the collagen because there is no genetic codon to specify the incorporation of hydroxyproline.

Chapter 22

1. The *N*-acetyl groups in the *de novo* pathway prevent the spontaneous cyclization of the semialdehyde intermediate.
3. One of the nitrogens enters the urea cycle via carbamoyl phosphate that comes from ammonia. The second urea nitrogen enters the urea cycle as part of aspartic acid that can come from transamination of oxaloacetate. Glutamate is the source of the amino group in the transamination.

5. Because of the importance of the urea cycle, the capacity to convert ornithine into arginine is obvious. Complete loss of the ability to produce ornithine (a catalyst or carrier in the urea cycle) would limit the organism's control over production of its nitrogen waste product.
7. Based on figure 22.11 Thr, Ala, Ser, Gly, Cys, Asn, Asp, Gln, Glu, His, Arg, Pro, Val, and Met are glucogenic; Lys, Trp, and Leu are ketogenic; and Phe, Tyr, and Ile are both ketogenic and glucogenic.
9. Because of its critical role in ATP production, a homozygotic defect in a gene for a protein involved in glycolysis, the citric acid cycle, or the electron transport chain probably leads to the death of the cell(s) soon after fertilization.
11. Pyridoxal phosphate forms a Schiff base (imine) with the glycine. A carbon-bound hydrogen is labile, and the resulting carbanion stabilized by resonance back into the pyridoxal phosphate. The carbanion approaches the carbonyl carbon of the succinyl-CoA. Following the elimination of the CoASH, the intermediate shown in figure 22.13 is formed. The intermediate then loses a CO_2, forming a carbanion that is resonance stabilized back into the pyridoxal phosphate.
13. We would predict an increased arginase activity in the liver of the untreated diabetic animal. The untreated diabetic animal synthesizes glucose primarily by hepatic gluconeogenesis utilizing amino acids derived from protein catabolism as the carbon source. The urea cycle activity must increase to accommodate the increased flux of amino groups removed from the amino acids. Arginase catalyzes the hydrolysis of arginine yielding urea plus ornithine and is the rate-limiting step in the urea cycle.
15. (a) L-glutathione is synthesized in successive steps, catalyzed by γ-glutamyl cysteine synthase and glutathione synthase. Glutathione synthesis is directed by the substrate specificity of these enzymes. (b) Decreased glutathione synthesis would increase the probability of oxidative damage to the cell.

Chapter 23

1. Carbamoyl phosphate synthase contributes to two processes: (a) the initial enzyme in the biosynthesis of pyrimidines and (b) a component in the synthesis of arginine biosynthesis or the urea cycle. In bacteria both of these processes occur within the same compartment. In human beings the carbamoyl phosphate synthase involved in the urea cycle is contained in the mitochondria; it is isolated from the cytosol counterpart that is involved in the biosynthesis of pyrmidines. Because the two carbamoyl phosphate synthases are in separate cellular compartments in human beings, control of the cytosol carbamoyl phosphate synthase by pyrimidine pathway products has no impact on the urea cycle.
3. Typically, the mechanism for amine addition involves a nucleophilic approach by a nitrogen, requiring a lone pair of electrons on the nitrogen. Ammonium ions are protonated at physiological pH and do not have a lone pair. The amide of glutamine is not protonated and carries a lone pair of electrons.
5. The phosphorylation step utilizing ATP converts the oxygen into a much better leaving group.
7. This oxygen becomes incorporated into an H_2O molecule.
9. Because Lesch-Nyhan patients lack the enzyme hypoxanthine-guanine phosphoribosyltransferase, they accumulate high levels of PRPP, which stimulates purine biosynthesis to high levels, leading to production of large amounts of uric acid. The brain may not have high levels of *de novo* purine biosynthesis and probably relies on the salvage pathway enzymes for its purine nucleotides.
11. The product would be 3-amino-2-methylpropanoic acid.

Chapter 24

1. In a liver cell, fatty acid synthesis takes place in the cytosol. It uses acetyl-CoA carboxylase and a large, multifunctional polypeptide fatty acid synthase. The first committed step is acetyl-CoA carboxylase, which is highly regulated by hormonal control; this results in the phosphorylation of the enzyme. Citrate is transported from the mitochondria and is used to generate acetyl-CoA and reducing power in the form of NADPH (NADPH is used in biosynthetic reactions instead of NADH). Citrate activates the carboxylase, but the end product, palmitate, inhibits the reaction. Fatty acid degradation occurs in the matrix of the mitochondria. A key point of regulation is on the uptake of the fatty acid into the matrix of the mitochondria. Malonyl-CoA, the product of the acetyl-CoA carboxylase, inhibits uptake and prevents the newly made palmitic acid from being degraded.

   Gluconeogenesis utilizes many of the glycolytic enzymes, yet three of these enzymes in glycolysis have large negative free energy changes in the direction of pyruvate formation. These reactions must be replaced in gluconeogenesis to make glucose formation thermodynamically favorable. Replacement allows glycolysis and gluconeogenesis to be thermodynamically favorable and at the same time permits the pathways to be independently regulated to avoid a futile cycle.

3. A hormone receptor must do two things if it is to function properly:
   (1) Distinguish the hormone from all other surrounding chemical signals and bind it with a very high affinity (K_d ranges from 1×10^{-7} M to 1×10^{-12} M).
   (2) Upon binding the hormone, undergo a conformational change into an active form that can then interact with other molecules that initiate the molecular events leading to the hormone's elicited response.

   Proteins are the only macromolecules that can exhibit this kind of behavior (specific binding and conformation change).
5. (1) A polypeptide signal sequence must be present if the protein is to be transported into the endoplasmic reticulum and subsequently secreted.
   (2) Additional polypeptide sequences are necessary for proper peptide chain folding (e.g., C peptide of insulin).
   (3) Cleavage allows control of hormones from inactive to active form (e.g., thyroxine).
   (4) Production of a number of different hormones from the same precursor allows coordinate production of several hormones. Specific cleavage by the cell allows control of which peptides are produced (e.g., cleavage of prepro-opiocortin to corticotropin, β-lipotropin, γ-lipotropin, α-MSH, β-MSH, γ-MSH, endorphin, and enkephalin).
   (5) A large precursor of the hormone can serve as a storage form (e.g., thyroglobulin).
7. The same amino acid, named as 5-oxoproline, is an intermediate in the γ-glutamyl cycle. The name pyroglutamate suggests formation involving dehydration via heat (fire) from glutamate.
9. Vitamin D can be considered both a hormone and a vitamin. Its mode of action is like that of other steroid hormones, and it is synthesized in the body. It can be given in the diet (e.g., in supplemented milk) and would then be called a vitamin.

Chapter 25

1. Avery, working with two different strains of pneumococcus, was able to show that a fraction isolated from the pathogenic S strain that transformed the nonpathogenic R strain was DNA. This transforming activity was not affected by RNase, proteases, or enzymes that degrade capsular polysaccharides but was destroyed by treatment with DNase. Purified DNA from S cells was able to transform R cells into S cells *in vitro.*
3. dpCpGpTpA or, in abbreviated form, CGTA.
5. (a) In the major groove of B-form DNA:
   Adenine N^7; N^6
   Cytosine N^4
   Guanine N^7, O^6
   Thymine O^4
   In the minor groove of B-form DNA:
   Adenine N^3
   Cytosine O^2
   Guanine N^3; N^2
   Thymine O^2
   (b) Although the numbers of electronegative atoms capable of forming hydrogen bonds in the major and minor groove are similar, access to those in the minor groove is hindered by ribose moiety.
7. The melting temperature (T_m) of DNA is affected by the base composition, with the G-C-rich DNA having a higher T_m than the A-T-rich DNA. The DNA samples could be heated in a spectrophotometer and the increase in absorbance of ultraviolet light (hyperchromism) could be plotted against temperature. The A-T-rich DNA would have a lower T_m than the G-C-rich DNA.
9. A simple approach would be to take small aliquots of the sample and treat them with the enzymes ribonuclease (RNase) or deoxyribonuclease (DNase). Digestion of the sample by one of these enzymes would indicate whether the sample is RNA or DNA. Another method is to treat a small sample with alkali, which degrades RNA to mononucleotides, but only denatures DNA to the single-stranded form. One way to detect whether these treatments had any effect on the nucleic acid is to subject it to electrophoresis on an agarose gel. Free nucleotides, or even small oligonucleotides, are not visible on agarose gels, whereas the original viral nucleic acid should yield one (or more) discrete high molecular weight bands.

   To determine whether the nucleic acid is single-stranded or double-stranded, it could be heated. A sharp increase in absorbance at 260 nm would indicate a double-stranded RNA or DNA; a broader melting curve would suggest a single-stranded nucleic acid. We could also analyze the nucleotide composition of the nucleic acid. Equivalence between A and T and between G and C would strongly suggest that the nucleic acid is double-stranded.

11. The 2′-OH group found on the ribose in RNA sterically prevents the B duplex from forming.
13. The differences in Ethidium (Et) binding capacities between linear duplex DNA and covalently closed circular DNA (cccDNA) can be understood in terms of the differences in topological constraints imposed on the two molecules. By intercalating between two adjacent base pairs, Et unwinds the double helix, which results in an increase in the length of the helix (pitch). For cccDNA, the conformational stress introduced by unwinding is compensated for by a change in tertiary structure of the molecule (i.e., supercoiling).

    At some point, the torsional stress caused by the positive supercoils will become energetically unfavorable and the tendency of the molecule will be toward winding, thus preventing further binding of Et. Linear duplex DNA does not undergo this torsional stress because it is not covalently closed and would be expected to have a greater binding capacity for Et.
15. The ratio of the histones in chromatin supports the model proposed for nucleosome structure. The core of the nucleosome is made up of an octamer of two molecules each of H2A, H2B, H3, and H4. The H1 histone seals off the nucleosome (i.e., only one H1 per nucleosome).

Chapter 26

1. If DNA replication were dispersive, each strand of each daughter molecule would have had an intermediate density in the experiment. The fact that after one generation, one strand of the DNA still had the same density as the parental (^{15}N) DNA served to further corroborate the conclusion that DNA replication is semiconservative.
3. As has been seen in many other biochemical reactions in the cell, the generation of pyrophosphate coupled with its hydrolysis by pyrophosphatase is the major driving force for DNA synthesis. The Mg^{2+} can bind to the transition state intermediate (trigonal bipyramidal intermediate) and stabilize it, lowering the energy of activation. The metal ion can also promote the reaction by charge shielding; the $Mg^{2+}NTP$ complex is the actual substrate, with the metal reducing the negative charge on the phosphate groups so as not to repel the electron pair of the attacking nucleophile.
5. The use of a primer increases fidelity in DNA synthesis by providing a more extensive stacked double helix to which the first few bases of DNA are added, allowing the proofreading exonuclease activity of the polymerase to evaluate more accurately the stability of the newly synthesized DNA. The reason that the primer is made of RNA and not DNA may be because RNA, even in a double-stranded nucleic acid, is readily recognized as being different from DNA (RNA–DNA duplexes adopt a different conformation because of the presence of the 2′-OH on the ribonucleotides.) A proofreading DNA polymerase recognizes the differences in the RNA primer and replaces it with DNA.
7. About 33 min would be required to replicate the chromosome because you have a bidirectional mode of replication with two replication forks. If you had multifork replication (one round of replication starting before the other finished), a 20-min division time would be possible.
9. 2.9×10^9 bp $\times$ 1 sec/60 bp $\times$ 1 h/3600 sec = 13,400 h

    The human genome would need at least 13,400 replication origins to be completely replicated in one hour.
11. DNA in eukaryotic chromosomes is complexed with histone proteins in complexes called nucleosomes. These DNA-protein complexes are disassembled directly in front of the replication fork. The nucleosome disassembly may be rate-limiting for the migration of the replication forks, as the rate of migration is slower in eukaryotes than prokaryotes. The length of Okazaki fragments is also similar to the size of the DNA between nucleosomes (about 200 bp). One model that would allow the synthesis of new eukaryotic DNA and nucleosome formation would be the disassembly of the histones in front of the replication fork and then the reassembly of the histones on the two duplex strands. Histone synthesis is closely coupled to DNA replication.
13. The SOS response is reversed when the protease activity of recA can no longer be activated because most or all of the damaged DNA has been repaired or eliminated and intact lexA protein levels begin to rise. Then lexA can act as a repressor, binding to the gene control regions of all of the genes it regulates, including its own and that of recA.
15. The recA protein is very important in DNA repair. An insult to the DNA leads to the activation of the

protease function of recA, which then cleaves lexA protein, turning on the genes in the SOS response. Once the DNA is repaired, the recA protease is inactivated and new lexA protein is made, repressing the DNA repair genes again. RecA mutations are totally deficient in homologous recombination, demonstrating the important role of the recA protein in this process. The purified recA protein will catalyze the exchange between duplex and single-stranded DNAs with the hydrolysis of ATP. Also, recA protein will form a complex between two circular helices if one helix is gapped on one strand. These functions of recA protein would place it at the hub of activities in recombination. RecA mutants would be very useful in genetic research because mutants generated would not be repaired, and in recombinant DNA cloning recombinational events between vector recombinant DNA and host genomic DNA would not occur.

Chapter 27

1. Spaces are introduced every 10 nucleotides for clarity: 5′-CAAAAAACGG ACCGGGTGTA CAACTTTTAC TATGGCGTGA CACCTAAATT ATAGGCAGAA ATAAGTACAT GACTATTGGG AGGAGCAGGA ACAAGTAGG-3′.
3. The frequency of occurrence of the sequence -CTGCAG- (a PstI site) in DNA that is 80% G + C is $0.4 \times 0.1 \times 0.4 \times 0.4 \times 0.1 \times 0.4$, or $(0.4)^4(0.1)^2$ = once every 3906 bp. An AAGCTT (HindIII) site should occur $(0.4)^2(0.1)^4$ = once every 62,500 bp.

   The *E. coli* genome composition is approximately 26% G, 26% C, 24% A, and 24% T. PstI would cleave this genome $(0.26)^4(0.24)^2$ = once every 3,800 bp, and HindIII, $(0.26)^2(0.24)^4$ = once every 4,500 bp.
5. There is no difference between the single-stranded ends generated by BamHI and by MboI. These ends are complementary, and they can anneal and be ligated together by DNA ligase. The resulting sequence contains an MboI site and can be cleaved by that enzyme. It has only a 25% probability of restoring a BamHI site, however, depending on the nucleotide located adjacent to the MboI site.
7. In order to isolate a 15-kb gene from a genome containing 3×10^9 bp, it would be necessary to isolate $3 \times 10^9/15{,}000$ or 200,000 fragments per genome, or 200,000 clones. We recommend that a library 3 to 10 times the minimum size should be prepared, to ensure a high probability that a given fragment will be represented at least once. In the case of the gene specified, the library should therefore contain between 6×10^5 and 2×10^6 clones.
9. (a) 2, 4
   (b) 1, 5
   (c) 1, 3
11. To reduce the hybridization stringency, researchers choose conditions that stabilize double-stranded DNA, allowing sequences that share only partial complementarity to form base pairs. Examples of such conditions include reducing the hybridization temperature and increasing the salt concentration in solution.
13. Chromosome jumping is a useful procedure to traverse long distances and skip troublesome regions of the genome (repetitious sequences, etc.). One approach with this technique is to digest the genome with rare cutting restriction enzymes (*Not*I recognizes an 8-base sequence and generates an average fragment size of 500 kb of DNA). The fragments are circularized with a small marker DNA between the ends. The marker DNA contains sequences necessary for cloning in λ. These circular DNA fragments are then cleaved with another restriction enzyme that produces fragments small enough to clone in λ. With this procedure only clones that contain the marker DNA (i.e., the ends of the original fragment) are isolated. This method would permit only one jump; thus another method used with this technique is called a linking library. The same sample of DNA is digested with a restriction enzyme that gives smaller fragments. These smaller fragments are then circularized with the same marker DNA used in the jumping library. The circular DNA is then digested with *Not*I to linearlize the fragments for insertion into the vector. The linking library carries sequences on both sides of the *Not*I restriction sites while the jumping library carries sequences from one side of two adjacent *Not*I sites.

Chapter 28

1. RNA and DNA polymerases catalyze the same reaction mechanistically, involving hydrolysis of a nucleotide triphosphate to release pyrophosphate and form a phosphodiester bond. In both cases, the order of nucleotide addition is specified by the template, and synthesis of the growing nucleic acid chain is in a 5′ to 3′ direction (the enzymes move in a 3′ to 5′ direction along the template strand). In addition to the obvious difference in substrates (RNA polymerase utilizes ribonucleotides, whereas DNA polymerase utilizes deoxyribonucleotides), these two enzymes differ in their requirements for initiating synthesis:

RNA polymerase initiates *de novo,* and does not require a primer, unlike DNA polymerase. Finally, RNA polymerase does not perform any proofreading activity, unlike DNA polymerases, which generally have a 3′ to 5′ exonuclease activity to remove misincorporated nucleotides.

3. mRNA has a short half-life (a couple of minutes); tRNA and rRNA are very stable (half-life measured in hours) and accumulate to make up most of the RNA in the cell (95%).
5. The bases within the base-paired region of each arm of the tRNA cloverleaf stack in a manner similar to the base stacking described in Chapter 25 for DNA. In addition to the base stacking within base-paired regions, there is also stacking of one helix on top of another in the tRNA molecule. In particular, the acceptor stem stacks with the TΨC stem and loop to form one nearly continuous stacked double helix. The anticodon stem and the D stem also stack on top of one another.
7. The D and T loops in tRNA interact with each other to form the tertiary structure, leaving only the anticodon with a single-stranded loop able to be cleaved by RNase.
9. The RNA polymerase binds to the same side of the duplex at the -10 and -35 regions with about two turns of the duplex helix between these boxes. This binding site can be determined by a variety of "footprint" experiments. DNA that is 5′-labeled can be mixed with the polymerase, and regions that are protected from digestion with an enzyme like DNase I can be determined on a sequencing gel. Sequences with tight contact with the RNA polymerase are observed as a series of blank spots in the sequencing ladder that look like footprints.
11. The 5′ terminus is capped (G^{m}pppX—), generating a guanosine nucleoside with 2′ and 3′-OH groups and a 5′-5′ pyrophosphate linkage. The other end of the mRNA has poly(A) added, so the 3′ end is adenosine. Therefore most of the mRNA in the cell has a guanosine and an adenosine 2′ and 3′-OH and no typical 5′-triphosphate ending (pppN—).
13. By flattening and widening the DNA minor groove, TATA binding protein may assist other factors in forming an open complexlike structure with separated strands. The bending induced by TATA binding protein would bring the DNA upstream and downstream of the TATA box closer together, promoting interactions between proteins bound to upstream elements and the start site of transcription.
15. If an intron or part of an intron containing a stop signal for translation was not removed, this mRNA would be longer but would yield a shorter polypeptide. Alternative splicing would remove the intron or use an alternative splice site, generating a shorter mRNA and a longer polypeptide.

Chapter 29

1. The difference in the mechanism of translation initiation in prokaryotes compared to eukaryotes has profound consequences for the strategy used to coordinate the expression of a set of genes in the two systems. In prokaryotes this coordination is achieved by organizing genes into transcription units that are transcribed to give polycistronic mRNAs. Each cistron within a polycistronic mRNA begins with a Shine-Dalgarno sequence and initiation AUG codon. In contrast, in eukaryotes, in which translation begins almost invariably (and exclusively) at the 5′-most AUG codon, monocistronic mRNAs are necessarily the order of the day. Genes that must be coordinately expressed are consequently not organized into transcriptional units, and coordination must be achieved in some other way.
3. This serine tRNA has inosine at the 5′ position of its anticodon, which can pair with U, C, or A. Thus the anticodon of this tRNA is most likely IGA (given in the correct 5′ to 3′ direction).
5. A number of variations have been found in the universal genetic code in genes located in mitochondria and chloroplast. These variations in the meaning of some of the code words represent divergences from the standard genetic code and not an independent origin of another genetic code. These divergences probably arose in these organelles because of the limited number of genes coded and requirements for the synthesis of ribosomes and tRNA. Clearly, something was unusual when only 24 types of tRNAs were found in mitochondria. Mitochondria do not use all 61 codons.
7. Met Val Glu Ile Arg Asp Thr His Leu Lys Lys Gln Ile Ala Phe Ter Ter
9. (5′)AAY TGG GCN CAR TGY AAY CC(3′). R is the abbreviation for a mixture of A or G (R = puRine), Y is the abbreviation for a mixture of C and T (Y = pYrimidine), and N represents a mixture of all four Nucleotides.
11. The x-ray crystal structure of EF-Tu bound to GDP and GTP is known. Based on this structure, many of the amino acids shared among IF-2, EF-Tu, EF-G, and RF-3 apparently are somehow involved in binding of GTP and GDP. Thus these proteins (and many others, including the proto-oncogene, Ras) share a similar GTP-binding domain.

13. Import into the endoplasmic reticulum requires an N-terminal signal sequence that contains a long stretch of hydrophobic amino acids. The mitochondrial transit peptide is a hydrophilic sequence rich in serine and threonine, with regularly spaced basic amino acids. Import into the ER requires the signal recognition particle and its receptor, but mitochondrial import does not require the SRP and presumably uses a different receptor. Import into mitochondria requires a membrane potential, but import into the ER does not.
15. GAA encodes a glutamic acid residue. If this glu is essential for the catalytic activity of the protein, any mutation (except GAA to GAG) will be harmful to the activity of the protein. If the glu residue is not essential for catalysis or proper folding, the possible mutations, in order of increasing potential severity, are:

GAA changes to	Glu changes to
GAG	Glu
GAT or GAC	Asp
CAA	Gln
GCA	Ala
AAA	Lys
GTA	Val
GGA	Gly
TAA	Terminator

Chapter 30

1. The existence of a repressor in *lac* operon regulation was first suggested by the results of studies of merodiploids. In merodiploids of the type i^+z^-/Fi^-z^+, Jacob and Monod were able to demonstrate that the i^+ (inducible) allele is dominant to the i^- (constitutive) allele when on the same chromosome (*cis*) or on a different chromosome (*trans*) with respect to the z^+ allele.
3. (a) Cells with the genotype $i^so^+z^+$ have a "super-repressor" i^s mutation, which causes the repressor to be insensitive to an inducer. Thus even though the operator and β-galactosidase loci are wild-type, no β-galactosidase will be produced in this mutant. On media containing X-gal, with or without IPTG, colonies will be white.
   (b) Cells with the genotype $i^so^cz^+$ still have the superrepressor, but the β-galactosidase gene is under the control of a constitutive operator, o^c. This mutation interferes with the ability of the repressor to bind and repress transcription. Therefore, β-galactosidase probably would be produced continuously. On X-gal, with or without IPTG, colonies will be blue.
   (c) Merodiploids with the genotype $i^so^cz^-/i^+o^+z^+$ would behave like the mutant described in part (a).
   (d) Merodiploid cells with the genotype $i^+o^cz^-/i^-o^+z^+$ would exhibit β-galactosidase regulation essentially identical to that of wild-type cells; on X-gal alone, colonies would be white. In the presence of an inducer, the β-galactosidase gene would be induced; on X-gal and IPTG, colonies would be blue.
5.

Strains	Uninduced Enz A	Uninduced Enz B	Induced Enz A	Induced Enz B
Haploid				
(1) $R^+O^+A^+B^+$	1	1	100	100
(2) $R^+O^cA^+B^+$	1–100	1–100	100	100
(3) $R^-O^+A^+B^+$	100	100	100	100
Diploid				
(4) $R^+O^+A^+B^+/R^+O^+A^+B^+$	2	2	200	200
(5) $R^+O^cA^+B^+/R^+O^+A^+B^+$	2–101	2–101	200	200
(6) $R^+O^+A^-B^+/R^+O^+A^+B^+$	1	2	100	200
(7) $R^-O^+A^+B^+/R^+O^+A^+B^+$	2	2	200	200

7. The explanation for the less than perfect match of most promoters to the consensus sequence is to be found in the need to regulate transcription. Transcriptional regulation is achieved in many instances by the selective improvement of the affinity of specific promoters for RNA polymerase. Such selective improvement is well illustrated in the case of regulation of the *lac* operon.
9. One possibility is the binding of the repressor nonspecifically to the DNA and then searching in one dimension (binding to the DNA and then sliding along until the promoter is reached). Also, a long section of DNA would not be randomly distributed but would form a loose ball of DNA that would define a domain much smaller than the solution in the test tube. When the repressor was released from the DNA it could more quickly find another strand of DNA to bind (effectively giving a much higher concentration of DNA).
11. The synthesis of ribosomal proteins is regulated in *E. coli* by translational regulation (i.e., free ribosomal proteins inhibit the translation of their own

mRNA). As long as rRNA is being made, these proteins bind to the rRNA and the translation of the ribosomal proteins continues. The genes for the ribosomal proteins are clustered in a number of operons that produce polycistronic mRNAs. One of the simplest operons is P_{L11}, which codes for proteins L1 and L11. L1 is the regulatory protein and can bind to the 23S rRNA or to the 5′ end of its own polycistronic mRNA. If the levels of L1 increase, it binds to its own mRNA and inhibits translation of both L1 and L11 proteins. This mechanism keeps the levels of L1 and L11 in register with the amount of rRNA. The other ribosomal proteins are regulated in a similar manner.

13. Translation is an amplification process, in that each molecule of ribosomal protein mRNA can yield many copies of the corresponding protein.
15. The genes encoded by bacteriophage lambda DNA are organized so that gene products required for the same function or process are clustered. Clustering facilitates regulation because all the gene products required at the same time can be induced simultaneously. Clustering also leads to more efficient organization and tighter packing of the bacteriophage genome.
17. The final proof of the two-site model came with the cocrystallization of the regulatory protein and its DNA-binding site. The protein binds on one side of the DNA in two adjacent major grooves. The hydrogen-bonding groups exposed in the major groove present many possibilities for interactions with the amino acid side chains of the protein.

Chapter 31

1. One definition is: *the DNA (or RNA) sequences necessary to produce a peptide (or RNA).* Some viruses have RNA as their genetic material and some genes do not code for a protein but make a functional RNA such as tRNA or rRNA.
3. No, this construct would not be activated by binding of *lexA* because *lexA* does not have a eukaryotic activation domain.
5. A deletion of *HMLE* would remove the element repressing transcription of HML_α at the "storage" location, and result in the constitutive expression of HML_α genes from this locus. The consequences of such constitutivity would depend on the mating type of the mutant. If HML_α were at the *MAT* locus (MAT_α), the arrangement characteristic of the α mating type, the cell exhibits the behavior expected of the α mating type. However, if HMR_a were at the *MAT* locus (MAT_a), the arrangement giving rise to the *a* mating type, the cell would resemble the diploid, expressing both *a* and α genes simultaneously, and thus be sterile.

   Mutants that began as α-type would become sterile at a rate conditioned by the frequency of transposition of the HMR_a to the *MAT* locus. In homothallic strains this frequency is very high, approximately once per cell division, while in heterothallic strains it is considerably lower. Deletion of the gene encoding $\alpha 2$ from MAT_α would result in a failure to inhibit expression of the *a*-specific genes. Haploid mutants that contained such a deletion would therefore resemble the diploid, and be sterile. If the $\alpha 2$ gene contained by the HML_α locus were similarly mutated, mutants of this type would be unable to switch to the α-type. However, transposition of HMR_a to the *MAT* locus would allow for expression of the *a* mating type. Again the frequency of transposition would condition the rate of switching between sterile and *a*-type, the frequency being far greater in homothallic strains than in heterothallic strains.
7. The results of the nuclear transplantation experiments of Briggs and King indicated that nuclei, to a certain stage in embryonic development, remain totipotent, in the sense that they can support the normal development of enucleated differentiating embryos. They found that blastula nuclei, although already "committed" to differentiated pathways, were capable of being "reprogrammed" by a proper environment to allow for some degree of dedifferentiation. Gastrula nuclei, in contrast, were found to support normal development only at a much reduced efficiency.
9. Histones bind to DNA by electrostatic interaction of basic amino acids (arginine and lysine) with the negative charge on the phosphate. These electrostatic (negative-positive charge) interactions are weakened by high salt, which will allow the histone proteins to disassociate from the DNA. Many regulatory proteins may initially interact in a nonspecific fashion with DNA until they locate their high-affinity binding sites. Once these proteins bind at their specific recognition sites, the major interaction is through hydrogen bonding and hydrophobic interactions. These hydrophobic interactions are stabilized by high salt.
11. MspI would cleave at every CCGG sequence in or near the globin genes, regardless of the cell type from which the DNA had been isolated because this enzyme is insensitive to methylation. Globin genes isolated from erythroblast cells would not differ from those of other tissues in their cleavage pattern with MspI.

HpaII does not recognize CCGG sequences when the second C is methylated. In tissues in which genes are *not* transcribed, CpG sequences tend to be methylated more often than in tissues where the same genes are transcribed. Thus HpaII is expected to recognize and cleave more sites in globin genes in erythroblast DNA than in DNA from other cell types (where the globin genes are not transcribed and the corresponding DNA is more highly methylated).

13. Most transcription factors have two-domain structures, one for DNA binding and another for protein binding. The DNA-binding domains involve various structural motifs that interact with the DNA in the major groove. The type of structure previously discussed in prokaryotes is the helix-turn-helix motif, which is also found in some eukaryotic transcription factors. A more common motif in eukaryotes is the zinc finger (zinc forms a tetrahedral complex with histidines and cysteines). Another type of structure found is the leucine zipper (a highly conserved stretch of amino acids with net basic charge followed by a region of four leucine residues at intervals of seven amino acids). All of these motifs found in these transcription factors involve interactions in the major groove of the DNA with either an α-helix segment (more common) or a two-stranded antiparallel β sheet.
15. One strategy employed to generate sufficient amounts of a product required at a specific stage of development is gene amplification. An alternative strategy used to provide large amounts of required products early in development is the accumulation of mRNAs or proteins prior to need.
17. Although the precise regulatory relationships between homeotic genes remain unclear, it is apparent that a hierarchy of interactions exists among gap, pair-rule, and segment polarity genes. Each gene class is influenced by the action of earlier acting genes that control larger units of pattern, and by some members of the same class. Thus gap genes that are expressed early influence the pattern of expression of pair-rule, (e.g., fushi terazu, *ftz*), segment polarity (e.g., engrailed, *en*) and homeotic genes (e.g., ultrabithorax, *ubx*), which are expressed later in a hierarchical fashion. Although segment polarity genes do not affect the expression of gap or pair-rule genes that occupy a higher position in the hierarchy, some gap genes, and presumably genes at other levels in the hierarchy, have been found to be mutually negative regulators of each other.

Glossary

A

A form. A duplex DNA structure with right-handed twisting in which the planes of the base pairs are tilted about 70° with respect to the helix axis.

Acetal. The product formed by the successive condensation of two alcohols with a single aldehyde. It contains two ether-linked oxygens attached to a central carbon atom.

Acetyl-CoA. Acetyl-coenzyme A, a high-energy ester of acetic acid that is important both in the tricarboxylic acid cycle and in fatty acid biosynthesis.

Actin. A protein found in combination with myosin in muscle and also found as filaments constituting an important part of the cytoskeleton in many eukaryotic cells.

Actinomycin D. An antibiotic that binds to DNA and inhibits RNA chain elongation.

Activated complex. The highest free energy state of a complex in going from reactants to products.

Active site. The region of an enzyme molecule that contains the substrate-binding site and the catalytic site for converting the substrate(s) into product(s).

Active transport. The energy-dependent transport of a substance across a membrane.

Adenine. A purine base found in DNA or RNA.

Adenosine. A purine nucleoside found in DNA, RNA, and many cofactors.

Adenosine diphosphate (ADP). The nucleotide formed by adding a pyrophosphate group to the 5′-OH group of adenosine.

Adenosine triphosphate (ATP). The nucleotide formed by adding yet another phosphate group to the pyrophosphate group on ADP.

Adenylate cyclase. The enzyme that catalyzes the formation of cyclic 3′,5′-adenosine monophosphate (cAMP) from ATP.

Adipocyte. A specialized cell that functions as a storage depot for lipid.

Aerobe. An organism that utilizes oxygen for growth.

Affinity chromatography. A column chromatographic technique that employs attached functional groups that have a specific affinity for sites on particular proteins.

Alcohol. A molecule with a hydroxyl group attached to a carbon atom.

Aldehyde. A molecule containing a doubly bonded oxygen and a hydrogen attached to the same carbon atom.

Alleles. Alternative forms of a gene.

Allosteric enzyme. An enzyme the active site of which can be altered by the binding of a small molecule at a nonoverlapping site.

Allosteric site. Location on an allosteric enzyme where the allosteric effector binds.

Aminotransferase. An enzyme that catalyzes the transfer of an amino group from an α-amino acid to an α-keto acid.

Amphibolic pathway. A metabolic pathway that functions in both catabolism and anabolism.

Anabolism. Metabolism that involves biosynthesis.

Anaerobe. An organism that does not require oxygen for maintenance or growth.

Anaplerotic reaction. An enzyme-catalyzed reaction that replenishes the intermediates in a cyclic pathway.

Angstrom (Å). A unit of length equal to 10^{-8} cm.

Anomers. The sugar isomers that differ in configuration about the carbonyl carbon atom. This carbon atom is called the anomeric carbon atom of the sugar.

Antibiotic. A natural product that inhibits bacterial growth (is bacteriostatic) and sometimes results in bacterial death (is bacteriocidal).

Antibody. A specific protein that interacts with a foreign substance (antigen) in a specific way.

Anticodon. A sequence of three bases on the transfer RNA that pair with the bases in the corresponding codon on the messenger RNA.

Antigen. A foreign substance that triggers antibody formation and is bound by the corresponding antibody.

Antiparallel β-pleated sheet (β sheet). A hydrogen-bonded secondary structure formed between two or more extended polypeptide chains.

Antiport. A protein system that can transport different molecules in opposite directions.

Apoactivator. A regulatory protein that stimulates transcription from one or more genes in the presence of a coactivator molecule.

Asexual reproduction. Growth and cell duplication that does not involve the union of nuclei from cells of opposite mating types.

Asymmetrical carbon. A carbon that is covalently bonded to four different groups.

Attenuator. A provisional transcription stop signal.

Autoradiography. The technique of exposing film in the presence of disintegrating radioactive particles. Used to obtain information on the distribution of radioactivity in a gel or a thin cell section.

Autoregulation. The process by which a gene regulates its own expression.

Autotroph. An organism that can form its organic constituents from CO_2.

Auxin. A plant growth hormone usually concentrated in the apical bud.

Auxotroph. A mutant that cannot grow on the minimal medium on which a wild-type member of the same species can grow.

Avogadro's number. The number of molecules in a gram molecular weight of any compound (6.022×10^{23}).

B

β bend. A characteristic way of turning an extended polypeptide chain in a different direction, involving the minimum number of residues.

β oxidation. Oxidative degradation of fatty acids that occurs by the successive oxidation of the β-carbon atom.

β sheet. A sheetlike structure formed by the interaction between two or more extended polypeptide chains.

B cell. One of the major types of cells in the immune system. B cells can differentiate to form memory cells or antibody-forming cells.

B form. The most common form of duplex DNA, containing a right-handed helix and about 10 (10.5 exactly) bp per turn of the helix axis.

Base analog. A compound, usually a purine or a pyrimidine, that differs somewhat from a normal nucleic acid base.

Base stacking. The close packing of the planes of base pairs, commonly found in DNA and RNA structures.

Bidirectional replication. Replication in both directions away from the origin, as opposed to replication in one direction only (unidirectional replication).

Bilayer. A double layer of lipid molecules with the hydrophilic ends oriented outward, in contact with water, and the hydrophobic parts oriented inward toward each other.

Bile salts. Derivatives of cholesterol with detergent properties that aid in the solubilization of lipid molecules in the digestive tract.

Biochemical pathway. A series of enzyme-catalyzed reactions that results in the conversion of a precursor molecule into a product molecule.

Bioluminescence. The production of light by a biochemical system.

Blastoderm. The stage in embryogenesis when a unicellular layer at the surface surrounds the yolk mass.

Bond energy. The energy required to break a bond.

Branchpoint. An intermediate in a biochemical pathway that can follow more than one route in subsequent steps.

Buffer. A conjugate acid–base pair that is capable of resisting changes in pH when acid or base is added to the system. This tendency is maximal when the conjugate forms are present in equal amounts.

C

cAMP. 3′,5′ cyclic adenosine monophosphate. The cAMP molecule plays a key role in metabolic regulation.

CAP. The catabolite gene activator protein, sometimes incorrectly referred to as the CRP protein. The latter term, in small letters (*crp*), should be used to refer to the gene but not to the protein.

Capping. Covalent modification involving the addition of a modified guanine group in a 5′-5″ linkage. It occurs only in eukaryotes, primarily on mRNA molecules.

Carbanion. A negatively charged carbon atom.

Carbohydrate. A polyhydroxy aldehyde or ketone.

Carboxylic acid. A molecule containing a carbon atom attached to a hydroxyl group and to an oxygen atom by a double bond.

Carcinogen. A chemical that can cause cancer.

Carotenoids. Lipid-soluble pigments that are made from isoprene units.

Catabolism. That part of metabolism concerned with degradation reactions.

Catabolite repression. The general repression of transcription of genes associated with catabolism that is seen in the presence of glucose.

Catecholamines. Hormones that are amino derivatives of catechol, for example, epinephrine or norepinephrine.

Catalyst. A compound that lowers the activation energy of a reaction without itself being consumed.

Catalytic site. The site of the enzyme involved in the catalytic process.

Catenane. An interlocked pair of circular structures, such as covalently closed DNA molecules.

Catenation. The linking of molecules without any direct covalent bonding between them, as when two circular DNA molecules interlock like the links in a chain.

cDNA. Complementary DNA, made *in vitro* from the mRNA by the enzyme, reverse transcriptase, and deoxyribonucleotide triphosphates.

Cell commitment. That stage in a cell's life when it becomes committed to a certain line of development.

Cell cycle. All of those stages that a cell passes through from one cell generation to the next.

Cell line. An established clone originally derived from a whole organism through a long process of cultivation.

Cell lineage. The pedigree of cells resulting from binary fission.

Cell wall. A tough outer coating found in many plant, fungal, and bacterial cells that accounts for their ability to withstand mechanical stress or abrupt changes in osmotic pressure. Cell walls always contain a carbohydrate component and frequently also a peptide and a lipid component.

Channeling. The direct transfer of a reaction intermediate from one enzyme to the next.

Chelate. A molecule that contains more than one binding site and frequently binds to another molecule through more than one binding site at the same time.

Chemiosmotic coupling. The coupling of ATP synthesis to an electrochemical potential gradient across a membrane.

Chimeric DNA. Recombinant DNA the components of which originate from two or more different sources.

Chiral compound. A compound that can exist in two forms that are nonsuperimposable images of one another.

Chlorophyll. A green photosynthetic pigment that is made of a magnesium dihydroporphyrin complex.

Chloroplast. A chlorophyll-containing photosynthetic organelle, found in eukaryotic cells, that can harness light energy.

Chromatin. The nucleoprotein fibers of eukaryotic chromosomes.

Chromatography. A procedure for separating chemically similar molecules. Segregation is usually carried out on paper or in glass or metal columns with the help of different solvents. The paper or glass columns contain porous solids with functional groups that have limited affinities for the molecules being separated.

Chromosome. A threadlike structure, visible in the cell nucleus during metaphase, that carries the hereditary information.

Chromosome puff. A swollen region of a giant chromosome; the swelling reflects a high degree of transcription activity.

***Cis* dominance.** Property of a sequence or a gene that exerts a dominant effect on a gene to which it is linked.

Cistron. A genetic unit that encodes a single polypeptide chain.

Citric acid cycle. *See* tricarboxylic acid (TCA) cycle.

Clone. One of a group of genetically identical cells or organisms derived from a common ancestor.

Cloning vector. A self-replicating entity to which foreign DNA can be covalently attached for purposes of amplification in host cells.

Closed system. A system that exchanges neither matter nor energy with its surroundings.

Coactivator. A molecule that functions in conjunction with a protein apoactivator. For example, cAMP is a coactivator of the CAP protein.

Codon. In a messenger RNA molecule, a sequence of three bases that represents a particular amino acid.

Coenzyme. An organic molecule that associates with enzymes and affects their activity.

Cofactor. A small molecule required for enzyme activity. It could be organic, like a coenzyme, or inorganic, like a metallic cation.

Competitive inhibition. An inhibitor that competes with the substrate for binding to the enzyme.

Complementary base sequence. For a given sequence of nucleic acids, the nucleic acids that are related to them by the rules of base pairing.

Configuration. The spatial arrangement of atoms in a molecule that can only be changed by breaking and reforming covalent bonds.

Conformation. The spatial arrangement of groups in a molecule that can be changed without breaking covalent bonds. Often molecules with the same configuration can have more than one conformation.

Consensus sequence. In nucleic acids, the "average" sequence that signals a certain type of action by a specific protein. The sequences actually observed usually vary around this average.

Constitutive enzymes. Enzymes synthesized in fixed amounts, regardless of growth conditions.

Cooperative binding. A situation in which the binding of one substituent to a macromolecule favors the binding of another. For example, DNA cooperatively binds histone molecules, and hemoglobin cooperatively binds oxygen molecules.

Coordinate induction. The simultaneous expression of two or more genes.

Cosmid. A DNA molecule with *cos* ends from λ bacteriophage that can be packaged *in vitro* into a virus for infection purposes.

Cot curve. A curve that indicates the rate of DNA–DNA annealing as a function of DNA concentration and time.

Covalent bond. A chemical bond that involves sharing electron pairs.

Cytidine. A pyrimidine nucleoside found in DNA and RNA.

Cytochromes. Heme-containing proteins that function as electron carriers in oxidative phosphorylation and photosynthesis.

Cytokinin. A plant hormone produced in root tissue.

Cytoplasm. The contents enclosed by the plasma membrane, excluding the nucleus.

Cytosine. A pyrimidine base found in DNA and RNA.

Cytoskeleton. The filamentous skeleton, formed in the cytoplasm, that is largely responsible for controlling cell shape.

Cytosol. The liquid portion of the cytoplasm, including the macromolecules but not including the larger structures, such as subcellular organelles or cytoskeleton.

D

D loop. An extended loop of single-stranded DNA displaced from a duplex structure by an oligonucleotide.

Dalton. A unit of mass equivalent to the mass of a hydrogen atom (1.66×10^{-24} g).

Dark reactions. Reactions that can occur in the dark, in a process that is usually associated with light, such as the dark reactions of photosynthesis.

***De novo* pathway.** A biochemical pathway that starts from elementary substrates and ends in the synthesis of a biochemical.

Deamination. The enzymatic removal of an amine group, as in the deamination of an amino acid to an α-keto acid.

Dehydrogenase. An enzyme that catalyzes the removal of a pair of electrons (and usually one or two protons) from a substrate molecule.

Deletion mutation. A mutation in which one or more nucleotides is removed from a region of the gene.

Denaturation. The disruption of the native folded structure of a nucleic acid or protein molecule; may be due to heat, chemical treatment, or change in pH.

Density-gradient centrifugation. The separation, by centrifugation, of molecules according to their density, in a gradient varying in solute concentration.

Dialysis. Removal of small molecules from a macromolecule preparation by allowing them to pass across a semipermeable membrane.

Diauxic growth. Growth on a mixture of two carbon sources in which one carbon source is used up before the other one is mobilized. For example, in the presence of glucose and lactose, *E. coli* utilizes the glucose before the lactose.

Difference spectra. Display comparing the absorption spectra of a molecule or an assembly of molecules in different states, for example, those of mitochondria under oxidizing or reducing conditions.

Differential centrifugation. Separation of molecules or organelles by sedimentation rate.

Differentiation. A change in the form and pattern of a cell and the genes it expresses as a result of growth and replication, usually during development of a multicellular organism. Also occurs in microorganisms (e.g., in sporulation).

Diploid cell. A cell that contains two chromosomes ($2N$) of each type.

Dipole. A separation of charge within a single molecule.

Directed mutagenesis. In a DNA sequence, an intentional alteration that can be genetically inherited.

Dissociation constant. An equilibrium constant for the dissociation of a molecule into two parts (e.g., dissociation of acetic acid into acetate anion and proton).

Disulfide bridge. A covalent linkage formed by oxidation between two SH groups either in the same polypeptide chain or in different polypeptide chains.

DNA. Deoxyribonucleic acid. A polydeoxyribonucleotide in which the sugar is deoxyribose; the main repository of genetic information in all cells and most viruses.

DNA cloning. The propagation of individual segments of DNA as clones.

DNA library. A mixture of clones, each containing a cloning vector and a segment of DNA from a source of interest.

DNA polymerase. An enzyme that catalyzes the formation of 3′-5′ phosphodiester bonds from deoxyribonucleotide triphosphates.

DNA supercoiling. The coiling of double helix DNA upon itself.

Domain. A segment of a folded protein structure showing conformational integrity. A domain can comprise the entire protein or just a fraction of the protein. Some proteins, such as antibodies, contain many structural domains.

Dominant. Describing an allele the phenotype of which is expressed regardless of whether the organism is homozygous or heterozygous for that allele.

Double helix. A structure in which two helically twisted polynucleotide strands are held together by hydrogen bonding and base stacking.

Duplex. Synonymous with double helix.

Dyad symmetry. Property of a structure that can be rotated by 180° to produce the same structure.

E

Ecdysone. A hormone that stimulates the molting process in insects.

Edman degradation. A systematic method of sequencing proteins, proceeding by stepwise removal of single amino acids from the amino terminal of a polypeptide chain.

Eicosanoid. Any fatty acid with 20 carbons.

Electron carrier. A protein or coenzyme that can reversibly gain and lose electrons and that serves the function of carrying electrons from one site to another.

Electron donor. A substance that donates electrons in an oxidation–reduction reaction.

Electrophoresis. The movement of particles in an electrical field. A commonly used technique for analysis of mixtures of molecules in solution according to their electrophoretic mobilities.

Elongation factors. Protein factors uniquely required during the elongation phase of protein synthesis. Elongation factor G (EF-G) brings about the movement of the peptidyl-tRNA from the A site to the P site of the ribosome.

Eluate. The effluent from a chromatographic column.

Embryo. Plant or animal at an early stage of development.

Enantiomers. Isomers that are mirror images of each other.

Endergonic reaction. A reaction with a positive free energy change.

Endocrine glands. Specialized tissues the function of which is to synthesize and secrete hormones.

Endonuclease. An enzyme that breaks a phosphodiester linkage at some point within a polynucleotide chain.

Endopeptidase. An enzyme that breaks a polypeptide chain at an internal peptide linkage.

Endoplasmic reticulum. A system of double membranes in the cytoplasm that is involved in the synthesis of transported proteins. The rough endoplasmic reticulum has ribosomes associated with it. The smooth endoplasmic reticulum does not.

End-product (feedback) inhibition. The inhibition of the first enzyme in a pathway by the end product of that pathway.

Energy charge. The fractional degree to which the AMP–ADP–ATP system is filled with high-energy phosphates (phosphoryl groups).

Enhancer. A DNA sequence that can bind protein factors that stimulate transcription at an appreciable distance from the site where it is located. It acts in either orientation and either upstream or downstream from the promoter.

Entropy. A thermodynamic measure of the randomness of a system.

Enzyme. A protein that contains a catalytic site for a biochemical reaction.

Epimers. Two stereoisomers with more than one chiral center that differ in configuration at one of their chiral centers.

Equilibrium. In chemistry the point at which the concentrations of two compounds are such that the interconversion of one compound into the other compound does not result in any change in free energy.

***Escherichia coli* (*E. coli*).** A gram negative bacterium commonly found in the vertebrate intestine. It is the bacterium most frequently used in the study of biochemistry and genetics.

Established cell line. A group of cultured cells derived from a single origin and capable of stable growth for many generations.

Ether. A molecule containing two carbons linked by an oxygen atom.

Eukaryote. A cell or organism that has a membrane-bound nucleus.

Excision repair. DNA repair in which a damaged region is replaced.

Excited state. An energy-rich state of an atom or a molecule, produced by the absorption of radiant energy.

Exergonic reaction. A chemical reaction that takes place with a negative change in free energy.

Exon. A segment within a gene that carries part of the coding information for a protein.

Exonuclease. An enzyme that breaks a phosphodiester linkage at one or the other end of a polynucleotide chain so as to release a single nucleotides or small oligonucleotides.

F

F factor. A large bacterial plasmid, known as the sex-factor plasmid because it permits mating between F^+ and F^- bacteria.

Facilitated diffusion. Diffusion of a substance across a membrane through a protein transporter.

Facultative aerobe. An organism that can use molecular oxygen in its metabolism but that also can live anaerobically.

Fatty acid. A long-chain hydrocarbon containing a carboxyl group at one end. Saturated fatty acids have completely saturated hydrocarbon chains. Unsaturated fatty acids have one or more carbon–carbon double bonds in their hydrocarbon chains.

Feedback inhibition. *See* end-product inhibition.

Fermentation. The energy-generating breakdown of glucose or related molecules by a process that does not require molecular oxygen.

Fingerprinting. The characteristic two-dimensional paper chromatogram obtained from the partial hydrolysis of a protein or a nucleic acid.

Fluorescence. The emission of light by an excited molecule in the process of making the transition from the excited state to the ground state.

Footprinting. A technique that results in a DNA sequence ladder in which part of the ladder is missing due to the binding of protein to the DNA before processing.

Frameshift mutations. Insertions or deletions of genetic material that lead to a shift in the translation of the reading frame. The mutation usually leads to nonfunctional proteins.

Free energy. That part of the energy of a system that is available to do useful work.

Furanose. A sugar that contains a five-member ring as a result of intramolecular hemiacetal formation.

Futile cycle. *See* pseudocycle.

G

G_1 phase. That period of the cell cycle in which preparations are being made for chromosome duplication, which takes place in the S phase.

G_2 phase. That period of the cell cycle between S phase and mitosis (M phase).

Gametes. The ova and the sperm, haploid cells that unite during fertilization to generate a diploid zygote.

Gel exclusion chromatography. A technique that makes use of certain polymers that can form porous beads with varying pore sizes. In columns made from such beads, it is possible to separate molecules, which cannot penetrate beads of a given pore size, from smaller molecules that can.

Gene. A segment of the genome that codes for a functional product.

Gene amplification. The duplication of a particular gene within a chromosome two or more times.

Gene splicing. The cutting and rejoining of DNA sequences.

General recombination. Recombination that occurs between homologous chromosomes at homologous sites.

Generation time. The time it takes for a cell to double its mass under specified conditions.

Genetic map. The arrangement of genes or other identifiable sequences on a chromosome.

Genome. The total genetic content of a cell or a virus.

Genotype. The genetic characteristics of an organism (distinguished from its observable characteristics, or phenotype).

Globular protein. A folded protein that adopts an approximately globular shape.

Gluconeogenesis. The production of sugars from nonsugar precursors, such as lactate or amino acids. Applies more specifically to the production of free glucose by vertebrate livers.

Glycogen. A polymer of glucose residues in 1,4 linkage and 1,6 linkage at branchpoints.

Glycogenic. Describing amino acids the metabolism of which may lead to gluconeogenesis.

Glycolipid. A lipid containing a carbohydrate group.

Glycolysis. The catabolic conversion of glucose to pyruvate with the production of ATP.

Glycoprotein. A protein linked to an oligosaccharide or a polysaccharide.

Glycosaminoglycans. Long, unbranched polysaccharide chains composed of repeating disaccharide subunits in which one of the two sugars is either *N*-acetylglucosamine or *N*-acetylgalactosamine.

Glycosidic bond. The bond between a sugar and an alcohol. Also the bond that links two sugars in disaccharides, oligosaccharides, and polysaccharides.

Glyoxylate cycle. A pathway that uses acetyl-CoA and two auxiliary enzymes to convert acetate into succinate and carbohydrates.

Glyoxysome. An organelle containing key enzymes of the glyoxylate cycle.

Goldman equation. An equation expressing the quantitative relationship between the concentrations of charged species on either side of a membrane and the resting transmembrane potential.

Golgi apparatus. A complex series of double-membrane structures that interact with the endoplasmic reticulum and that serve as a transfer point for proteins destined for other organelles, the plasma membrane, or extracellular transport.

Gram molecular weight. For a given compound, the weight in grams that is numerically equal to its molecular weight.

Ground state. The lowest electronic energy state of an atom or a molecule.

Growth factor. A substance that must be present in the growth medium to permit cell proliferation.

Growth fork. The region on a DNA duplex molecule where synthesis is taking place. It resembles a fork in shape because it consists of a region of duplex DNA connected to a region of unwound single strands.

Guanine. A purine base found in DNA or RNA.

Guanosine. A purine nucleoside found in DNA and RNA.

H

Hairpin loop. A single-stranded complementary region that folds back on itself and base-pairs into a double helix.

Half-life. The time required for the disappearance of one half of a substance.

Haploid cell. A cell containing only one chromosome of each type.

Heavy isotopes. Forms of atoms that contain greater numbers of neutrons (e.g., ^{15}N, ^{13}C).

Helix. A spiral structure with a repeating pattern.

Heme. An iron–porphyrin complex found in hemoglobin and cytochromes.

Hemiacetal. The product formed by the condensation of an aldehyde with an alcohol; it contains one oxygen linked to a central carbon in a hydroxyl fashion and one oxygen linked to the same central carbon by an ether linkage.

Henderson-Hasselbach equation. An equation that relates the pK_a to the pH and the ratio of the proton acceptor (A^-) and the proton donor (HA) species of a conjugate acid–base pair.

Heterochromatin. Highly condensed regions of chromosomes that are not usually transcriptionally active.

Heteroduplex. An annealed duplex structure between two DNA strands that do not show perfect complementarity. Can arise by mutation, recombination, or the annealing of complementary single-stranded DNAs.

Heteropolymer. A polymer containing more than one type of monomeric unit.

Heterotroph. An organism that requires preformed organic compounds for growth.

Heterozygous. Describing an organism (a heterozygote) that carries two different alleles for a given gene.

Hexose. A sugar with a six-carbon backbone.

High-energy compound. A compound that undergoes hydrolysis with a high negative standard free energy change.

Histones. The family of basic proteins that is normally associated with DNA in most cells of eukaryotic organisms.

Holoenzyme. An intact enzyme containing all of its subunits with full enzymatic activity.

Homeobox. A conserved sequence of 180 bp encoding a protein domain found in many eukaryotic regulatory proteins.

Homologous chromosomes. Chromosomes that carry the same pattern of genes but not necessarily the same alleles.

Homopolymer. A polymer composed of only one type of monomeric building block.

Homozygous. Describing an organism (a homozygote) that carries two identical alleles for a given gene.

Hormone. A chemical substance made in one cell and secreted so as to influence the metabolic activity of a select group of cells located elsewhere in the organism.

Hormone receptor. A protein that is located on the cell membrane or inside the responsive cell and that interacts specifically with the hormone.

Host cell. A cell used for growth and reproduction of a virus.

Hybrid (or chimeric) plasmid. A plasmid that contains DNA from two different organisms.

Hydrogen bond. A weak attractive force between one electronegative atom and a hydrogen atom that is covalently linked to a second electronegative atom.

Hydrolysis. The cleavage of a molecule by the addition of water.

Hydrophilic. Preferring to be in contact with water.

Hydrophobic. Preferring not to be in contact with water, as is the case with the hydrocarbon portion of a fatty acid or phospholipid chain.

I

Ion-exchange resin. A polymeric resinous substance, usually in bead form, that contains fixed groups with positive or negative charge. A cation exchange resin has negatively charged groups and is therefore useful in exchanging the cationic groups in a test sample. The resin is usually used in the form of a column, as in other column chromatographic systems.

Isoelectric pH. The pH at which a protein has no net charge.

Isomerase. An enzyme that catalyzes an intramolecular rearrangement.

Isomerization. Rearrangement of atomic groups within the same molecule without any loss or gain of atoms.

Isoprene. The hydrocarbon 2-methyl-1,3-butadiene, which in some form serves as the precursor for many lipid molecules.

Isozymes. Multiple forms of an enzyme that differ from one another in one or more properties.

K

K_m. *See* Michaelis constant.

Ketogenic. Describing amino acids that are metabolized to acetoacetate and acetate.

Ketone. A functional group of an organic compound in which a carbon atom is double-bonded to an oxygen. Neither of the other substituents attached to the carbon is a hydrogen. Otherwise the group would be called an aldehyde.

Ketone bodies. Refers to acetoacetate, acetone, and β-hydroxybutyrate made from acetyl-CoA in the liver and used for energy in nonhepatic tissues.

Ketosis. A condition in which the concentration of ketone bodies in the blood or urine is unusually high.

Kilobase. One thousand bases in a DNA molecule.

Kinase. An enzyme catalyzing phosphorylation of an acceptor molecule, usually with ATP serving as the phosphate (phosphoryl) donor.

Kinetochore. A structure that attaches laterally to the centromere of a chromosome; it is the site of chromosome tubule attachment.

Krebs cycle. *See* tricarboxylic acid (TCA) cycle.

L

Lampbrush chromosome. Giant diplotene chromosome found in the oocyte nucleus. The loops that are observed are the sites of extensive gene expression.

Law of mass action. The finding that the rate of a chemical reaction is a function of the product of the concentrations of the reacting species.

Leader region. The region of an mRNA between the 5′ end and the initiation codon for translation of the first polypeptide chain.

Lectins. Agglutinating proteins usually extracted from plants.

Ligase. An enzyme that catalyzes the joining of two molecules together. In DNA it joins 3′-OH to 5′ phosphates.

Linkers. Short oligonucleotides that can be ligated to larger DNA fragments, then cleaved to yield overlapping cohesive ends, suitable for ligation to other DNAs that contain comparable cohesive ends.

Linking number. The net number of times one polynucleotide chain crosses over another polynucleotide chain. By convention, right-handed crossovers are given a plus designation.

Lipid. A biological molecule that is soluble in organic solvents. Lipids include steroids, fatty acids, prostaglandins, terpenes, and waxes.

Lipid bilayer (*see* Bilayer). Model for the structure of the cell membrane based on the hydrophobic interaction between phospholipids.

Lipopolysaccharide. Usually refers to a unique glycolipid found in Gram negative bacteria.

Lyase. An enzyme that catalyzes the removal of a group to form a double bond, or the reverse reaction.

Lysogenic virus. A virus that can adopt an inactive (lysogenic) state, in which it maintains its genome within a cell instead of entering the lytic cycle. The circumstances that determine whether a lysogenic (temperate) virus adopts an inactive state or an active lytic state are often subtle and depend on the physiological state of the infected cell.

Lysosome. An organelle that contains hydrolytic enzymes designed to break down proteins that are targeted to that organelle.

Lytic infection. A virus infection that leads to the lysis of the host cell, yielding progeny virus particles.

M

M phase. That period of the cell cycle when mitosis takes place.

Meiosis. Process in which diploid cells undergo division to form haploid sex cells.

Membrane transport. The facilitated transport of a molecule across a membrane.

Merodiploid. An organism that is diploid for some but not all of its genes.

Mesosome. An invagination of the bacterial cell membrane.

Messenger RNA (mRNA). The template RNA carrying the message for protein synthesis.

Metabolic turnover. A measure of the rate at which already existing molecules of the given species are replaced by newly synthesized molecules of the same type. Usually isotopic labeling is required to measure turnover.

Metabolism. The sum total of the enzyme-catalyzed reactions that occur in a living organism.

Metamorphosis. A change of form, especially the conversion of a larval form to an adult form.

Metaphase. That stage in mitosis or meiosis when all of the chromosomes are lined up on the equator (i.e., an imaginary line that bisects the cell).

Micelle. An aggregate of lipids in which the polar head groups face outward and the hydrophobic tails face inward; no solvent is trapped in the center.

Michaelis constant (K_m). The substrate concentration at which an enzyme-catalyzed reaction proceeds at one-half maximum velocity.

Michaelis-Menten equation (also known as the Henri-Michaelis-Menten equation). An equation relating the reaction velocity to the substrate concentration of an enzyme.

Microtubules. Thin tubules, made from globular proteins, that serve multiple purposes in eukaryotic cells.

Mismatch repair. The replacement of a base in a heteroduplex structure by one that forms a Watson-Crick base pair.

Missense mutation. A change in which a codon for one amino acid is replaced by a codon for another amino acid.

Mitochondrion. An organelle, found in eukaryotic cells, in which oxidative phosphorylation takes place. It contains its own genome and unique ribosomes to carry out protein synthesis of only a fraction of the proteins located in this organelle.

Mitosis. The process whereby replicated chromosomes segregate equally toward opposite poles prior to cell division.

Mixed-function oxidases. Enzymes that use molecular oxygen to oxidize two different molecules simultaneously, usually a substrate and a coenzyme.

Mobile genetic element. A segment of the genome that can move as a unit from one location on the genome to another, without any requirement for sequence homology.

Molecularity of a reaction. The number of molecules involved in a specific reaction step.

Monolayer. A single layer of oriented lipid molecules.

Mutagen. An agent that can bring about a heritable change (mutation) in an organism.

Mutagenesis. A process that leads to a change in the genetic material that is inherited in subsequent generations.

Mutant. An organism that carries an altered gene or change in its genome.

Mutarotation. The change in optical rotation of a sugar that is observed immediately after it is dissolved in aqueous solution, as the result of the slow approach of equilibrium of a pyranose or a furanose in its α and β forms.

Mutation. The genetically inheritable alteration of a gene or group of genes.

Myofibril. A unit of thick and thin filaments in a muscle fiber.

Myosin. The main protein of the thick filaments in a muscle myofibril. It is composed of two coiled subunits (M_r about 220,000) that can aggregate to form a thick filament that is globular at each end.

N

Nascent RNA. The initial transcripts of RNA, before any modification or processing.

Negative control. Regulation of the activity by an inhibitory mechanism.

Negative feedback. Regulation of a reaction or a pathway by the end product.

Nernst equation. An equation that relates the redox potential to the standard redox potential and the concentrations of the oxidized and reduced form of the couple.

Nitrogen cycle. The passage of nitrogen through various valence states, as the result of reactions carried out by a wide variety of different organisms.

Nitrogen fixation. Conversion of atmospheric nitrogen into a form that can be converted by biochemical reactions to an organic form. This reaction is carried out by a very limited number of microorganisms.

Nitrogenous base. An aromatic nitrogen-containing molecule with basic properties. Such bases include purines and pyrimidines.

Noncompetitive inhibitor. An inhibitor of enzyme activity the effect of which is not reversed by increasing the concentration of substrate molecule.

Nonsense mutation. A change in the base sequence that converts a sense codon (one that specifies an amino acid) to one that specifies a stop (a nonsense codon). There are three nonsense codons.

Northern blotting. *See* Southern blotting.

Nuclease. An enzyme that cleaves phosphodiester bonds of nucleic acids.

Nucleic acids. Polymers of the ribonucleotides or deoxyribonucleotides.

Nucleohistone. A complex of DNA and histone.

Nucleolus. A spherical structure visible in the nucleus during interphase. The nucleolus is associated with a site on the chromosome that is involved in ribosomal RNA synthesis.

Nucleophilic group. An electron-rich group that tends to attack an electron-deficient nucleus.

Nucleosome. A complex of DNA and an octamer of histone proteins in which a small stretch of the duplex is wrapped around a molecular bead of histone.

Nucleotide. An organic molecule containing a purine or pyrimidine base, a five-carbon sugar (ribose or deoxyribose), and one or more phosphate groups.

Nucleus. In eukaryotic cells, the centrally located organelle that encloses most of the chromosomes. Minor amounts of chromosomal substance are found in some other organelles, most notably the mitochondria and the chloroplasts.

O

Okazaki fragment. A short segment of single-stranded DNA that is an intermediate in DNA synthesis. In bacteria, Okazaki fragments are 1,000–2,000 bases in length; in eukaryotes, 100–200 bases in length.

Oligonucleotide. A polynucleotide containing a small number of nucleotides. The linkages are the same as in a polynucleotide; the only distinguishing feature is the small size.

Oligosaccharide. A molecule containing a small number of sugar residues joined in a linear or a branched structure by glycosidic bonds.

Oncogene. A gene of cellular or viral origin that is responsible for rapid, unruly growth of animal cells.

Operon. A group of contiguous genes that are coordinately regulated by two *cis*-acting elements: A promoter and an operator. Found only in prokaryotic cells.

Optical activity. The property of a molecule that leads to rotation of the plane of polarization of plane-polarized light when the latter is transmitted through the substance. Chirality is a necessary and sufficient property for optical activity.

Organelle. A subcellular membrane-bound body with a well-defined function.

Osmotic pressure. The pressure generated by the mass flow of water to that side of a membrane-bound structure that contains the higher concentration of solute molecules. A stable osmotic pressure is seen in systems in which the membrane is not permeable to some of the solute molecules.

Oxidation. The loss of electrons from a compound.

Oxidative phosphorylation. The formation of ATP as the result of the transfer of electrons to oxygen.

Oxido-reductase. An enzyme that catalyzes oxidation–reduction reactions.

P

Palindrome. A sequence of bases that reads the same in both directions on opposite strands of the DNA duplex (e.g., GAATTC).

Pentose. A sugar with five carbon atoms.

Pentose phosphate pathway. The pathway involving the oxidation of glucose-6-phosphate to pentose phosphates and further reactions of pentose phosphates.

Peptide. An organic molecule in which a covalent amide bond is formed between the α-amino group of one amino acid and the α-carboxyl group of another amino acid, with the elimination of a water molecule.

Peptide mapping. Same as fingerprinting.

Peptidoglycan. The main component of the bacterial cell wall, consisting of a two-dimensional network of heteropolysaccharides running in one direction, cross-linked with polypeptides running in the perpendicular direction.

Periplasm. The space between the inner and outer membranes of a bacterium.

Permease. A protein that catalyzes the transport of a specific small molecule across a membrane.

Peroxisomes. Subcellular organelles that contain flavin-requiring oxidases and that regenerate oxidized flavin by reaction with oxygen.

Phenotype. The observable trait(s) that result from the genotype in cooperation with the environment.

Phenylketonuria. A human disease caused by a genetic deficiency in the enzyme that converts phenylalanine to tyrosine. The immediate cause of the disease is an excess of phenylalanine. The condition can be alleviated by a diet low in phenylalanine.

Pheromone. A hormonelike substance associated with insects that acts as an attractant.

Phosphodiester. A molecule containing two alcohols esterified to a single molecule of phosphate. For example, the backbone of nucleic acids is connected by 5′-3′ phosphodiester linkages between the adjacent individual nucleotide residues.

Phosphogluconate pathway. Another name for the pentose phosphate pathway. This name derives from the fact that 6-phosphogluconate is an intermediate in the formation of pentoses from glucose.

Phospholipid. A lipid containing charged hydrophilic phosphate groups; a component of cell membranes.

Phosphorolysis. Phosphate-induced cleavage of a molecule. In the process the phosphate becomes covalently linked to one of the degradation products.

Phosphorylation. The formation of a phosphate derivative of a biomolecule.

Photoreactivation. DNA repair in which the damaged region is repaired with the help of light and an enzyme. The lesion is repaired without excision from the DNA.

Photosynthesis. The biosynthesis that directly harnesses the chemical energy resulting from the absorption of light. Frequently used to refer to the formation of carbohydrates from CO_2 that occurs in the chloroplasts of plants or the plastids of photosynthetic microorganisms.

Pitch length (or pitch). The number of base pairs per turn of a duplex helix.

p*K*. The negative logarithm of the equilibrium constant.

Plaque. A circular clearing on a lawn of bacterial or cultured cells, resulting from cell lysis and production of phage or animal virus progeny.

Plasma membrane. The membrane that surrounds the cytoplasm.

Plasmid. A circular DNA duplex that replicates autonomously in bacteria. Plasmids that integrate into the host genome are called episomes. Plasmids differ from viruses in that they never form infectious nucleoprotein particles.

Polar group. A hydrophilic (water-loving) group.

Polar mutation. A mutation in one gene that reduces the expression of a gene or genes distal to the promoter in the same operon.

Polarimeter. An instrument for determining the rotation of polarization of light as the light passes through a solution containing an optically active substance.

Polyamine. A hydrocarbon containing more than two amino groups.

Polycistronic messenger RNA. In prokaryotes, an RNA that contains two or more cistrons; note that only in prokaryotic mRNAs can more than one cistron be utilized by the translation system to generate individual proteins.

Polymerase. An enzyme that catalyzes the synthesis of a polymer from monomers.

Polynucleotide. A chain structure containing nucleotides linked together by phosphodiester (5′-3′) bonds. The polynucleotide chain has a directional sense, with a 5′ and a 3′ end.

Polynucleotide phosphorylase. An enzyme that polymerizes ribonucleotide diphosphates. No template is required.

Polypeptide. A linear polymer of amino acids held together by peptide linkages. The polypeptide has a directional sense, with an amino- and a carboxyl-terminal end.

Polyribosome (polysome). A complex of an mRNA and two or more ribosomes actively engaged in protein synthesis.

Polysaccharide. A linear or branched-chain structure containing many sugar molecules linked by glycosidic bonds.

Porphyrin. A complex planar structure containing four substituted pyrroles covalently joined in a ring and frequently containing a central metal atom. For example, heme is a porphyrin with a central iron atom.

Positive control. A system that is turned on by the presence of a regulatory protein.

Posttranslational modification. The covalent bond changes that occur in a polypeptide chain after it leaves the ribosome and before it becomes a mature protein.

Primary structure. In a polymer, the sequence of monomers and the covalent bonds.

Primer. A structure that serves as a growing point for polymerization.

Primosome. A multiprotein complex that catalyzes synthesis of RNA primer at various points along the DNA template.

Prochiral molecule. A nonchiral molecule that may react with an enzyme so that two groups that have a mirror image relationship to each other are treated differently.

Prokaryote. A unicellular organism that contains a single chromosome, no nucleus, no membrane-bound organelles, and has characteristic ribosomes and biochemistry.

Promoter. The region of the gene that signals RNA polymerase binding and the initiation of transcription.

Prophage. The silent phage genome. Some prophages integrate into the host genome; others replicate autonomously. The prophage state is maintained by a phage-encoded repressor.

Prophase. The stage in meiosis or mitosis when chromosomes condense and become visible as refractile bodies.

Proprotein. A protein that is made in an inactive form, so that it requires processing to become functional.

Prostaglandin. An oxygenated eicosanoid that has a hormonal function. Prostaglandins are unusual hormones in that they usually have effects only in that region of the organism where they are synthesized.

Prosthetic group. Synonymous with coenzyme except that a prosthetic group is usually more firmly attached to the enzyme it serves.

Protamines. Highly basic, arginine-rich proteins found complexed to DNA in the sperm of many invertebrates and fish.

Protein subunit. One of the components of a complex multicomponent protein.

Protein targeting. The process whereby proteins following synthesis are directed to specific locations.

Proteoglycan. A protein-linked heteropolysaccharide in which the heteropolysaccharide is usually the major component.

Protist. A relatively undifferentiated organism that can survive as a single cell.

Proton acceptor. A functional group capable of accepting a proton from a proton donor molecule.

Proton motive force (Δp). The thermodynamic driving force for proton translocation. Expressed quantitatively as $\Delta G_{H^+}/F$ in units of volts.

Protooncogene. A cellular gene that can undergo modification to a cancer-causing gene (oncogene).

Pseudocycle. A sequence of reactions that can be arranged in a cycle but that usually do not function simultaneously in both directions. Also called a futile cycle because the net result of simultaneous functioning in both directions would be the expenditure of energy without accomplishing any useful work.

Pulse-chase. An experiment in which a short labeling period is followed by the addition of an excess of the same, unlabeled compound to dilute out the labeled material.

Purine. A heterocyclic ring structure with varying functional groups. The purines adenine and guanine are found in both DNA and RNA.

Puromycin. An antibiotic that inhibits polypeptide synthesis by competing with aminoacyl-tRNA for the ribosomal binding site A.

Pyranose. A simple sugar containing the six-member pyran ring.

Pyrimidine. A heterocyclic six-member ring structure. Cytosine and uracil are the main pyrimidines found in RNA, and cytosine and thymine are the main pyrimidines found in DNA.

Pyrophosphate. A molecule formed by two phosphates in anhydride linkage.

Q

Quaternary structure. In a protein, the way in which the different folded subunits interact to form the multisubunit protein.

R

R group. The distinctive side chain of an amino acid.

R loop. A triple-stranded structure in which RNA displaces a DNA strand by DNA–RNA hybrid formation in a region of the DNA.

Rapid-start complex. The complex that RNA polymerase forms at the promoter site just before initiation.

Recombination. The transfer to offspring of genes not found together in either of the parents.

Redox couple. An electron donor and its corresponding oxidized form.

Redox potential (E). The relative tendency of a pair of molecules to release or accept an electron. The standard redox potential ($E°$) is the redox potential of a solution containing the oxidant and reductant of the couple at standard concentrations.

Regulatory enzyme. An enzyme in which the active site is subject to regulation by factors other than the enzyme substrate. The enzyme frequently contains a nonoverlapping site for binding the regulatory factor that affects the activity of the active site.

Regulatory gene. A gene the principal product of which is a protein designed to regulate the synthesis of other genes.

Renaturation. The process of returning a denatured structure to its original native structure, as when two single strands of DNA are reunited to form a regular duplex, or the process by which an unfolded polypeptide chain is returned to its normal folded three-dimensional structure.

Repair synthesis. DNA synthesis following excision of damaged DNA.

Repetitive DNA. A DNA sequence that is present in many copies per genome.

Replica plating. A technique in which an impression of a culture is taken from a master plate and transferred to a fresh plate. The impression can be of bacterial clones or phage plaques.

Replication fork. The Y-shaped region of DNA at the site of DNA synthesis; also called a growth fork.

Replicon. A genetic element that behaves as an autonomous replicating unit. It can be a plasmid, phage, or bacterial chromosome.

Repressor. A regulatory protein that inhibits transcription from one or more genes. It can combine with an inducer (resulting in specific enzyme induction) or with an operator element (resulting in repression).

Resonance hybrid. A molecular structure that is a hybrid of two structures that differ in the locations of some of the electrons. For example, the benzene ring can be drawn in two ways, with double bonds in different positions. The actual structure of benzene is a blending of these two equivalent structures.

Restriction-modification system. A pair of enzymes found in most bacteria (but not eukaryotic cells). The restriction enzyme recognizes a certain sequence in duplex DNA and makes one cut in each unmodified DNA strand at or near the recognition sequence. The modification enzyme methylates (or modifies) the recognition sequence, thus protecting it from the action of the restriction enzyme.

Reverse transcriptase. An enzyme that synthesizes DNA from an RNA template, using deoxyribonucleotide triphosphates.

Rho factor. A protein involved in the termination of transcription of some messenger RNAs.

Ribose. The five-carbon sugar found in RNA.

Ribosomal RNA (rRNA). The RNA parts of the ribosome.

Ribosomes. Small cellular particles made up of ribosomal RNA and protein. They are the site, together with mRNA, of protein synthesis.

RNA (ribonucleic acid). A polynucleotide in which the sugar is ribose.

RNA polymerase. An enzyme that catalyzes the formation of RNA from ribonucleotide triphosphates, using DNA as a template.

RNA splicing. The excision of a segment of RNA, followed by a rejoining of the remaining fragments.

Rolling-circle replication. A mechanism for the replication of circular DNA. A nick in one strand allows the 3′ end to be extended, displacing the strand with the 5′ end, which is also replicated, to generate a double-stranded tail that can become larger than the unit size of the circular DNA.

S

S phase. The period during the cell cycle when the chromosome is replicated.

Salting in. The increase in solubility that is displayed by typical globular proteins on the addition of small amounts of certain salts such as ammonium sulfate.

Salting out. The decrease in protein solubility that occurs when salts such as ammonium sulfate are present at high concentrations.

Salvage pathway. A family of reactions that permits nucleosides or purine and pyrimidine bases resulting from the partial breakdown of nucleic acids, to be reutilized in nucleic acid synthesis.

Satellite DNA. A DNA fraction the base composition of which differs from that of the main component of DNA, as revealed by the fact that it bands at a different density in a CsCl gradient. Usually repetitive DNA or organelle DNA.

Scissile. Capable of being cut smoothly or split easily.

Second messenger. A diffusible small molecule, such as cAMP, that is formed at the inner surface of the plasma membrane in response to a hormonal signal.

Secondary structure. In a protein or a nucleic acid, any repetitive folded pattern that results from the interaction of the corresponding polymeric chains.

Semiconservative replication. Duplication of DNA in which the daughter duplex carries one old strand and one new strand.

Sigma factor. A subunit of bacterial RNA polymerase that recognizes specific sites on DNA for initiation of RNA synthesis.

Signal sequence. A sequence in a protein that serves as a signal to guide the protein to a specific location.

Signal transduction. The process by which an extracellular signal is converted into a cellular response.

Single-copy DNA. A region of the genome the sequence of which is present only once per haploid complement.

Somatic cell. Any cell of an organism that cannot contribute its genes to a subsequent generation.

SOS system. A set of DNA repair enzymes and regulatory proteins that regulate their synthesis so that maximum synthesis occurs when the DNA is damaged.

Southern blotting. A method for detecting a specific DNA restriction fragment, developed by Edward Southern. DNA from a gel electrophoresis pattern is blotted onto nitrocellulose paper; then the DNA is denatured and fixed on the paper. Subsequently, the pattern of specific sequences in the Southern blot can be determined by hybridization to a suitable probe and autoradiography. A northern blot is similar, except that RNA is blotted instead onto the nitrocellulose paper.

Splicing. *See* RNA splicing.

Sporulation. Formation from vegetative cells of metabolically inactive cells that can resist extreme environmental conditions.

Stacking energy. The energy of interaction that favors the face-to-face packing of purine and pyrimidine base pairs.

Steady-state. In enzyme kinetic analysis, the time interval when the rate of reaction is approximately constant with time. The term is also used to describe the state of a living cell in which the concentrations of many molecules are approximately constant because of a balancing between their rates of synthesis and breakdown.

Stem cell. A cell from which other cells stem or arise by differentiation.

Stereoisomers. Isomers that are nonsuperimposable mirror images of each other.

Steroids. Compounds that are derivatives of a tetracyclic structure composed of a cyclopentane ring fused to a substituted phenanthrene nucleus.

Structural domain. An element of protein tertiary structure that recurs in many structures.

Structural gene. A gene encoding the amino acid sequence of a polypeptide chain.

Structural protein. A protein that serves a structural function.

Subunit. Individual polypeptide chains in a protein.

Supercoiled DNA. Supertwisted, covalently closed duplex DNA.

Suppressor gene. A gene that can reverse the phenotype of a mutation in another gene.

Suppressor mutation. A mutation that restores a function lost by an initial mutation and that is located at a site different from the initial mutation.

Svedberg unit (S). The unit used to express the sedimentation constant *s*: $1\ S = 10^{-13}$ s. The sedimentation constant *s* is proportional to the rate of sedimentation of a molecule in a given centrifugal field and is related to the size and shape of the molecule.

Synapse. The chemical connection for communication between two nerve cells or between a nerve cell and a target cell such as a muscle cell.

Synapsis. The pairing of homologous chromosomes, seen during the first meiotic prophase.

T

Tandem duplication. A duplication in which the repeated regions are immediately adjacent to one another.

TCA cycle. *See* tricarboxylic acid cycle.

Template. A polynucleotide chain that serves as a surface for the absorption of monomers of a growing polymer and thereby dictates the sequence of the monomers in the growing chain.

Termination factors. Proteins that are exclusively involved in the termination reactions of protein synthesis on the ribosome.

Terpenes. A diverse group of lipids made from isoprene precursors.

Tertiary structure. In a protein or nucleic acid, the final folded form of the polymer chain.

Tetramer. Structure resulting from the association of four subunits.

Thioester. An ester of a carboxylic acid with a thiol or mercaptan.

Thymidine. One of the four nucleosides found in DNA.

Thymine. A pyrimidine base found in DNA.

Topoisomerase. An enzyme that changes the extent of supercoiling of a DNA duplex.

Transamination. Enzymatic transfer of an amino group from an α-amino acid to an α-keto acid.

Transcription. RNA synthesis that occurs on a DNA template.

Transduction. Genetic exchange in bacteria that is mediated via phage.

Transfection. An artificial process of infecting cells with naked viral DNA.

Transfer RNA (tRNA). Any of a family of low-molecular-weight RNAs that transfer amino acids from the cytoplasm to the template for protein synthesis on the ribosome.

Transferase. An enzyme that catalyzes the transfer of a molecular group from one molecule to another.

Transformation. Genetic exchange in bacteria that is mediated via purified DNA. In somatic cell genetics the term is also used to indicate the conversion of a normal cell to one that grows like a cancer cell.

Transgenic. Describing an organism that contains transfected DNA in the germ line.

Transition state. The activated state in which a molecule is best suited to undergoing a chemical reaction.

Translation. The process of reading a messenger RNA sequence for the specified amino acid sequence it contains.

Transport protein. A protein the primary function of which is to transport a substance from one part of the cell to another, from one cell to another, or from one tissue to another.

Tricarboxylic acid (TCA) cycle. The cyclical process whereby acetate is completely oxidized to carbon dioxide and water, and electrons are transferred to NAD^+ and FAD. The TCA cycle is localized to the mitochondria in eukaryotic cells and to the plasma membrane in prokaryotic cells. Also called the Krebs cycle.

Trypsin. A proteolytic enzyme that cleaves peptide chains next to the basic amino acids arginine and lysine.

Tryptic peptide mapping. The technique of generating a chromatographic profile characteristic of the fragments resulting from trypsin enzyme cleavage of the protein.

Tumorigenesis. The mechanism of tumor formation.

Turnover number. The maximum number of molecules of substrate that can be converted to product per active site per unit time.

U

Ultracentrifuge. A high-speed centrifuge that can attain speeds up to 60,000 rpm and centrifugal fields of 500,000 times gravity. Useful for characterizing and separating macromolecules.

Uncoupler. A substance that uncouples phosphorylation of ADP from electron transfer; for example, 2,4-dinitrophenol.

Unidirectional replication. *See* bidirectional replication.

Unwinding proteins. Proteins that help to unwind double-stranded DNA during DNA replication.

Urea cycle. A metabolic pathway in the liver that leads to the synthesis of urea from amino groups and CO_2. The function of the pathway is to convert the ammonia resulting from catabolism to a nontoxic form, which is subsequently secreted.

UV irradiation. Electromagnetic radiation with a wavelength shorter than that of visible light (200–390 nm). Causes damage to DNA (mainly pyrimidine dimers).

V

van der Waals forces. Refers to two types of interactions, one attractive and one repulsive. The attractive forces are due to favorable interactions among the induced instantaneous dipole moments that arise from fluctuations in the electron charge densities of neighboring nonbonded atoms. Repulsive forces arise when noncovalently bonded atoms come too close together.

Viroids. Pathogenic agents, mostly of plants, that consist of short (usually circular) RNA molecules.

Virus. A nucleic acid–protein complex that can infect and replicate inside a specific host cell to make more virus particles.

Vitamin. A trace organic substance required in the diet of some species. Many vitamins are precursors of coenzymes.

W

Watson-Crick base pairs. The type of hydrogen-bonded base pairs found in DNA, or comparable base pairs found in RNA. The base pairs are A-T, G-C, and A-U.

Wild-type gene. The form of a gene (allele) normally found in nature.

Wobble. A proposed explanation for base-pairing that is not of the Watson-Crick type and that often occurs between the 3′ base in the codon and the 5′ base in the anticodon.

X

X-ray crystallography. A technique for determining the structure of molecules from the x-ray diffraction patterns that are produced by crystalline arrays of the molecules.

Y

Ylid. A compound in which adjacent, covalently bonded atoms, both having an electronic octet, have opposite charges.

Z

Z form. A duplex DNA structure in which the usual type of hydrogen bonding occurs between the base pairs but in which the helix formed by the two polynucleotide chains is left-handed rather than right-handed.

Zwitterion. A dipolar ion with spatially separated positive and negative charges. For example, most amino acids are zwitterions, having a positive charge on the α-amino group and a negative charge on the α-carboxyl group but no net charge on the overall molecule.

Zygote. A cell that results from the union of haploid male and female sex cells. Zygotes are diploid.

Zymogen. An inactive precursor of an enzyme. For example, trypsin exists in the inactive form trypsinogen before it is converted to its active form, trypsin.

Credits

Portions of this text have been adapted from Geoffrey Zubay, *Genetics.* To be published.

Molecular Modeling

Molecular Graphics The molecular graphics photos listed below were developed by Michael Pique at the SCRIPPS Research Institute using software by Yng Chen, Michael Connolly, Michael Carson, Alex Shah, and AVS, Inc. Images copyright 1994 by the SCRIPPS Research Institute. Visualization advice was provided by Holly Miller, Wake Forest University Medical Center.

Atomic Coordinates All atomic coordinates for the molecular graphics images were obtained from the Protein Data Bank at Brookhaven National Laboratory, Upton, N.Y.

4.21: Phosphoglycerate kinase: entry 3PGK of July 1992. T. N. Bryant, et al. (To be published). **7.7**: Thermolysin: entry 1TLP of June 1987. D. E. Tronrud, et al., *European Journal of Biochemistry,* 157:261, 1986. **8.6a:** Trypsin: entry 3PTB of Sept. 1982. M. Marquart, et al., *ACTA Crystallography,* Section B, 39:480, 1983: Chymotrypsin: entry 4CHA of Nov. 1984. H. Tsukada & D. M. Blow, *Journal of Molecular Biology,* 184:703, 1985; Elastase: entry 2EST of Mar. 1986. D. L. Hughes, et al., *Journal of Molecular Biology,* 162:645, 1982. **8.6b**: Trypsin: entry 3PTB of Sept. 1982. M. Marquart, et al., *ACTA Crystallography,* Section B, 39:480, 1983; Subtilisin: entry 1SBC of May 1988. D. J. Neidhart & G. A. Petsko, (To be published). **8.7a-b**: Trypsin: entry 3PTB of Sept. 1982. M. Marquart, et al., *ACTA Crystallography,* Section B, 39:480, 1983. **8.19a-b**: Ribonuclease A: entry 6RSA of Feb. 1986. B. Borah, et al., *Biochemistry,* 24:2058, 1985. **8.20a-b**: Triose phosphate isomerase: entry 1TIM of Sept. 1976. D. W. Banner, et al., *Biochemical & Biophysical Research Communications,* 72:146, 1976. **9.9**: Phosphofructokinase: entry 4PFK of Jan. 1988. P. R. Evans, et al., *Philosophical Transactions of the Royal Society of London,* 293:53, 1981. **21.4a-b**: Glutamine synthetase: entry 2GLS of May 1989. M. M. Yamashita, et al., *Journal of Biological Chemistry,* 264:17681, 1989.

Chapter 1

1.1: From Sylvia S. Mader, *Inquiry into Life,* 7th edition. Copyright © 1994 Wm. C. Brown Communications, Inc., Dubuque, Iowa. All Rights Reserved. Reprinted by permission.
1.10: From R. E. Dickerson and I. Geis, *The Structure and Action of Proteins,* Benjamin/Cummings, Menlo Park, Calif., 1969. Illustration copyright by Irving Geis. Reprinted by permission.
1.15: From R. E. Dickerson and I. Geis, *The Structure and Action of Proteins,* Benjamin/Cummings, Menlo Park, Calif., 1969. Coordinates courtesy of D. C. Phillips, Oxford. Illustration copyright by Irving Geis. Reprinted by permission.

Chapter 3

3.6: From R. H. Haschenmeyer and A. E. V. Haschenmeyer, *A Guide to Study by Physical and Chemical Methods,* John Wiley & Sons, New York, 1973.

Chapter 4

4.2: From R. E. Dickerson and I. Geis, *The Structure and Action of Proteins,* Benjamin/Cummings, Menlo Park, Calif., 1969. Illustration copyright by Irving Geis. Reprinted by permission.
4.5: From R. E. Dickerson and I. Geis, *The Structure and Action of Proteins,* Benjamin/Cummings, Menlo Park, Calif., 1969. Illustration copyright by Irving Geis. Reprinted by permission.
4.24a: From Lansing M. Prescott, John P. Harley, and Donald A. Klein, *Microbiology,* 2d edition. Copyright © 1993 Wm. C. Brown Communications, Inc., Dubuque, Iowa. All Rights Reserved. Reprinted by permission.
4.25: From Lansing M. Prescott, John P. Harley, and Donald A. Klein, *Microbiology,* 2d edition. Copyright © 1993 Wm. C. Brown Communications, Inc., Dubuque, Iowa. All Rights Reserved. Reprinted by permission.
4a: From C. H. Bamford et al., *Synthetic Polypeptides,* Academic Press, Orlando, Fla., 1956.

Chapter 5

5.15: From John W. Hole, Jr., *Human Anatomy and Physiology,* 5th edition. Copyright © 1990 Wm. C. Brown Communications, Inc., Dubuque, Iowa. All Rights Reserved. Reprinted by permission.

Chapter 6

6.2: From E. J. Cohn and J. T. Edsall, *Proteins, Amino Acids, and Peptides as Ions and Dipolar Ions,* Reinhold, New York, 1942.

Chapter 7

7.5: From *Enzyme Structure and Mechanism,* 2d edition, by Alan Ferscht. Copyright © 1985 by W. H. Freeman and Company. Reprinted with permission.
7.3: From *Enzyme Structure and Mechanism* by Alan Ferscht. Copyright © 1985 by W. H. Freeman and Company. Reprinted with permission.

Chapter 9

9.20: From N. B. Madsen in *The Enzymes,* 3d edition, vol. XVII, ed. by P. D. Boyer and E. G. Krebs, Academic Press, New York, 1986.

Chapter 16

16.5: Reproduced with permission from B. Alberts, D. Bray, J. Lewis, M. Raff, K. Roberts, and J. D. Watson, *Molecular Biology of the Cell* (New York: Garland Publishing, 1989).

Chapter 17

17.3: Adapted from M. K. Jain and R. C. Wagner, *Introduction to Biological Membranes,* John Wiley & Sons, New York, 1980.
17.18: From J. T. Segrest and L. D. Kohn, Protein-lipid interactions of the membrane penetrating MN-glycoprotein from the human erythrocyte, *Protides of the Biological Fluids,* 21st colloquium, ed. by J. Peeters, Pergamon Press, New York, 1973.
17.4: Adapted from M. K. Jain and R. C. Wagner, *Introduction to Biological Membranes,* John Wiley & Sons, New York, 1980.
17.23: From B. Alberts et al., *Molecular Biology of the Cell,* 2d edition, Garland Publishing, New York, 1989.

Supplement 1

S1.3: Adapted from S. W. Kuffler and J. G. Nicholls, *From Neuron to Brain,* Sinauer Associates, Sunderland, Mass., 1976.

Chapter 25

25.16: From Geoffrey Zubay, *Genetics,* Benjamin/Cummings, Menlo Park, Calif., 1987. Reprinted by permission of the author.
25.18b: From Robert F. Weaver and Philip W. Hedrick, *Basic Genetics.* Copyright © 1991 Wm. C. Brown Communications, Inc., Dubuque, Iowa. All Rights Reserved. Reprinted by permission.

Chapter 26

26.16: From Robert F. Weaver and Philip W. Hedrick, *Basic Genetics.* Copyright © 1991 Wm. C. Brown Communications, Inc., Dubuque, Iowa. All Rights Reserved. Reprinted by permission.
26.21: From Geoffrey Zubay, *Genetics,* Benjamin/Cummings, Menlo Park. Calif., 1987. Reprinted by permission of the author.
26.22: From Geoffrey Zubay, *Genetics,* Benjamin/Cummings, Menlo Park, Calif., 1987. Reprinted by permission of the author.

Chapter 27

27.4: From Robert F. Weaver and Philip W. Hedrick, *Basic Genetics.* Copyright © 1991 Wm. C. Brown Communications, Inc., Dubuque, Iowa. All Rights Reserved. Reprinted by permission.

Chapter 28

28.2: From Geoffrey Zubay, *Genetics,* Benjamin/Cummings, Menlo Park, Calif., 1987. Reprinted by permission of the author.

28.5: From Robert F. Weaver and Philip W. Hedrick, *Genetics,* 2d edition. Copyright © 1992 Wm. C. Brown Communications, Inc., Dubuque, Iowa. All Rights Reserved. Reprinted by permission.
28.6: From Robert F. Weaver and Philip W. Hedrick, *Genetics.* Copyright © 1989 Wm. C. Brown Communications, Inc., Dubuque, Iowa. All Rights Reserved. Reprinted by permission.
28.8: From Geoffrey Zubay, *Genetics,* Benjamin/Cummings, Menlo Park, Calif., 1987. Reprinted by permission of the author.

Chapter 29
29.1: From Robert F. Weaver and Philip W. Hedrick, *Basic Genetics.* Copyright © 1991 Wm. C. Brown Communications, Inc., Dubuque, Iowa. All Rights Reserved. Reprinted by permission.
29.8: From Robert F. Weaver and Philip W. Hedrick, *Genetics,* 2d edition. Copyright © 1992 Wm. C. Brown Communications, Inc., Dubuque, Iowa. All Rights Reserved. Reprinted by permission.
29.23: From *Molecular Cell Biology,* 2d edition, by Darnell, Lodish, and Baltimore. Copyright © 1990 by Scientific American Books, Inc. Reprinted with permission of W. H. Freeman and Company.

Chapter 30
30.25: From C. Branden and J. Tooze, *Introduction to Protein Structure,* Garland Publishing, 1991, p. 102. Copyright 1991 by Garland Publishing, New York. Reprinted by permission.
30.28: From C. Branden and J. Tooze, *Introduction to Protein Structure,* Garland Publishing, New York, 1991, p. 109. Adapted from unpublished diagrams, courtesy of S. Phillips. Reprinted by permission.
30.29: From C. Branden and J. Tooze, *Introduction to Protein Structure,* Garland Publishing, New York, 1991, p. 105. Adapted from R.-g. Zhang et al., The crystal structure of trp aporepressor at 1.8 Å shows how binding tryptophan enhances DNA affinity, *Nature* 327:591, 1987. Reprinted by permission.

Chapter 31
31.10: From Geoffrey Zubay, *Genetics,* Benjamin/Cummings, Menlo Park, Calif., 1987. Reprinted by permission of the author.
31.13: From Geoffrey Zubay, *Genetics,* Benjamin/Cummings, Menlo Park, Calif., 1987. Reprinted by permission of the author.
31.21: From C. Branden and J. Tooze, *Introduction to Protein Structure,* Garland Publishing, New York, 1991, p. 126. Adapted from C. R. Vinson, P. B. Sigler, and S. L. McKnight, Scissors-grip model for DNA recognition by a family of leucine zipper proteins, *Science* 246:911, 1989. Copyright 1989 by the AAAS. Reprinted by permission.
31.24: Adapted from A. P. Bird, Gene reiteration and gene amplification, *Cell Biology,* vol. 3, *Gene Expression: The Production of RNAs,* ed. by L. Goldstein and D. M. Prescott, Academic Press, New York, 1980.

Supplement 3
S3.6: From Geoffrey Zubay, *Genetics,* Benjamin/Cummings, Menlo Park, Calif., 1987. Reprinted by permission of the author.
S3.7: From Geoffrey Zubay, *Genetics,* Benjamin/Cummings, Menlo Park, Calif., 1987. Reprinted by permission of the author.
S3.2: From Geoffrey Zubay, *Genetics,* Benjamin/Cummings, Menlo Park, Calif., 1987. Reprinted by permission of the author.
S3.10: From Geoffrey Zubay, *Genetics,* Benjamin/Cummings, Menlo Park, Calif., 1987. Reprinted by permission of the author.
S3.16: From Geoffrey Zubay, *Genetics,* Benjamin/Cummings, Menlo Park, Calif., 1987. Reprinted by permission of the author.

Supplement 4
S4.0: © Scientific American, Inc., George V. Kelvin. Reprinted by permission.
S4.1: Reproduced with permission from B. Alberts, D. Bray, J. Lewis, M. Raff, K. Roberts, and J. D. Watson, *Molecular Biology of the Cell* (New York: Garland Publishing, 1989).
S4.5: From Robert F. Weaver and Philip W. Hedrick, *Genetics,* 2d edition. Copyright © 1992 Wm. C. Brown Communications, Inc., Dubuque, Iowa. All Rights Reserved. Reprinted by permission.
S4.6: From Geoffrey Zubay, *Genetics,* Benjamin/Cummings, Menlo Park, Calif., 1987. Reprinted by permission of the author.

Index

Note: Page numbers followed by F indicate illustrations; page numbers followed by T indicate tables.